MACROMOLECULES
An Introduction to Polymer Science

CONTRIBUTORS

F. A. Bovey
M. J. Bowden
T. W. Kwei
L. D. Loan
S. Matsuoka
F. H. Winslow

MACROMOLECULES
An Introduction to Polymer Science

Edited by

F. A. BOVEY

and

F. H. WINSLOW

BELL LABORATORIES
MURRAY HILL, NEW JERSEY

ACADEMIC PRESS   New York   San Francisco   London   1979

A Subsidiary of Harcourt Brace Jovanovich, Publishers

ACADEMIC PRESS, INC.
111 Fifth Avenue, New York, New York 10003

United Kingdom Edition published by
ACADEMIC PRESS, INC. (LONDON) LTD.
24/28 Oval Road, London NW1   7DX

Library of Congress Cataloging in Publication Data

Main entry under title:

Macromolecules.

    Includes bibliographies.
    1. Macromolecules.   I. Bovey, Frank Alden, Date
II. Winslow, Field Howard, Date
QD381.M352   547'.7   78–20041
ISBN 0–12–119755–7

PRINTED IN THE UNITED STATES OF AMERICA

79 80 81 82   9 8 7 6 5 4 3 2 1

CONTENTS

CHAPTER 4. **Macromolecules in Solution**

T. K. KWEI

CHAPTER 5. **Macromolecules in the Solid State: Morphology**

F. A. BOVEY

CHAPTER 6. **Physical Behavior of Macromolecules**

S. MATSUOKA AND T. K. KWEI

CHAPTER 7. **Reactions of Macromolecules**

L. D. LOAN AND F. H. WINSLOW

CHAPTER 8. **Biological Macromolecules**

F. A. BOVEY

LIST OF CONTRIBUTORS

Numbers in parentheses indicate the pages on which authors' contributions begin.

F. A. BOVEY (1, 207, 317, 445), Bell Laboratories, Murray Hill, New Jersey 07974

M. J. BOWDEN (23), Bell Laboratories, Murray Hill, New Jersey 07974

T. K. KWEI (207, 273, 339), Bell Laboratories, Murray Hill, New Jersey 07974

L. D. LOAN (409), Bell Laboratories, Murray Hill, New Jersey 07974

S. MATSUOKA (339), Bell Laboratories, Murray Hill, New Jersey 07974

F. H. WINSLOW (1, 409), Bell Laboratories, Murray Hill, New Jersey 07974

FOREWORD

Polymer science and technology have been among the most important areas of discovery during the twentieth century. Much of the vitality of macromolecular research and development has come from the diversity of interests, skills, and viewpoints with which they have been pursued in science and engineering.

Chemists, physicists, engineers, designers, market developers, and production workers have all made important contributions to the creation, provision, and use of the range of plastics, rubbers, adhesives, fibers, and finishes that play so large a part in modern civilization.

We now know that the polymer molecule is the common factor in all of this. But bringing diverse themes together and combining the interests in polymers has been a historic role of teachers and researchers in universities and other basic science laboratories. Thus, the dominant part played by macromolecules in all living things—plants, animals, and single cells—has extended the community of interests into a realm reaching from agriculture through biomedicine to zoology.

Books about polymers go beyond the usual expectations for scientific and technical monographs. In an age characterized by a flood of information, writers are therefore challenged to communicate with a vast potential readership with diverse viewpoints and experience and to find a common language with which to express their ideas. The present work draws upon the editors' and authors' special knowledge of the uses of polymeric matter and upon their participation in an institution where much of the century's materials science and engineering began, and where modern principles of solids and their mechanical, electromagnetic, and thermodynamic properties were established. The editors have, themselves, been associated with a major coordinated application of polymers in communication and information processing. These projects range from the replacement of lead as a sheath in electrical cable to meeting the most exacting requirements for dielectrics for transoceanic communication; from the design of films for circuit stabilization to the preparation of adjuncts to light guides and optical circuitry. In one way or another these polymers—plastics, rubbers, and adhesives—are crucial to our enterprise, which over the past three decades has provided five to ten percent of the yearly national investment in capital facilities.

Thus, these editors and their collaborators can write for a principal component of the macromolecular constituency—the users and adapters and those who must live with the results of polymer synthesis and fabrication. Such a position stimulates discovery and development of polymers in ways that complement the more conventional view of the producer. For instance, this will involve especially the effects of processing on properties. Our particular applications include extensive high-speed fabrication of polymers in virtually all forms except fibers. Accordingly, from the community in which these authors have worked have come original findings of significance concerning the effect of thermal history on the mechanical, electrical, and chemical properties of thermoplastics. Similarly, strong interest in polymer stability under a variety of atmospheric and other exposures has been stimulated by the work of these authors and their associates.

The user aspect of polymer science and engineering, typical of the communications and information processing industries, has led to the establishment of significant principles. Among those noted in this volume are: establishment of the qualities of three-dimensional nets of molecular dimensions, which stimulated the modern light scattering concepts of Debye and which are now used to modify natural and synthetic rubbers; early determinations of entropy consequences in the melting of polar polymers, such as polyesters and polyamides; and novel electrical and mechanical properties of densely netted hydrocarbons showing a continuous transition between highly insulating and highly semiconducting states. These are a few examples of how user interests in special functions can interact with basic understanding.

Thus, although the range of books on polymers is now very extensive, the present volume provides a different and useful vision of both the known and the expected.

W. O. BAKER

PREFACE

This book arose from the need for a text to accompany an introductory course in polymer science at Bell Laboratories. The authors are all staff members in this organization. The book provides a fairly complete treatment of all phases of macromolecular chemistry and physics at the undergraduate or first year graduate level. It is primarily fundamental rather than technological in orientation.

Because there are now a number of textbooks in this field, one may well ask what special virtues can be claimed for this one. We feel that it covers several areas more completely and perhaps also more authoritatively than most other general introductory treatments. Polymerization kinetics is treated in Chapter 2 with unusual completeness. Chapter 3 includes a discussion of the rotational isomeric state method of calculation of polymer chain dimensions, which will provide the reader with at least a survey of this important discipline, not normally described in introductory texts. An entire chapter (Chapter 7) is devoted to chemical reactions and degradation of macromolecules, usually treated very cursorily if at all. Finally, biological molecules are discussed with exceptional thoroughness in Chapter 8.

It appears customary to introduce a general book on macromolecular science by pointing out the great importance of this field and by, at the same time, deploring the inadequate training which most chemists receive in it. We heartily agree with both these sentiments and can only hope that this volume may contribute toward finding a remedy for any educational shortcomings.

We thank Dr. A. E. Tonelli for material assistance with Chapter 3, Dr. H. D. Keith and Mr. F. J. Padden for helpful discussions and many of the illustrations in Chapter 5, and Dr. W. H. Starnes, Jr. for several helpful comments on Chapter 7. We are also grateful to many authors for permission to use numerous illustrations and figures appearing throughout the text.

MACROMOLECULES
An Introduction to Polymer Science

Chapter 1

THE NATURE OF MACROMOLECULES

F. A. BOVEY AND F. H. WINSLOW

1.1. Introduction

In the early part of this century organic chemists were still confining their attention to compounds that could be readily distilled or crystallized. When substances were encountered that could not be purified by conventional procedures, they were promptly thrown away. Many of the discarded materials were *polymers*, which in those days were assumed to be impure aggregates of small molecules held together by colloidal forces. Although it may seem odd now, this concept was well entrenched when Staudinger (1920) proposed that rubber and other polymers were actually composed of giant molecules that he called *macromolecules* (Staudinger and Fritschi, 1922). At first virtually no one agreed with him. In fact, his hypothesis met with stubborn criticism all through the next decade. Meanwhile, bits of evidence corroborating the macromolecular theory began to accumulate from x-ray studies, molecular weight measurements, and similar sources. After 1930 even the most persistent doubts faded away rapidly with the introduction of numerous new polymers that gradually revolutionized the plastics, textile, rubber, and related industries.

1

With the exception of the Staudinger studies, nearly all pioneering work on synthetic macromolecules was done in industrial laboratories. But as academic interest grew, research turned toward the natural polymers that form so much of our food, clothing, and shelter as well as the basic structures and functions of all living things.

1.2. Formation of Macromolecules

The term polymer is derived from Greek words meaning "many parts." They are prepared by a process known as polymerization, which involves the chemical combination of many small chemical units known as *monomers* ("single parts"). The repeating units in a polymer molecule may be either single atoms as in sulfur molecules or groups of atoms such as the methylene units, $-CH_2-$, in polyethylene[†]:

$$-S-S-S-S-S-S-S-S-$$

$$-CH_2CH_2CH_2CH_2CH_2CH_2CH_2CH_2-$$

The total number of repeating units in a polymer is called the *degree of polymerization* or DP.

As shown schematically in Fig. 1.1a, some polymers have a *linear* or threadlike structure. Others are *branched* or *cross-linked* in three-dimensional networks. Still others have less common shapes resembling combs, stars, or ladders. Polymers having flexible linear or branched structures are *thermoplastic*; that is, they can be molded or extruded at elevated temperatures and pressures. In contrast, the cross-linked *thermosetting* resins are permanently rigid materials.

Homopolymers consist of only one type of repeating unit, whereas *copolymers* (Fig. 1.1b) are composed of two or more different monomer units arranged in either random or alternating sequences. A few copolymers possess *block* or *graft* structures with relatively long sequences of one repeating unit bonded to similar sequences of another.

Polymer formation involves either *chain* or *step* reactions. Earlier, the terms *addition* and *condensation*, respectively, were used for describing these processes. One of the main differences between these mechanisms is outlined in Fig. 1.2. Once a chain reaction is *initiated*, monomer molecules add in rapid succession to the reactive end group of the growing polymer chain until it *terminates* and becomes unreactive. In the growth or *propagation*

[†] The commercial polymer is derived from ethylene, $CH_2=CH_2$. It has the same repeat unit as polymethylene made from diazomethane, CH_2N_2.

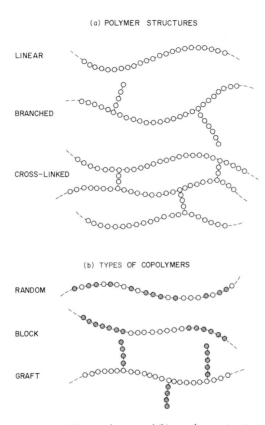

(a) POLYMER STRUCTURES

LINEAR

BRANCHED

CROSS-LINKED

(b) TYPES OF COPOLYMERS

RANDOM

BLOCK

GRAFT

FIG. 1.1.   (a) Homopolymer and (b) copolymer structures.

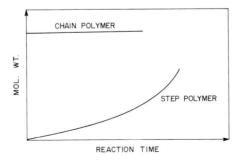

FIG. 1.2.   Dependence of molecular weight on reaction time for chain and step polymerization processes.

step, several thousand monomer units add one at a time to a growing chain in a time interval of less than one second. Thus, at any stage in a chain polymerization, the reaction mixture consists almost entirely of only two species: monomer and very large molecules.

In step reactions any pair of molecules can react, generally with elimination of a small molecule such as water. Under these circumstances, monomer disappears at the outset and the polymer molecules grow throughout the reaction at a slow, steady rate. The mechanisms and kinetics of chain and step reactions are described in detail in Chapter 2.

Macromolecular materials prepared by either process usually contain a broad range of molecular sizes. In chain reactions, this results from the random nature of the initiation and termination processes. In the step reactions, it results from the random interactions of all species, the rates of which are independent of molecular weight. The distribution of molecular weights varies widely and can be determined by methods described in Chapters 2 and 4.

1.3. Structure of Macromolecules

Vinyl chloride, $CH_2{=}CHCl$, is an ethylene derivative in which one hydrogen is replaced by a chlorine atom. Since the resulting monomer is unsymmetrical, it could in principle form a polymer chain comprising some combination of (a) head-to-head or (b) head-to-tail arrangements. In fact, the

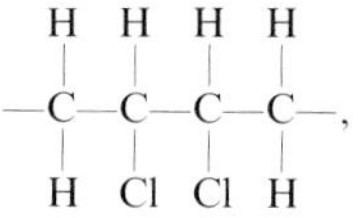

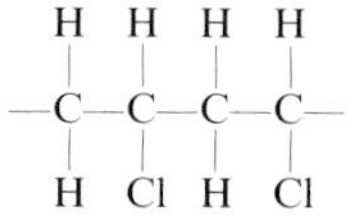

head-to-tail structure is strongly predominant in poly(vinyl chloride) and in most other vinyl polymers. Poly(vinyl fluoride), the most noteworthy exception, has about 16% of its units reversed:

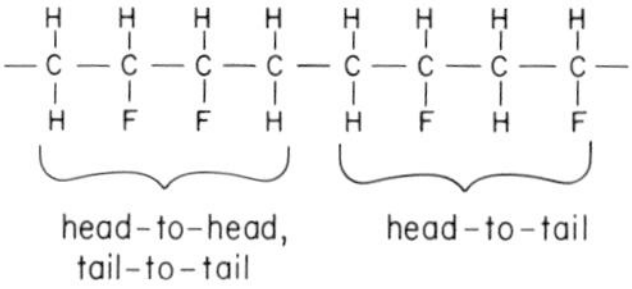

Vinyl monomers of this type may also exhibit large differences in *stereochemical configuration*, i.e., the relative handedness of successive monomer units may vary. The simplest regular arrangments of successive groups along a chain are the *isotactic* structure (Fig. 1.3a), in which all R substituents are located on the same side of the zigzag plane representing the chain stretched out in an all-trans conformation; and the *syndiotactic* arrangement, in which the R groups alternate from side to side (Fig. 1.3b). In the *atactic* arrangement the R groups appear at random on either side of the zigzag plane (Fig. 1.3c). These isomeric forms cannot be interconverted by rotating R groups about the carbon–carbon bonds of the main chain.

Another type of isomerism occurs in polymers having unsaturation in the main chain. Since carbon atoms linked together by double bonds are not free

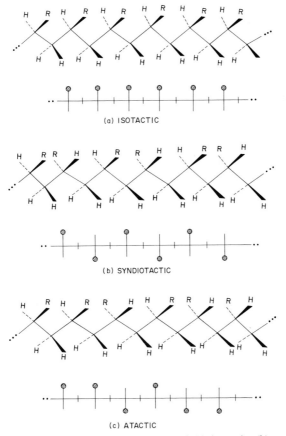

FIG. 1.3. Schematic representation of structures of (a) isotactic, (b) syndiotactic, and (c) atactic chains.

to rotate about the chain axis, repeating units in polymers such as poly-isoprene can exist as two different *geometrical* isomers:

$$CH_3 \diagdown \quad \diagup H$$
$$C{=}C$$
$${\sim}CH_2 \diagup \quad \diagdown CH_2{\sim}$$

$$CH_3 \diagdown \quad \diagup CH_2{\sim}$$
$$C{=}C$$
$${\sim}CH_2 \diagup \quad \diagdown H$$

cis   *trans*
natural rubber   gutta percha

polyisoprenes

The trans polymer is a semicrystalline plastic whereas the cis form is normally a rubber at room temperature.

1.4. Morphology and Mechanical Properties

Homopolymers of regular structure usually crystallize. The extent of crystallization is limited by irregularities in the molecular structure. Thus, ordinary polystyrene, which is nearly atactic, is amorphous, whereas isotactic polystyrene readily crystallizes. Again, branched polyethylene, produced at high pressure by a free radical process (Chapter 2, Section 2) is about 60% crystalline at room temperature, whereas linear polyethylene, formed at low pressure by heterogeneous catalysis (Chapter 2, Section 2), may be over 90% crystalline.

1.4.1. MELTING AND GLASS TEMPERATURES

Crystalline polymers, like all other crystalline compounds, show melting points T_m. All polymers, whether crystalline or amorphous, also exhibit *second order transitions*, the most important of which is the glass transition T_g. This transition is evidenced by a change in slope of a plot of specific volume versus temperature (Fig. 1.4). Changes in mechanical and other properties may occur at the same temperature (Fig. 1.5). This behavior results from the abrupt onset of extensive molecular motion, chiefly rotation of the polymer chains, as the temperature is raised (or the "freezing" of such motion as the temperature is lowered). These motions are inhibited in the glassy state, in which the viscosity is so high that the specific volume cannot reach its equilibrium value in a practical time span. This subject is discussed in detail in Chapter 6. Values of T_m and T_g for a selected group of polymers are shown in Table 1.1.

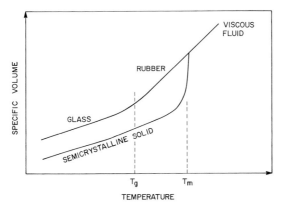

FIG. 1.4. Specific volume as a function of temperature for glassy and semicrystalline polymers.

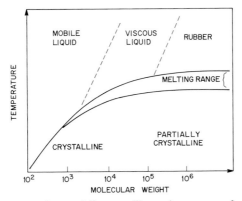

FIG. 1.5. Melting curves for partially crystalline polymers as a function of molecular weight.

1.4.2. INTERCHAIN FORCES

Many important polymer properties, such as melting point, solubility, and viscosity, depend on the interaction between adjacent molecules. These forces are typically one or two orders of magnitude less than the strength of the covalent bonds holding each molecule together, but have themselves a range of strength of about an order of magnitude. The strongest secondary interactions involve hydrogen bonds. These account for the high softening temperature of nylon and the intractability of cellulose. Less polar carbon–chlorine bonds provide interactions that are somewhat weaker yet sufficient

TABLE 1.1

TRANSITION TEMPERATURES OF SELECTED POLYMERS

Polymer	Chain unit	T_m (°C)	T_g (°C)	Remarks
(a) Vinyl and Diene Polymers				
Polyethylene, high pressure (branched)	$+CH_2CH_2)_{\overline{n}}$	~115	—	
linear	$+CH_2CH_2)_{\overline{n}}$	135	−125(?)	T_g value controversial
Polypropylene, isotactic	$+CH{-}CH_2)_{\overline{n}}$ CH_3	165	−10	
atactic		—	−20	
Polybutene-1, isotactic	$+CH{-}CH_2)_{\overline{n}}$ CH_2CH_3	142	−20	
Polyisobutene	$+C(CH_3)_2CH_2)_{\overline{n}}$	—	−73	
Polystyrene, atactic	$+CHCH_2)_{\overline{n}}$ ⬡	—	100	
isotactic		240	100	Several crystalline forms
Poly(vinyl chloride)	$+CHClCH_2)_{\overline{n}}$	—	81	
Poly(vinylidene chloride)	$+CCl_2CH_2)_{\overline{n}}$	190	—	

Polymer	Structure		T_m	T_g	
Poly(vinyl fluoride)	$-\!(CHFCH_2)_n\!-$		200	-20	T_g value controversial
Poly(vinylidene fluoride)	$-\!(CF_2CH_2)_n\!-$		171	-45	
Poly(trifluorochloroethylene)	$-\!(CF_2CFCl)_n\!-$		218	45	
Poly(tetrafluoroethylene)	$-\!(CF_2CF_2)_n\!-$		327	—	
Poly(methyl methacrylate), isotactic	CH_3 $-\!(C\!-\!CH_2)_n\!-$ CO_2CH_3		160	~45	
syndiotactic			200	115	
atactic			—	105	
Poly(methyl acrylate)	$-\!(CHCH_2)_n\!-$ CO_2CH_3		—	6	
Poly(vinyl alcohol)	$-\!(CHCH_2)_n\!-$ OH		258	85	
Poly(vinyl acetate)	$-\!(CH\!-\!CH_2)_n$ $OCOCH_3$		—	28	
85:15 Vinyl chloride: vinyl acetate copolymer			—	45	

TABLE 1.1 (*Continued*)

Polymer	Chain unit	T_m(°C)	T_g(°C)	Remarks
Poly-*cis*-1,4-isoprene (natural rubber)	CH_3, H / $C=C$ / $-CH_2$, $-CH_2\!-\!_n$		−73	
Poly-*trans*-1,4-isoprene (gutta percha)	CH_3, $-CH_2\!-\!_n$ / $C=C$ / $-CH_2$, H		−58	
Poly-*cis*-1,4-butadiene	H, H / $C=C$ / $-CH_2$, $-CH_2\!-\!_n$	6	−108	
Poly-*trans*-1,4-butadiene	H, $-CH_2\!-\!_n$ / $C=C$ / $-CH_2$, H	148	−18	
Poly-1,2-butadiene, isotactic	$-CH\!-\!CH_2\!-\!_n$ / $CH=CH_2$	125	−4	
syndiotactic		154	—	
75:25 Butadiene: styrene copolymer		—	−70	

(b) Ring Opening and Condensation Polymers

Polyoxymethylene	$-OCH_2\!-\!_n$	195	−85	

Poly(ethylene oxide)	$+CH_2-CH_2-O+_n$	66	-67
Poly(propylene oxide)	$+CH-CH_2-O+_n$ CH_3	75	-75
Poly(styrene oxide)	$+CH-CH_2-O+_n$ (phenyl)	149	37
Polyhexamethylene adipamide (nylon 66)	$+NH(CH_2)_6NHCO(CH_2)_4CO+_n$	265	49
Poly(ethylene terephthalate)	$+O-CH_2CH_2-OCO-\bigcirc-CO+_n$	265	69
Poly(4,4-isopropylidene diphenylene carbonate)	(structure with CH_3, C, CH_3, O, C=O)	267	149
Poly(dimethylsiloxane)	$+Si-O+_n$ CH_3 / CH_3	—	-123

to make poly(vinyl chloride) stiff and hard even though it is essentially noncrystalline.

The chain segments in nonpolar polymers such as polyethylene are held together by weak dispersion forces common to all polymers. As the melting curves in Fig. 1.5 indicate, cumulative secondary forces are large even in nonpolar polymers. The melting point at first rises with increasing molecular weight but then levels off, showing that melting of high polymer fractions depends more on attractive forces between chain segments than on interactions between discrete molecules. The melting range reflects the molecular weight distribution in a typical polyethylene.

1.4.3. MECHANICAL STRENGTH

Simple tensile tests provide the most common means for measuring important mechanical properties of polymers. For example, the initial slope of each stress–strain curve in Fig. 1.6 gives the Young's modulus or stiffness of the material. Polystyrene, which is brittle, breaks at low elongation. At a slightly higher elongation, branched polyethylene yields and forms a neck. In the region where the neck forms, crystallites realign and molecular chains slide by one another to orient in the draw direction. The cross section of the neck is much smaller than that of the original specimen, and although the overall stress decreases, the oriented material may have a tensile strength more than tenfold greater than that of the unoriented polymer. However, the drawn material is much weaker perpendicular to the draw direction. The strain at rupture varies inversely as the molecular weight of the specimen and the rate of extension. The area under the curve represents the work required for rupture. The initial modulus of vulcanized natural rubber is low but stress-induced crystallization markedly increases the modulus at strains

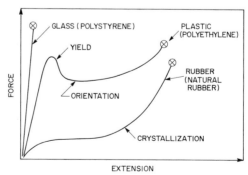

FIG. 1.6.   Stress–strain curves for glassy, semicrystalline, and rubbery polymers.

above 400%. As the rubber is stretched it releases heat, but when stress is removed the specimen absorbs heat and contracts as thermal motion of molecular chains leads to greater disorder or increased entropy without significant change in volume. This rubbery behavior is typical of all long-chain polymers with (1) freely rotating repeating units that are (2) slightly cross-linked and (3) have weak interchain forces. Stretching causes permanent distortion of unvulcanized (noncross-linked) rubber. As might be expected dense cross-linking forms "hard rubber," an intractable *thermoset* resin.

1.5. Biological Macromolecules and Biopolymers

One can readily defend the proposition that of all the classes of macromolecules discussed in this book, biological polymers are by far the most significant since they are essential constituents of living cells. The term *biopolymers* is not confined to macromolecules actually occurring in nature but also includes synthetic polymers of simpler but analogous structure, often useful as models for the behavior of their natural counterparts.

It is convenient and useful to consider biopolymers under three headings: (a) polypeptides and proteins, (b) polynucleotides, and (c) polysaccharides. Polypeptides and proteins share the common structural unit of α-amino acids joined by peptide (CONH) bonds:

$$NH_2-\underset{\underset{R_1}{|}}{CH}-\underset{\underset{O}{\|}}{C}\left[NH-\underset{\underset{R_2}{|}}{CH}-CONH-\underset{\underset{R_3}{|}}{CH}-\underset{\underset{O}{\|}}{C}\cdots\right]_n OH$$

The synthesis of polypeptides will be discussed in Chapter 8. Small naturally occurring polypeptides (n usually not exceeding ~ 10) are of great biological importance as hormones and antibiotics, but will not be discussed in this book. *Proteins* assume many forms and functions, from structural types (collagen, keratin, elastin) to those having oxygen-carrying properties (myoglobin, hemoglobin), electron transfer functions (cytochrome c, ferredoxin), or highly specific enzymatic (i.e., catalytic) functions. Twenty different amino acids, corresponding to 20 different R side chains, permit a vast number of sequence permutations. Chain lengths may also vary. In contrast to synthetic polymers, the sequences and chain lengths are not random but highly specific for a given protein although they may vary somewhat from one species to another. The conformations of chains show a high degree of variability from protein to protein and are now generally thought to be a direct consequence of the particular amino acid sequences; again, they are quite specific for a given type of protein.

Polysaccharides may also vary in structure but these variations take the form of differing linkages between the basic units, most commonly glucose, as well as differing side chains. They show a distribution of chain lengths. Among the most studied are *cellulose*, the structural polysaccharide; *starch*, a medium for energy storage; *dextran*, synthesized by certain bacteria; and *chitin*, which is a structural unit of cell walls and of insect and crustacean exoskeletons.

Nucleic acid chains are composed of sugar units connected by phosphate links, with purine and pyrimidine bases attached to the sugar units. The base–sugar–phosphate unit is known as a *nucleotide unit*:

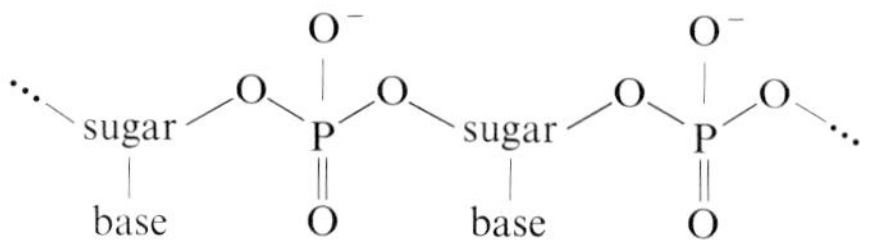

Two classes of nucleic acids are recognized: ribonucleic acids or RNA and deoxyribonucleic acids or DNA. DNA occurs in cell nuclei and is the genetic material that transmits information for protein synthesis from one generation to the next. (In certain viruses, RNA plays this role.) The DNA chains, as is now well known, have a double-helical conformation.

The structure and conformation of biological polymers will be discussed in greater detail in Chapter 8.

1.6. Historical Perspective

More than a century ago Thomas Graham (1861) noted distinct differences in the diffusion rates of various compounds in solution. Crystalline compounds such as sugar diffused more rapidly than those that did not crystallize. He coined the word *colloid* (from a Greek word *kolla*, meaning glue) to describe gelatin and similar compounds that diffused slowly in solution and did not pass through parchment paper. The notion quickly caught the fancy of chemists accustomed to thinking in terms of small molecules, and as studies of colloids became more popular, an impression grew that under proper conditions most materials consisted of large impure aggregates of small structural units held together by peculiar "association" forces. Some scientists even went so far as to postulate a fourth or "colloidal" state of matter in addition to the familiar gaseous, liquid, and solid states. Eventually, the fallacy of applying the colloidal theory to macromolecules was recognized; but another 70 years passed before the macromolecular concept prevailed. Since early studies of macromolecules have been described in detail by Flory (1953) and Mark (1967), we review here only a few selected highlights in the history of polymer science.

1.6.1. EARLY STUDIES OF RUBBER

In 1511 Pietro Martyre d'Anghiera described a peculiar bouncing ball used by Aztec Indians in their favorite game (Wolf, 1964). More than two centuries later, in 1770, Joseph Priestley named the resilient material "india rubber" because he found it to be much better than bread crumbs for rubbing out unwanted pencil marks. However, caoutchouc, or natural rubber as it is more generally known today, had limited commercial importance before Charles Goodyear discovered sulfur vulcanization[†] in 1839.

One of the first chemists to show an interest in polymers was an Englishman named Greville Williams (1860) who pyrolyzed natural rubber and obtained a new compound he called isoprene. He reported that "the composition of caoutchouc coincides with that of isoprene, as found by analysis, to a degree which is remarkable . . . ." Although the experiment indicated to him that isoprene was a basic building block of rubber, he and his successors had only a vague notion of what a polymer molecule was really like. His experiments were later confirmed independently by Gustave Bouchardat, William Tilden, and Otto Wallach. Tilden (1892) reported making rubberlike products from isoprene and claimed that "the artificial rubber unites with sulfur in the same way as ordinary rubber, forming a tough elastic compound." But as time passed he became less enthusiastic about the prospects for a synthetic rubber. Finally, in 1907, he concluded that synthetic rubber was commercially impossible at any price. Three years later Francis Mathews in England and Carl Dietrich Harries in Germany discovered almost simultaneously that sodium metal catalyzed the rapid polymerization of isoprene, and in the same year Lebedev in Russia obtained a rubberlike product from butadiene.

Natural rubber was particularly difficult to explain by colloidal theories since the forces between small hydrocarbon molecules in the liquid or solid state were exceptionally weak. Harries (1905) resorted to Thiele's theory of partial valences in a vain attempt to explain the chemical structure. In his view, rubber consisted of the cyclic dimer or higher cyclic homologs of isoprene linked together at double bond sites in contiguous rings. The idea was rather short-lived because in 1922 Herman Staudinger hydrogenated

---

[†] The word vulcanization was actually proposed by a friend of Thomas Hancock; Hancock was an English pioneer in rubber processing who independently discovered the sulfur vulcanization process in 1843.

rubber and obtained a saturated paraffinic product that lacked a sharp melting point, was noncrystalline, formed viscous solutions, and could not be distilled without decomposition. The retention of polymeric characteristics in the absence of unsaturation reinforced Staudinger's belief (1920) that rubber molecules had the same valence structure as volatile hydrocarbons but were larger in size by several orders of magnitude.

Three main factors have been cited by Mark (1973) as tipping the scales in favor of the macromolecular theory. First, x-ray studies by Katz (1925) revealed a fibrelike diagram for stretched rubber, and similar work led Meyer and Mark (1928) to propose a semicrystalline structure for long-chain molecules in cellulose. Second, dilute solution viscosity (Staudinger, 1928), ultracentrifugation (Svedberg, 1926), and other methods were developed for measuring molecular weights and molecular weight distributions of polymers. The narrow molecular weight distributions found for proteins were particularly difficult to explain by the colloidal theory. Summer (1926) succeeded in crystallizing an enzyme, urease, despite the fact that colloids were considered noncrystallizable. Finally, Wallace Carothers (1931) and his group at du Pont secured the argument for the macromolecular concept through their extensive research in the synthesis and characterization of polyesters, polyamides, and vinyl polymers.

The interest of the du Pont Company in these materials was stimulated by some studies on acetylenic compounds reported in 1925 by Father Julius Nieuwland of the University of Notre Dame. Soon after du Pont contracted to take over development of these materials, Carothers discovered a method for adding hydrogen chloride to vinyl acetylene in the presence of cuprous chloride. The resulting 2-chlorobutadiene polymerized far more rapidly than isoprene to form polychloroprene, an oil and ozone resistant rubber which du Pont began marketing in 1931 under the trade name Duprene, which was later changed to Neoprene.

Meanwhile the German and Russian workers had shifted the focus of their polymerization studies from isoprene to butadiene. At first the Germans experimented with "Buna" rubbers made from butadiene using sodium as a catalyst, as the name implies. However, by 1930 Edward Tschunkur of I. G. Farben, Leverkusen, had developed an aqueous emulsion system for making two new rubbers with superior properties: Buna-S and Buna-N, which are butadiene copolymers with styrene and acrylonitrile, respectively.

In 1940 the US Government became concerned over its rubber supply and made plans to produce GR-S (Government Rubber-Styrene), a general purpose rubber similar to Buna-S. Full-scale production began in December 1943 at Institute, West Virginia. Although total US synthetic rubber production in 1940 was less than 2600 tons, it grew rapidly during World War II, reaching 760,000 tons in 1946.

German chemists had also developed a process for making polyisobutylene in 1932. Since the saturated polymer could not be vulcanized, William Sparks and Robert Thomas at the Standard Oil Development Company decided in 1937 to modify the synthesis. They added about 2% of a diene to the iso-butylene and obtained a vulcanizable, ozone resistant rubber with low permeability to air. Commercial production of butyl rubber began in 1940 for use primarily in the manufacture of inner tubes for automobile tires.

1.6.2. POLYETHYLENE AND OTHER VINYL POLYMERS

Rapid advances were also made in fundamental studies of macromolecules during the 1930s. Staudinger and Steinhofer (1935) demonstrated the existence of a head-to-tail arrangement of repeating units in polystyrene, with phenyl groups attached to alternate carbon atoms along the chain, and Marvel and Levesque (1938) found the same head-to-tail structures in poly(vinyl methyl ketone) and in poly(vinyl chloride) (Marvel et al., 1939). About the same time Flory (1937) outlined the fundamental principles of vinyl polymerization. He introduced the concept of hydrogen transfer in radical polymerizations and showed how it led to termination of chain growth and occasionally to chain branching in accordance with the laws of chance (see Chapter 2, Section 2.2.2).

Chance also has played a prominent role in the discovery and development of polyethylene and related polymers. The polyethylene story (Perrin, 1953; Hunter, 1955) has special historical significance because it illustrates how experiment has repeatedly outstripped theory in the advance of polymer science.

During the course of a research program on the effect of pressure on organic reactions, British chemists at Imperial Chemical Industries Ltd. isolated in March 1933 a waxlike substance that they concluded was an ethylene polymer. Attempts to repeat the experiment failed until nearly three years later, in December 1935, they finally succeeded in "rediscovering" polyethylene. In doing so they learned that their original success resulted from the fortuitous presence of oxygen, a contaminant that catalyzed the polymerization. Thus, polyethylene was discovered twice by the same group, the first time by accident resulting from defective equipment.

The versatility and enormous economic value of the polymer was immediately apparent. Early tests turned up several unexpected properties, such as a tendency to develop a spherulitic structure and to undergo cracking under complex stress (Richards, 1946; Hopkins et al., 1950). Also, the unprotected polymer degraded rapidly outdoors (Pross and Black, 1950) or at elevated temperatures even though it is a paraffin, which by definition should be inert.

With the advent of infrared analysis (Fox and Martin, 1940), the British found to their surprise that polyethylene molecules were not strictly linear chains but instead contained a small number of branches. Despite Flory's earlier proposal the origin of the branches was not entirely clear until Roedel (1953) accounted for them by a "backbiting" mechanism of hydrogen transfer (see Chapter 3, Section 3.8).

Meanwhile, Karl Ziegler *et al.* (1955) at the Max-Planck Institute for Coal Research, in the course of developing an antiknock additive for gasoline, discovered a new catalyst for making linear polyethylene at room temperature and atmospheric pressure. The catalyst was attached by coordination to the growing chain end and made the reaction more selective by avoiding hydrogen transfer. The resulting chains were linear.

Giulio Natta *et al.* (1955) in Milan extended the use of Ziegler's catalysts to other α-olefins. He found that propylene yielded a mixture of a rubber and a high melting plastic. To his astonishment, both the rubber and the plastic were found to consist of molecules of nearly identical size. The difference in properties was traced to the spatial arrangement of the methyl groups on alternate carbon atoms in the polypropylene chain (Section 1.1). When all repeating units in the chain had the same configuration, the polymer molecules had a regular helical conformation and formed a semicrystalline plastic. This type of isomerism had been suggested earlier by Staudinger (1932), by Huggins (1944), and by Schildknecht *et al.* (1948). If the molecules consisted of repeating units with a random mixture of configurations, they formed a rubber. Natural polymers often have a regular helical structure. Thus, Natta demonstrated for the first time that it was possible for man to mimic nature and make polymers with stereospecific structures like those of cellulose and proteins (see Section 1.5). Production of synthetic natural rubber with a Ziegler-type catalyst was announced first by the B. F. Goodrich Co. in 1954. Almost simultaneously the Firestone Co. reported development of the same polymer using lithium catalyst systems.

Another surprising result followed from the discovery of linear polyethylene. In 1955, Jaccodine obtained lozenge-shaped crystals of the polymer from dilute solution. Two years later, Andrew Keller (1957) demonstrated that in this solution-crystallized material the chains were not extended zigzags as in short-chain paraffins but were folded back and forth on one another like fire hose. Although the fold spacing was only about 10^{-6} cm, the broad crystal faces were often large enough to see under an optical microscope. Actually, Keller was not the first to report chain folding in polymer crystals. Nearly 20 years earlier, Keith Storks (1938) using electron diffraction had observed this phenomenon in gutta percha, the plastic form of polyisoprene. At the time, gutta percha was the only known polymer that readily showed this behavior. Since gutta percha was a notoriously unstable polymer of natural origin, the Storks report was promptly ignored and almost forgotten.

1.6.3. NYLONS AND SILICONES

Finally, brief mention should be made of two polymers with curious histories. The development of silicones is especially relevant because it indicates how the commercial value of these and similar substances escaped the notice of many early chemists. Frederick S. Kipping and his students at the University of Nottingham, England, studied the chemistry of organo-silicon compounds over a span of 45 years. As Rochow (1951) has observed, Kipping overlooked the importance of the oily or gluelike substances he frequently synthesized, being mainly concerned with isolating and characterizing pure compounds. In fact, he ended his 1937 Bakerian Lecture at Cornell University with these words: "Most if not all of the known types of organic derivatives of silicon have now been considered and it may be seen how few they are in comparison with those which are entirely organic; as moreover the few which are known are very limited in their reactions, the prospect of any immediate and important advance in this section of organic chemistry does not seem to be very hopeful." Commercial production of poly(dimethylsiloxane) (see Table 1.1) and other silicone polymers began about six years later.

The du Pont Co. became interested in polymers in 1927 and began a research program aimed at developing synthetic fibers (Bolton, 1942). A year later Carothers joined the group, and initiated studies that eventually led to the first commercial production of nylon 66 (see Table 1.1) in 1939. Although the original announcement claimed that the nylon "was made from coal, air, and water," it is more precise to say that it is a derivative of hexamethylenediamine and adipic acid. Evidently no attempts were made to develop a related polyamide, nylon 6,

$$-(NH(CH_2)_5\underset{\displaystyle \underset{O}{\|}}{C})_n-$$

because Carothers had concluded (Achilladelis, 1970) that ε-caprolactam was not readily polymerizable. As soon as the du Pont patents were filed in 1937, Paul Schlack at the Badische Anilin und Soda Fabrik (BASF) in Germany began studying caprolactam and achieved the synthesis of nylon 6 in late January 1938. Commercial production of the polymer commenced in 1943.

After World War II, the polymer industry burgeoned as new applications and improved methods of stabilization and processing of these materials were found. Staudinger's pioneering contributions (Yarsley, 1967) were belatedly honored by a Nobel Prize in 1953 and since then the award has been presented to Ziegler and Natta in 1963 and to Flory in 1974. Several others have received Nobel Awards for studies of biopolymers.

Highlights in the commercial development of various familiar rubbers and plastics are listed in Table 1.2.

TABLE 1.2

MILESTONES IN POLYMER HISTORY

Date[a]	Polymer	Pioneer(s)
1844[b]	Vulcanized rubber	Goodyear, Hancock
1870	Cellulose nitrate	Hyatt
1909	Phenolics	Baekeland
1919	Cellulose acetate	C. and H. Dreyfus
1927	Acrylics	Röhm
	Alkyds	Kienle
1929	Thiokol rubber	Patrick
1930	Polystyrene	Staudinger[c]
	Styrene–butadiene rubber	Tschunkur
1931	Poly(vinyl chloride)	Klatte, Semon
	Polychloroprene	Carothers
1939	Melamines	Henkel
	Branched polyethylene	Fawcett and Gibson
	Nylon 66	Carothers
1941	Polytetrafluoroethylene	Plunkett[d]
1943	Butyl rubber	Sparks and Thomas
	Nylon 6	Schlack
	Silicones	Rochow, Hyde, Andrianov
1944	Poly(ethylene terephthalate)	Whinfield and Dickson
1946	ABS resins	
1947	Epoxies	Castan[c], Greenlee
1950	Polyurethanes	Bayer[c]
1956	Linear polyethylene	Ziegler, Hogan
1957	Polypropylene	Natta
1959[c]	Polyacetals	Staudinger, McDonald
1959	Polycarbonates	Schnell, Fox
1964	Ionomers	Rees
	Polyimides	Stroog
1966	Polyphenylene oxide	Hay
	Aramids	Morgan[e]

[a] Dates refer to the approximate time of commercial introduction. U.S. dates for commercial introduction were somewhat later for the following polymers: cellulose acetate (1924), acrylics (1931), poly(vinyl chloride) (1936), polystyrene (1937), branched polyethylene (1943), and polyurethanes (1953).

[b] Although Goodyear discovered rubber vulcanization in 1839, another five years were required to perfect and patent his invention.

[c] Staudinger studied polystyrene and the polyacetals during the 1920s. Bayer's work on polyurethane began in 1937 in Germany. A year later Castan commenced his studies of epoxies in Switzerland.

[d] Polytetrafluoroethylene was discovered quite by accident in 1938 while Roy Plunkett and Jack Rebok were studying gaseous refrigerants.

[e] Commercial production of meta-substituted aramids began in 1966. The para-substituted aramids were introduced in 1971.

REFERENCES

Achilladelis, B. (1970). *Chem. Ind. (London)* 1549.
Bolton, E. K. (1942). *Ind. Eng. Chem.* **34**, 53.
Carothers, W. H. (1931). *Chem. Rev.* **8**, 353.
Flory, P. J. (1937). *J. Am. Chem. Soc.* **59**, 241.
Flory, P. J. (1953). "Principles of Polymer Chemistry." Cornell Univ. Press, Ithaca, New York.
Fox, J. J., and Martin, A. E. (1940). *Proc. R. Soc. London Ser. A* **175**, 211.
Graham, T. (1861). *Phil. Trans. R. Soc. (London) Ser. A.* **151**, 183.
Harries, C. D. (1905). *Ber.* **38**, 3935.
Hopkins, I. L., Baker, W. O., and Howard, J. B. (1950). *J. Appl. Phys.* **21**, 206.
Huggins, M. L. (1944). *J. Am. Chem. Soc.* **66**, 1991.
Hunter, E. (1955). *Chem. Ind. (London)* 396.
Jaccodine, R. (1955). *Nature (London)* **176**, 305.
Katz, I. R. (1925). *Naturwissenschaften* **13**, 410.
Keller, A. (1957). *Phil. Mag.* (8) **2**, 1171.
Mark, H. (1967) Polymers—Past, present, future, "Proc. Robert A. Welch Foundation, Polymers" (W. O. Milligan, ed.), Vol. 10, Chapter 2. Houston, Texas.
Mark, H. (1973). *J. Chem. Educ.* **50**, 757.
Marvel, C. S., and Levesque, C. L. (1938). *J. Am. Chem. Soc.* **60**, 280.
Marvel, C. S., Sample, J. H., and Roy, M. F. (1939). *J. Am. Chem. Soc.* **61**, 3241.
Meyer, K. H., and Mark, H. (1928). *Ber.* **61**, 593.
Natta, G. *et al.* (1955). *J. Am. Chem. Soc.* **77**, 1708.
Perrin, M. W. (1953). *Research* **6**, 111.
Pross, A. W., and Black, R. M. (1950). *J. Chem. Soc. Ind. (London)* **69**, 113.
Richards, R. B. (1946). *Trans. Faraday Soc.* **42**, 10.
Rochow, E. G. (1951). "An Introduction to the Chemistry of Silicones," 2nd ed., p. 77. Wiley, New York.
Roedel, M. J. (1953). *J. Am. Chem. Soc.* **75**, 6110.
Schildknecht, C. E., Gross, S. J., Davidson, H. R., Lambert, J. M., and Zoss, A. O. (1948). *Ind. Eng. Chem.* **40**, 2104.
Staudinger, H. (1920). *Ber.* **53**, 1073.
Staudinger, H. (1928). *Ber.* **61B**, 2427.
Staudinger, H. (1932). "Die Hochmolecularen Organischen Verbindungen." Springer-Verlag, Berlin and New York.
Staudinger, H., and Fritschi, J. (1922). *Helv. Chem. Acta* **6**, 705.
Staudinger, H., and Steinhofer, A. (1935). *Ann. Chem.* **517**, 35.
Storks, K. H. (1938). *J. Am. Chem. Soc.* **60**, 1753.
Sumner, J. B. (1926). *J. Biol. Chem.* **69**, 435.
Svedberg, T. (1926). *Z. Phys. Chem.* **121**, 65.
Tilden, W. (1892). *Chem. News* **65**, 265.
Williams, C. G. (1860). *Proc. R. Soc. London Ser. A* **10**, 516.
Wolf, R. F. (1964). *Rubber World* (10), 64.
Yarsley, V. E. (1967). *Chem. Ind. (London)* 250.
Ziegler, K., Holzkamp, E., Breil, H., and Martin, H. (1955). *Angew. Chem.* **67**, 426.

Chapter 2

FORMATION OF MACROMOLECULES

M. J. BOWDEN

2.1. Introduction

It was seen in Chapter 1 that most macromolecules are made up of repeat sequences of structural units derived from simple molecules called monomers. Polyethylene, for example, consists of repeat structural units of the type $-(CH_2-CH_2)_x$. These in turn derive from the monomer ethylene, which has the structure $CH_2=CH_2$. The synthesis of polyethylene is formally represented by the equation

$$xCH_2 = CH_2 \longrightarrow -(CH_2-CH_2)_{\overline{x}}$$

23

Obviously, if a molecule is to be classified as a monomer, it must be capable of bonding to two or more molecules, i.e., its functionality must be ≥ 2. There are three basic ways of achieving the necessary difunctionality (Billingham and Jenkins, 1972):

1. Opening a double bond;
2. Opening a ring;
3. Using molecules bearing two reactive functional groups.

This chapter will be concerned with the mechanism of these various processes and with the derivation of kinetic expressions that govern the rate at which they occur.

2.2. Vinyl and Diene Polymerization

2.2.1. Nature of Addition Polymerization

The difunctionality necessary to form a linear macromolecule from a vinyl monomer is achieved by opening the double bond. Such reactions invariably proceed by a chain reaction mechanism. Polymerization usually begins with the addition of an initiator I, which is capable of producing an active species R*. The active species then reacts with the monomer M to produce an activated molecule by opening the double bond.

$$I \longrightarrow R^*, \qquad R^* + M \longrightarrow RM^*$$

Polymerization then proceeds by successive addition of further monomer molecules to the activated end of the growing chain. Thus the mechanism is termed *addition* or *chain polymerization* and polymers from this type of reaction are called *addition polymers.*

The addition of monomer molecules to the activated chain end is known as *propagation* and this reaction will proceed until the reactive center is destroyed by one of a number of *termination* reactions or until the supply of monomer is exhausted. Addition polymerization may then be represented as follows:

$$
\left.
\begin{array}{l}
I \longrightarrow R^* \\
R^* + M \longrightarrow RM^*
\end{array}
\right\} \quad \text{initiation}
$$

$$
\left.
\begin{array}{l}
RM^* + M \longrightarrow RMM^* \\
RMM^* + M \longrightarrow RMMM^* \\
\qquad \text{or in general} \\
R(M)_x M^* + M \longrightarrow R(M)_{x+1} M^*
\end{array}
\right\} \quad \text{propagation}
$$

$$
R(M)_x M^* \longrightarrow \text{inactive polymer} \} \quad \text{termination}
$$

2.2.1.1. General Considerations of Polymerizability

The elementary processes in polymerization reactions are in principle reversible and subject to the laws of thermodynamics. Polymerization will be possible only if the free energy difference ΔG between monomer and polymer is negative. Since ΔG depends only on the initial and final states of the system, it is independent of the chemical nature of the intermediates, whether they are free radicals or ionic species. The chemical nature of the intermediate is extremely important however when considering the kinetic feasibility of a reaction. Whereas the polymerization of a wide variety of monomers is feasible from the thermodynamic standpoint (ΔG negative), very specific reaction conditions may be required to actually accomplish a particular polymerization. Indeed, many simple molecules such as acetone, benzene, and acetonitrile, formerly thought of only as solvents, can now be classified as "monomers" since they can now be polymerized in the presence of specific initiators or catalysts. The same applies to the polymerization of the α-olefins, which only became possible with the development of specific coordination catalysts.

2.2.1.2. Initiation

The reactive initiating species R^* must be capable of adding to the π bond of the monomer. When R^* is a free radical species $R \cdot$, it produces homolytic opening of the π bond with the resultant formation of a new free radical at the propagating end. Similarly, when the reactive species is electrophilic R^+ or nucleophilic R^-, heterolytic opening of the π bond occurs with the formation of a carbenium ion or carbanion, respectively, at the propagating end. It is the nature of the reactive chain end that determines the polymerization mechanism. Thus, free radical initiators give rise to free radical polymerization. Similarly, electrophilic and nucleophilic initiators give rise to cationic and anionic polymerization mechanisms, respectively.

2.2.1.3. Effect of Substituents

As was noted in Section 2.2.1.1, the nature of the intermediates—radicals, cations or anions—can have a profound effect on the kinetic feasibility of a particular polymerization. Whereas styrene can be polymerized by all three mechanisms, methacrylate esters can only be polymerized by radical and anionic mechanisms, isobutene can only be polymerized by a cationic process, and vinyl chloride only by a free radical reaction.

The actual mechanism by which a monomer polymerizes depends on the nature (inductive and resonance characteristics) of the substituents around the double bond and their number and spatial arrangement. When the spatial arrangement of substituents provides severe steric crowding around the active center, the monomer is not normally homopolymerizable by any method (Minoura, 1969). For example, homopolymers cannot normally be made from 1,2-disubstituted ethylenes although dialkoxy-substituted ethylenes will homopolymerize. In this latter case the steric strain is relieved by the flexible nature of the C–O–C linkage.

The susceptibility of a monomer to a particular polymerization mechanism depends primarily on the degree of stabilization of the active propagating center afforded by the substituents. Almost all substituents are able to stabilize a free radical species by resonance delocalization of the radical over two or more atoms, as for example in the polymerization of methyl methacrylate:

$$R \cdot + CH_2 {=} \underset{\underset{OCH_3}{|}}{\overset{CH_3}{\underset{|}{\overset{|}{C}}}} {\overset{|}{\underset{C=O}{}}} \longrightarrow R{-}CH_2{-}\underset{\underset{OCH_3}{|}}{\overset{CH_3}{\underset{|}{\overset{|}{C}\cdot}}}\underset{C=O}{} \longleftrightarrow R{-}CH_2{-}\underset{\underset{OCH_3}{|}}{\overset{CH_3}{\underset{||}{\overset{|}{C}}}}\underset{C{-}O\cdot}{}$$

Consequently, most vinyl monomers are polymerizable by free radical initiators.

On the other hand, monomers undergoing anionic polymerization require electron-withdrawing substitutents such as cyano or ester groups to facilitate attack by an anionic species and to stabilize the new propagating carbanion formed. Thus, in addition to being polymerizable by free radical initiators, methyl methacrylate is also susceptible to attack by anionic initiators since the ester group can also stabilize the propagating anionic species by resonance:

$$R^- + CH_2 {=} \underset{\underset{OCH_3}{|}}{\overset{CH_3}{\underset{|}{\overset{|}{C}}}} {\overset{|}{\underset{C=O}{}}} \longrightarrow R{-}CH_2{-}\underset{\underset{OCH_3}{|}}{\overset{CH_3}{\underset{|}{\overset{|}{C}^-}}}\underset{C=O}{} \longleftrightarrow R{-}CH_2{-}\underset{\underset{OCH_3}{|}}{\overset{CH_3}{\underset{||}{\overset{|}{C}}}}\underset{C{-}O^-}{}$$

Similar resonance stabilization is afforded by phenyl and alkenyl groups even though these groups tend to be electron donating, e.g., in styrene polymerization

$$R^- + CH_2{=}CH \longrightarrow R{-}CH_2{-}\bar{C}H \longleftrightarrow R{-}CH_2{-}CH$$

Carbenium ions are stabilized by electron-donating substituents. Such groups increase the electron density at the double bond, thereby facilitating attack by a cationic species, and also stabilize the resulting propagating carbenium ion by resonance delocalization of the positive charge. Vinyl compounds containing such substituents as alkyl, alkoxy, phenyl, and alkenyl groups may therefore be expected to polymerize cationically. For example, the following resonance structures may be written for vinyl ethers:

$$R^+ + CH_2 \!\!=\!\! \underset{\underset{OR}{|}}{\overset{\overset{H}{|}}{C}} \longrightarrow R\!-\!CH_2\!-\!\underset{\underset{OR}{|}}{\overset{\overset{H}{|}}{C^+}} \longleftrightarrow R\!-\!CH_2\!-\!\underset{\underset{\overset{OR}{+}}{||}}{\overset{\overset{H}{|}}{C}}$$

Halogens are electron withdrawing by inductive effects and electron donating by resonance effects, but since both effects are weak neither cationic nor anionic polymerization is appreciably facilitated. In systems where the monomer is highly polarized, radical polymerization does not take place. Thus, aldehydes and ketones are polymerized only by anionic and cationic initiators.

2.2.1.4. Structural Arrangement of Monomer Units

Consideration of the propagation reaction involving the addition of a monosubstituted or disubstituted vinyl monomer to the active species leads to the conclusion that there are two possible addition modes, referred to as head-to-tail (I) and head-to-head tail-to-tail (II) (Chapter 1, Section 1.3):

$$\sim\!\!\sim\!\! CH_2\!-\!\underset{\underset{Y}{|}}{\overset{\overset{X}{|}}{C^*}} + CH_2\!\!=\!\!\underset{\underset{Y}{|}}{\overset{\overset{X}{|}}{C}} \longrightarrow \sim\!\!\sim\!\! CH_2\!-\!\underset{\underset{Y}{|}}{\overset{\overset{X}{|}}{C}}\!-\!CH_2\!-\!\underset{\underset{Y}{|}}{\overset{\overset{X}{|}}{C^*}}$$

I

$$\sim\!\!\sim\!\! CH_2\!-\!\underset{\underset{Y}{|}}{\overset{\overset{X}{|}}{C^*}} + CH_2\!\!=\!\!\underset{\underset{Y}{|}}{\overset{\overset{X}{|}}{C}} \longrightarrow \sim\!\!\sim\!\! CH_2\!-\!\underset{\underset{Y}{|}}{\overset{\overset{X}{|}}{C}}\!-\!\underset{\underset{Y}{|}}{\overset{\overset{X}{|}}{C}}\!-\!CH_2^*$$

II

In practice, steric and resonance effects almost invariably produce head-to-tail addition. For example, if the active center is a free radical species, the propagating radical I is more stable than II because of the resonance stability provided by the substituents X and Y. Particularly when the X and Y substituents are relatively bulky, there will be substantial steric hindrance to the approach of a monomer to give structure II compared with the approach at the unsubstituted carbon to give structure I. In certain cases, when these substituents are small and do not have large resonance stabilizing effects, head-to-head propagation may occur. For example, high resolution ^{19}F

NMR studies have demonstrated the existence of about 16% head-to-head placements in poly(vinyl fluoride) (Wilson and Santee, 1965).

It may be noted that head-to-head polymers can be prepared by special synthetic techniques. For example, Tanaka and Vogl (1974) prepared head-to-head poly(methyl acrylate) by esterification of an alternating copolymer of ethylene and maleic anhydride:

$$\text{www}CH_2\text{—}CH_2\text{—}CH\text{—}CH\text{ www} \xrightarrow{CH_3OH/H_2SO_4} \text{www}CH_2\text{—}CH_2\text{—}CH\text{—}CH\text{ www} \xrightarrow{CH_3N_2}$$

with the groups $O{=}C$ and $COOH$, OCH_3 pendant, and the final product:

$$\text{www}CH_2\text{—}CH_2\text{—}CH\text{——}CH\text{ www}$$
$$O{=}C \qquad C{=}O$$
$$OCH_3 \quad OCH_3$$

2.2.2. FREE RADICAL ADDITION POLYMERIZATION

Free radical addition polymerization is the most commonly used method for the preparation of polymers from a wide variety of vinyl and diene monomers. Of all the addition polymerization processes, it is the most widely studied and the best understood. A complete understanding of the mechanism requires a detailed analysis of the various steps in the chain process, viz., initiation, propagation, and termination. These are discussed in the following sections.

2.2.2.1. Initiation

This is the first step in the chain process. It requires the generation of free radicals capable of reacting with monomer according to the reaction

$$R\cdot + CH_2{=}\underset{X}{\overset{H}{C}} \longrightarrow R\text{—}CH_2\text{—}\underset{X}{\overset{H}{C}}\cdot$$

Initiating free radicals may be generated by several methods. They may be formed directly from the monomer by exposure to heat (thermal polymerization) or to ultraviolet or high-energy radiation. Usually, they are generated via decomposition of some other molecule termed an initiator although sometimes referred to rather inaccurately as a catalyst.

Some monomers, notably styrene and methyl methacrylate, will undergo spontaneous or thermal polymerization in the absence of other initiating species or in the absence of radiation. (It may be noted that a pure thermal mechanism is often difficult to establish since trace impurities such as peroxides often prove to be the true initiator in such systems.) Various mechanisms have been proposed to explain the observed kinetic features, but as yet the true mechanism remains obscure (Allen and Patrick, 1974).

Free radicals are produced when the monomer is exposed to high-energy radiation or irradiated with ultraviolet light (provided it has sufficiently strong absorption in the ultraviolet region). The latter is referred to as photopolymerization. Under high-energy irradiation, ionization of the monomer occurs with the formation of a cationic molecular species (Chapiro, 1962; Wilson, 1974)

$$M \rightsquigarrow M^+ + e^-$$

together with molecular excitation

$$M \rightsquigarrow M^*$$

These species, produced in the primary act, can undergo a variety of reactions that result in the formation of radical, cationic, and anionic species. Thus propagation by all three mechanisms is possible although it is usually only at low temperatures that the ionic species are sufficiently stable to initiate polymerization. Consequently, most radiation-induced polymerizations carried out at room temperature and above proceed by a free radical mechanism.

Since the energy of ultraviolet light with $\lambda > 200$ nm is insufficient to cause ionization, free radical mechanisms prevail although the exact mechanism of initiation, as in thermal initiation, is somewhat obscure.

The common method of initiating vinyl polymerization is by decomposition of an added initiator. Decomposition may be effected either by heat or radiation or by bimolecular reactions, such as one-electron transfer processes that occur in certain oxidation–reduction systems. The field has been reviewed extensively by Bevington (1961), North (1966) and O'Driscoll and Ghosh (1969).

The most widely used method is the thermal decomposition of organic compounds such as peroxides, hydroperoxides, and azo compounds. These compounds contain labile bonds (bond energy ~ 25–40 kcal/mole) that undergo homolytic dissociation over a temperature range characteristic of the particular initiator. Benzoyl peroxide undergoes thermal cleavage of the peroxy bond according to the reaction

This reaction is a two-stage process, although in the presence of monomer, no CO_2 formation is observed, indicating that it is the intermediate benzoyl radicals that start polymerization.

Hydroperoxides undergo similar reactions, e.g., cumyl hydroperoxide thermally decomposes according to the reaction

$$C_6H_5-\underset{\underset{CH_3}{|}}{\overset{\overset{CH_3}{|}}{C}}-O-OH \longrightarrow C_6H_5-\underset{\underset{CH_3}{|}}{\overset{\overset{CH_3}{|}}{C}}-O\cdot + \cdot OH$$

Azo compounds decompose with the liberation of a molecule of nitrogen. Azobisisobutyronitrile (AIBN) is typical of this class of initiator and is one of the most widely used both in bulk and in solution polymerization:

$$CH_3-\underset{\underset{CN}{|}}{\overset{\overset{CH_3}{|}}{C}}-N=N-\underset{\underset{CN}{|}}{\overset{\overset{CH_3}{|}}{C}}-CH_3 \longrightarrow 2CH_3-\underset{\underset{CN}{|}}{\overset{\overset{CH_3}{|}}{C}}\cdot + N_2$$

The initiator chosen for a particular polymerization will mainly depend on the temperature at which polymerization is to be carried out. Table 2.1 shows some of the more common peroxy and azo initiators and the temperature range over which they decompose at a rate suitable for polymerization.

The majority of these initiators are soluble in organic solvents. In aqueous solution polymerizations and in emulsion systems, water-soluble initiators such as hydrogen peroxide and potassium persulfate are used. The latter forms a sulfate anion radical on decomposition according to the reaction

$$(O_3S-O-O-SO_3)^{2-} \longrightarrow 2SO_4^-\cdot$$

The decomposition of peroxy and azo compounds can also be induced photochemically. Unlike thermal decomposition for which the activation energy is of the order of 30 kcal/mole, the activation energy for photochemical initiation is approximately zero and so polymerization can be initiated at much lower temperatures. Other compounds with bonds that are too strong to undergo thermal homolysis can be cleaved photolytically and are efficient initiators of polymerization under ultraviolet irradiation. Compounds of this type include benzoin (and its derivatives) and disulfides:

$$RSSR \xrightarrow{h\nu} 2RS\cdot$$

There are other means of reducing the activation energy of the peroxide decomposition reaction, thereby markedly increasing the rate of decomposi-

TABLE 2.1

INITIATORS FOR RADICAL POLYMERIZATION

Name	Formula	Useful polymerization temperature (°C)[a]
Benzoyl peroxide	C_6H_5—C(=O)—O—O—C(=O)—C_6H_5	40–90
t-Butyl peroxide	CH_3—C(CH_3)(CH_3)—O—O—C(CH_3)(CH_3)—CH_3	80–150
t-Butyl hydroperoxide	CH_3—C(CH_3)(CH_3)—O—OH	100–170
sec-Butyl peroxydicarbonate	$CH(CH_3)(C_2H_5)$—O—C(=O)—O—O—C(=O)—O—$CH(CH_3)(C_2H_5)$	15–60
Azobisisobutyronitrile	CH_3—C(CH_3)(CN)—N=N—C(CH_3)(CN)—CH_3	25–80
Potassium persulfate	KO—S(=O)(=O)—O—O—S(=O)(=O)—OK	40–80

[a] Range over which the rate constant for initiator decomposition varies between $\sim 10^{-4}$ sec and 10^{-7} sec^{-1}.

tion at low temperature. The addition of a reducing agent results in radical formation from an oxidation–reduction reaction between the two components. Metal ions in their reduced state such as Fe^{2+} or Cu^+ are commonly used, e.g.,

$$C_6H_5\text{—C}(CH_3)(CH_3)\text{—O—OH} + Fe^{2+} \longrightarrow C_6H_5\text{—C}(CH_3)(CH_3)\text{—O}\cdot + OH^- + Fe^{3+}$$

Another reaction typical of this type of initiation is the reaction between ferrous ion and hydrogen peroxide (Fenton's reagent)

$$Fe^{2+} + H-O-O-H \longrightarrow HO\cdot + Fe^{3+} + OH^-$$

A large variety of hydroperoxides and peroxides can be used as well as certain inorganic peroxides such as the persulfate ion

$$Fe^{2+} + S_2O_8^{2-} \longrightarrow Fe^{3+} + SO_4^- + SO_4^{2-}$$

There is also a wide choice of reducing agents. Various metal cations in low valence states such as Cr^{2+}, V^{2+}, and T_i^{3+} have been used. In general, such systems are water soluble and so find use as aqueous emulsion polymerization initiators. In organic media, certain amines can be used as reducing agents, e.g., dimethyl aniline–benzoyl peroxide has been extensively investigated as an initiator of low-temperature polymerization (O'Driscoll and Ghosh, 1969). Metal ions may be used as reductants in nonaqueous media and can be introduced as naphthenate salts—for example, benzoyl peroxide with copper naphthenate

The initiating systems discussed so far may be thought of as somewhat conventional. Many of them are used industrially even though in some cases the exact mechanistic details of the initiation process are not known.

In recent years, many novel initiating systems have been reported based on the use of various organometallic compounds (Inoe, 1969; Bamford, 1974). Organozinc, cadmium, boron, and aluminum compounds in conjunction with oxygen as a coinitiator are active initiators for the polymerization of vinyl compounds at low temperatures. The mechanism involves formation of an organometallic peroxide although the initiating radical does not appear to be produced by simple decomposition of the peroxide. For example, in the case of initiation by the trialkyl boron–oxygen system, the following mechanism has been postulated (Barney et al., 1966)

$$R_3B + O_2 \longrightarrow R_2B-O-O-R$$

$$R_2BOOR + 2R_3B \longrightarrow 2R\cdot + R_2BOBR_2 + R_2BOR$$

Initiation by transition metal carbonyls and transition metal chelates has been extensively studied. It was found that the carbonyls of chromium, molybdenum, tungsten, manganese, rhenium, cobalt, and nickel are initiators for free radical polymerization of methyl methacrylate in the presence of an organic halide, notably carbon tetrachloride, at 80 or 100°C. The process of radical formation in these systems involves electron transfer from transition metal to halide with the metal assuming a higher oxidation state and the halide generally splitting into an ion and a radical fragment. The following initiation scheme may be written for the $Mn_2(CO)_{10}$–CCl_4 system (Bamford and Denyer, 1966). The trichloromethyl radical is the initiating species.

$$Mn_2(CO)_{10} \rightleftharpoons Mn_2(CO)_9 + CO$$
$$Mn_2(CO)_9 + CCl_4 \longrightarrow complex + CO$$
$$complex \longrightarrow \cdot CCl_3 + complex$$

Various chelates such as the acetyl acetonate derivatives of manganese and cobalt will initiate the polymerization of styrene at temperatures of 100–180°C. Kinetic evidence and copolymerization studies reveal that the polymerization is initiated by a free radical process and end group analysis gives evidence for the generation of initiating radicals from the ligands

$$(acac)_2Mn \underset{O=C}{\overset{O-C}{\diagup}} \overset{\diagup CH_3}{\underset{\diagdown CH_3}{CH}} \rightarrow (acac)_2Mn^{II} + \overset{O=C}{\underset{O=C}{}} \overset{\diagup CH_3}{\underset{\diagdown CH_3}{CH\cdot}}$$

2.2.2.2. Propagation

Propagation involves the successive addition of monomer units to the active radicals generated in the initiation step.

$$R-CH_2-\overset{H}{\underset{X}{C}}\cdot + CH_2=\overset{H}{\underset{X}{C}} \longrightarrow R-CH_2-\overset{H}{\underset{X}{C}}-CH_2-\overset{H}{\underset{X}{C}}\cdot$$

or in general

$$R(CH_2-\overset{H}{\underset{X}{C}})_xCH_2-\overset{H}{\underset{X}{C}}\cdot + CH_2=\overset{H}{\underset{X}{C}} \longrightarrow R(CH_2-\overset{H}{\underset{X}{C}})_{x+1}CH_2-\overset{H}{\underset{X}{C}}\cdot$$

As discussed in Section 2.2.1.4, addition occurs in the head-to-tail configuration. In dienes with isolated double bonds, both bonds can react independently of each other and cross-linking can result. In certain cases, cooperative addition reactions of one double bond to the other on the same molecule lead to polymers with cyclic repeating units in a process known as *cyclopolymerization* (McCormick and Butler, 1972; Solomon, 1975). This addition mode is particularly important if 5- or 6-membered rings are formed.

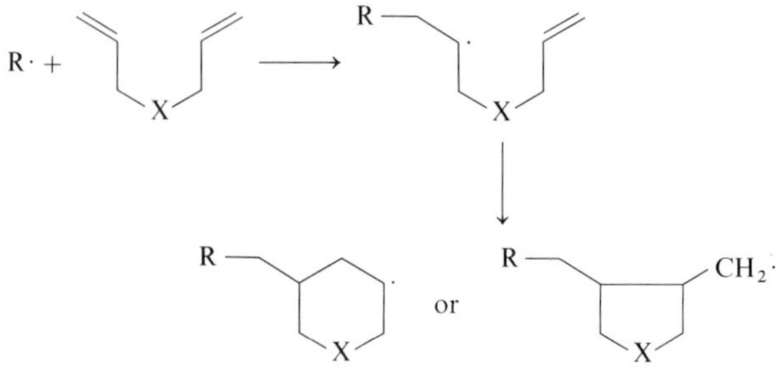

$R \cdot + $ [diene structure with X] $\longrightarrow$ [radical structure with R and X]

$\downarrow$

[6-membered ring structure with R and X] or [5-membered ring structure with R, X and $CH_2\cdot$]

Thermodynamic reasoning suggests that the 6-membered ring should form in preference to the 5-membered ring because of the greater stability of the secondary radical produced. However, a number of systems are known to give 5-membered rings, suggesting that the ring structure may be determined by kinetic factors rather than thermodynamic factors (Solomon, 1975).

In the case of conjugated dienes such as 1,3-butadiene, propagation may proceed by several modes (Chapter 1, Section 1.3):

(i)   1,4 addition with both cis and trans geometric isomers being possible;
(ii) 1,2 addition.

The various structures are shown below:

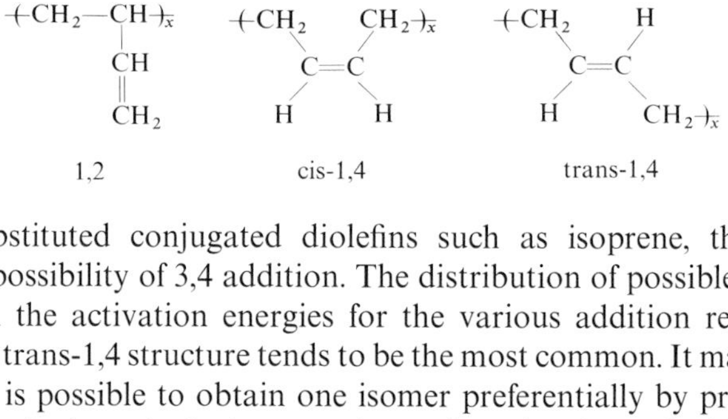

$\begin{array}{ccc}
\cdots(CH_2{-}CH)_x{\cdots} & \cdots(CH_2 \quad CH_2)_x{\cdots} & \cdots(CH_2 \quad H) \\
\quad | & \quad \diagdown \diagup & \quad \diagdown \diagup \\
\quad CH & \quad C{=}C & \quad C{=}C \\
\quad \| & \diagup \quad \diagdown & \diagup \quad \diagdown \\
\quad CH_2 & H \quad\quad H & H \quad\quad (CH_2)_x{\cdots} \\
\\
1,2 & cis\text{-}1,4 & trans\text{-}1,4
\end{array}$

With substituted conjugated diolefins such as isoprene, there is the additional possibility of 3,4 addition. The distribution of possible structures depends on the activation energies for the various addition reactions; in general, the trans-1,4 structure tends to be the most common. It may be noted here that it is possible to obtain one isomer preferentially by prealignment of the molecules in an inclusion complex such as that formed with urea. The monomer is ordered in the canals of the crystal complex so that propagation takes place in only one way. Butadiene, for example, forms a clathrate complex with urea. Polymerization is stereospecific giving a crystalline polymer with 99% trans-1,4 structure (White, 1960).

2.2.2.3.   Termination

The process whereby the activity of the growing chain end is destroyed is known as termination. Free radicals may be terminated by a variety of mechanisms both first and second order in reactive species. These have been reviewed by North and Postlethwaite (1969). Second-order processes are by far the most important where the two principal modes of termination are

(a) Combination

$$\sim\!\!\sim\!\!\sim CH_2-\underset{\underset{X}{|}}{\overset{\overset{H}{|}}{C}}\cdot \;+\; \cdot\underset{\underset{X}{|}}{\overset{\overset{H}{|}}{C}}-CH_2\sim\!\!\sim\!\!\sim \;\longrightarrow\; \sim\!\!\sim\!\!\sim CH_2-\underset{\underset{X}{|}}{\overset{\overset{H}{|}}{C}}-\underset{\underset{X}{|}}{\overset{\overset{H}{|}}{C}}-CH_2\sim\!\!\sim\!\!\sim$$

(b) Disproportionation

$$\sim\!\!\sim\!\!\sim CH_2-\underset{\underset{X}{|}}{\overset{\overset{H}{|}}{C}}\cdot \;+\; \cdot\underset{\underset{X}{|}}{\overset{\overset{H}{|}}{C}}-CH_2\sim\!\!\sim\!\!\sim \;\longrightarrow\; \sim\!\!\sim\!\!\sim CH_2-CH_2 \;+\; \underset{\underset{X}{|}}{\overset{\overset{H}{|}}{C}}\!\!=\!\!CH\sim\!\!\sim\!\!\sim$$

The latter takes place by a hydrogen transfer reaction and results in the formation of two molecules, one of which has an unsaturated end group. The mode of termination can be determined by measuring the number average molecular weight in conjunction with quantitative end group determination using polymers prepared with radioactively labeled initiators such as ^{14}C-containing AIBN. Termination by combination will lead to two radioactive end groups per chain and termination by disproportionation will lead to only one.

In contrast to combination, disproportionation is an activated process since it requires transfer of a hydrogen atom. Hence the rate of the disproportionation reaction is temperature dependent and becomes increasingly important as the temperature is increased. The temperature at which it does become important depends on the monomer, e.g., termination during polymerization of methyl methacrylate at 60–70°C is almost entirely by disproportionation, whereas with styrene and acrylonitrile it is negligible over this temperature range and termination by combination predominates (Bamford *et al.*, 1959).

In processes that remove radicals via entanglement or occlusion in an impermeable polymer, termination corresponds to a unimolecular loss of radical activity. Such a physical deactivation process occurs in the bulk polymerization of acrylonitrile since the polymer is insoluble in the monomer and precipitates as a solid in which neither propagation or termination can occur further. As might be expected, this situation gives rise to different kinetic expressions and the reader is referred to other texts for further discussion of this point (Jenkins, 1967).

2.2.2.4. *Overall Kinetics*

Now that the various steps of the chain reaction are known, kinetic expressions for the rate of reaction, molecular weight, and molecular weight

distribution may be derived. Consider a system in which radicals are generated by the decomposition of a chemical initiator in a homogeneous reaction
medium. Rate expressions for each of the kinetic processes of the chain
reaction may be written as follows:

Initiation

$$I \xrightarrow{k_d} 2R\cdot$$

$$R\cdot + M \xrightarrow{k_i} RM\cdot$$

Rate expression

$$R_i = \frac{d[R\cdot]}{dt} = -2\left(\frac{d[I]}{dt}\right) = 2fk_d[I] \tag{2.1}$$

The decomposition of initiator into primary radicals is considerably slower
than the addition of the primary radical to monomer and thus is the rate
determining step in the initiation sequence. In fact, the rate constant for
decomposition is often so small that very little of the initiator may be consumed in polymerizing to complete conversion, so that $[I]$ can be regarded
as constant.

The factor f is termed the *initiator efficiency* and is included in the rate
expression to take account of the fact that not all of the radicals produced
initiate a chain. Reduced efficiency is attributed to cage effects in which
radicals formed in the initiation reaction have a finite probability of recom-
bining before they can diffuse through the cage of surrounding liquid
molecules and initiate polymerization. Values for f usually range between
0.6 and 1.

The factor of 2 in the rate expression arises from the fact that two radicals
are produced from the decomposition of one initiator molecule.

Propagation

$$RM\cdot + M \xrightarrow{k_p} RM_2\cdot$$

$$RM_2\cdot + M \xrightarrow{k_p} RM_3\cdot$$

$$RM_{\dot{x}} + M \xrightarrow{k_p} RM_{\dot{x}+1}$$

Rate expression

$$R_p = -\frac{d[M]}{dt} = k_i[R\cdot][M] + k_p[M]\sum[M_{\dot{x}}] = k_p[M][M\cdot] \tag{2.2}$$

In this rate expression, it is assumed that the radical reactivity expressed
by the propagation rate constant k_p is independent of the molecular weight
of the growing chain. Hence $[M\cdot]$ represents the total concentration of all
chain radicals. This assumption is consistent with the concept of a flexible,

freely moving chain able to undergo collisions with monomer molecules. The number of monomer molecules consumed by initiation is small compared with the number consumed during propagation, and hence the contribution of the former to the rate of monomer loss is omitted in the final expression and the polymerization rate is given simply by the rate of propagation, Eq. (2.2).

Termination   (i)   Combination

$$RM_{\dot{x}} + RM_{\dot{y}} \xrightarrow{k_{tc}} P_{x+y}$$

Rate expression

$$R_t = -\frac{d[R\cdot]}{dt} = 2k_{tc}[M\cdot]^2 \qquad (2.3)$$

(ii)   Disproportionation

$$RM_{\dot{x}} + RM_{\dot{y}} \xrightarrow{k_{td}} P_x + P_y$$

Rate expression

$$R_t = -\frac{d[R\cdot]}{dt} = 2k_{td}[M\cdot]^2 \qquad (2.4)$$

Although the two termination processes are different and result in quite different molecular weights, the kinetics of radical disappearance are the same so that the termination step can be expressed by

$$RM_{\dot{x}} + RM_{\dot{y}} \xrightarrow{k_t} P$$

for which

$$R_t = -\frac{d[R\cdot]}{dt} = 2k_t[M\cdot]^2 \qquad (2.5)$$

where

$$k_t = k_{tc} + k_{td} \qquad (2.6)$$

The factor of 2 in the rate expression is included by convention to indicate that two molecules are involved in the termination reaction.

In order to solve these equations in a convenient but realistic fashion, use is made of the so called *steady-* or *stationary-state assumption* that requires that the concentration of radicals rapidly increase to a constant steady-state value, i.e., the rate of change of radical concentration should quickly approach and then remain zero during the course of polymerization. This implies that the rate of radical production must equal the rate of radical termination so that

$$2fk_d[I] = 2k_t[M\cdot]^2 \qquad (2.7)$$

whence

$$[M\cdot] = \left(f\frac{k_d}{k_t}[I] \right)^{1/2} \tag{2.8}$$

Substituting for $[M\cdot]$ in the propagation rate expression Eq. (2.2), the expression for the rate of polymerization becomes

$$R_p = -\frac{d[M]}{dt} = k_p \left(f\frac{k_d}{k_t}[I] \right)^{1/2}[M] \tag{2.9}$$

Equation (2.9) is the rate expression for polymerization under homogeneous conditions and has been shown to hold in the early stages of polymerization for a large number of free radical polymerizations. The equation predicts a first-order dependence on monomer concentration and a square-root dependence on initiator concentration, the latter being a direct consequence of the bimolecular nature of the termination reaction.

2.2.2.5. Absolute Rate Constants

It is frequently of interest to know the absolute magnitude of the individual rate constants k_d, k_p, and k_t since knowledge of these rate constants would enable one to correlate the reactivity of the species involved with its chemical structure.

While the rate constant for the initiation reaction can be estimated in favorable cases from experiments with inhibitors or by measuring the rate of decomposition of the initiator, the determination of k_p and k_t is far from straightforward. In principle, monomer disappearance is predominantly due to the propagation reaction, particularly when the degree of polymerization is large, in which case the rate of polymerization R_p is given by Eq. (2.2). Hence, measurement of R_p and the radical concentration $[M\cdot]$ for a given monomer concentration would yield a value for k_p. Unfortunately in most common polymerization systems, the concentration of free radicals is very low ($\sim 10^{-8}\,M$) and is beyond the sensitivity of most methods of measurement.

Analysis of the steady-state kinetic expression does allow the determination of the ratio of k_p to k_t as k_p^2/k_t. Rearrangement of Eq. (2.9) gives

$$k_p^2/k_t = R_p^2/f k_d[I][M]^2 \tag{2.10}$$

$$= 2R_p^2/R_i[M]^2 \tag{2.11}$$

from which k_p^2/k_t may be calculated since the variables R_p, R_i, and $[M]$ in Eq. (2.11) can be measured.

However, it is not possible to separate these two rate constants. Various relationships can be derived, but it always happens that there is one more unknown than there are relationships between the unknowns. In order to determine these rate constants separately, recourse to nonstationary state kinetics involving measurement of radical lifetimes is necessary.

The average lifetime τ of the kinetic chain during stationary-state conditions is given by

$$\tau = \frac{\text{steady state radical concentration}}{\text{steady state rate of radical disappearance}} = \frac{[M \cdot]_s}{2k_t[M \cdot]_s^2} = \frac{1}{2k_t[M \cdot]_s}$$

(2.12)

Substituting for $[M \cdot]_s$ by means of Eq. (2.2), the expression for τ becomes

$$\tau = k_p[M]/2k_t R_p$$

(2.13)

Thus any measurement of τ in conjunction with R_p will yield the ratio k_p/k_t, and comparison with a previously determined value of k_p^2/k_t immediately allows evaluation of the individual constants.

The most widely used method of obtaining radical lifetimes is by the intermittent illumination of a photosensitized reaction. In the rotating sector method first employed by Burnett and Melville (1947), a rotating sector, which may be supposed to have equal-sized closed and cut out portions, is placed between the light source and the polymerization vessel so that each period of irradiation Δt is followed by an equally long period of darkness.

If I_0 is the intensity of the source, the rate of polymerization at full illumination can be shown to be given by Eq. (2.14).

$$R_p = k_p[M](\Phi I_0 \varepsilon [I]/k_t)^{1/2}$$

(2.14)

where Φ is the quantum efficiency, I_0 is the incident light intensity, $[I]$ is the photoinitiator concentration and ε is the molar absorptivity (extinction coefficient) for the monomer at the wavelength of the absorbing radiation. For photolysis of the pure monomer without added initiator, this expression becomes

$$R_p = k_p[M]^{3/2}(\Phi I_0 \varepsilon/k_t)^{1/2}$$

(2.15)

In either case, R_p is directly proportional to the half power of the incident intensity.

The sector is now set rotating at a rate such that the duration of the light flash Δt is much longer than the radical lifetime. The radical concentration rapidly increases to its stationary-state value, and polymerization takes place at a rate equal to $k_o I_0^{1/2}$, where k_o is the apparent overall rate constant. During the dark period, the rate of polymerization will be essentially zero since the time interval over which the radicals decay is very short compared with the length of the dark period. The overall rate of polymerization will therefore be given by

$$R_{p(\text{slow sector speed})} = \tfrac{1}{2}k_o I_0^{1/2}$$

(2.16)

Now suppose that the sector is greatly speeded up so that the flashes are considerably shorter than τ. Radicals initiated during a light interval will continue to grow through several successive light and dark intervals. The radical concentration neither rises nor falls to the extreme values experienced during slow sector rotation and the effect approaches that produced by steady illumination having an intensity one-half that of the unsectored illumination, i.e.,

$$R_{\text{p(fast sector speed)}} = k_o(\tfrac{1}{2}I_0)^{1/2} \tag{2.17}$$

These two cases are illustrated in Fig. 2.1 as a plot of radical concentration or rate of polymerization against time. So long as the intensity exponent is not unity, the average rates for high and low sector speeds will differ, the change from one value to the other occurring when the dark time is approximately equal to the radical lifetime. From Eqs. (2.16) and (2.17) it is apparent that

$$R_{\text{p(slow)}} = 0.707\, R_{\text{p(fast)}} \tag{2.18}$$

In a typical experiment, the average rate of polymerization is measured as a function of the speed of sector rotation and plotted as the rate ratio of the average polymerization rate (R_p (average)) to the polymerization rate corre-

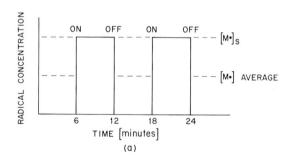

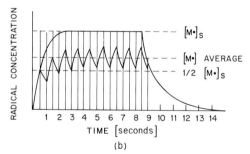

FIG. 2.1. Variation of radical concentration with time for intermittent UV irradiation. (a) $\Delta t \gg \tau$, $[\text{M}\cdot]_{\text{av}} = \tfrac{1}{2}[\text{M}\cdot]_{\text{s}}$. (b) $\Delta t \leq \tau$, $[\text{M}\cdot]_{\text{av}} > \tfrac{1}{2}[\text{M}\cdot]_{\text{s}}$ (Vollmert, 1973).

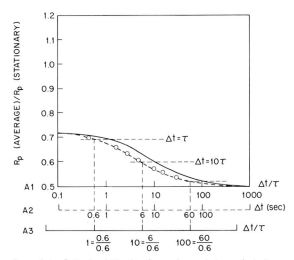

FIG. 2.2. Semilog plot of R_p (ave)/R_p (stationary) vs. $\Delta t/\tau$ and Δt:(————), theoretical curve (abscissa A_1, log scale); (–○–), experimental curve (abscissa A_2, log scale), arbitrarily chosen; A_3 = abscissa A_1, displaced until both curves coincide (Vollmert, 1973).

sponding to the stationary-state radical concentration (R_p (stationary)) against log Δt (Fig. 2.2). The shape of this curve may be calculated theoretically (Briers et al., 1926) as a function of $\log(\Delta t/\tau)$. Since the abscissa of the theoretical function is $\log \Delta t - \log \tau$, the value of τ may be calculated by overlaying the experimental and theoretical curves and determining the displacement of the abscissa that would be required to make both curves coincide (Fig. 2.2). Knowing the value of τ, the value of k_p and k_t can then be determined.

Values of the propagation and termination rate constants for several monomers are tabulated in Table 2.2. It can be seen that values of k_p are

TABLE 2.2

ABSOLUTE RATE CONSTANTS FOR PROPAGATION AND
TERMINATION OF SELECTED VINYL MONOMERS AT $25°C^a$

Monomer	$k_p \times 10^{-2}$ (liter/mole sec)	$k_t \times 10^{-7}$ (liter/mole sec)
Acrylonitrile	19.1	14.5
Methyl acrylate	15.8	2.75
Methyl methacrylate	2.6	1.05
Styrene	0.44	2.37
Vinyl chloride	62	55.0
Vinyl acetate	10.0	2.95

^{a} Data are selected from Polymer Handbook (Brandrup and Immergut, 1975).

usually in the range of 10^2-10^4 liters/mole sec indicating that propagation is very rapid with polymer being formed virtually instantaneously. Values of k_t generally fall within the range of 10^6-10^8 liters/mole sec, i.e., significantly greater than k_p, resulting in rapid termination, short radical lifetime and hence, as was indicated earlier, low steady-state radical concentrations. Although $k_t \gg k_p$, the concentration of radicals is small, and hence a significant number of propagation steps take place before termination intervenes.

2.2.2.6. *Deviations from Standard Kinetics*

Equation (2.9) predicts that the rate of polymerization should be proportional to monomer concentration and to the square root of initiator concentration. This relationship has been confirmed for polymerization of many monomers, particularly for low degrees of conversion. Figure 2.3 shows excellent agreement with the predicted half-order dependence on initiator concentration for the polymerization of styrene and methyl methacrylate.

As with any kinetic scheme, there are always exceptions and deviations and although the above kinetic relationships hold in a large number of cases, there are many cases in which they do not. Equation (2.9) implies for example that the initiator efficiency is independent of [M] and [I], an assumption that is not always true, particularly at low values of [M]. This again is attrib-

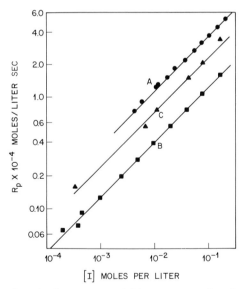

Fig. 2.3. Initial polymerization rate versus initiator concentration: A, methyl methacrylate with AIBN at 50°C; B, styrene with benzoyl peroxide at 60°C; C, methyl methacrylate with benzoyl peroxide at 50°C (Ham, 1967, after Mayo, 1959).

uted to the cage effect in which radicals cannot all diffuse away before being destroyed by recombination. This effect would clearly be more pronounced as the likelihood of a radical meeting a monomer molecule is diminished by dilution.

Deviations from the expected monomer exponent of unity have been reported, the order varying in some cases with monomer concentration Similarly, initiator exponents both greater and smaller than 0.5 have been reported. These effects have been attributed to deviations in the termination reaction, such as primary radical termination, degradative chain transfer, and diffusion effects and to deviations in the propagation reaction, such as propagation by both "hot" and thermalized polymer radicals and formation of complexes between polymer radicals, solvent, and monomer. Aspects of these different factors are discussed in reviews by Smith (1967), Scott and Senogles (1973), and Bamford (1975).

2.2.2.6.1. Autoacceleration. The realization that termination is probably diffusion controlled from the *beginning* of polymerization has only been fully realized recently and is due in large part to the work of North (1966). Deviations resulting from diffusional control of termination are very apparent however in some systems when polymerization is carried to high conversion. Figure 2.4 shows curves for the polymerization of methyl methacrylate as a function of monomer concentration. Provided the monomer concentration is less than 40%, approximately first-order kinetics are observed. At higher monomer concentrations, there is considerable deviation from first-order kinetics, the point of departure depending on the monomer concentration. This deviation, which is manifested by a marked increase in rate of polymerization, is also accompanied by a sharp increase in molecular weight and breadth of the molecular weight distribution.

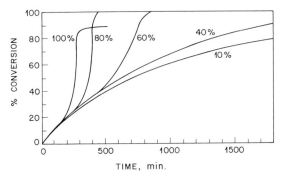

FIG. 2.4. Extent of conversion as a function of time for the benzoyl peroxide catalyzed polymerization of methyl methacrylate in benzene at 50°C. The different curves are for various concentrations of monomer in solvent (Schultz and Haborth, 1948).

The phenomenon is known as *autoacceleration* or the *Trommsdorf* or *gel* effect (Norrish and Smith, 1942; Trommsdorf *et al.*, 1948). It is caused by a decrease in the termination rate constant as conversion increases. The termination reaction is controlled by the rate of diffusion together of the polymeric macroradicals. As conversion increases, the viscosity of the medium also increases with a consequent decrease in the rate of diffusion of the macroradicals, i.e., the kinetics become controlled by diffusion processes rather than by chemical processes. On the other hand, monomer molecules are small enough to diffuse more readily through the reaction medium and this coupled with the lower value for k_p implies that the maintenance of the concentration of monomer at a suitable equilibrium level in the environment of the radical is less subject to diffusion processes. At very high conversion ($>95\%$), where the polymerization mixture approaches a glass at temperatures less than 90°C, the conversion curve levels off and it is necessary to increase the temperature above the glass transition temperature of the polymer (105°C) in order to achieve 100% conversion. This may reflect the onset of diffusion control of the propagation step, or as suggested by Allen and Patrick (1963), microheterogeneity may develop in the polymer gel at such high conversions with the result that polymer radicals become trapped in monomer-free regions.

Since such vinyl polymerization reactions are exothermic, the rapid increase in polymerization rate associated with autoacceleration can lead to thermal runaway of the reaction and even to thermal explosion. In certain instances, polymerization can be autocatalytic from the start of the reaction. This usually occurs in systems where termination is restricted as might occur when a polymer precipitates from the reaction medium. Under such conditions, radicals can be occluded in the precipitated heterogeneous phase where their reduced mobility results in a very long lifetime. Bulk polymerization of acrylonitrile and vinyl chloride are examples of this type of behavior.

2.2.2.6.2. Effect of Temperature. The effect of temperature on the rate of polymerization and molecular weight is of prime importance in polymer production. Quantitatively, temperature effects are complex since R_p depends on a combination of three rate constants, k_d, k_p, and k_t. If it is assumed that each rate constant can be expressed by an Arrhenius-type expression, viz.,

$$k = Ae^{-E/RT} \tag{2.19}$$

where A is the collision frequency factor and E the activation energy, then the temperature dependence of the ratio $k_p(k_d/k_t)^{1/2}$ in Eq. (2.9) can be obtained by combining the three separate Arrhenius expressions:

$$k_p(k_d/k_t)^{1/2} = A_p(A_d/A_t)^{1/2}\exp -(E_p + E_d/2 - E_t/2) \tag{2.20}$$

The overall activation energy for polymerization E_a is therefore given by

$$E_a = E_p + E_d/2 - E_t/2 \qquad (2.21)$$

For the case under consideration, viz., initiation by thermal decomposition of a chemical initiator, E_d is of the order of 30–40 kcal/mole for the chemical initiators commonly used. The E_p values for most monomers lie within the range 5–10 kcal/mole, and E_t is generally in the range of 2–5 kcal/mole. Consequently, the overall activation energy is approximately 20 kcal/mole and hence there should be a two- to threefold increase in polymerization rate for each 10°C increase in temperature. It should be noted that the magnitude of E_a depends primarily on the value of E_d and hence on the method of initiation. For photochemical- or radiation-induced polymerization, the value of E_d is zero since initiation results from the interaction of a photon with the monomer. Consequently, the overall activation energy will be only a few kilocalories per mole and R_p will not be very sensitive to an increase in temperature.

Although the absolute value of E_a may depend on the method of initiation, it is usually positive, and hence R_p should increase with temperature in the manner depicted in curve I, Fig. 2.5.

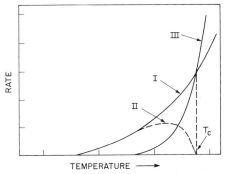

FIG. 2.5. Variation of propagation rate (curve I), overall polymerization rate (curve II), and depropagation rate (curve III) with temperature.

It is observed however that R_p passes through a maximum and then decreases with increasing temperature (curve II), ultimately becoming zero at a temperature that is characteristic of the monomer system and the prevailing conditions of monomer concentration and pressure (Dainton and Ivin, 1958). This temperature is known as the *ceiling temperature* T_c and is the temperature above which it is impossible to form a long chain polymer.

The origin of this phenomenon lies in the reversibility of the propagation reaction, i.e., the propagation step should correctly be written as

$$RM_{\dot{x}} + M \underset{k_{dp}}{\overset{k_p}{\rightleftarrows}} RM_{\dot{x}+1}$$

where k_{dp} is the rate constant for depropagation. Equation (2.2) thus becomes

$$-d[M]/dt = (k_p[M] - k_{dp})[M\cdot] \tag{2.22}$$

The temperature dependence of the rate constants will be given by their Arrhenius expressions $k_p = A_p \exp(-E_p/RT)$ and $k_{dp} = A_{dp}\exp(-E_{dp}/RT)$, and the heat of polymerization ΔH_p will be given by $E_p - E_{dp}$. As already noted, E_p for most monomers lies within the range of 5–10 kcal/mole. Further, the heat of polymerization is negative (exothermic) with values generally ranging from 10 to 20 kcal/mole. Consequently, $E_{dp} \gg E_p$, and the rate of depropagation will increase more rapidly than the rate of propagation as the temperature is raised (curve III, Fig. 2.5). The ceiling temperature then is that temperature at which the rates of propagation and depropagation become equal and the net rate of polymer production is zero. Equation (2.22) becomes

$$-d[M]/dt = 0 = (k_p[M] - k_{dp})[M\cdot] \tag{2.23}$$

so that

$$k_p/k_{dp} = K = 1/[M] \tag{2.24}$$

where K is the equilibrium constant. The thermodynamic implications of the ceiling temperature will be examined in Section 2.2.2.11.

For many vinyl monomers, the ceiling temperature is far higher (equilibrium lies well to the right) than the temperature normally encountered for polymerization. For example, T_c for polymerization of pure styrene is 310°C. Nevertheless, the T_c values of many monomers lie in the temperature range over which polymerizations are usually performed. In those circumstances the failure of a given unsaturated compound to polymerize may be due to the fact that the polymerization temperature is above the ceiling temperature under the prevailing conditions. It follows that it may simply be a matter of lowering the temperature in order to achieve polymerizability. Such an approach can have the effect of making the reaction *kinetically* difficult. For example, the ceiling temperature for the polymerization of α-methylstyrene in bulk is 61°C. Hence polymerization must be initiated at somewhat lower temperatures and since the propagation reaction is hindered by the two bulky substituents on the α-carbon atom, the overall polymerization rate is very slow.

2.2.2.7. *Molecular Weight and Control by Transfer*

So far the kinetic analysis has been concerned with expressions for the polymerization rate in terms of reactant concentrations and kinetic parameters of the various steps in the chain process. It is also possible to derive important information concerning the degree of polymerization (extent to which propagation takes place before some termination process intervenes). This problem is frequently compounded in practice by the phenomenon of chain branching.

One must distinguish between the term chain used in the molecular sense and chain used as a kinetic concept. The kinetic chain length v is defined as

$$v = \frac{\text{total number of monomer units consumed}}{\text{total number of initiator fragments produced}} \tag{2.25}$$

$$= -\int_0^t \frac{d[M]}{dt}\, dt \Big/ \int_0^t \frac{d[R\cdot]}{dt}\, dt \tag{2.26}$$

In a system that has attained the stationary state, both integrands may be constant over a short but significant period of time t. Thus

$$v = \left(-\frac{d[M]}{dt} \cdot t\right) \Big/ \left(\frac{d[R\cdot]}{dt} \cdot t\right) = -\frac{d[M]}{dt} \Big/ \frac{d[R\cdot]}{dt} \quad \text{or} \quad R_p/R_i \tag{2.27}$$

Substituting for R_p and R_i using Eqs. (2.2) and (2.1) yields

$$v = k_p[M][M\cdot]/2f k_d[I] \tag{2.28}$$

Since $R_i = R_t$ under stationary-state conditions, Eq. (2.28) may also be written as

$$v = k_p[M][M\cdot]/2k_t[M\cdot]^2$$
$$= k_p[M]/2k_t[M\cdot] \tag{2.29}$$

By eliminating $[M\cdot]$ in Eq. (2.29) by means of Eq. (2.8), the expression for kinetic chain length becomes

$$v = k_p[M]/2(f k_d k_t[I])^{1/2} \tag{2.30}$$

This equation illustrates an important characteristic of free radical addition polymerization, viz., that the kinetic chain length is inversely proportional to the square root of the initiator concentration. Increasing the initiator concentration leads to smaller size polymer molecules.

In this simplified reaction scheme the kinetic chain length is directly related to the molecular chain length or mean degree of polymerization $\bar{X}_n$.

Termination by combination involves the production of one polymer molecule from two kinetic chains, i.e.,

$$\bar{X}_n = 2v \tag{2.31}$$

Termination by disproportionation (or by unimolecular termination) involves the production of one polymer molecule per kinetic chain, i.e.,

$$\bar{X}_n = v \tag{2.32}$$

If only a fraction x_d of the macromolecules are formed in disproportionation reactions, then

$$\bar{X}_n = 2v/(1 + x_d). \tag{2.33}$$

In all instances, the number average molecular weight $\bar{M}_n$ will be given by

$$\bar{M}_n = \bar{X}_n M_0 \tag{2.34}$$

where M_0 = molecular weight of the monomer.

The mean kinetic chain length of a radical polymerization can be determined by radiochemical assay of the polymers using radioactively labeled initiators. By this means the number of initiator fragments per unit weight of polymer can be calculated, and hence the number of monomer molecules polymerized by each fragment can be determined. It is frequently found that the polymer molecular weight is much lower than that predicted on the basis of a termination mechanism consisting of combination or disproportionation. This effect has been attributed to *chain transfer*, which is a process by which a propagating chain is prematurely terminated by some species from which a new initiating radical is generated (Flory, 1953). It is envisaged that the propagating active end abstracts either a hydrogen or some other atom, e.g., chlorine, from some species in the system, thus terminating itself but regenerating the active site on the new molecule. The reaction may be depicted as

$$\text{\small{wwww}} CH_2 \overset{\overset{\displaystyle H}{|}}{\underset{\underset{\displaystyle X}{|}}{C}} \cdot + AY \longrightarrow \text{\small{wwww}} CH_2 \overset{\overset{\displaystyle H}{|}}{\underset{\underset{\displaystyle X}{|}}{C}} - Y + A \cdot$$

where AY can be monomer, solvent, initiator, polymer, or some other species deliberately added to the system and referred to as a *chain transfer agent*.

The effect of transfer reactions on the rate of polymerization depends on the subsequent reactivity of the radical A· towards the monomer relative to that of the propagating radicals. If the reactivity is equal, R_p will be unaffected although the molecular weight of the polymer will be reduced.

The extent of branching in a polymer is very dependent on the nature of the chain transfer reaction. Transfer reactions between the propagating radical and dead polymer are important and become increasingly probable as conversion increases

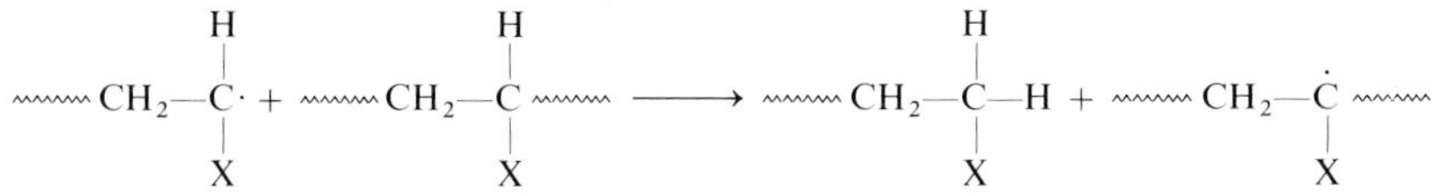

In this way a radical is produced on the backbone from which propagation can ensue, resulting in the production of a long chain branch. An interesting variation of this phenomenon takes place in the high-pressure polymerization of ethylene. The polymer has many butyl and amyl branches along the chain (about 10–20 per $1000CH_2$ groups) that are believed to be formed by intramolecular chain transfer or "back biting" reactions (see Chapter 3, Section 3.8.2). The presence of the short branches reduces the ability of the chains to pack into crystallites and thus strongly influences the physical properties of the polymer.

Intermolecular chain transfer leading to long chain branches can lead to cross-linking, particularly at high conversion. This occurs, for example, in the bulk polymerization of acrylates. The transfer constant of the polymer toward its own radicals is high and cross-linked insoluble polymer is usually obtained. The addition of a specific chain transfer agent to control molecular weight is frequently practised in industrial polymerization where chain length must be controlled to facilitate ease of processing. The most common agents in industrial practice are mercaptans.

2.2.2.8. *Kinetics of Chain Transfer Processes*

The effects of transfer reactions can readily be incorporated into the kinetic scheme developed in Section 2.2.2.4. The following scheme may be written (neglecting transfer to polymer)

Initiation

$$I \xrightarrow{\ k_d\ } 2R\cdot,$$

$$R\cdot + M \xrightarrow{\ k_i\ } RM\cdot$$

$$R_i = 2f k_d[I] \tag{2.35}$$

Propagation

$$RM_{\dot{x}} + M \xrightarrow{\ k_p\ } RM_{\dot{x}+1}$$

$$R_p = k_p[M][M\cdot] \tag{2.36}$$

Transfer to solvent or chain transfer agent

$$RM_{\dot{x}} + S \xrightarrow{\ k_{trS}\ } P_x + S\cdot$$

$$R_{trS} = k_{trS}[M\cdot][S] \tag{2.37}$$

Transfer to monomer

$$RM_{\dot{x}} + M \xrightarrow{k_{trM}} P_x + M\cdot$$

$$R_{trM} = k_{trM}[M\cdot][M] \qquad (2.38)$$

Transfer to initiator

$$RM_{\dot{x}} + I \xrightarrow{k_{trI}} P_x + I\cdot$$

$$R_{trI} = k_{trI}[M\cdot][I] \qquad (2.39)$$

Termination (by coupling)

$$RM_{\dot{x}} + RM_{\dot{y}} \xrightarrow{k_t} P_{x+y}$$

$$R_t = 2k_{tc}[M\cdot]^2 \qquad (2.40)$$

Since the radical concentration remains unaffected by transfer and assuming the rate constant for the addition of M to S·, M·, and I· is of the same magnitude as k_p, then R_p will be given by Eq. (2.9).

The quantitative effects on $\bar{X}_n$ can also be readily derived. For polymerization under stationary-state conditions, $\bar{X}_n$ is defined by

$$\bar{X}_n = \frac{\text{rate of polymerization}}{\text{rate of production of pairs of chain ends}} \qquad (2.41)$$

$$= \frac{\text{rate of polymerization}}{\sum \text{rate of all reactions leading to dead polymer}} \qquad (2.42)$$

$$= \frac{k_p[M][M\cdot]}{k_{tc}[M\cdot]^2 + k_{trS}[M\cdot][S] + k_{trM}[M\cdot][M] + k_{trI}[M\cdot][I]} \qquad (2.43)$$

or

$$\frac{1}{\bar{X}_n} = \frac{k_{tc}[M\cdot]}{k_p[M]} + \frac{k_{trS}[S]}{k_p[M]} + \frac{k_{trM}}{k_p} + \frac{k_{trI}[I]}{k_p[M]} \qquad (2.44)$$

Substituting for $[M\cdot]$ in Eq. (2.44) using Eq. (2.2), one obtains

$$\frac{1}{\bar{X}_n} = \frac{k_{tc}R_p}{k_p^2[M]^2} + \frac{k_{trS}[S]}{k_p[M]} + \frac{k_{trM}}{k_p} + \frac{k_{trI}[I]}{k_p[M]} \qquad (2.45)$$

(It is usual to consider the inverse degree of polymerization since this function exhibits a linear dependence on the rate of polymerization.) The propensity for chain transfer is generally tabulated in terms of a transfer constant C defined by

$$C_S = \frac{k_{trS}}{k_p}, \qquad C_M = \frac{k_{trM}}{k_p}, \qquad C_I = \frac{k_{trI}}{k_p} \qquad (2.46)$$

where C_S, C_M, and C_I are the chain transfer constants to solvent, monomer, and initiator, respectively. Thus, Eq. (2.45) becomes

$$\frac{1}{\overline{X}_n} = \frac{k_{tc}R_p}{k_p^2[M]^2} + C_S\frac{[S]}{[M]} + C_M + C_I\frac{[I]}{[M]} \qquad (2.47)$$

This equation provides a ready means of evaluating the various transfer constants. In the absence of solvent or transfer agent, $C_S = 0$ and Eq. (2.46) reduces to

$$\frac{1}{\overline{X}_n} = \frac{k_{tc}R_p}{k_p^2[M]^2} + C_M + C_I\frac{[I]}{[M]} \qquad (2.48)$$

Hence a plot of $1/\overline{X}_n$ vs. R_p with varying initiator concentration ([M] is held constant) should give a linear plot over the region of low initiator concentration where the contribution of chain transfer to initiator is negligible, i.e., $C_I = 0$. The intercept yields the value of C_M and the slope the value of $k_{tc}/k_p^2[M]^2$, from which k_{tc}/k_p^2 can be determined since [M] is constant and is known. Figure 2.6 shows the dependence of the degree of polymerization of styrene on polymerization rate for several initiating systems. It can be seen that the curve for AIBN does not deviate from linearity and indicates chain transfer to this initiator is negligible. Chain transfer to initiator is also known as *induced decomposition* and the absence of induced decomposition

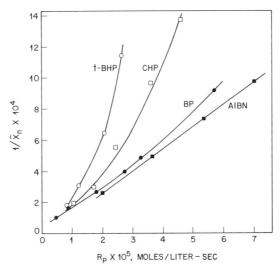

FIG. 2.6. Dependence of the degree of polymerization on the polymerization rate for several initiating systems. *t*BHP—*t*-butyl hydroperoxide; CHP—cumyl hydroperoxide; BP—benzoyl peroxide; AIBN—azobisisobutyronitrile (Baysal and Tobolsky, 1952).

TABLE 2.3
TRANSFER CONSTANTS TO MONOMERS
AT $60°C^a$

Monomer	$C_M \times 10^4$
Acrylonitrile	0.26–0.3
Methyl acrylate	0.036–0.325
Methyl methacrylate	0.07–0.18
Styrene	0.6–1.37
Vinyl acetate	1.75–2.8

[a] Data selected from Polymer Handbook
(Brandrup and Immergut, 1975).

for AIBN makes it a preferred initiator for kinetic studies. Once C_M is known, C_I can be determined by similar rearrangement of Eq. (2.48). Typical values for C_M and C_I are shown in Tables 2.3 and 2.4.

The situation where a chain transfer agent is deliberately added either as a solvent or as an added compound (regulator) represents a special case. If the polymerization is carried out under conditions where $R_p/[M]^2$ is constant (by adjusting the initiator concentration throughout the course of the reaction), and where R_{trI} is negligible (by using a low concentration of initiator or using an initiator such as AIBN, the transfer constant of which is negligible, then Eq. (2.47) may be written

$$1/\bar{X}_n = (1/\bar{X}_n)_0 + C_S[S]/[M] \tag{2.49}$$

TABLE 2.4
CHAIN TRANSFER CONSTANTS TO INITIATORS[a]

	C_I		
Initiator	$\sim\sim\sim CH_2-\overset{\displaystyle H}{\underset{\displaystyle COOCH_3}{C}}\cdot$	$\sim\sim\sim CH_2-\overset{\displaystyle CH_3}{\underset{\displaystyle COOCH_3}{C}}\cdot$	$\sim\sim\sim CH_2-\overset{\displaystyle H}{\underset{\displaystyle \bigcirc}{C}}\cdot$
Azobisisobutyronitrile	0	0	0
t-Butyl peroxide	0.00047 (65°C)	0 (20°C)	0.00023–0.0013
t-Butyl hydroperoxide	0.01	$1.27 \times [I]$	0.035
Benzoyl peroxide	0.0246	0–0.02	0.048–0.101

[a] Values for C_I are at 60°C unless otherwise noted. Data are selected from Polymer Handbook (Brandrup and Immergut, 1975).

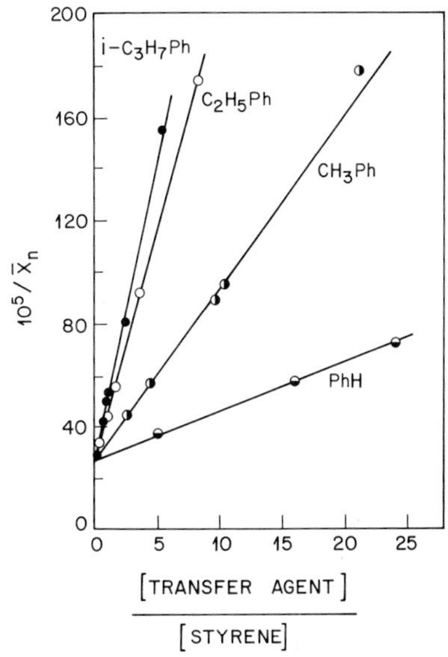

FIG. 2.7. Plots of $1/\bar{X}_n$ versus $[S]/[M]$ showing the effect of various chain transfer agents on the degree of polymerization of styrene at 100°C (Gregg and Mayo, 1947).

where $(1/\bar{X}_n)_0$ is the value of $1/\bar{X}_n$ in the absence of chain transfer agent. Thus a plot of $1/\bar{X}_n$ vs. $[S]/[M]$ should yield a straight line with a slope equal to C_S. Such plots are shown in Fig. 2.7 for several chain transfer agents in styrene polymerization. Transfer constants for several solvents and additives are given in Table 2.5. The magnitude of C_S depends on the nature of the solvent or additive and the chain radical. It is also temperature dependent. The reason that mercaptans are generally preferred as regulators is that their chain transfer constants are large and only small amounts are needed to obtain a considerable decrease in the degree of polymerization.

2.2.2.9 Molecular Weight Distribution

So far, only the average degree of polymerization has been considered, whereas in fact polymer molecules have a broad range of molecular weights referred to as a *molecular weight distribution*. In order to calculate the distribution of molecular weights that results from addition polymerization, it is necessary to make several assumptions:

(a) the concentration of monomer remains constant throughout the polymerization (this condition requires stopping the reaction before any appreciable amount of monomer has been used or else continually adding monomer during polymerization);

(b) the rate of initiation is considered to be constant over the entire time of polymerization; and

TABLE 2.5

TRANSFER CONSTANTS TO SOLVENTS AND ADDITIVES[a]

$C_S \times 10^4$

Solvent	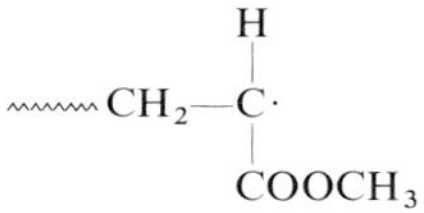 $\sim CH_2-\overset{H}{\underset{COOCH_3}{C}}\cdot$	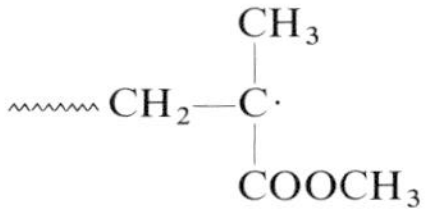 $\sim CH_2-\overset{CH_3}{\underset{COOCH_3}{C}}\cdot$	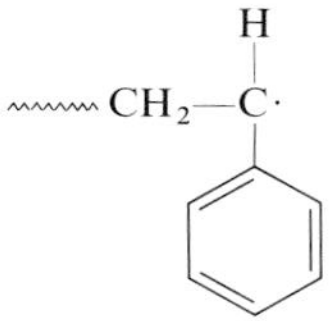 $\sim CH_2-\overset{H}{\underset{C_6H_5}{C}}\cdot$	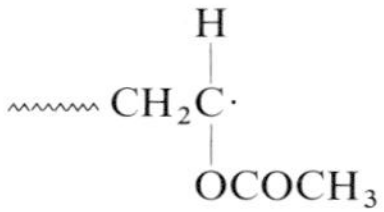 $\sim CH_2-\overset{H}{\underset{OCOCH_3}{C}}\cdot$
Acetone	0.622–1.1	0.225–0.275	<0.5–4.1 (60°C)	1.5–12 (60°C)
Benzene	0.326–0.45	0.075–0.24	0.06–0.156	1.07–20 (60°C)
Carbon tetrabromide	4100 (60°C)	3300	23,000	28,740 (60°C)
Carbon tetrachloride	1.25–1.55	2.393–3.3	133	800–10,700 (60°C)
Carbon tetrachloride	1.25–1.55	2.393–3.3	133	800–10,700 (60°C)
Chloroform	2.1–2.5	1.13–1.9	0.5–0.92	125–170 (60°C)
Cyclohexane	0.027–1.2	0.10	0.066–0.156	6.6–100 (60°C)
Toluene	1.775–2.7	0.29–0.91	0.15–0.313	17.8–69 (60°C)
Triethylamine	400 (60°C)	8.3–1900	1.4–7.5 (60°C)	370 (60°C)
n-Butyl mercaptan	16,900 (60°C)	6,700 (60°C)	210,000–250,000 (60°C)	480,000 (60°C)
n-Dodecyl mercaptan	—	—	148,000–190,000 (60°C)	—

[a] Data represent the range of values listed in Polymer Handbook (Brandrup and Immergut, 1975). All values are for 80°C unless otherwise noted.

(c) termination of growth is unimolecular in polymer, i.e., termination occurs exclusively by disproportionation or chain transfer to a single species S, which may be solvent or monomer.

Most derivations of molecular weight distribution employ a statistical approach based on kinetic parameters and this method will be followed here. A probability factor p is defined as the probability that a given radical will propagate rather than terminate. The following question may then be asked: What is the probability P of forming an x-mer after $x - 1$ propagation steps and one termination or transfer? This probability is simply $p^{x-1}(1 - p)$ and since probability is analogous to number fraction, one may write

$$P = n_x = p^{x-1}(1 - p) \tag{2.50}$$

where n_x is the fraction of molecules of degree of polymerization x. It is equal to N_x/N_T where N_x is the number of molecules of degree of polymerization x and N_T is the total number of molecules in the system. Thus

$$N_x = N_T p^{x-1}(1 - p) \tag{2.51}$$

This distribution is referred to as the most probable distribution. The molecular weight averages $\bar{X}_n$ and $\bar{X}_w$ of the distribution may be calculated from the expressions

$$\bar{X}_n = \Sigma N_x x / \Sigma N_x \tag{2.52}$$

$$\bar{X}_w = \Sigma W_x x / \Sigma W_x = \Sigma(N_x x)x / \Sigma N_x x \tag{2.53}$$

where W_x is the weight of molecules of degree of polymerization x. Substituting for N_x from Eq. (2.51) in Eqs. (2.52) and (2.53), it is possible to show that

$$\bar{X}_n = 1/(1 - p) \tag{2.54}$$

and

$$\bar{X}_w = (1 + p)/(1 - p) \tag{2.55}$$

The probability factor p can be related to kinetic parameters in terms of the rate constants for propagation, transfer, and termination in the form of the relationship

$$p = \frac{R_p}{R_p + R_{tr} + R_t} \tag{2.56}$$

$$= \frac{k_p[M][M\cdot]}{k_p[M][M\cdot] + k_{trs}[M\cdot][S] + 2k_{td}[M\cdot]^2} \tag{2.57}$$

$$= \frac{k_p[M]}{k_p[M] + k_{trs}[S] + 2k_{td}[M\cdot]} \tag{2.58}$$

Substitution for $[\mathrm{M}\cdot]$ using the steady-state approximation, Eq. (2.8), yields

$$p = \frac{k_p[\mathrm{M}]}{k_p[\mathrm{M}] + 2k_{td}(fk_d/k_{td}[\mathrm{I}])^{1/2} + k_{trs}[\mathrm{S}]} \tag{2.59}$$

$$= \frac{k_p[\mathrm{M}]}{k_p[\mathrm{M}] + 2(fk_dk_{td}[\mathrm{I}])^{1/2} + k_{tr}[\mathrm{S}]} \tag{2.60}$$

Hence

$$\bar{X}_n = \frac{1}{1-p} = \frac{k_p[\mathrm{M}] + 2(fk_dk_{td}[\mathrm{I}])^{1/2} + k_{trs}[\mathrm{S}]}{k_{trs}[\mathrm{S}] + 2(fk_dk_{td}[\mathrm{I}])^{1/2}} \tag{2.61}$$

In the absence of transfer, the mean kinetic chain length is great so that $k_p[\mathrm{M}] \gg 2(fk_dk_{td}[\mathrm{I}])^{1/2}$ and $k_{trs} = 0$. Equation (2.61) then reduces to

$$\bar{X}_n = k_p[\mathrm{M}]/2f(k_dk_{td}[\mathrm{I}])^{1/2} \tag{2.62}$$

which is identical to the expression for $\bar{X}_n$ derived in Sec. 2.2.2.7. For most addition polymerizations, propagation is highly favored over termination and transfer (this must be true if polymer of high molecular weight is to be produced), i.e., $R_p \gg R_t + R_{tr}$. Hence $p \approx 1$ from which it follows that $\bar{X}_w/\bar{X}_n \approx 2$ (assuming that termination is first order). The value of $\bar{X}_w/\bar{X}_n$ is a measure of the breadth of the distribution. Expressions may also be derived for the molecular weight distribution produced when there is termination by combination. Specifically,

$$N_x = (1-p)^2(x-1)p^{x-2} \tag{2.63}$$

so that substitution for N_x in Eqs. (2.52) and (2.53) yields

$$\bar{X}_n = 2/(1-p) \tag{2.64}$$

$$\bar{X}_w = (2+p)/(1-p) \tag{2.65}$$

The breadth of the distribution is therefore given by

$$\bar{X}_w/\bar{X}_n = (2+p)/2 \tag{2.66}$$

Thus as $p \to 1$, $\bar{X}_w/\bar{X}_n \to 1.5$ so that the distribution produced by coupling is somewhat narrower than that produced by disproportionation.

2.2.2.10. Inhibition and Retardation

It has been seen how chain transfer processes lower the kinetic chain length by prematurely terminating the propagating chain by abstraction of a hydrogen atom or some other species from another molecule. No reduction in polymerization rate will be observed provided the reactivity of the new radical formed toward monomer molecules is the same as that of the propagating radicals. Obviously, if its reactivity is lower, then the rate of polymerization will be similarly reduced. Such a process is called *degradative chain transfer* or more generally *retardation*. In some cases retardation is so effective as to produce complete *inhibition* of polymerization. Thus in the presence of an *inhibitor*, the rate of polymerization will be completely suppressed until such time as the inhibitor is totally consumed at which point polymerization will commence at a rate determined by the kinetics discussed in Section 2.2.2.4. In the presence of a retarder, polymerization is not suppressed but rather proceeds at a much reduced rate from that predicted by Eq. (2.9). Mechanistically there is no difference between an inhibitor and a retarder—it is simply a matter of degree. These processes are shown schematically in Fig. 2.8.

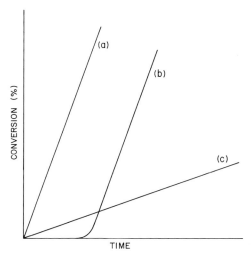

FIG. 2.8. Schematic curves showing the effect of inhibitors and retarders on the rate of polymerization: (a) no additives present, (b) with inhibitor present, and (c) with retarder present (Billingham and Jenkins, 1972).

A variety of compounds have been shown to inhibit or retard vinyl polymerization and an equal variety of mechanisms postulated to explain their action. The commonly used inhibitors appear to function by reacting in some manner with initiator radicals or oligomeric propagating radicals to to yield a species of low reactivity, and hence a reduced tendency for propagation. These inhibitors include phenols, quinones, aromatic nitro and nitroso compounds, halogenated compounds, amines, and thiols (Goldfinger et al., 1967).

Several of these compounds inhibit or retard polymerization by a radical abstraction or chain transfer mechanism in which the product radical is unreactive toward monomer. Secondary and tertiary amines are believed to act by this mechanism. For example, diphenylamine retards vinyl acetate polymerization, presumably by the reaction

$$\text{---}CH_2\text{---}\underset{\underset{OCOCH_3}{|}}{\overset{\overset{H}{|}}{C}}\cdot + (C_6H_5)_2NH \longrightarrow \text{---}CH_2\text{---}\underset{\underset{OCOCH_3}{|}}{\overset{\overset{H}{|}}{C}}\text{---}H + (C_6H_5)_2N\cdot$$

yielding an unreactive diphenylnitrogen radical that may terminate another chain. Many substituted phenols such as *tert*-butyl catechol are also believed to function by this mechanism.

Certain metal salts such as $FeCl_3$ inhibit polymerization of many vinyl monomers (Walling, 1957). Inhibition presumably results from electron transfer reactions of the type

$$\text{---}CH_2\text{---}\underset{\underset{X}{|}}{\overset{\overset{H}{|}}{C}}\cdot + Fe^{3+} \longrightarrow \text{---}CH_2\text{---}\underset{\underset{X}{|}}{\overset{\overset{H}{|}}{C}}^+ + Fe^{2+}$$

and

$$\text{---}CH_2\text{---}\underset{\underset{X}{|}}{\overset{\overset{H}{|}}{C}}^+ + Cl^- \left\langle \begin{array}{l} \text{---}CH_2\text{---}\underset{\underset{X}{|}}{\overset{\overset{H}{|}}{C}}\text{---}Cl \\[2ex] \text{---}CH\text{=}CHX + HCl \end{array} \right.$$

In addition to abstraction or transfer reactions, many compounds retard or inhibit polymerization via radical addition reactions to produce a radical, which again is unreactive toward the monomer. Quinones are probably the most important class of inhibitors of this type. Benzoquinone completely stops the polymerization of monomers such as vinyl acetate, styrene, and methyl methacrylate. The precise termination mechanism is not entirely understood. Price (1946) has suggested an initial polymer radical attack on the aromatic nucleus:

Other mechanisms have been proposed that involve radical attack on the oxygen molecule:

$$RM_{\dot{x}} + O = \left\langle \bigcirc \right\rangle = O \longrightarrow RM_x - O - \left\langle \bigcirc \right\rangle - O\cdot$$

An additional complication with these systems is a nonstoichiometric relation between the number of kinetic chains terminated and the number of quinone molecules consumed. The precise radical–quinone stoichiometry depends on the fate of the inhibitor radical produced and on whether such radicals are completely active, dimerize, disproportionate among themselves or with other polymer radicals or even undergo further growth to give copolymers (George, 1967).

The extent of retardation of inhibition is determined entirely by the reactivity of the radical toward the monomer; hence, as might be expected, a compound that might be an inhibitor for one monomer may have no effect upon another. For example, aromatic di- and trinitro compounds act as inhibitors in vinyl acetate polymerization, as retarders for styrene poly-merization, and have no effect on methacrylate or acrylate polymerizations.

Molecular oxygen is also an effective inhibitor for vinyl polymerization. It reacts with chain radicals to form the relatively unreactive peroxy radical. In some cases, notably methyl methacrylate and styrene, reaction proceeds further by addition to monomer producing an alternating copolymer (Bovey and Kolthoff, 1947; Barnes et al., 1950). The peroxides formed in the presence of oxygen can decompose at elevated temperatures and initiate polymeriza-tion. Many reports of pure thermal initiation can be attributed to the presence of trace peroxides in the system. Indeed, some commercial processes for ethylene polymerization involve initiation by trace amounts of oxygen presumably through the decomposition of peroxide species so formed.

Stable free radicals are also effective inhibitors for polymerization. While they are far too stable to initiate polymerization, they are active enough to react with free radicals. Diphenylpicrylhydrazyl, DPPH, is perhaps the best example of this class of inhibitor. It has the structure

and is an extremely efficient inhibitor, being capable of producing induction periods in the polymerization of most vinyl monomers when present in concentrations as low as $10^{-4} M$. The stoichiometry between the number of chains terminated and the number of DPPH radicals consumed is 1:1, thereby making this compound very useful for quantitative measurements.

A special type of degradative chain transfer occurs if the monomer itself acts as a retarder or inhibitor. This phenomenon is shown by allylic monomers and is known as *autoinhibition*. The hydrogen atom attached to the allylic carbon atom (carbon atom α to the double bond) is readily abstracted by the propagating radical producing a resonance-stabilized allylic radical. The resonance stabilization of the allylic radical results in the transfer reaction being thermodynamically favored over the propagation reaction.

$$\text{\~\~CH}_2\text{--}\overset{\displaystyle \text{H}}{\underset{\displaystyle \text{CH}_2\text{X}}{\text{C}}}\cdot \; + \; \text{CH}_2\text{=CH--CH}_2\text{X}$$

$$\overset{k_{trM}}{\longrightarrow} \quad \text{\~\~CH}_2\text{--}\overset{\displaystyle \text{H}}{\underset{\displaystyle \text{CH}_2\text{X}}{\text{C}}}\text{--H} \; + \; \underset{\displaystyle \dot{\text{C}}\text{H}_2\text{--CH=CHX}}{\text{CH}_2\text{=CH--}\dot{\text{C}}\text{HX}}$$

$$\overset{k_p}{\longrightarrow} \quad \text{\~\~CH}_2\text{--}\overset{\displaystyle \text{H}}{\underset{\displaystyle \text{CH}_2\text{X}}{\text{C}}}\text{--CH}_2\text{--}\overset{\displaystyle \text{H}}{\underset{\displaystyle \text{CH}_2\text{X}}{\text{C}}}\cdot$$

Since $k_{trM} \gg k_p$, the degree of polymerization is very small and to all intents and purposes, polymerization is inhibited. A similar mechanism may also explain why α-olefins such as propylene and butene do not undergo free radical polymerization. In such cases, a change in polymerization mechanism is necessary to form high molecular weight polymer. The polymerization of α-olefins was not achieved until the discovery of the Ziegler–Natta or coordination-type catalysts (Section 2.2.5).

2.2.2.11. Thermodynamics of Polymerization

It was seen in Section 2.2.1.1 that the thermodynamic feasibility of polymerization is determined by the magnitude of the free energy difference (ΔG) between the monomer and the polymer. When ΔG is negative, polymerization is thermodynamically favored and high polymer will be produced provided a suitable kinetic pathway is available. The magnitude of ΔG is determined by two factors, viz., the enthalpy change ΔH and entropy change ΔS for the polymerization, through the relationship

$$\Delta G = \Delta H - T\,\Delta S \tag{2.67}$$

where T is the absolute temperature.

Some typical values for ΔH and ΔS are shown in Table 2.6. It is evident from this table that values for ΔH vary over a wide range, e.g., 8 kcal/mole for

TABLE 2.6
HEATS AND ENTROPIES OF POLYMERIZATION OF
SELECTED MONOMERS AT 25°C[a]

Monomer	$-\Delta H_{1c}$(kcal/mole)	$-\Delta S_{1c}$ cal/deg mole
Ethylene	22.7	24.0
Propylene	20.5	27.7
Styrene	16.7	25.0
α-Methylstyrene	8.4	24.8
Acrylonitrile	18.4	26.0
Methyl acrylate	18.8	—
Methyl methacrylate	13.5	28.0
Vinyl chloride	22.9	—
Vinyl acetate	21.0	26.2
Tetrafluoroethylene	37.2	26.8

[a] Data are taken from Polymer Handbook (Brandrup and Immergut, 1975).

α-methylstyrene to 37 kcal/mole for tetrafluoroethylene. These variations arise mainly from the following causes (Joshi and Zwolinski, 1969):

(1) Energy difference between monomer and polymer resulting from either resonance stabilization (delocalization) due to conjugation–hyperconjugation or change of bond type (hybridization);
(2) Steric strain in the polymer as a result of bond stretching, bond angle deformation, or interaction between nonbonded atoms;
(3) Steric strain in the monomer; and
(4) Differences in extent of hydrogen bonding and perhaps also of strong dipole interactions in monomer and polymer.

On the other hand, the ΔS values are relatively insensitive to structural variations. When entropy changes are examined in terms of the component changes of translation, rotation, and vibration, it is found that on polymerization, the loss of external rotational entropy nearly balances the gain in vibrational and rotational entropy so that the entropy change accompanying polymerization reflects essentially the loss of the monomer's translational entropy (Dainton and Ivin, 1958).

Under normal conditions of temperature, the exothermicity of the reaction generally has an overriding effect on the entropy term so that ΔG will be negative. As the polymerization temperature increases, the degree of counteraction provided by the $T \Delta S$ term increases until eventually a condition is reached where $\Delta H = T \Delta S$ and hence $\Delta G = 0$. At this temperature, the system is in equilibrium and it corresponds to the ceiling temperature T_c

discussed in Section 2.2.2.6.2. Thus, T_c in thermodynamic terms is defined by

$$T_c = \Delta H / \Delta S \tag{2.68}$$

where ΔH and ΔS are the heat and enthalpy changes for the polymerization process under the prevailing conditions. Kinetically, it was attributed to the propagation–depropagation equilibrium reaction

$$RM_{\dot{x}} + M \underset{k_{dp}}{\overset{k_p}{\rightleftarrows}} RM_{\dot{x}+1}$$

for which the free energy change per mole of monomer added to the reactive intermediates is given by

$$\Delta G = \Delta G^\circ + RT \ln \frac{[RM_{\dot{x}+1}]}{[RM_{\dot{x}}][M]} \tag{2.69}$$

$$= \Delta G^\circ + RT \ln \frac{1}{[M]} \tag{2.70}$$

where ΔG° is the free energy change for polymerization when both monomer and polymer are in appropriate standard states. At equilibrium, $\Delta G = 0$, hence

$$\Delta G^\circ = -RT \ln \frac{1}{[M]_c} = RT \ln[M]_c \tag{2.71}$$

$1/[M]_c$ is the equilibrium constant for the propagation–depropagation reaction (cf. Eq. (2.24)). Further, since

$$\Delta G^\circ = \Delta H^\circ - T \Delta S^\circ \tag{2.72}$$

Equation (2.71) may be rearranged to give

$$T_c = \frac{\Delta H^\circ}{\Delta S^\circ + R \ln[M]_c} \tag{2.73}$$

Thus from a knowledge of the standard enthalpies and entropies of polymerization, it is possible to calculate the ceiling temperature for polymerization corresponding to any value of the monomer concentration. Values of T_c for the polymerization of several monomers are listed in Table 2.7.

It should be noted that a dead polymer molecule may be quite stable above its ceiling temperature. Degradation by depolymerization first requires formation of a propagating macroradical since it is to this species that the propagation-depropagation equilibrium pertains. The production of such a macroradical by chain breaking can be easily achieved by thermal or radiation means and depolymerization can then follow until a monomer concentration corresponding to $[M]_c$ for the particular temperature is

TABLE 2.7
CEILING TEMPERATURE FOR
POLYMERIZATION OF PURE
LIQUID MONOMER

Monomer	$T_c(°C)$
α-methylstyrene	61
Methyl methacrylate	220
Styrene	310

reached. For example, poly(methyl methacrylate) undergoes thermal degradation by a depolymerization mechanism. However, for many polymers, other reactions intervene so that the overall degradation mechanism is extremely complex (Conley, 1970).

2.2.2.12. Polymerization Conditions

Polymerization reactions are classified as homogeneous or heterogeneous depending on the number of phases in the system. In homogeneous polymerization, all the reactants, viz., monomer, initiator, and solvent, are mutually soluble and compatible with the polymer. Examples of this type of process include bulk polymerization of methyl methacrylate and solution polymerization of styrene in benzene. Some systems become heterogeneous due to precipitation of the polymer as it is formed, e.g., polyacrylonitrile is insoluble in its own monomer and precipitates during bulk polymerization. On the other hand, if dimethylsulfoxide is used as a solvent, the polymer remains in solution and the polymerization process remains homogeneous.

Other polymerization systems, notably suspension and emulsion polymerization, consist of the monomer dispersed as droplets in an inert suspending medium and are completely heterogeneous. The major difference between these two processes is the locus of polymerization.

2.2.2.12.1. Bulk Polymerization.

Bulk polymerization is inherently the simplest of all polymerization processes. It involves adding a small amount of initiator to the pure monomer and heating to a temperature where the initiator breaks down to give free radicals. The absence of other additives results in a highly pure product. There are however several disadvantages that limit the applicability of this process. Due to the high exothermicity of vinyl polymerization, thermal control of the reaction may be difficult. This is particularly true of commercial processing, which requires reasonable rates of polymerization in large volumes of reactants. Under such conditions, the rate of thermal transfer is insufficient to remove the heat of polymerization. Autoacceleration effects further compound the problem and in addition lead to broadened molecular weight distributions. Consequently, bulk polymerization is restricted to those monomers with somewhat low reactivities and heats of polymerization such as styrene and methyl methacrylate. Bulk polymerization is well suited to the production of cast sheet, e.g., production of acrylic cast sheet was expected to reach 650 million pounds in 1977 (Wood, 1975). Still, in order to avoid problems associated with a rapid rise in polymerization temperature, process conditions often require castings to be polymerized slowly over a period of many hours at relatively low temperatures. One interesting application of bulk polymerization is in the manufacture of dental acrylics. A slurry is prepared containing some 50–60% of polymer that is swollen by monomer (to which initiator has been added) to form a "dough." The dough is then molded and polymerization completed by heating the mold.

The problems associated with bulk processing may be compounded in those cases where the polymer precipitates, although continuous bulk processes have been developed, such as the Pechiney–St. Gobain process for vinyl chloride, which has proved satisfactory. Polymerization of many monomers can also be initiated in bulk below the monomer melting point. This process is referred to as *solid-state polymerization* (Tabata, 1968, 1969). High-energy radiation has been the principal method of initiation and both radical and ionic mechanisms have been observed. In spite of a vast amount of work conducted in this field, it continues to be only of academic interest.

2.2.2.12.2. Solution Polymerization. Many of the problems of bulk polymerization are eliminated by diluting the monomer with a suitable solvent. The viscosity of the medium does increase but not nearly as drastically as it does in bulk polymerization. Thermal control is much easier because of improved heat transfer. In fact, removal of the heat of polymerization can easily be facilitated by carrying out the reaction at the reflux temperature of the solvent where the heat is removed as latent heat of vaporization. The major disadvantage of the process is the difficulty of removing and handling large quantities of solvents. Nevertheless, solution polymerization is often used industrially, especially for the preparation of relatively small quantities of special polymers such as the preparation of polymeric resists for microelectronic device fabrication (Thompson, 1974).

2.2.2.12.3. Suspension Polymerization. Suspension polymerization is a heterogeneous process in which a water-insoluble monomer is dispersed in water in the form of droplets with diameters varying from 0.01 to 0.5 cm. The suspension is stabilized by mechanical agitation and with a protective colloid or dispersant such as poly(vinyl alcohol). The dispersant prevents the aggregation of partially polymerized monomer droplets at the stage when they are sticky. Polymerization is initiated by means of a monomer soluble initiator so that the locus of polymerization is in the individual droplets. The whole process may therefore be viewed as being made up of many separate bulk polymerizations and as a consequence, the kinetics of polymerization are the same as those for bulk and solution polymerization. Since water is the continuous phase, the viscosity remains constant and the large surface area of the individual particles ensures good heat transfer. This method is widely used for the commercial production of many polymers and copolymers.

2.2.2.12.4. Emulsion Polymerization. This is also a heterogeneous process and, like suspension polymerization, takes place with water as the continuous phase. In this case however, the monomer droplets are dispersed in the aqueous phase by means of an emulsifying agent or surfactant and a stable emulsion is produced. Water-soluble initiators are used so that the initial locus of polymerization is in the aqueous phase. The polymer is thus formed as a latex of polymer particles stabilized by the emulsifying agent. The diameter of the particles is of the order of $0.05-0.2\,\mu m$, considerably less than the diameter of the particles from suspension polymerization. The process has many of the advantages of suspension polymerization such as low viscosity and ease of thermal control. In addition, the latex may be directly usable as in paints, and soft or tacky polymers may be conveniently prepared and handled. It is extremely difficult however to remove the various additives such as emulsifiers and coagulants from the polymer and the high residual impurity level may degrade the polymer properties.

2.2.2.13. Kinetics of Emulsion Polymerization

Unlike homogeneous bulk, solution, and suspension polymerization that exhibit the kinetic features discussed in Section 2.2.2.4, the kinetics of emulsion polymerization are relatively complicated. In order to better appreciate the kinetics, it is instructive to examine the physical picture of emulsion polymerization, which is based on the qualitative description of the process by Harkins (1947).

When a relatively water-insoluble monomer such as styrene is emulsified in water with an anionic soap (emulsifier) such as sodium stearate, three phases result:

(1) the aqueous phase in which a small amount of monomer and emulsifier is dissolved;

(2) emulsified monomer droplets ($\sim 10^{12}/ml$) with a diameter of the order of $0.5-1.0\,\mu m$, and

(3) soap micelles saturated with monomer ($\sim 10^{18}/ml$).

Soap micelles are aggregates of about 50–100 soap molecules radially ordered with respect to each other so that the hydrophilic COO^- group forms the outer surface. Micelles begin to form when the emulsifier concentration reaches a particular level (~ 0.1–0.3%) known as the critical micelle concentration. Monomer molecules are solubilized within the micelles.

When a water-soluble initiator such as potassium persulfate is added to this system, radicals are formed in the aqueous phase by decomposition of the persulfate ions

$$S_2O_8^{2-} \xrightarrow{\ k_d\ } 2SO_4^-.$$

The rate of production of sulfate ion radicals expressed as the number of radicals formed per cubic centimeter of water per second is given by

$$R = 2N_{Avo}k_d[I] \tag{2.74}$$

where N_{Avo} is Avogadro's number, $[I]$ is the initiator concentration, and k_d is the initiator decomposition rate constant. Reaction of these primary radicals with monomer molecules present in the aqueous phase produces short chain radicals that are captured by the micelles and emulsified monomer droplets at rates proportional to their surface areas. Calculations show that the radicals will be captured by the organic phase before they are large enough to precipitate. Further, the high surface area of the micelles relative to the monomer droplets (the micellar surface area is approximately 60 times larger than the monomer droplet surface area) means that initiation takes place primarily in the micelles and any polymerization in the monomer droplets can be ignored.

Three regions of kinetic interest may be distinguished. In the early stage of reaction (stage I), radicals diffuse into the micelles and initiate polymerization, at which point the micelles are said to become latex particles. As polymerization proceeds, the micelles continue to grow by addition of monomer from the aqueous phase, which is in turn replenished from the monomer droplets. The monomer droplets therefore serve as a reservoir of monomer for the polymerizing latex particles.

The driving force for diffusion of monomer into the latex particle is the free energy of mixing, which is countered by the interfacial free energy change of the system due to the increase in surface area of the particle on swelling. Consequently, the latex particles can only absorb a limited amount of monomer until at saturation swelling characterized by the equilibrium volume fraction of monomer φ_M, the interfacial free energy change and the free energy of mixing balance (Gardon, 1968). Typical values of φ_M are 0.6–0.8 for monomers that are soluble in their polymer, e.g., $\varphi_{styrene}$ is 0.6. In such cases, swelling measurements indicate that diffusion of monomer from the droplet is not rate limiting and that the monomer is likely to be uniformly distributed within the latex particle. Thus the monomer concentration in the particle is considered to be constant (and equal to the saturation swelling) as long as there are monomer droplets present in the aqueous phase.

As polymerization proceeds, the latex particles continue to grow. The surface area increases and therefore more emulsifier is required for stabilization. Emulsifier molecules are absorbed from the solution until a point is finally reached where the concentration of emulsifier in solution is less than the critical micelle concentration and as a result, the inactive micelles (those in which polymerization has not yet been initiated) collapse. Thus no more latex particles will be initiated. This point corresponds to the end of stage I and usually occurs at about 2–15% conversion. The physical picture of stage I is represented schematically in Fig. 2.9.

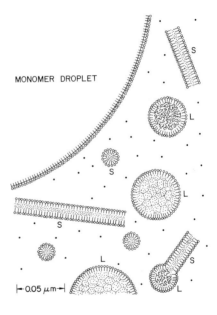

FIG. 2.9.   Schematic illustration of emulsion polymerization during stage I. S = soap micelle;   L = latex   particle;   • = monomer molecules (Vollmert, 1973).

At the end of this nucleation period, micellar soap is no longer present in solution, and as a result there is a rise in the surface tension of the latex. During stage II, the latex particles formed during stage I continue to grow. Since the number of particles and the concentration of monomer in each particle are constant, the overall rate of polymerization is essentially constant. Finally at a conversion of ∼ 50–80%, the monomer droplets disappear and the latex particles contain all the unreacted monomer. This corresponds to the commencement of stage III, which is characterized by a decreasing rate of polymerization as the monomer in the swollen latex particle is used up. Conversions approaching 100% are usually obtained and the final polymer particles have diameters of the order of 0.05–0.2 μm, i.e., intermediate in size between the initial micelles and the monomer droplets.

In homogeneous polymerization, the rate equation for polymerization is given by (cf. Eq. (2.2))

$$-d[M]/dt = k_p[M][M\cdot] \tag{2.75}$$

where [M] and [M·] are the molar concentrations of monomer and free radicals. In emulsion polymerization, each particle is a separate miniscule reactor containing on average $\bar{n}$ radicals and monomer at a saturation molar concentration M_L in the latex particle. Applying Eq. (2.75) to this model one may write

$$R_p = k_p M_L (N_P \bar{n}/N_{Avo}) \tag{2.76}$$

where R_p is the polymerization rate in moles per cubic centimeter of water, N_P is the number of particles per cubic centimeter containing an average number of radicals $\bar{n}$, and N_{Avo} is Avogadro's number. Other authors express the monomer concentration in the swollen particle in terms of the volume fraction φ_M defined by Eq. (2.77)

$$\varphi_M = (M_0 M_L/d_m)10^{-3} \tag{2.77}$$

where M_0 is the molecular weight of the monomer and d_m its density. The rate of polymerization in terms of the volume of polymer produced is given by

$$R'_p = k'_p \varphi_M (d_m/d_p) N_P \bar{n}/N_{Avo} \tag{2.78}$$

where k'_p is in units of cubic centimeter per mole per second. In order to predict the overall rate of polymerization, one must derive expressions for $\bar{n}$, the average number of radicals per particle, and N_P, the number of particles. The following analysis centers around the constant rate period observed during stage II.

According to the Harkins model, the number of latex particles remains constant during this stage. As might be expected, the emulsifier has an important effect on particle formation, influencing it by the area it occupies at the interface between water and the monomer-containing phase. This area is defined by the parameter A according to Eq. (2.79).

$$A = N_{Avo} a_e [E] \tag{2.79}$$

where [E] is the emulsifier concentration, and a_e the area per molecule. Calculations have led to the following expressions for the total number of particles N_P, the root-mean-cube average particle radius r_{rmc}, and the volume of polymer formed per unit volume of water when particle nucleation is completed P_{cr} (Gardon, 1975):

$$N_P = 0.208 A^{0.6}(R/K)^{0.4} \tag{2.80}$$

$$r_{rmc} = 1.05(m/w)^{0.333}(d_{water}/d_p)^{0.333} A^{-0.2}(K/R)^{0.133} \tag{2.81}$$

$$P_{cr} = 0.209 A^{1.2}(K/R)^{0.2}(1 - \varphi_M) \tag{2.82}$$

In these expressions, R is the initiation rate defined by Eq. (2.74), m/w is the volume ratio of monomer to water, and K is the volume growth rate constant obtained from Eq. (2.78) by setting $\bar{n} = 1$:

$$K = \tfrac{3}{4}\pi(k_p/N_{Avo})(d_m/d_p)\varphi_M/(1 - \varphi_M) = dr^3/dt \tag{2.83}$$

The theory predicts constancy of particle numbers when the reaction proceeds beyond a conversion corresponding to P_{cr} which, as has already been noted, usually occurs at a few percent conversion. Constancy of particle numbers has been observed for the polymerization of such monomers as styrene, methyl methacrylate, and vinyl acetate. Further support for this postulate has come from polymerization studies in seed lattices where polymerization takes place in already preformed particles. This is not meant to imply a general acceptance of the constancy of particle numbers postulate. Monomers that show a degree of water solubility can show wide deviations from theoretical predictions. Such deviations are believed to result from particle initiation via precipitation of polymer produced by initiation in the aqueous phase (as opposed to micellar initiation). One would therefore expect wide variation in the predicted exponents of emulsifier and initiator concentration on particle number. With styrene the theoretical exponents of 0.6 and 0.4 are in excellent agreement with experimental observations, but with other monomers, the number of polymer particles displays a dependence on emulsifier concentration anywhere from zero to third order. Wide variations are also observed in the initiator exponent. Many of these systems however violate the basic theoretical assumptions, and no universal explanation can be given based on any single theory.

The other parameter of importance in Eqs. (2.76) and (2.78) is $\bar{n}$ the average number of radicals per particle. Most calculations of $\bar{n}$ rest on the application of steady-state conditions to stage II. The steady-state condition implies that the rate of generation of N_n-type particles (N particles containing $\bar{n}$ radicals) can be equated to their rate of disappearance. N_n-type particles are created by three processes:

(1) first-order entry of single free radicals into N_{n-1}-type particles;
(2) first-order exit of single free radicals from N_{n+1}-type particles;
(3) mutual termination in N_{n+2}-type particles.

The same three processes occurring in N_n-type particles lead to their disappearance. The following recursion formula (Bovey *et al.*, 1955) may be written

$$N_{n-1}\frac{\rho'}{N} + N_{n+1}k_0 a_i\left[\frac{(n+1)}{v}\right] + N_{n+2}k_t\left[\frac{(n+2)(n+1)}{v}\right]$$

$$= N_n\frac{\rho'}{N} + N_n k_0 a_i\left(\frac{n}{v}\right) + N_n k_t\left[\frac{n(n-1)}{v}\right] \tag{2.84}$$

where N is the number of particles containing n radicals, v is the particle volume, a_i is the particle surface area, ρ' is the rate of radical entry into the particles, k_0 is the rate constant for transfer out of the particles, and k_t is the rate constant for termination. This recursion formula may be simplified to

$$\alpha'N_{n-1} + mN_{n+1}(n+1) + N_{n+2}(n+2)(n+1) = N_n(\alpha' + mn + n)(n-1)) \tag{2.85}$$

where $\alpha' = \rho'v/Nk_t$ (a parameter indicative of radical initiation rate and particle size) and $m = k_0 a_i/k_t$ (a parameter indicative of the rate of radical transfer out of the particle relative to the rate of radical termination within the particle). Stockmayer (1957) obtained a general solution of the recursion formula in the form

$$\bar{n} = a/4\{I_{-m}(a)/I_{(1-m)}(a)\} \qquad \text{for} \quad m \le 1 \tag{2.86}$$

$$\bar{n} = -(m-1)/2 + a/4\{I_{m-2}(a)/I_{m-1}(a)\} \qquad \text{for} \quad m \ge 1 \tag{2.87}$$

where $\bar{n}$ is the average number of radicals per particle, I_{-m}, I_{1-m}, I_{m-2}, and I_{m-1} are Bessel functions of the first kind of order a where $a = (8\alpha')^{1/2}$. Under conditions where radicals do not transfer out of the particle ($m = 0$) Eq. (2.86) becomes

$$\bar{n} = (a/4)\{I_0(a)/I_1(a)\} \tag{2.88}$$

so that the rate of polymerization in this instance becomes

$$R_p = \frac{k_p}{N_{\text{Avo}}} M_L N_P \left(\frac{\alpha'}{2}\right)^{1/2} \frac{I_0(a)}{I_1(a)} \tag{2.89}$$

or in terms of volume fraction

$$R'_p = \frac{k'_p}{N_{\text{Avo}}} \varphi_M \frac{d_m}{d_p} N_P \left(\frac{\alpha'}{2}\right)^{1/2} \frac{I_0(a)}{I_1(a)} \tag{2.90}$$

Figure 2.10 shows a plot of Z^* defined as $I_0(a)/I_1(a)$ as a function of a. It may be seen that Z^* is approximately unity for $a \ge 6$ which corresponds to a relatively small limiting value of $\bar{n} \geq 1.5$. Under such conditions, it may be shown that the observed kinetics are those of bulk polymerization, and no kinetic advantage would be won by having a large number of small particles.

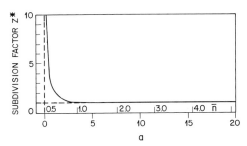

FIG. 2.10. The subdivision factor Z^* as a function of the parameter a or the average number of free radicals per particle $\bar{n}$ (Alexander and Napper, 1971).

Interest centers however on the value of Z^* as $a \rightarrow 0$. This situation corresponds to slow radical initiation rates, or small particle size or both. Under such conditions, $\bar{n} \rightarrow 0.5$ and the rate of polymerization becomes

$$R_p = (k_p/N_{\mathrm{Avo}})M_L(N_P/2) \tag{2.91}$$

or in terms of volume fraction

$$R_p' = (k_p'/N_{\mathrm{Avo}})\varphi_M(d_m/d_p)(N_P/2) \tag{2.92}$$

This result was first derived by Smith and Ewart (1948). It corresponds to the physical situation in which radicals are generated in the aqueous phase and diffuse one by one into the particles. Transfer of radicals out of the particles is negligible, and entry of a second radical into a particle containing a free radical results in instant termination. Thus, when the steady state is reached, each particle grows with one radical in it for half of the time and the other half of the time it is dead. Since the rate of radical entry into the particles is constant at any time, half of the particles contain one radical and the other half contains none.

Combining Eqs. (2.80), (2.83), and (2.92) yields

$$R_p' = 0.185k_p\varphi_M A(d_m/d_p N_{\mathrm{Avo}})^{0.6}R(1 - \varphi_M)^{0.4} \tag{2.93}$$

as the overall rate expression.

If the assumption of instantaneous termination is justified, the molecular weight of the polymer produced during stage II is readily calculable. Since radicals are being produced at a rate R radicals per second per cubic centimeter and are absorbed by N polymer particles per cubic centimeter, then on average one polymer particle absorbs R/N_P radicals per second or one radical is absorbed every N_P/R seconds. This term is designated β and is the frequency with which a radical enters a particle. It represents the average length of undisturbed chain growth. The number of monomer molecules that add to a chain radical per second is given by $k_p M_L$ so that

$$\bar{X}_n = k_p M_L \beta = N_P k_p M_L/R \tag{2.94}$$

In terms of volume fraction,

$$\bar{X}_n = k_p' d_m \varphi_M N_P/R \tag{2.95}$$

The expressions denoted by Eqs. (2.94) and (2.95) illustrate one of the important characteristics of emulsion polymerization, viz., that in contrast to other types of radical polymerization, an increase in the polymerization rate is also accompanied by an increase in the degree of polymerization. The prime factor determining the polymerization rate and $\bar{X}_n$ is the number of particles produced, which in turn is determined by the initiation rate and emulsifier concentration. Thus for a given constant initiation rate, R_p and $\bar{X}_n$ can be markedly increased by increasing the emulsifier concentration.

An examination of the literature shows that the behavior of no single monomer is wholly explained by the theory. The best agreement with Smith–Ewart kinetics is observed with water-insoluble monomers such as styrene. One major source of deviation is the actual value of $\bar{n}$. Smith and Ewart's equations were developed on the basis of $\bar{n} = \frac{1}{2}$, but with certain monomers there is quite rapid transfer out of the latex particle. This appears to be true of systems in which the monomer is insoluble in its polymer and possibly stems from the fact that radicals are localized near the surface of the polymer. In such cases, it is likely that polymerization proceeds in an adsorbed layer on the surface of the particles. Such a model forms the basis of another mechanism of emulsion polymerization proposed by Medvedev, which explains reasonably well the kinetics of polymerization of such monomers (Vanderhoff, 1969).

At higher conversions with monomers, such as styrene and methyl methacrylate, and probably with most monomers that are solvents for their polymers, deviations in the opposite sense occur and radical contents of 2–10 per particle may be present. The actual extent of deviation is governed by the particle size. During styrene polymerization for example, $\bar{n}$ increases from 0.5 to 0.6 at 90% conversion for 0.7 μm particles, but increases to more than 2 at an equivalent conversion for 1.4 μm particles (Gerrens, 1959). Thus as conversion and particle size increase, deviation from Smith–Ewart kinetics will increase and the kinetics will approach those expected for bulk polymerization. Gardon (1975) has developed formulas to take into account the dependence of $\bar{n}$ on conversion.

Equations (2.96) and (2.97) describe the time dependence of the polymer yield P and the conversion dependence of the average number of radicals per particle:

$$P = At^2 + Bt \qquad (2.96)$$

$$\bar{n} = 0.5[1 + 4AP/B^2]^{0.5} \qquad (2.97)$$

where A and B are specific rate expressions related to bulk and ideal Smith–Ewart kinetics, respectively. It is obvious from these equations that the assumption of instantaneous termination requires $4AP/B^2 \ll 1$, a situation which prevails at low initiator and high surfactant concentration. For small particle sizes, the Bt term dominates and conversion is linear with time. For large particles, the At^2 term dominates so that the square root of conversion varies linearly with time. The conversion versus time curves thus exhibit a curvature that is convex to the time axis and this is associated with significant radical accumulation. Values of A and B for styrene and methyl methacrylate and values for $\bar{n}$ are given in Table 2.8.

TABLE 2.8

KINETIC PARAMETERS FOR METHYL METHACRYLATE AND STYRENE[a]

	Methyl methacrylate at 40°C ($m/w = 35/65$)		Styrene at 60°C ($m/w = 40/60$)
Sodium laurylsulfate (%)	0.63	2.58	0.67
Potassium persulfate (%)	0.23	0.23	0.24
$B \times 10^5$	5.48	7.55	10.2
$A \times 10^9$	2.29	7.45	10.9
$\bar{n}$ 25% conversion	0.58	0.63	—
$\bar{n}$ 50% conversion	—	—	1.3
$\bar{n}$ 60% conversion	0.96	1.25	—
$\bar{n}$ 80% conversion	2.90	2.34	—

[a] From Gardon (1968).

2.2.3. CATIONIC ADDITION POLYMERIZATION

Cationic polymerization proceeds by attack on the monomer by an acidic (electrophilic) species, resulting in heterolytic splitting of the double bond to produce a carbenium ion. This species has been termed a carbonium ion in the past but, as Olah (1972) has pointed out, this terminology should be reserved for pentavalent positive carbocations derived from CH_5^+ in accordance with accepted nomenclature. In contrast, the propagating species in cationic polymerization is a trivalent carbocation, the parent of which is

CH_3^+ and for which the term carbenium ion is preferred (Kennedy, 1975; Vogl, 1975). This primary initiation process is followed by successive addition of monomer molecules leading to the formation of a macroion, which is stabilized (terminated) by any one of several termination reactions.

Whether or not a particular monomer will polymerize by a cationic mechanism depends very much on the stability of the carbenium ion formed in the initiation step. Although ions of low stability (high energy) would be expected to react with olefinic double bonds and thereby sustain propagation, they are invariably not formed or else are easily destroyed. Only an ion of fairly high stability will form and have a reasonable lifetime for growth by propagation. For example, the polar splitting of the double bond of isobutene yields a relatively stable tertiary carbenium ion whose isomerization is energetically unfavorable and, at low temperatures, even impossible. Thus isobutene can be polymerized to high molecular weight polymer under the right reaction conditions. On the other hand, propene under the same reaction conditions yields a low molecular weight oil because the secondary propyl carbenium ion is much more reactive and unstable than the tertiary butyl cation so that other terminating side reactions occur before any significant propagation can take place (Zlamal, 1969).

It is not surprising then that the tendency for a monomer to polymerize by a cationic mechanism depends on the nature of the substituents attached to the double bond. It is found that cationic polymerization of vinyl monomers is essentially limited to those monomers with electron-donating substituents such as 1:1 dialkyl, alkoxy, phenyl, or vinyl groups. Such groups can stabilize the cationic propagating species by resonance as, for example, in the polymerization of vinyl ethers, where the alkoxy group allows the delocalization of the positive charge.

$$\text{www}CH_2-\underset{\underset{R}{\overset{\mid}{O}}}{\overset{\overset{H}{\mid}}{C^+}} \longleftrightarrow \text{www}CH_2-\underset{\underset{R}{\overset{\|}{O^+}}}{\overset{\overset{H}{\mid}}{C}}$$

III

Electron-donating substituents also enhance the electron density at the carbon–carbon double bond, i.e., they enhance the nucleophilicity of the monomer and since the monomer acts as an electron donor (nucleophile) in the course of cationic polymerization, the presence of such groups greatly facilitates its bonding to the carbenium ion.

Besides nucleophilicity, there are other important factors that determine the susceptibility of a monomer to cationic attack, among them being steric and polarity effects. For instance, diisobutene (IV), in which a CH_3 group of isobutene is replaced by a neopentyl group, undergoes only dimerization under the influence of BF_3 under conditions where isobutene polymerizes easily. Such experimental observations can only be explained by the steric hindrance resulting from the more bulky neopentyl substituent.

$$CH_2 = \underset{\underset{\underset{\underset{CH_3}{|}}{CH_3-C-CH_3}}{\underset{|}{CH_2}}}{\overset{\overset{CH_3}{|}}{C}} \longrightarrow \text{dimers and trimers}$$

IV

In contrast to the precise kinetic studies on free radical polymerization, there is a paucity of information relating to the kinetics of cationic polymerization, since it is extremely difficult in most instances to obtain reproducible kinetic data. Rates of polymerization are frequently uncontrollably high, and the reactions are extremely sensitive to the presence of small amounts of impurities.

One additional complicating feature of the mechanism is the importance of the reaction medium. The reactive intermediates in cationic polymerization usually are not free carbenium ions but exist as ion pairs, with the propagating carbenium ion being closely associated with its *counter* or *gegen ion* throughout its lifetime. In solvents with relatively high dielectric constants (the absolute value is still low since solvents of polarity high enough to solvate the ions interact with the initiators), the ion pair tends to be separated, resulting in an increase in the rate of polymerization. This increase in polymerization rate arises from an increasing contribution to propagation by free ions that propagate faster than ion pairs. The increase in separation between the carbenium ion and its counter ion facilitates monomer insertion for chain growth so that the propagation rate constant is increased.

2.2.3.1.   Initiation

The generation of a carbenium ion from a vinyl monomer containing an electron-donating substituent may be accomplished by two methods: (1) addition of a cation to the monomer, and (2) removal of an electron from the monomer to form a radical cation.

Initiating systems of the first type are generally strong Lewis acids. Protonic acids such as sulfuric acid and perchloric acid can be used to initiate polymerization via addition of a proton to the monomer, although their catalytic activity is very dependent on the reaction conditions such as temperature and polarity of the reaction medium. The initiation reaction may be represented as shown where $C^+ \cdots A^-$ represents the carbenium

$$HA + CH_2 {=} \underset{\underset{Y}{|}}{\overset{\overset{X}{|}}{C}} \longrightarrow CH_3 {-} \underset{\underset{Y}{|}}{\overset{\overset{X}{|}}{C}}{}^+ \cdots A^-$$

ion–counter ion paired species. The ability of an acid to initiate polymerization depends markedly on the nucleophilicity of the acid anion. If this counter ion is highly nucleophilic, termination by direct coupling or combination will occur, producing a stable covalent bond

$$HA + CH_2 {=} \underset{\underset{Y}{|}}{\overset{\overset{X}{|}}{C}} \longrightarrow CH_3 {-} \underset{\underset{Y}{|}}{\overset{\overset{X}{|}}{C}}A$$

Consequently, acids such as perchloric and sulfuric, the anions of which are reasonably non-nucleophilic, are in general better initiators than the simple halogenated acids such as HCl. However, the reactivity of the anion can be strongly suppressed by solvation, thereby enhancing the tendency of the monomer to undergo polymerization. Initiator efficiency thus depends on the polarity of the medium. For example, hydrogen chloride induces the polymerization of styrene in polar solvents, while in nonpolar solvents the reaction does not proceed (Zlamal, 1969). The halide ion is quite nucleophilic and will couple with the carbenium ion unless it is solvated, as occurs in the polar medium.

The actual monomer involved is also of great importance. Aliphatic olefins including isobutene polymerize with difficulty with all acids. In these systems, the carbenium ions formed by protonation are extremely reactive and very often the overall reaction is the addition of the acid across the double bond. Substituted styrenes such as α-methylstyrene display a high tendency to polymerize in the presence of protonic acids, presumably because of the greater stability of the carbenium ion and the diminished tendency to react with the counter ion.

There are a variety of compounds, particularly the halides of group III and IV metals, which display marked activity as initiators. These compounds comprise the Friedel–Crafts catalysts. Examples are BF_3, $AlCl_3$, and $SnCl_4$, all of which are strong Lewis acids. Boron trifluoride is one of the most active initiators of cationic polymerization (Plesch, 1963). It is found however that BF_3 will not usually initiate polymerization if the monomer has been scrupulously dried, although addition of a trace amount of water results in rapid polymerization. This observation illustrates an interesting point concerning this class of initiator, viz., that they usually require a second component termed a coinitiator or cocatalyst in order to become active. Direct initiation by the metal halide alone would require rearrangement of the weak π-complex that forms between the monomer and the initiator (Nakane *et al.*, 1962). This rearrangement is energetically unfavorable because of the high electron density at the double bond. It is therefore necessary to reduce the increased electron density at the double bond of the asymmetrically substituted olefin by interacting with an acceptor molecule, viz., the coinitiator.

$$
\begin{array}{c}
CH_2 \\
\parallel \\
C \\
X \quad Y
\end{array}
+ BF_3 \longrightarrow
\begin{array}{c}
CH_2 \\
\| \!\!-\!\! BF_3 \\
C \\
X \quad Y
\end{array}
\xrightarrow[\text{unfavorable}]{\text{energetically}}
\begin{array}{c}
\bar{B}F_3 \\
| \\
CH_2 \\
| \\
{}^+C \\
X \quad Y
\end{array}
$$

(charge complex)

$\downarrow H_2O$

$$
\begin{array}{c}
CH_3 \\
| \\
C^+ \cdots BF_3OH^- \\
X \quad Y
\end{array}
\longleftarrow
\left[
\begin{array}{c}
\quad\; H \cdots \cdots H \\
CH_2 \qquad\; O \\
| \qquad\qquad | \\
C \cdots \cdots \cdots BF_3 \\
X \quad Y
\end{array}
\right]
$$

The coinitiator is usually mixed with the initiator prior to the addition of the monomer, in which case a complex protonic acid probably forms first, which can attack the monomer directly.

$$ BF_3 + H_2O \longrightarrow H^+ BF_3OH^- $$

$$
H^+ BF_3OH^- +
\begin{array}{c}
CH_2 \\
\parallel \\
C \\
X \quad Y
\end{array}
\longrightarrow
\left[
\begin{array}{c}
CH_2 \\
\| \!\!-\!\! H^+ BF_3OH^- \\
C \\
X \quad Y
\end{array}
\right]
$$

$\downarrow$

$$
\begin{array}{c}
CH_3 \\
| \\
C^+ \cdots BF_3OH^- \\
X \quad Y
\end{array}
$$

Other effective coinitiators include alcohols, organic acids, and even organic hydrocarbons.

The common feature of all these systems is that they contain an active hydrogen. The electronegative component of the coinitiator is attached to the Lewis acid, thereby liberating the proton, which is the effective initiator. The coinitiator need not necessarily be protogenic. For example, $SnCl_4$ functions as an initiator in the presence of organic halides. The product of the interaction is a carbenium ion that is capable of initiating polymerization

$$SnCl_4 + RCl \rightleftharpoons R^+SnCl_5^-$$

$$R^+SnCl_5^- + CH_2{=}\underset{\underset{Y}{|}}{\overset{\overset{X}{|}}{C}} \longrightarrow RCH_2{-}\underset{\underset{Y}{|}}{\overset{\overset{X}{|}}{C^+}} \cdots SnCl_5^-$$

Most polymerizations exhibit a maximum rate at a particular ratio of initiator to coinitiator. Above this optimum ratio the rate decreases. This behavior is found especially with water and is referred to as *coinitiator poisoning*. The actual concentration at which maximum efficiency is observed varies with the nature of the initiator, coinitiator, monomer, solvent, and other reaction conditions. This effect may be attributed to competition of coinitiator with monomer for the complex protonic acid formed in the initial equilibrium, the former yielding an oxonium ion salt whose reactivity toward the monomer is substantially lower than that of the complex protonic acid.

$$BF_3 + H_2O \longrightarrow H^+BF_3OH^- \xrightarrow{H_2O} [H_3O^+][BF_3OH^-]$$

Alternatively, the water may act to terminate the chain by the reaction

$$\text{\small\leftwavy} CH_2\underset{\underset{Y}{|}}{\overset{\overset{X}{|}}{C^+}} \cdots BF_3OH^- + H_2O \longrightarrow \text{\small\leftwavy} CH_2{-}\underset{\underset{Y}{|}}{\overset{\overset{X}{|}}{C}}{-}OH + H^+BF_3OH^-$$

The study of initiation with alkyl halides and common Lewis acids such as $SnCl_4$ and $TiCl_4$ has to be carried out under high vacuum anhydrous conditions because of the extreme moisture sensitivity of these systems. Progress in this field of coinitiation with cationogens was made by the discovery and development of certain new alkyl aluminum halide systems such as AlR_2X/RX. These initiator–coinitiator systems are much less sensitive to protogenic impurities and consequently better suited to experimentation in an inert gas atmosphere. Other advantages of these alkyl aluminum halide systems include ease of catalyst preparation, concentration control, improved uniformity, and reproducibility. In addition to alkyl and aryl halides, protonic acids may also be used as coinitiators with these systems. The following scheme represents the mechanism of initiation of isobutene polymerization with $(C_2H_5)_2AlCl–HCl$ (Kennedy and Gilham, 1972):

$$(C_2H_5)_2AlCl + HCl \longrightarrow H^+[Al(C_2H_5)_2Cl_2]^-$$

$$H^+[Al(C_2H_5)_2Cl_2]^- + CH_2{=}\underset{\underset{CH_3}{|}}{\overset{\overset{CH_3}{|}}{C}} \longrightarrow CH_3{-}\underset{\underset{CH_3}{|}}{\overset{\overset{CH_3}{|}}{C^+}} \cdots [Al(C_2H_5)_2Cl_2]^-$$

While the phenomenon of coinitiation with protogenic and cationogenic compounds is well documented, it is not entirely clear whether a Lewis acid initiator is effective in the complete absence of a coinitiator. Certain monomer-initiator systems apparently do not require a third component (coinitiator). It has been proposed (Kennedy, 1972) that initiation in such systems occurs by the removal of an allylic hydrogen atom in the form of a hydride ion from the nucleophilic monomer by a strong Lewis acid

$$-C{=}C{-}C{-}H + MX_n \longrightarrow -C{=}C{-}C^+ \cdots MX_nH^-$$

where MX_n denotes a Lewis acid such as $AlBr_3$ and $TiCl_4$. The ultimate success or failure of initiation appears to depend on the acidity of the Friedel–Crafts halide relative to that of the monomer since the reaction is akin to an acid–base reaction in which the monomer is the base (nucleophile) and the Friedel–Crafts halide, the acid (electrophile). This process has been termed (Kennedy, 1972) self-initiation. It should be pointed out, however, that there is no direct proof for the self-initiation mechanism. One problem is that it is probably a slow process, particularly in nonpolar media, and is easily masked by more facile initiation mechanisms, e.g., coinitiation by trace protogenic or cationogenic impurities.

An important class of cationogenic initiators are the stable carbenium ion salts such as hexachloroantimonate ($SbCl_6^-$) salts of cycloheptatrienyl ($C_7H_7^+$) and triphenylmethyl (Ph_3C^+) cations, which have been shown to be efficient initiators for the polymerization of certain reactive olefins (Ledwith, 1969). Certain alkyl or acylperchlorates and tetrafluoroborates may also be included in this group, which appear to function by direct addition of the cation to the monomer. The fact that these cations are stable means that they react rapidly only with the more reactive vinyl monomers, viz., those having electron-releasing substituents such as vinyl ethers. Initiation involves direct addition of the carbocation.

$$Ph_3C^+SbCl_6^- + CH_2{=}\underset{\underset{OR}{|}}{\overset{\overset{H}{|}}{C}}$$

$$\downarrow$$

$$Ph_3C{-}CH_2{-}\underset{\underset{OR}{|}}{\overset{\overset{H}{|}}{C^+}} \cdots SbCl_6^-$$

Cationogenic initiators do not always initiate polymerization by direct addition across the double bond. Carbenium ions are produced by hydride ion abstraction in certain systems (cf. Section 2.3.2) or by electron transfer from the olefin to the cation as occurs for example in the initiation of N-vinylcarbazole polymerization by stable aminium salts (Ledwith and Sherrington, 1974). The cation radical produced probably dimerizes, producing a growing dication.

Other initiation systems include iodine, which is an effective initiator of the more reactive vinyl monomers such as p-methoxystyrene and alkyl vinyl ether. Initiation proceeds via the reaction

$$2I_2 \rightleftharpoons I^+I_3^-$$

Radiation may also be used to initiate cationic polymerization. In the far ultraviolet ($\lambda < 150\,\text{nm}$), the quantum energy may exceed the ionization energy, in which case the absorbing molecule will be ionized and the radical cation so formed may initiate cationic polymerization. This has been demonstrated for isobutene at $120°\text{K}$ using light of shorter than a critical wavelength of about 160 nm (Vermal et al., 1964).

High-energy radiation such as γ rays or electrons can initiate polymerization, the exact mechanism depending on the monomer, presence of impurities, and polymerization conditions. Irradiation produces a radical cation (Tabata, 1969) which is then free to propagate via a radical mechanism at one end and a cationic mechanism at the other. In the absence of impurities and at very low temperature, the ionic mechanism predominates because of the much higher value of the propagation rate constant. The cationic propagation reaction is, however, susceptible to small amounts of impurities. For example, styrene polymerizes under γ irradiation by a cationic mechanism only if the monomer has been scrupulously dried. The cationic propagation is completely suppressed by traces of water, in which case a radical mechanism prevails (Tabata, 1969). One advantage of radiation initiation is that the propagation characteristics of free cations can be investigated in the absence of problems involving ion-pair formation.

The second method of initiation involves transfer of an electron from the monomer to a suitable acceptor in an initiation process referred to as *one electron transfer* or *charge transfer* initiation (Gaylord, 1970). In some cases, these reactions occur spontaneously on the interaction of the donor and acceptor. The product of such an interaction may require the input of thermal or other energy before polymerization is initiated.

$$\text{monomer} + \text{acceptor} \longrightarrow \begin{bmatrix} \text{charge transfer} \\ \text{complex} \end{bmatrix} \xrightarrow{\text{dissociation}} \text{products} \xrightarrow{\text{monomer}} \text{polymer}$$

The electron-donating characteristics of the monomer result from the presence of nucleophilic substituents or heteroatoms containing unshared electrons. Various organic acceptors have been used such as *p*-chloranil and tetracyanoethylene. Also ferric, cupric, ceric, and other metal salts in such aprotic solvents as dioxane and benzene have been used as well as certain cations, cation radicals, and carbenium ion salts.

Just how the charge transfer complex dissociates thereby initiating polymerization is a matter of some speculation. The monomer N-vinyl carbazole has been extensively studied in this regard. With tetracyanoethylene as the acceptor, Bawn *et al.* (1965) believe that the complex dissociates to form a monomer radical-cation in which the acceptor fragment becomes the counterion.

This species is then responsible for initiating polymerization by a cationic mechanism. A similar mechanism is believed to prevail in photo- and radiation-induced decomposition of charge transfer complexes of monomers such as α-methyl styrene with tetracyanobenzene (Irie *et al.*, 1975).

Similar electron-transfer processes take place at the electrodes of an electrochemical cell, and hence it is possible to initiate ionic polymerization by electrochemical initiation. It is the usual practice to add an electrolyte to the system to carry the cell current. Electrode processes are usually single-electron transfers that can give rise to radicals, anions, or cations. Discharge of cations at a cathode forms reactive radicals or free atoms. The former can initiate free radical polymerization, while the latter may undergo electron exchange to form the monomer radical anion (see Section 2.2.4.1). At the anode, the discharged anion may react with monomer to initiate radical polymerization or undergo electron exchange with the monomer to form the monomer-radical cation.

$$E^- \longrightarrow E\cdot + e^-$$

$$E\cdot + M \underset{\longrightarrow}{\overset{\longrightarrow}{\longleftarrow}} \begin{array}{l} EM\cdot \\ E^- + M^{\ddagger} \end{array}$$

In general, the kinetics of these processes are complex and many aspects of the mechanism are not well understood (Tsukamoto and Vogl, 1971).

2.2.3.2. Propagation

The propagation reaction involves the successive insertion of monomer molecules into the partial bond between the propagating species and its counterion.

$$\underset{Y}{\overset{X}{CH_3\!-\!C^+}} \cdots A^- + \underset{Y}{\overset{X}{CH_2\!=\!C}} \longrightarrow \underset{Y}{\overset{X}{CH_3\!-\!C}}\!-\!CH_2\!-\!\underset{Y}{\overset{X}{C^+}} \cdots A^-$$

or in general

$$CH_3\!-\!\overset{X}{\underset{Y}{C}}\!\!\left(CH_2\!-\!\overset{X}{\underset{Y}{C}}\right)_{\!\!x}\!\!CH_2\!-\!\overset{X}{\underset{Y}{C^+}} \cdots A^- + \overset{X}{\underset{Y}{CH_2\!=\!C}} \longrightarrow CH_3\!-\!\overset{X}{\underset{Y}{C}}\!\!\left(CH_2\!-\!\overset{X}{\underset{Y}{C}}\right)_{\!\!x+1}\!\!CH_2\!-\!\overset{X}{\underset{Y}{C^+}} \cdots A^-$$

This concept of a propagating carbenium ion species is adequate for the majority of cationic polymerizations. However, the presence of the counter ion results in considerable complexity of the propagation step. The bond between the carbenium ion and the counterion may possess a high degree of covalency, or there may be varying degrees of ionic character ranging from contact ion pairs to completely solvated free ions (Winstein and Robinson, 1958).

$$M\!-\!A \;\rightleftharpoons\; M^+A^- \;\rightleftharpoons\; M^+\|A^- \;\rightleftharpoons\; M^+ + A^-$$

| covalent | contact ion pair | solvent-separated ion pair | free solvated ions |

Since the rate constant for propagation by free ions is much greater than that for ion pair propagation, the rate of propagation will be markedly dependent on the degree of association. This in turn is determined by such factors as the nature of the counterion, solvent polarity, and temperature. As might be expected, propagation is faster in a polar solvent than in a nonpolar solvent for a given counterion. For example, in the low dielectric constant ($\varepsilon = 2.3$) carbon tetrachloride, k_p for styrene initiated by $HClO_4$ is 0.0012 liter/mole sec, whereas in the high dielectric constant ($\varepsilon = 9.7$) 1,2-dichloroethane, k_p is 17.0 liter/mole sec (Pepper and Reilly, 1962).

Values of the propagation rate constant in cationic polymerization are frequently lower than those for free radical polymerization. However, the concentration of growing chain ends in cationic systems is considerably larger ($\sim 10^{-3}$–10^{-4} mole/liter versus 10^{-8} mole/liter for free radical systems) so that the overall polymerization rates are much larger.

Cationic polymerization is further complicated by the fact that propagation can involve at least two different species—free ions and ion pairs, so that the observed rate constant can be regarded as being the sum of these two contributions

$$k_p = \alpha k_{pf}^+ + (1 - \alpha)k_{pp}^\pm \tag{2.98}$$

where α is the degree of dissociation of ion pairs into free ions and k_{pf}^+ and k_{pp} are the propagation rate constants of free ions and ion pairs, respectively. It has not been possible to resolve these rate constants for cationic polymerization of vinyl monomers. Determination of the rate constants requires a living polymerization technique (see anionic polymerization in Section 2.2.4.5), which eliminates the complication of initiation and termination. It has been possible to determine the value of k_{pf}^+ under conditions where only free ions are present, as in radiation-induced polymerization. Values range between 10^3–10^9 liter/mole sec depending on monomer and polymerization conditions (Ledwith and Sherrington, 1974).

In certain cases, the presence of ions or ion pairs cannot be detected until the polymerization is almost complete, leading to some doubt as to whether carbenium ions are involved in the propagation step. Such cases are referred to as *pseudocationic polymerization* (Plesch, 1963). The polymerization of styrene initiated by perchloric acid in methylene chloride has been widely studied. In this system covalent ester species participate in the equilibrium

$$St^+ + ClO_4^- \rightleftharpoons St^+ \cdots ClO_4^- \rightleftharpoons StOClO_3$$

free ions   ion pairs   covalent ester

Propagation is visualized as occurring through repetitive monomer insertion into the ester bond. Further work remains to be done, however, to establish the exact propagation mechanism.

The propagation reaction may be further complicated by the occurrence of intramolecular rearrangement reactions in which the reactive species rearranges to an energetically preferred, i.e., more thermodynamically stable species prior to each addition of a monomer molecule. These isomerization reactions occur by two basic mechanisms (Tsukamoto and Vogl, 1971): (1) isomerization by bond or electron shift and (2) isomerization by material transport. An example of the former occurs in the polymerization of non-conjugated aliphatic dienes. These monomers can undergo cyclopoly-merization in which the intermolecular propagation step is preceded by an intramolecular cyclization reaction, resulting in the formation of a different carbenium ion (cf. Section 2.2.2.2). Also, dialdehydes suitable for the formation of 5- and 6-membered rings have been shown to undergo cyclo-polymerization (Tsukamoto and Vogl, 1971).

Isomerization by material transport, e.g., by hydride ion shift, is particu-larly important during the polymerization of certain olefins. Initiation of 3-methyl-1-butene by coordination catalysts yields the conventional 1,2 polymer. However, initiation by Lewis acids at low temperature, e.g., $AlCl_3$ at $-130°C$, results in the production of high molecular weight 1,3 polymer as a result of a 1,2 hydride ion shift generating a thermodynamically favored propagating radical (Kennedy et al., 1964a).

Initiation at temperatures above $-100°C$ yields a rubbery product thought to be a random copolymer of 1,3 and 1,2 units. The importance of such reactions depends upon the relative rates of propagation and isomerization and on the stabilization of the propagating and rearranged carbenium ions.

Other isomerization reactions include carbenium ion shift as in the polymerization of 3,3'-dimethyl-1-butene at temperatures less than $-130°C$ (Kennedy *et al.*, 1964b).

$$CH_2{=}CH \xrightarrow{R^+} R{-}CH_2{-}\overset{+}{CH} \longrightarrow \underset{H \quad CH_3}{\overset{CH_3 \ CH_3}{RCH_2{-}C{-}C^+}} \longrightarrow 1,3 \text{ polymer}$$

(with $CH_3{-}\underset{CH_3}{\overset{|}{C}}{-}CH_3$ substituents)

Both hydride and methide migration are believed to take place consecutively in the polymerization of 4,4-dimethyl-1-pentene (Sartori *et al.*, 1971). Propagation is believed to proceed by

$$CH_2{=}CH \xrightarrow{R^+} RCH_2{-}\overset{+}{CH} \xrightarrow[\sim H]{} RCH_2{-}CH_2$$

$$\downarrow \sim CH_3$$

$$RCH_2{-}CH_2$$
$$\overset{|}{CH}{-}CH_3$$
$$\underset{+}{CH_3{-}C{-}CH_3}$$

The driving force of the hydride shift is provided by the *t*-butyl group and that of the methide migration by the higher stability of the tertiary cation.

Halide ion migration has been shown to take place in the polymerization of 3-chloro-3-methyl-1-butene

$$CH_2{=}CH \xrightarrow{R^+} RCH_2{-}CH^+ \xrightarrow{\sim Cl^-} RCH_2{-}\underset{Cl \quad CH_3}{CH{-}C^+}$$

About 50% of the polymer is made of such units formed by a 1,3 chloride ion shift. The remaining units are simple 1,2 addition products (Kennedy *et al.*, 1966).

2.2.3.3. Transfer

There are a wide variety of transfer reactions that can occur in cationic polymerization. Some of them are general and apply to any monomer susceptible to attack by electrophilic reagents, while others are specific to certain monomers. Proton transfer to monomer is one of the most common chain-terminating reactions and most often is the molecular weight controlling reaction.

$$\text{wwwCH}_2\!-\!\overset{\overset{\text{X}}{|}}{\underset{\underset{\text{Y}}{|}}{\text{C}}}^{\!+}\cdots\text{A}^- + \text{CH}_2\!=\!\overset{\overset{\text{X}}{|}}{\underset{\underset{\text{Y}}{|}}{\text{C}}} \longrightarrow \text{wwwCH}\!=\!\overset{\overset{\text{X}}{|}}{\underset{\underset{\text{Y}}{|}}{\text{C}}} + \text{CH}_3\!-\!\overset{\overset{\text{X}}{|}}{\underset{\underset{\text{Y}}{|}}{\text{C}}}^{\!+}\cdots\text{A}^-$$

Another type of chain transfer to monomer is that involving hydride ion abstraction

$$\text{wwwCH}_2\!-\!\overset{\overset{\text{X}}{|}}{\underset{\underset{\text{Y}}{|}}{\text{C}}}^{\!+}\cdots\text{A}^- + \text{CH}_2\!=\!\overset{\overset{\text{X}}{|}}{\underset{\underset{\text{Y}}{|}}{\text{C}}} \longrightarrow \text{wwwCH}_2\!-\!\overset{\overset{\text{X}}{|}}{\underset{\underset{\text{Y}}{|}}{\text{C}}}\!-\!\text{H} + \text{CH}_2\!=\!\overset{\overset{\text{X}'^+\cdots\text{A}^-}{|}}{\underset{\underset{\text{Y}}{|}}{\text{C}}}$$

where X'^+ represents an alkyl group from which a hydride ion has been abstracted. Both reactions involving chain transfer to monomer are kinetically indistinguishable, but one yields an unsaturated group and the other a saturated end group.

One of the simplest transfer reactions is the spontaneous regeneration of the protonic catalyst species.

$$\text{CH}_2\!-\!\overset{\overset{\text{X}}{|}}{\underset{\underset{\text{Y}}{|}}{\text{C}}}^{\!+}\cdots\text{A}^- \longrightarrow \text{CH}\!=\!\overset{\overset{\text{X}}{|}}{\underset{\underset{\text{Y}}{|}}{\text{C}}} + \text{HA}$$

This process is referred to as chain transfer to counterion. The ejected proton is a powerful electrophile and can readily add to another monomer molecule, thereby initiating a new polymer chain.

Transfer by abstraction can also take place involving anions other than the hydride ion, e.g., aromatic hydrocarbons and relatively simple halide molecules can also act as transfer agents, an important consideration in view of the fact that many solvents belong to these groups of compounds. Thus in the polymerization of isobutene initiated by BF_3/H_2O with an alkyl halide as the solvent, transfer with the incorporation of a halogen atom into the polymer may take place (Kennedy et al., 1963).

$$\text{wwwCH}_2\!-\!\overset{\overset{\text{CH}_3}{|}}{\underset{\underset{\text{CH}_3}{|}}{\text{C}}}^{\!+}\cdots\text{BF}_3\text{OH}^- + \text{RCl} + \text{CH}_2\!=\!\overset{\overset{\text{CH}_3}{|}}{\underset{\underset{\text{CH}_3}{|}}{\text{C}}} \longrightarrow$$

$$\text{wwwCH}_2\!-\!\overset{\overset{\text{CH}_3}{|}}{\underset{\underset{\text{CH}_3}{|}}{\text{C}}}\!-\!\text{Cl} + \text{RCH}_2\!-\!\overset{\overset{\text{CH}_3}{|}}{\underset{\underset{\text{CH}_3}{|}}{\text{C}}}^{\!+}\cdots\text{BF}_3\text{OH}^-$$

When the anion donor is present in the polymer backbone, an active center will be generated on the polymer backbone leading to branching.

$$\sim\!\!\sim CH_2\!-\!\overset{\displaystyle H}{\underset{\displaystyle Y}{\overset{\displaystyle |}{\underset{\displaystyle |}{C}}}}{}^{+}\cdots A^{-} \; + \; \sim\!\!\sim CH_2\!-\!\overset{\displaystyle H}{\underset{\displaystyle Y}{\overset{\displaystyle |}{\underset{\displaystyle |}{C}}}}\!\!\sim\!\!\sim \; \longrightarrow$$

$$\sim\!\!\sim CH_2\!-\!\overset{\displaystyle H}{\underset{\displaystyle Y}{\overset{\displaystyle |}{\underset{\displaystyle |}{C}}}}\!\!-\!H \; + \; \sim\!\!\sim CH_2\!-\!\overset{\displaystyle A^{-}}{\underset{\displaystyle Y}{\overset{\displaystyle \cdots}{\underset{\displaystyle |}{C}}}}{}^{+}\!\!\sim\!\!\sim \quad \xrightarrow{\text{monomer}}$$

$$\sim\!\!\sim CH_2\!-\!\underset{\displaystyle Y}{\overset{\displaystyle |}{\underset{\displaystyle |}{C}}}\!\!\sim\!\!\sim \qquad \text{branched polymer}$$

2.2.3.4. Termination

Most termination reactions in cationic polymerization involve transfer reactions. True termination in which the activity of the chain carrier is lost without regeneration of an active centre is very rare. For example, water is an effective terminator of polymerization, possibly via the transfer reaction

$$\sim\!\!\sim CH_2\!-\!\overset{\displaystyle X}{\underset{\displaystyle Y}{\overset{\displaystyle |}{\underset{\displaystyle |}{C}}}}{}^{+}\cdots A^{-} + H_2O \; \longrightarrow \; \sim\!\!\sim CH_2\!-\!\overset{\displaystyle X}{\underset{\displaystyle Y}{\overset{\displaystyle |}{\underset{\displaystyle |}{C}}}}\!\!-\!OH + H^{+}A^{-}$$

The result is a regeneration of the protogenic species that can lead to reinitiation. Of course, with water present in excess, the transfer reaction will predominate and lead to ill-defined oligomeric products containing hydroxyl end groups.

Other types of basic additives, e.g., amines, ethers, and sulfides are effective terminators because ammonium, oxonium, and sulfonium ions are more stable than carbenium ion species. For example, termination by amines probably involves the formation of an unreactive stable quarternary ion:

$$\sim\!\!\sim CH_2\!-\!\overset{\displaystyle X}{\underset{\displaystyle Y}{\overset{\displaystyle |}{\underset{\displaystyle |}{C}}}}{}^{+}\cdots A^{-} + :NR_3 \; \longrightarrow \; \sim\!\!\sim CH_2\!-\!\overset{\displaystyle X}{\underset{\displaystyle Y}{\overset{\displaystyle |}{\underset{\displaystyle |}{C}}}}\!\!-\!{}^{+}NR_3A^{-}$$

If, in the absence of impurities, true termination does not occur, then with monomer present as the sole nucleophile, polymerization should continue until the monomer is totally depleted. Although transfer reactions may occur, the carbenium ion should still be capable of initiating polymerization at the same rate on the addition of an equal aliquot of fresh monomer. In practice, such "living" polymerizations are rarely observed in the cationic polymerization of vinyl monomers. This is due primarily to extreme difficulties in removing impurities from the system.

Termination by combination of the propagating carbenium ion with its counterion can occur if the latter is sufficiently nucleophilic to form a covalently bonded species. For example, in the polymerization of styrene initiated by perchloric acid, some covalent perchlorate species are formed

$$\text{wwww}\ CH_2\overset{+}{-}CH \cdots \bar{Cl}O_4 \longrightarrow \text{wwww}\ CH_2-CH-O-ClO_3$$

A similar termination mechanism is observed in the trifluroacetic acid catalyzed polymerization of styrene

$$\text{wwww}\ CH_2\overset{+}{-}CH \cdots O-\underset{\overset{\|}{O}}{C}-CF_3^- \longrightarrow \text{wwww}\ -CH_2-CH-O-\underset{\overset{\|}{O}}{C}-CF_3$$

Alternatively, the propagating cation may combine with an anionic fragment from the counterion. For example, the predominant terminal groups produced in the polymerization of isobutene with BF_3/H_2O are olefinic linkages formed by proton elimination (transfer to monomer) or hydroxyl groups formed by combination with the coinitiator

$$\text{wwww}\ CH_2-\underset{\underset{CH_3}{|}}{\overset{\overset{CH_3}{|}}{C^+}} \cdots BF_3OH^- \longrightarrow \text{wwww}\ CH_2-\underset{\underset{CH_3}{|}}{\overset{\overset{CH_3}{|}}{C}}-OH + BF_3$$

A similar reaction occurs with tetrachloroaluminum as the counterion

$$\text{wwww}\ CH_2-\underset{\underset{CH_3}{|}}{\overset{\overset{CH_3}{|}}{C^+}} \cdots AlCl_4^- \longrightarrow \text{wwww}\ CH_2-\underset{\underset{CH_3}{|}}{\overset{\overset{CH_3}{|}}{C}}-Cl + AlCl_3$$

The successful polymerization of isobutene to high molecular weight polymer requires very low temperatures ($\sim -100°C$), where the rate of these deleterious side reactions is low.

The transfer reaction involving hydride abstraction from the monomer may result in termination if the new cationic species is sufficiently stabilized so that it is unable to reinitiate polymerization. Such a mechanism has been postulated to occur in the polymerization of propene where the resultant cation can be allylically stabilized.

$$\text{wwww CH}_2-\overset{+}{\text{CH}} + \text{CH}_2\text{=CH} \longrightarrow \text{wwww CH}_2-\text{CH}_2 + \overset{\delta^+}{\text{CH}_2}\overset{\cdots\cdots\cdots}{-\text{CH}}\overset{\delta^+}{-\text{CH}_2}$$
$$\quad\quad\quad |\quad\quad\quad\quad |\quad\quad\quad\quad\quad\quad\quad\quad\quad |$$
$$\quad\quad\quad \text{CH}_3\quad\quad\quad\text{CH}_3\quad\quad\quad\quad\quad\quad\quad\quad\text{CH}_3$$

This does not seem to be an important termination mechanism for isobutene although in the presence of n-alkenes such as propene and butene, the polymerization of isobutene is terminated (Kennedy and Squires, 1967). In the initiator system, $AlCl_3/CH_3Cl$, termination is believed to involve hydride transfer from the n-alkene

$$\quad\quad\quad\text{CH}_3\quad\quad\quad\quad\quad\quad\quad\quad\quad\quad\quad\quad\text{CH}_3$$
$$\quad\quad\quad |\quad\quad\quad\quad\quad\quad\quad\quad\quad\quad\quad\quad\quad |$$
$$\text{wwww CH}_2-\overset{+}{\text{C}}\cdots\text{AlCl}_4^- + \text{CH}_2\text{=CH} \longrightarrow \text{wwww CH}_2-\text{CH} + \overset{\delta^+}{\text{CH}_2}\overset{\text{AlCl}_4^-}{-\text{CH}}\overset{\delta^+}{-\text{CHR}}$$
$$\quad\quad\quad |\quad\quad\quad\quad\quad\quad\quad |\quad\quad\quad\quad\quad\quad\quad\quad\quad |$$
$$\quad\quad\quad\text{CH}_3\quad\quad\quad\quad\text{CH}_2\text{R}\quad\quad\quad\quad\quad\text{CH}_3$$

2.2.3.5. Kinetics

The development of valid kinetic expressions may be complicated by several factors. The rate of initiation is frequently very rapid and, in the possible absence of true termination reactions, this results in a nonstationary-state condition. The situation is further complicated by high reaction rates, dependence on the reaction medium, difficult experimental conditions, and the sensitivity of the process to trace amounts of polar impurities. Hence it is impossible to set up a generally valid reaction scheme comprising a sequence of elementary reactions as was done with radical polymerization (Section 2.2.2.4). However, it is possible to derive specific kinetic expressions assuming a particular sequence of reactions, and solutions to the kinetic expressions may be found if the stationary-state assumption is made. Consider for example, the polymerization of a monomer M by a Lewis acid initiator C in conjunction with a coinitiator RH. It will be assumed that transfer to monomer occurs and that a unimolecular termination reaction takes place such as combination of the propagating carbenium ion with its counterion. The reaction scheme and the rate expressions may be summarized as follows:

Initiation

$$C + RH \rightleftharpoons H^+CR^-, \quad H^+CR^- + M \xrightarrow{k_i} HM^+ \cdots CR^-$$

$$d[M^+]/dt = R_i = k_i[M][H^+CR^-] = k_iK[M][C][RH] \quad (2.99)$$

where K is the equilibrium constant for the formation of the protonated complex.

Propagation

$$HM^+ \cdots CR^- + M \xrightarrow{k_p} HM_2^+ \cdots CR^-$$

$$HM_x^+ \cdots CR^- + M \xrightarrow{k_p} HM_{x+1}^+ \cdots CR^-$$

$$-d[M]/dt = R_p = k_p[M][M^+] \qquad (2.100)$$

where $[M^+]$ represents the total concentration of all propagating ion pairs.

Transfer to monomer

$$HM_x^+ \cdots CR^- + M \xrightarrow{k_{trM}} P_x + HM^+ \cdots CR^-$$

$$R_{tr} = k_{trM}[M^+][M] \qquad (2.101)$$

Termination

$$HM_x^+ \cdots CR^- \xrightarrow{k_t} HM_x\!-\!CR \quad \text{or} \quad P_x$$

$$-d[M^+]/dt = R_t = k_t[M^+] \qquad (2.102)$$

Assuming that the steady-state condition holds, $R_i = R_t$,

$$k_i K[M][C][RH] = k_t[M^+] \qquad (2.103)$$

whence

$$[M^+] = (k_i K/k_t)[M][C][RH] \qquad (2.104)$$

Substituting for $[M^+]$ in Eq. (2.100), the rate of polymerization becomes

$$R_p = (k_p k_i K/k_t)[M]^2[C][RH] \qquad (2.105)$$

The rate of polymerization should therefore be second order in monomer and first order in initiator concentration. This is in contrast to radical polymerization, which is first order in monomer and half-order in initiator concentration.

Several systems are known that follow this type of kinetic expression, e.g., polymerization of styrene by stannic chloride (Pepper, 1949). Kinetic expressions can be derived for other situations (Zlamal, 1969) but it is again pointed out that it is impossible to assign in advance any system under study to any one particular kinetic formulation, since almost every cationic system possesses its own unique features.

2.2.3.6. *Molecular Weight and Control by Transfer*

Both termination and transfer reactions in cationic polymerization are first order with respect to growing chains. Assuming the steady-state condition, and recalling that the degree of polymerization of the polymer being formed is determined by the ratio of the rate of chain growth to the sum of the rates of all reactions leading to the formation of dead polymer, $\bar{X}_n$ for the polymerization model developed above will be given by

$$\bar{X}_n = \frac{k_p[M][M^+]}{k_t[M^+] + k_{trM}[M^+][M]} \tag{2.106}$$

$$= \frac{k_p[M]}{k_t + k_{trM}[M]} \tag{2.107}$$

Thus the degree of polymerization is determined by the balance between the rates of propagation and chain breaking reactions and not, as was the case for free radical polymerization, on the initiator concentration. Cross-termination cannot occur since all chain ends are positively charged; hence increasing the initiator concentration does not increase the probability of termination. If spontaneous termination predominates over transfer,

$$\bar{X}_n = (k_p/k_t)[M] \tag{2.108}$$

and for the case where transfer predominates

$$\bar{X}_n = k_p/k_{trM} = C_M \tag{2.109}$$

where C_M is the chain transfer constant to monomer.

2.2.3.7. Energetics and Its Relationship to Molecular Weight Control

The thermodynamic changes in the enthalpy (ΔH) and entropy (ΔS) of polymerization merely depend on the initial (monomer) and final state (polymer) of the system and are independent of the path by which the reaction proceeds. Thus values for ΔH and ΔS will be essentially independent of the initiation mechanism.

The effect of temperature on the overall rate of polymerization can be determined from a consideration of Eq. (2.105). In this case the overall activation energy E_a will be given by

$$E_a \approx E_p + E_i - E_t \tag{2.110}$$

where E_p, E_i, and E_t are the activation energies for propagation, initiation, and termination, respectively.

The overall energies of activation usually fall within the range of $-10-+15\,\text{kcal/mole}$ (Sawada, 1969). A negative value is frequently observed indicating that the polymerization rate increases as the temperature is decreased. However, the sign and value of E_a vary from one monomer to another and even for the same monomer, the value of E_a may vary considerably. For example, the activation energy for cationic polymerization of styrene can vary from -8.5 to $+14$ kcal/mole depending on the aforementioned parameters (Odian, 1970). The variations in E_a are a consequence of the differences in the initiator/coinitiator system and the solvating power of the reaction medium on the individual activation energies of Eq. (2.110). The transfer or termination reaction involves complex ion pair rearrangements in a low dielectric medium with the consequence that E_t is higher than that of propagation, usually by about 2–3 kcal/mole. Thus a negative activation energy will result when the activation energy for initiation is near zero. This is often the case since initiation usually involves the approach of a neutral molecule to an ion. Since $E_t > E_p$, consideration of Eqs. (2.108) and (2.109) indicates that the activation energy for degree of polymerization is also negative (values usually range from -3 to -7 kcal/mole) so that $\bar{X}_n$ is increased by lowering the temperature.

2.2.3.8. Practical Techniques

Because of the difficulty in producing polymers with high molecular weight at controlled polymerization rates there has been limited commercial interest in cationic polymerization of vinyl monomers. The polymerization of isobutene to yield polyisobutene, as well as its copolymerization with a small amount (3%) of isoprene to give butyl rubber, is the only example of a major industrial process. Some other polymers such as poly(vinyl ethers) are prepared on a smaller scale. Generally, one only resorts to cationic polymerization if the monomer cannot be polymerized by other means.

Cationic polymerization reactions are usually carried out in solution and at low temperatures (in order to minimize transfer reactions). Rigorous drying of monomers, solvents, and catalysts is essential and polymerizations are usually carried out under high vacuum conditions. Plesch (1963) has defined the standards regarded as acceptable in experimental work on cationic polymerization and given details of the techniques required to achieve them.

2.2.4. ANIONIC ADDITION POLYMERIZATION

Anionic polymerization proceeds by attack on the monomer of a basic (nucleophilic) species resulting in the heterolytic splitting of the double bond to produce a carbanion followed by propagation of this ion. As one might expect, there are many similarities between an anionic and cationic polymerization mechanism. Thus, the stability of the carbanion produced in the initiation reaction is enhanced by the presence of electron-withdrawing substituents such as cyano, nitro, phenyl, or carbonyl groups. These groups stabilize the propagating anionic species by resonance thereby facilitating polymerization. For example, the following resonance structures (V) may be written for the propagating species from acrylonitrile

$$\text{wwww CH}_2\text{—}\overset{\overset{\displaystyle H}{|}}{\underset{\underset{\displaystyle N}{\overset{\displaystyle |||}{C}}}{C}}:^- \longleftrightarrow \text{wwww CH}_2\text{—}\overset{\overset{\displaystyle H}{|}}{\underset{\underset{\displaystyle \ddot{N}^-}{\overset{\displaystyle ||}{C}}}{C}}$$

V

As with cationic polymerization, the propagating species is not a free ion but exists as an ion pair with the degree of association between the carbanion and its counterion again dependent on the nature of the monomer, initiator, and dielectric or solvating power of the medium. The counterions in anionic polymerization are usually small metal ions that can be easily solvated. Thus, the nature of the solvent affords valuable control over both ion–ion pair equilibria and counterion reactivity. Such is not the case in cationic polymerization, which is limited by the greater intrinsic reactivity of the carbenium ion and the difficulty in solvating anions—large anions in particular. In contrast to cationic polymerization, many anionic polymerization reactions are devoid of termination reactions. Polymerization continues until all the monomer is exhausted and in the presence of highly pure reactants and solvents, the anionic end groups will have an indefinite lifetime. Such polymers are referred to as *living polymers* (Szwarc, 1968) and if one were to recharge the system with fresh monomer, polymerization would resume. A suitably chosen terminator deliberately added to a living polymer system can form polymers with specific end groups such as carboxyl or hydroxyl groups.

2.2.4.1. *Initiation*

The generation of a carbanion from a vinyl monomer containing an electron-withdrawing substituent may be accomplished in two different ways. The first mode of initiation involves the addition of an anion (base) to the double bond of the monomer:

$$M^+B^- + CH_2\!=\!\overset{\overset{\displaystyle X}{|}}{\underset{\underset{\displaystyle Y}{|}}{C}} \longrightarrow B\text{—}CH_2\text{—}\overset{\overset{\displaystyle X}{|}}{\underset{\underset{\displaystyle Y}{|}}{C}}^- \cdots M^+$$

where the $C^- \cdots M^+$ bond may have varying character ranging from partially covalent to completely ionic.

The most common initiators of this type are the alkyl and aryl derivatives of alkali metals. Organolithium initiators have been widely used since they are readily prepared by reaction of the metal with alkyl or aryl chlorides and are soluble in the hydrocarbon solvents used in their preparation. The alkyl and aryl derivatives of the other alkali metals are insoluble under these conditions and are difficult to purify. They also attack ether and aromatic solvents more readily than do lithium compounds. The reactivity of organo-lithium compounds varies greatly with their structure and is in part determined by the basicity of the negative ion. For example, fluorenyllithium will initiate methyl methacrylate but not styrene, while benzyllithium will initiate both monomers. In many cases, however, the order of reactivity is counter to the order of inherent basicity. Methyl-, ethyl-, and butyllithium for example are much slower in initiating polymerization of styrene, butadiene, or isoprene in hydrocarbon media than the less basic benzyllithium. This effect is attributed to the tendency of these compounds, the lithium alkyls in particular, to associate in hydrocarbon solvents (Smid, 1969). Aggregates range from dimers to hexamers depending on the initiator. For example, n-butyllithium dissolved in benzene exists in solution as an aggregate of six molecules. The rate of initiation is found to obey the expression (Worsfold and Bywater, 1960)

$$R_i = kC_{n\text{-BuLi}}^{1/6}[M] \qquad (C_{n\text{-BuLi}} > 10^{-4} M) \qquad (2.111)$$

where $C_{n\text{-BuLi}}$ represents the total initial concentration of all forms of the initiator. This observed relationship between the reaction order of the initiator and the reciprocal of the degree of association can be explained by assuming that only the unassociated organolithium species can react with monomer and that this species is in rapid equilibrium with the associated form

$$(n\text{-C}_4\text{H}_9\text{Li})_6 \; \rightleftharpoons \; 6\, n\text{-C}_4\text{H}_9\text{Li}$$

However, a survey of the literature (Morton and Fetters, 1975) shows that initiation orders for n-butyllithium range between one sixth and first order and vary depending on whether aliphatic or aromatic solvents are used. Further, the one step dissociation to six unassociated n-butyllithium molecules has been questioned on energetic grounds (Brown, 1970). Mechanisms involving reaction of monomer with associated organolithium species can account for the low fractional orders that are observed.

The order of initiator efficiency is very sensitive to the solvent composition. In polar solvents where the aggregates are completely broken down, the initiation rate shows a first-order dependence on initiator concentration. Further, the order of reactivity in such solvents follows the order of inherent basicity with the more basic alkyllithium initiators being more reactive than the aryllithium compounds.

Other base initiator systems that have been used are the alkali metal amides ($NaNH_2$, KNH_2), alkoxides (RONa), and Grignard reagents (C_2H_5MgBr). The basicity of the initiator required to initiate polymerization depends on the electronegativity of the substituents X and Y (p. 93). The relative electronegativities of some selected substituents are $-CN > -COOR > -C_6H_5 > -CH=CH_2$. Hence nonpolar monomers like styrene or butadiene require relatively strong bases such as metal alkyls for initiation. Weaker bases such as fluorenyllithium, while unable to initiate polymerization of nonpolar monomers, will polymerize methyl methacrylate, whereas acrylonitrile can be polymerized by relatively weak bases such as sodium methoxide.

The second mode of initiation involves electron transfer from an electron donor to a monomer. This type of process is based upon the ability of the alkali metals to supply electrons to double bonds. Free metals may be employed as solutions in liquid ammonia or certain ether solvents or as suspensions in inert solvents. In liquid ammonia, alkali metals dissolve with the formation of a solvated electron that imparts a deep blue color to the solution

$$Li + NH_3 \longrightarrow Li^+(NH_3) + e^-(NH_3)$$

The solvated electron is then transferred to the monomer to form a radical anion. Not all ammonia solutions of alkali metals initiate polymerization by electron transfer. Potassium for example is believed to form potassium amide, which initiates polymerization by addition of amide ion (Overberger *et al.*, 1960).

When the metal is present as a dispersion, the reaction is heterogeneous and involves a direct one-electron transfer from the alkali metal to the monomer to form a radical anion

$$M^0 + CH_2=\overset{\overset{\displaystyle X}{|}}{\underset{\underset{\displaystyle Y}{|}}{C}} \longrightarrow \cdot CH_2-\overset{\overset{\displaystyle X}{|}}{\underset{\underset{\displaystyle Y}{|}}{C}}{}^- \cdots M^+$$

The anion then dimerizes to a dianion forming a polymer chain capable of growing at both ends

$$M^+ \cdots {}^-\overset{\overset{\displaystyle X}{|}}{\underset{\underset{\displaystyle Y}{|}}{C}}-CH_2\cdot + \cdot CH_2-\overset{\overset{\displaystyle X}{|}}{\underset{\underset{\displaystyle Y}{|}}{C}}{}^- \cdots M^+ \longrightarrow M^+ \cdots {}^-\overset{\overset{\displaystyle X}{|}}{\underset{\underset{\displaystyle Y}{|}}{C}}-CH_2-CH_2-\overset{\overset{\displaystyle X}{|}}{\underset{\underset{\displaystyle Y}{|}}{C}}{}^- \cdots M^+$$

The heterogeneous nature of this initiation process makes the alkali metal initiators rather inefficient and less reliable than the soluble complexes formed between alkali metals and aromatic hydrocarbons extensively studied by Szwarc and his co-workers (Szwarc, 1956; Szwarc *et al.*,

1956). These complexes, e.g., sodium naphthalene, act readily as initiators of anionic polymerization by transferring an electron to the monomer (provided of course that the monomer has a similar or higher electron affinity). Production of the complex involves a one-electron transfer from sodium to the naphthalene in a suitable coordinating solvent.

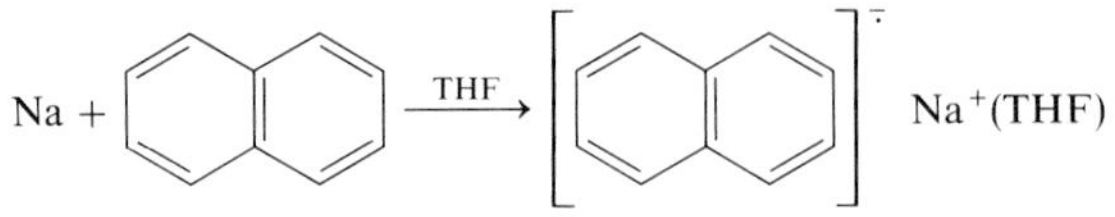

(green)

The position of the equilibrium depends on the electron affinity of the hydrocarbon and on the donor properties of the solvent. The solvation of the cation is an important driving force in the complete formation of the radical-anion. Suitable solvents include tetrahydrofuran and diethyl ether.

On the addition of a monomer such as styrene, an electron-transfer reaction takes place from the complex to styrene, forming the styryl radical anion, which rapidly dimerizes to form the dianion

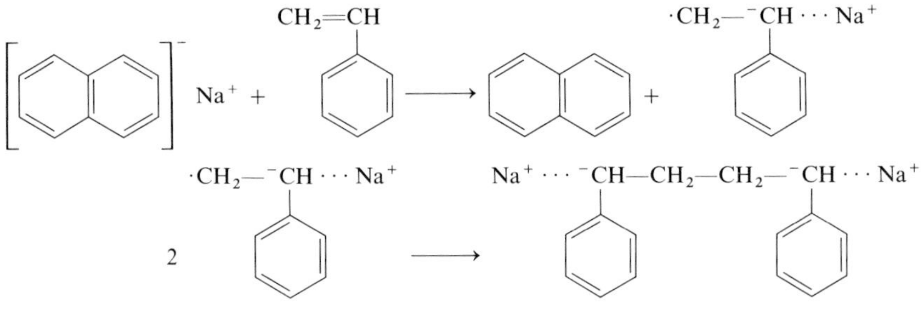

(red)

This reaction is accompanied by a color change from green to red, the latter being characteristic of the polystyryl anion.

Electron-transfer reactions in the cathode compartment of an electro-chemical cell can initiate anionic polymerization. Styrene, α-methylstyrene, and isoprene, for example, have been polymerized using alkali metal tetra-alkyl aluminates or tetraphenyl borates as conductive electrolytes (Yamazaki, 1969). Electrochemical initiation is a convenient way of initiating living anionic polymerization since the monomer solutions can be easily purged of impurities by passing current until the color of the living propagating anion appears (Funt et al., 1966).

An electron-transfer process must occur producing dianions as described earlier. This may occur by addition of a electron at the cathode to form the monomer radical anion or indirectly by reduction of the electrolyte cation to the metal, which itself may transfer an electron to the monomer. A terminating agent is produced in the anode compartment requiring a sintered disk between the two compartments to prevent diffusion of the terminating agent into the cathode area. Generally the kinetics of these processes are quite complex and many aspects of the mechanism are imperfectly understood.

2.2.4.2. Propagation

Propagation involves the successive insertion of monomer molecules into the partial bond between the propagating species and its counterion

$$
\begin{array}{c}
\quad\quad X \quad\quad\quad\quad X \quad\quad\quad\quad\quad\quad X \quad\quad\quad X \\
\quad\quad | \quad\quad\quad\quad | \quad\quad\quad\quad\quad\quad | \quad\quad\quad | \\
B-CH_2-C^- \cdots M^+ + CH_2{=}C \longrightarrow B-CH_2-C-CH_2-C^- \cdots M^+ \\
\quad\quad | \quad\quad\quad\quad | \quad\quad\quad\quad\quad\quad | \quad\quad\quad | \\
\quad\quad Y \quad\quad\quad\quad Y \quad\quad\quad\quad\quad\quad Y \quad\quad\quad Y
\end{array}
$$

or in general

$$
\begin{array}{c}
\quad\quad X \quad\quad\quad X \quad\quad\quad\quad\quad X \quad\quad\quad\quad\quad X \quad\quad\quad X \\
\quad\quad | \quad\quad\quad | \quad\quad\quad\quad\quad | \quad\quad\quad\quad\quad | \quad\quad\quad | \\
B(CH_2-C)_x CH_2-C^- \cdots M^+ + CH_2{=}C \longrightarrow B(CH_2-C)_{x+1} CH_2-C^- \cdots B^+ \\
\quad\quad | \quad\quad\quad | \quad\quad\quad\quad\quad | \quad\quad\quad\quad\quad | \quad\quad\quad | \\
\quad\quad Y \quad\quad\quad Y \quad\quad\quad\quad\quad Y \quad\quad\quad\quad\quad Y \quad\quad\quad Y
\end{array}
$$

As with cationic polymerization, the presence of the counterion complicates the propagation step. The bond between the metal and the alkyl group may possess a high degree of covalency as occurs with lithium compounds or there may be varying degrees of ionic character. Although the Winstein ion pair dissociation scheme was originally devised to account for the carbenium ion-forming reactions, it can be applied to carbanion-forming systems as well (Cram, 1965).

$$
R-\text{metal} \rightleftharpoons R^- {}^+\text{Metal} \rightleftharpoons R^-\|{}^+\text{Metal} \rightleftharpoons R^- + {}^+\text{Metal}
$$

covalent	contact ion pair	solvent-separated ion pair	free solvated ions

Thus factors such as the nature of the counterion, solvent polarity, and temperature, which have a marked influence on the propagation rate constant for cationic polymerization, will similarly affect the propagation rate constant for anionic polymerization. As an example, lithium counterion tends to form covalent bonds in nonpolar solvents and so propagation is slower than when cesium is the counterion.

The propagation step is further complicated by association phenomena involving the growing active centers. In nonpolar solvents, growing polystyrene chains occur in the form of ion pair dimers with either lithium or sodium as the counterion. With potassium, dissociation of the dimers occurs at high dilution, whereas with cesium, no association can be measured over a wide range of concentration (Bywater, 1975). The reaction kinetics changes from one-half (Li^+, Na^+) to first order (Cs^+) with respect to active center concentration with the potassium compound showing an order which increases from one-half to one with dilution. This behavior suggests that it is the monomeric form that is active in monomer addition since the monomeric and dimeric forms are in mobile equilibrium, e.g., with n-butyl lithium, the following equilibrium reaction occurs

$$(n\text{-}C_4H_9\text{---}M_x^- \cdots Li^+)_2 \underset{}{\overset{K_d}{\rightleftharpoons}} 2\, n\text{-}C_4H_9\text{---}M_x^- \cdots Li^+$$

Only in cases where a first-order dependence is observed can an absolute propagation rate constant for the ion pair be measured. In the case of lithium and sodium, only a complex rate constant equivalent to $k_p^{\mp}(K_d/2)$ can be determined, where K_d is the dimer dissociation constant.

Even higher aggregates are found in polyisopropenyl lithium and polybutadienyl lithium in solvents such as cyclohexane and heptane, where evidence suggests the presence of tetrameric species (Worsfold and Bywater, 1964).

The association phenomena observed in nonpolar solvents tend to disappear in polar solvents such as ethers, in which the major species present is the ion pair. As the dielectric constant of the solvent increases, a further development is possible, viz., increasing separation of the ion pair to form free ions. The rate constant for propagation by the free ion is considerably greater than that for the ion pair, and although the concentration of free ions may be small for initiator concentrations normally used in polymerization, their high reactivity can exert an overriding effect on the polymerization rate. This effect is examined further in Section 2.2.4.5.

For polar monomers, additional complicating features arise. The initiator and sometimes the growing active centers can react with the polar group on both the monomer and polymer to produce unwanted side reactions. Organometallic compounds react readily with ester, ketone, and halide groups. In the case of methyl methacrylate initiated with organolithium compounds, two basic reactions take place between initiator and monomer

The relative importance of the two steps depends on the initiator, solvent, and temperature. The propagation reaction is further complicated in hydrocarbon solvents by the formation of an inactive cyclic trimer (VI)

VI

It is thought that a slow rate of monomer addition to this cyclic trimer may account for the broad molecular weight distributions observed. In ether solvents, the polymerization is considerably simpler. The molecular weight distribution of the isolated polymer becomes much narrower and characteristics approaching that of a living polymer system are attained.

2.2.4.3. Transfer

A growing polymer chain may react with another species in such a way that its anion is transferred to a new species and the growing chain is terminated.

Since this latter reaction involves proton transfer to the active center, it may terminate the reaction if R^- is too stable to initiate a new chain. Small amounts of cyclopentadiene, for example, will inhibit the polymerization of isoprene due to the formation of the cyclopentadienyl anion, which will not initiate isoprene polymerization. On the other hand, in the polymerization of styrene initiated by KNH_2 in liquid NH_3, transfer to the ammonia solvent results in chain termination and the production of amide anion that can reinitiate polymerization (Higginson and Wooding, 1952).

Hydrocarbon solvents such as toluene also function as chain transfer agents. Low molecular weight polydienes of various microstructures are prepared commercially by an anionic chain transfer polymerization using toluene as the chain transfer agent and lithium as the counterion (Kamienski et al., 1973a,b). As expected, the molecular weight distributions are broad ($M_w/M_n \approx 2$), reflecting the random nature of the transfer step.

Proton transfer from monomer has been observed in the polymerization of acrylonitrile (Ottolenghi and Zilkha, 1963).

$$\text{CH}_2\text{CH}^-\cdots\text{M}^+ + \text{CH}_2{=}\text{CH} \longrightarrow \text{CH}_2{-}\text{CH}_2 + \text{CH}_2{=}\text{C}^-\cdots\text{M}^+$$

$$\underset{\text{CN}}{|}\qquad\underset{\text{CN}}{|}\qquad\qquad\underset{\text{CN}}{|}\qquad\underset{\text{CN}}{|}$$

Transfer reactions to monomer or polymer via possible hydride ion transfer may also occur giving an unsaturated chain end

$$\text{CH}_2{-}\overset{X}{\underset{Y}{\text{C}}}{}^-\cdots\text{M}^+ + \text{CH}_2{=}\overset{X}{\underset{Y}{\text{C}}} \longrightarrow \text{CH}{=}\overset{X}{\underset{Y}{\text{C}}} + \text{CH}_3{-}\overset{X}{\underset{Y}{\text{C}}}{}^-\cdots\text{M}^+$$

2.2.4.4. Termination

As was pointed out earlier, termination reactions may be virtually absent in anionic polymerization, particularly when extreme care is taken to remove impurities from the system. The polymeric carbanions are, however, rapidly terminated on exposure to air as they react rapidly with water, oxygen, or carbon dioxide. For example, reaction with carbon dioxide produces a new anion that is too stable to undergo further propagation (Altares *et al.*, 1965)

$$\text{CH}_2{-}\overset{X}{\underset{Y}{\text{C}}}{}^-\cdots\text{M}^+ + \text{CO}_2 \longrightarrow \text{CH}_2{-}\overset{X}{\underset{Y}{\text{C}}}{-}\text{C}\overset{O}{\underset{O^-\text{M}^+}{\diagup}}$$

In the presence of acid, a polymer with carboxyl end groups will result.

Polymerization may be deliberately terminated at the conclusion of an experiment by addition of proton donors such as an alcohol or water to produce polymers with saturated end groups.

$$\text{CH}_2{-}\overset{X}{\underset{Y}{\text{C}}}{}^-\cdots\text{M}^+ + \text{ROH} \longrightarrow \text{CH}_2{-}\overset{X}{\underset{Y}{\text{C}}}{-}\text{H} + \text{RO}^-\text{M}^+$$

This ability to determine the nature of the end groups is extremely important. The technique is most often applied to polymerization of living polymers having difunctional end groups, where termination will produce the same reactive group on both ends of the chain.

Another mode of termination is provided by isomerization reactions. Solutions of living polystyrene in unreactive solvents gradually lose their activity over a period of days or weeks as a result of isomerization of the carbanion to a more stable form (Spach *et al.*, 1962). The termination reaction is a two-step process involving hydride elimination to produce an unsaturated end group followed by abstraction of the allylic hydrogen of the unsaturated end group to yield an unreactive diphenyl allyl anion

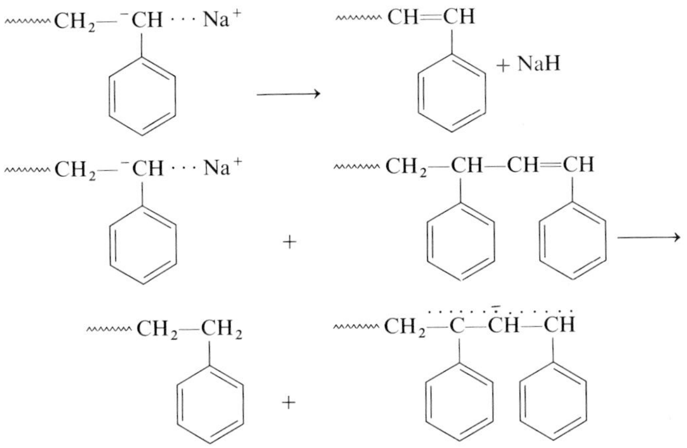

The rate of these isomerization reactions depends on the nature of the solvent, monomer, and counterion. In ether solvents, the above isomerization reactions proceed at a faster rate at room temperature than in hydrocarbon solvents. In addition, a true termination reaction occurs in the form of ether cleavage by the growing anions. These processes are kinetically limiting, particularly for dienes where they effectively compete with propagation, making measurement of kinetic parameters virtually impossible.

2.2.4.5. Kinetics

As with cationic polymerization, the exact mechanism of anionic polymerization is frequently complex, and it is correspondingly impossible to derive an overall kinetic expression based on a generally valid reaction scheme comprising a sequence of elementary reactions. Kinetic expressions can be derived assuming a particular reaction sequence and have been found to be valid for certain systems. For example, the polymerization of styrene in liquid ammonia initiated by potassium amide has been extensively studied (Higgenson and Wooding, 1952), and the kinetics found to be explicable in terms of the following sequence of reactions:

Initiation

$$KNH_2 \xrightleftharpoons{K} K^+ + NH_2^-, \qquad NH_2^- + M \xrightarrow{k_i} H_2N-M_1^-$$

$$\frac{d[M^-]}{dt} = R_i = k_i[M][NH_2^-] = \frac{k_i K[M][KNH_2]}{[K^+]} \qquad (2.112)$$

Propagation

$$H_2N\text{—}M_1^- + M \xrightarrow{k_p} H_2N\text{—}M_2^-$$

$$H_2N\text{—}M_x^- + M \xrightarrow{k_p} H_2N\text{—}M_{x+1}^-$$

$$-\frac{d[M]}{dt} = R_p = k_p[M^-][M] \tag{2.113}$$

Transfer to solvent

$$H_2N\text{—}M_x^- + NH_3 \xrightarrow{k_{trS}} H_2N\text{—}M_x H + NH_2^-$$

$$-\frac{d[M^-]}{dt} = R_{tr} = k_{trS}[M^-][NH_3] \tag{2.114}$$

Assuming the steady-state condition holds, $R_i = R_{trS}$, so that

$$\frac{k_i K[M][KNH_2]}{[K^+]} = k_{trS}[M^-][NH_3] \tag{2.115}$$

whence

$$[M^-] = \frac{k_i K[M][KNH_2]}{k_{trS}[NH_3][K^+]} \tag{2.116}$$

and the rate of polymerization becomes

$$R_p = \frac{k_p k_i K[M]^2[KNH_2]}{k_{trS}[NH_3][K^+]} \tag{2.117}$$

Under conditions where there is no externally added potassium ions, $[K^+] = [NH_2^-]$, and

$$K = \frac{[K^+]^2}{[KNH_2]}$$

so that $[K^+] = K^{1/2}[KNH_2]^{1/2}$. Substituting for $[K^+]$ in Eq. (2.117), the expression for R_p becomes

$$R_p = \frac{k_p k_i K^{1/2}[M]^2[KNH_2]^{1/2}}{k_{trS}[NH_3]} \tag{2.118}$$

The rate is therefore second order in monomer, and one-half order in initiator. Part of the reason for the simplified kinetics in this case is due to the fact that propagation involves only free ions due to the high dielectric constant and solvating power of ammonia.

The kinetics of living polymer systems has been of particular interest since the absence of termination and a fast initiation rate compared to propagation rate enable the propagation reaction to be studied independently of the other competing processes. The rate of polymerization in such nonterminating systems is therefore given by

$$R_p = k_p[M][M^-] \tag{2.119}$$

The kinetics of living polymer systems have been studied with particular emphasis on organolithium polymerization. Unlike the aromatic complexes of the alkali metals, these compounds are soluble in a variety of solvents including hydrocarbons and ethers. Most studies on such systems have demonstrated a first-order dependence on monomer concentration.

The order with respect to concentration of growing chains varies considerably depending on the nature of the solvent, monomer, and counterion. As has already been noted, growing chains tend to associate in hydrocarbon solvents when lithium is the counterion, giving rise to orders less than one. In polar media such as tetrahydrofuran, a first-order dependence is observed for organolithium polymerizations of isoprene and butadiene (Morton et al., 1961) and for the polymerization of styrene initiated by sodium naphthalene (Geacintov et al., 1962). The latter polymerization proved to be somewhat complex in that the rate constant calculated for the propagation reaction was very dependent on the concentration of growing chains at which it was measured. This arises from the fact that propagation is actually sustained by two reactive species, the ion pair and the free ion, the concentration of the latter increasing with dilution (Bhattacharyya et al., 1964, 1965; Hostalka and Schulz, 1965). If k_{pf}^- is the rate constant for propagation by free ions and $k_{pp}^\mp$ the rate constant for ion pairs, the overall rate of polymerization may be written as the sum of the two contributing propagation reactions

$$M_x\!-\!M^- + M \xrightarrow{k_{pf}^-} M_{x+1}\!-\!M^-$$

$$M_x\!-\!M^-X^+ + M \xrightarrow{k_{pp}^\mp} M_{x+1}\!-\!M^-X^+$$

so that

$$R_p = k_{pf}^-[M^-][M] + k_{pp}^\mp[M^-X^+][M] \tag{2.120}$$

The two propagating species are in equilibrium so that

$$M^-X^+ \underset{}{\overset{K_d}{\rightleftharpoons}} M^- + X^+$$

for which the dissociation constant K_d is given by

$$K_d = \frac{[M^-][X^+]}{[M^-X^+]} = \frac{[M^-]^2}{[M^-X^+]} \quad \text{for} \quad [M^-] = [X^+] \tag{2.121}$$

Substitution for $[M^-]$ in Eq. (2.120) yields

$$R_p = k_{pf}^- K_d^{1/2} [M^-X^+]^{1/2}[M] + k_{pp}^\mp [M^-X^+][M] \qquad (2.122)$$

$$= \left\{ \frac{k_{pf}^- K_d^{1/2}}{[M^-X^+]^{1/2}} + k_{pp}^\mp \right\} [M^-X^+][M] \qquad (2.123)$$

If the extent of dissociation into free ions is small, $[M^-X^+]$ can be approximated by the total concentration of living ends (i.e., by the initiator concentration). Thus the observed rate constant k_p in Eq. (2.123) becomes

$$k_p = k_{pp}^\mp + \frac{k_{pf}^- K_d^{1/2}}{[LE]^{1/2}} \qquad (2.124)$$

where $[LE]$ is the total concentration of living ends. Hence a plot of the experimentally determined k_p values versus $[LE]^{-1/2}$ should be linear, giving a value of $k_{pp}^\mp$ at the intercept. The value of k_{pf}^- can be obtained from the slope of the line provided K_d is known. The latter can be determined from conductivity measurements. These measurements have been made for the styrene/tetrahydrofuran system (Fig. 2.11), where it was found that $k_{pf}^- \sim 6.5 \times 10^4$ liter/mole sec and $k_{pp}^\mp \sim 80$ liter/mole sec. Values for $k_{pp}^\mp$

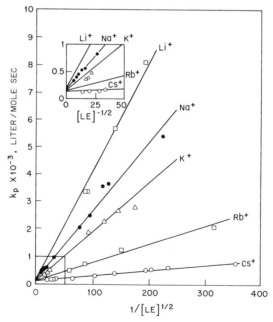

FIG. 2.11. The experimentally observed rate constant k_p of living polystyrene polymerization in tetrahydrofuran as a function of $1/[LE]^{1/2}$. The results are given for lithium, sodium, potassium, rubidium, and cesium salts (Szwarc, 1975).

actually ranged between 160 liter/mole sec for lithium counterion and 22 liter/mole sec for cesium counterion. In view of the vast difference in propagation rate constants, it is hardly surprising that parameters such as solvent polarity and type of counterion, which determine the relative amounts of free ions and ion pairs, markedly effect the overall rate of polymerization.

2.2.4.6. Molecular Weight Distribution

It is interesting to speculate on the conditions that are necessary to produce a polymer in which all the chains are of equal length ($M_w/M_n = 1$). Such a system is said to be monodisperse and should be obtained if

(1) all growing chains are initiated instantaneously
(2) there is equal probability of growth of all chains
(3) there is a complete absence of transfer or termination reactions
(4) depropagation is very slow compared to propagation.

In view of these considerations, it is hardly surprising that the molecular weight distributions predicted by virtually all polymerization mechanisms are reasonably broad (polydisperse). For instance, free radical polymerization is characterized by a slow rate of initiation followed by competition between propagation, transfer, and termination reactions. Chains initiated later in the polymerization may not necessarily grow to the same length as those produced earlier owing to monomer depletion and/or possible auto-acceleration effects. As a consequence, the distribution resulting from free radical polymerization is characterized by values of M_w/M_n ranging from 1.5 to quite high values such as 10 or more.

In the case of some homogeneous anionic systems under conditions where spontaneous termination is avoided through a judicious choice of experimental conditions, the reaction conditions necessary to form monodisperse polymer can almost be achieved. The number average degree of polymerization is given simply by the ratio of the concentrations of monomer to living ends

$$\bar{X}_n = [\text{M}]/[\text{M}^-] \tag{2.125}$$

$$= nb[\text{M}]/[\text{I}] \tag{2.126}$$

where b is the fractional conversion to polymer, $[\text{M}]$ and $[\text{I}]$ are the concentrations of monomer and initiator, respectively, and $n = 1$ or 2 depending on whether initiation is by dianions or monoanions. The resultant molecular weight distribution approaches the Poisson distribution as predicted by Flory (1940), for which the number fraction of x-mers n_x is given by

$$n_x = e^{-k}k^{x-1}/(x-1)! \tag{2.127}$$

where k is the number of monomer molecules reacted per initiator molecule. The weight to number average chain length ratio is given by

$$\bar{X}_w/\bar{X}_n = 1 + (\bar{X}_n/(\bar{X}_{n+1})^2) \tag{2.128}$$

which can be approximated by

$$\bar{X}_w/\bar{X}_n = 1 + 1/\bar{X}_n \tag{2.129}$$

Thus for any but very low molecular weight polymer, the size distribution will be very narrow with $\bar{X}_w/\bar{X}_n$ being close to unity. This has been found in many systems (Morton and Fetters, 1975) and the technique has been widely used to prepare polymers with very sharp molecular weight distributions useful as standards for gel permeation chromatography (GPC); see Chapter 3, Section 3.

2.2.4.7. Steric Control of Propagation

The nature of the solvent and counterion not only influences the rate at which a particular monomer adds to an activated chain end, but also the stereochemistry of the chain. This is most readily demonstrated in the polymerization of isoprene which, as discussed in Section 2.2.2.2, may enter into the polymer in any one of four configurations, viz., 1,2, 3,4, cis-1,4, and trans-1,4. In bulk or in a nonpolar solvent such as pentane, anionic polymerization of this monomer initiated by alkali metals results in polymers containing different proportions of the various isomeric unit structures depending on the particular metal (Parry, 1974). Of particular interest is the lithium-initiated polymer, which is almost 100% cis-1,4-polyisoprene. This effect of counterion on chain structure is characteristic of ionic mechanisms and indicates that the carbanion exists in the form of an ion pair structure with its counterion. Similar effects are shown by the alkali alkyl initiators, lithium alkyls exerting a strong stereoregulating influence. The degree of stereoregularity is also highly dependent on the nature of the solvent. Polar solvents in particular drastically reduce the degree of stereoregularity. For example, initiation of isoprene with lithium metal in tetrahydrofuran gives a product whose microstructure is approximately 16% 1,2, 51% 3,4, and 33% trans-1,4 units. This effect of solvent is further demonstrated in the case of the anionic polymerization of methyl methacrylate initiated by 9-fluorenyllithium. An isotactic polymer is obtained in toluene, a stereoblock in ether, and syndiotactic polymer in tetrahydrofuran (Goode et al., 1960).

These effects are qualitatively understandable in terms of the bond between the carbanion and counterion, the exact nature of which depends on the chemical structure of the catalyst and the reaction medium. When conditions are such as to allow extensive coordination between the propagating chain end, initiator, and monomer species, then the configuration of this "coordination complex" apparently allows entry of monomer molecules into the polymer chain only in a stereospecific manner. For example, cyclic transition complexes (VII) have been postulated to account for the cis-1,4

VII

mode of addition in the lithium catalyzed polymerization of isoprene in nonpolar solvents (Stearns and Forman, 1959). Polymerization via coordination mechanisms will be discussed further in Section 2.2.5.

2.2.4.8. Energetics of Polymerization and Its Relationship to Molecular Weight Control

The overall activation energy for living polymerization systems is simply the activation energy of the propagation step. Actual values vary depending on the nature of the solvent, cation, and the activation energies of free ion and ion pair propagation but are usually on the order of a few kilocalories per mole. Thus the rate of polymerization increases slowly with increasing temperature.

The proposed mechanism of anionic polymerization involving ion pairs and free ions means, however, that one does not measure a true activation energy. Activation parameters have been determined for both free ion and ion pair propagation reactions of styrene in several solvents (Shimomura et al., 1967a). The activation energy of free ion propagation E_{pf}^- was 5–6 kcal/mole and was independent of the solvent and counterion. On the other hand, the activation energy for ion pair propagation $E_{pp}^{\mp}$ was much lower and under certain conditions (depending on solvent and temperature) negative. This dependence of $E_{pp}^{\mp}$ on reaction parameters can be rationalized in terms of two kinds of propagation ion pairs, contact and solvent separated, in rapid equilibrium with each other

$$R^- \, {}^+metal \underset{}{\overset{K_d}{\rightleftarrows}} \quad R^- \| \, {}^+metal$$
$$\text{contact ion} \qquad \text{solvent-separated}$$
$$\text{pair} \qquad\qquad \text{ion pair}$$

Under conditions where the solvent-separated ion pair is much more reactive than the contact ion pair, it can be shown (Smid, 1969) that

$$E_{pp}^{\mp} = E_{ps}^{\mp} + \Delta H_d/(1 + K_d) \qquad (2.130)$$

where $E_{ps}^{\mp}$ is the activation energy for propagation of the solvent-separated ion pair, and K_d and ΔH are the equilibrium constant and enthalpy change for the equilibrium reaction. Accordingly, $E_{pp}^{\mp}$ will be negative for $\Delta H_d/1 + K_d > E_{ps}^{\mp}$, and thus the sign of the activation energy will depend on ΔH_d and K_d. Variation of K_d with temperature may result in a change from a negative to a positive activation as is observed in the polymerization of styrene in 1,2-dimethoxyethane (Fig. 2.12) with sodium as the counterion at about $0°C$ (Shimomura et al., 1967b).

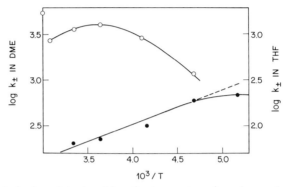

FIG. 2.12. Arrhenius plots describing the temperature dependence of propagation of $S^- \cdots Na^+$ ion pairs in (●) tetrahydrofuran (THF) and (○) dimethoxyethane (DME) (Szwarc, 1975).

When there is only one propagating species in a living polymerization system, the degree of polymerization is independent of the propagation rate constant, being dependent only on the relative amounts of initiator and monomer. A Poisson distribution is obtained in which $\bar{X}_n$ should be insensitive to changes in temperature. When propagation proceeds by more than one propagating species, a Poisson-type molecular weight distribution may not always be possible. The effect can be overcome by adding an inert common ion salt such as sodium tetraphenylboride to the polymerization medium to suppress the dissociation of ion pairs into free ions.

A fascinating demonstration of the ceiling temperature phenomenon may be seen in the anionic polymerization of α-methylstyrene. Addition of sodium naphthalene to a solution of α-methylstyrene in tetrahydrofuran at 50–60°C results in the development of the dark red color characteristic of the polystyryl anion. However, there is no evolution of heat nor any increase in viscosity. On cooling to $-40°C$, polymer is formed and the viscosity of the solution increases. If the solution is warmed again, the poly(α-methylstyrene) depolymerizes and the viscosity drops markedly. Recooling causes polymerization to take place again. Thus extensive propagation to long chain polymer can only occur when the polymerization temperature is well below the ceiling temperature corresponding to the particular monomer concentration.

2.2.4.9. Practical Techniques

The stringent control of purity of monomers, initiators, and solvents required for cationic polymerization is also needed for anionic polymerization. Typically, the monomer is added gradually throughout polymerization to avoid excessive heat buildup. In practice therefore, the rate of polymerization is determined by the rate of addition of monomer, which in turn is limited by the rate at which the heat of polymerization can be removed. Morton and Fetters (1975) have reviewed experimental techniques for anionic polymerization.

Although anionic polymerization reactions are very sensitive to impurities, the reactions are somewhat easier to carry out on a large scale because such factors as surface to volume ratio are smaller. As a consequence, several polymers, notably certain synthetic thermoplastic elastomers, are made by anionic polymerization techniques on an industrial scale with good control of polymer properties.

2.2.5. COORDINATION ADDITION POLYMERIZATION

It is instructive to reflect for a moment on the configuration of the polymers produced by the mechanisms that have been discussed so far. The polymerization of a monosubstituted ethylene $CH_2=CHR$ leads to a polymer in which every other carbon atom can be considered to be asymmetric as shown in the polymer structure VIII,

$$\text{www}\ CH_2-\underset{\underset{R}{|}}{\overset{\overset{H}{|}}{C^*}}\ \text{www}$$

VIII

where C^* is the asymmetric carbon atom. The asymmetry is due to the fact that C^* is bonded to four different substituents, viz., H, R, and two polymer chain segments of different lengths. However, C^* is not truly asymmetric since this term implies optical activity, which is determined only by the first few atoms of the substituents attached to C^*. But although the polymer molecule VIII will not exhibit optical activity, C^* is a site of steric isomerization in polymerization in that it can exhibit either of two configurations and thus give rise to the isotactic, syndiotactic, or atactic structures discussed in Chapter 1.

Most of the polymerization mechanisms that have been discussed so far involve freely propagating species that undergo both modes of addition at rates determined by the energetics of the propagation reaction. In free radical addition polymerization, the difference in activation enthalpy between syndiotactic and isotactic placement leads to an increasing tendency toward syndiotacticity with decreasing polymerization temperature (Fordham, 1959). Since free radical polymerizations are usually carried out at moderately high temperatures, most of the resulting polymers are consequently atactic. It is of interest, however, that polymerization of methyl methacrylate in the presence of preformed isotactic poly(methyl methacrylate) serving as a polymer matrix results in a syndiotactic polymer. This type of polymerization is referred to as a stereospecific matrix or template polymerization. The driving force for the stereoselectivity of the template polymerization is believed to be the tendency of isotactic and syndiotactic poly(methyl methacrylate) to form stereocomplexes (Gons et al., 1975).

The tendency to form atactic polymer is also observed in ionic polymerization in which free propagating species are involved. Again, syndiotactic placement is increasingly favored as the polymerization temperature is lowered. However, some anionic polymerizations are quite stereoregular as a result of coordination between the propagating chain end, counterion, and incoming monomer. The coordination complex allows entry of the monomer molecules only in a stereospecific manner.

The number of such instances of stereospecific polymerization was very limited prior to 1954 and polymers in commercial use were atactic. In that year, a new class of polymerization catalysts, which possessed powerful stereoregulating properties, was discovered by Ziegler. These catalysts were to revolutionize polymer chemistry, for they not only facilitated the synthesis of polymers from the α-olefins, which hitherto could not be polymerized by any means, but they also gave polymers with well-defined stereochemical structures (Ziegler et al., 1955). The most important applications of these catalysts are the polymerization of nonpolar olefins such as ethylene and propene, and the conjugated diolefins, especially butadiene (Boor, 1974). Ethylene for example can be polymerized at low pressures (~ 1–10 atm) and low temperatures (50–75°C) to a linear, highly crystalline polymer. The absence of branches (see Chapter 3, Section 3.8.2) results in a higher melting point and superior mechanical properties compared to the product of high pressure free radical polymerization. Both the isotactic and syndiotactic forms of polypropene can be readily synthesized, whereas previously this monomer had not been polymerizable by conventional free radical or ionic means.

The catalysts discovered by Ziegler and further developed by Natta are commonly referred to as Ziegler–Natta catalysts (Reich and Schindler, 1966).

These catalyst systems appear to function by a coordination complex between the catalyst, growing chain, and incoming monomer. Hence this process is referred to as *coordination addition polymerization*.

2.2.5.1. Catalysts

The Ziegler–Natta catalysts are formed by combining under an inert atmosphere a metal alkyl or hydride of a group I–III metal in the periodic table with a transition metal salt (usually halides) of groups IV–VIII. Of the latter, the halides of titanium, vanadium, chromium, and zirconium have been extensively studied. The catalysts are prepared from solutions of their components in a suitable inert solvent such as hexane. On mixing the components, the active catalyst may be formed as a solution, but more commonly it is formed as a heterogeneous dispersion or slurry. Polymerization is carried out by the addition of monomer to this dispersion or solution. Table 2.9 shows some typical homogeneous and heterogeneous catalyst systems for olefin polymerization.

TABLE 2.9

TYPICAL HOMOGENEOUS AND HETEROGENEOUS
CATALYST SYSTEMS FOR OLEFIN POLYMERIZATION

Transition metal compound	Metal alkyl
Homogeneous catalysts	
VCl_4, V(acetylacetonate)$_3$	R_2AlCl
VCl_4, $VOCl_3$	$(C_6H_5)_4Sn + AlBr_3$
$(C_2H_5)_2TiCl_2$	$(CH_3)_3Al$
Heterogeneous catalysts	
$TiCl_4$	$(Et)_3Al$
$TiCl_3$	Et_2AlCl
VCl_3	R_3Al

Not all the catalyst combinations are equally reactive or selective, and wide differences in reaction rate and stereospecificity are found with different catalyst combinations for a given monomer. For example, the heterogeneous catalyst formed from diethylaluminum chloride and titanium trichloride polymerizes propene at 50°C to a highly isotactic polymer, whereas the soluble system formed by replacing titanium trichloride with vanadium tetrachloride polymerizes propene at −78°C to a highly syndiotactic polymer. Further developments in Ziegler–Natta catalysts are aimed primarily at greater polymerization efficiencies. These have mainly involved modifications to the transition metal component and have resulted in a more than twentyfold increase in efficiency (Weissermel *et al.*, 1975). For example, highly active catalysts are obtained by reaction of magnesium alkoxides with tetravalent titanium chlorides prior to activation with aluminum alkyls. During the course of catalyst preparation, the original crystal structure of $TiCl_4$ is completely destroyed as new crystalline species are formed.

At the same time that Ziegler and Natta were carrying out their research on the two-component catalyst systems, it was found that certain one-component systems in which there is no organometallic activator also showed catalytic activity towards ethylene and α-olefins. These systems consist of transition metal oxides such as chromium oxide and molybdenum oxides physically dispersed on another material called the *support*. Typical supports include alumina, silica, titania, zirconia, and thoria. The catalysts must be activated before use by exposure to air at temperatures ranging from 400 to 1000°C for several hours. These catalysts are mainly employed for the polymerization of ethylene; the resultant polymer has an essentially linear structure that is free from the branching characteristic of the free radically produced polymer (Chapter 3, Section 3.8.2). They are not good catalysts for the higher olefins and do not possess good stereoregulating properties. For example, the supported chromium oxide catalysts developed by Phillips Petroleum Co. (Clark *et al.*, 1956) convert ethylene to a highly crystalline polymer but the homopolymerization of propene results in a polymer of low stereoregularity which is, therefore, only partially crystalline. The higher olefins such as 1-butene, 1-pentene, and 1-hexene give polymers ranging from tacky semisolids to viscous liquids (Witt, 1974).

A variety of one-component catalyst systems based on organometallic compounds of transition metals have been reported in recent years (Yermakov and Zakharov, 1975; Ballard, 1975). Research in this area has been stimulated by advances in the understanding of the mechanism of the Ziegler–Natta two-component system. As will be seen later, an essential feature of the formation of the propagating centers in such systems is the alkylation of the transition metal ions by the organometallic cocatalyst. For this reason, the search for individual organometallic compounds that possess catalytic activity in olefin polymerization is of great interest. Several relatively stable organometallic compounds of transition metals have been shown to be active catalysts for the polymerization of olefin and vinyl monomers. They fall into two main groups: σ-organometallic compounds and π-allyl compounds. The former include compounds of transition metals such as zirconium and titanium with such ligands as $-CH_2Ph$, $-CH_2Si(CH_3)_2$, and $-CH_2C(CH_3)_3$. Numerous active π-allyl compounds have been reported. For example, the polymerization of ethylene to linear polyethylene occurs in the presence of π-allyl compounds of zirconium, titanium, and chromium, among others. Nickel-based catalysts, e.g., π-allyl nickel halides, are particularly efficient catalysts for the conversion of conjugated dienes to polymers of high steric order.

Also falling into the category of one-component catalysts are the subhalides of transition metals in which there is no organic component at all. Titanium dichloride is the most active system of this type.

2.2.5.2. Polymerization Mechanism

The formation of a coordination complex between the propagating chain end, incoming monomer, and catalyst site results in orientation of the monomer with respect to the growing chain end so that each monomer adds in a sterically controlled fashion. In the case of ethylene, complexation of the active end removes the possibility of the backbiting transfer reaction that characterizes the free radical propagation and hence an essentially linear polymer results. Termination is also negligible, since the length of the polymer molecules is limited primarily by chain transfer reactions to monomer or catalyst.

A large majority of mechanistic studies have been concerned with Ziegler–Natta catalysis. While it is easy to generalize on the nature of the mechanism, answers to questions concerning the chemical nature of the propagating species and the mechanism of complexation have been debated since the first discoveries were made. It is important to realize that not all Ziegler–Natta combinations operate by coordination mechanisms. Polymerization by conventional free radical, anionic, or cationic mechanisms may be induced by the individual catalyst components or products of their interaction, and may take place with or without stereoregularity. This appears to be true for many of the polar vinyl monomers, where in most of the examples reported the catalyst serves as a source of conventional cationic, anionic, or radical initiating species. When the components of the catalyst are strong electron acceptors, e.g., $AlEtCl_2$ or $TiCl_4$, vinyl monomers such as vinyl ethers can be polymerized via a cationic mechanism. Similarly, anionic polymerization at the base metal–carbon bond is also possible when the monomer is polymerizable by the metal alkyl component of the Ziegler–Natta catalyst, e.g., tert-butyl acrylate or isoprene by n-butyllithium in the n-butyllithium/$TiCl_4$ combination.

By far the most important application of these systems is the polymerization of nonpolar olefins and conjugated diolefins taking place through formation of coordinated complexes. The reaction mechanism is believed to be ionic in character and may be classified according to the charge distribution in the transition state of the coordinated complex formed in the propagation reaction. The terms anionic coordination or cationic coordination have been used depending on whether the monomer on insertion is negatively or positively charged with respect to the metal. It is generally accepted that the polymerization mechanism involves an insertion reaction that is based on a coordinated anionic species in which the coordinated monomer enters the chain through a polarized catalyst–polymer bond, $M^{\delta+}-R^{\delta-}$. The rate determining step is assumed to be the preliminary co-ordination of the monomer to the active site of the catalyst followed by a nucleophilic attack by the incipient carbanion at the end of the polymer chain on the monomer double bond. These reaction steps may be represented schematically as

$$\text{\small{mmmm}}\ CHR\overset{\delta-}{-}\overset{}{C}H_2 \cdots\cdots \overset{\delta+}{M_T}$$
$$R\overset{\delta+}{-}\overset{}{C}H\overset{\delta-}{=}\overset{}{C}H_2$$
$$\text{IX}$$

where M_T is the active catalyst center. The anionic character of this mechanism is in accord with the reactivity sequence ethylene > propene > butene − 1, which is opposite to the cationic polymerization sequence: isobutene > propene > ethylene. Additional evidence favoring a coordinated anionic mechanism is reviewed by Berger et al. (1969). A coordinated cationic mechanism has been proposed for certain monomer systems, such as vinyl ethers. For example, a cationic insertion mechanism has been postulated (Reich and Schindler, 1966) for the polymerization of isobutyl vinyl ether by i-$Bu_3Al/VCl_3-\frac{1}{3}AlCl_3$ to a highly stereoregular polymer. Discussion in the rest of this chapter centers around the coordinated anionic mechanism.

Proposals for coordinate anionic polymerization may be further distinguished according to whether the transition metal (M_T) or the base metal center (M_B) or a bimetallic complex involving both centers is considered to be the active site of chain growth. An illustration of these mechanisms is the polymerization of propene by the heterogeneous catalyst formed by reaction of titanium trichloride with triethylaluminum. Both bimetallic active sites (X) and monometallic sites (XI) have been postulated as the active site

$$\text{X} \qquad\qquad \text{XI}$$

(Natta and Mazzanti, 1960; Cossee, 1960) with the bulk of experimental evidence favoring the latter. Isotactic polypropene, for example, has been formed in the presence of transition metal compounds alone. Other olefins can also be polymerized with certain transition metal compounds in the absence of added metal alkyl. The presence of the latter has a marked influence only on catalyst activity (where activity is taken to mean polymerization rate) and to some extent on stereospecificity. Further, reactivity ratios in copolymerization are markedly dependent on the nature of the transition metal with little dependence on the nature of the organometallic component (Karol and Carrick, 1961). These studies suggest that all the essential features of the stereospecific polymerization are present in the surface of the transition metal halide.

The chemical structure of the monometallic active species is believed to result from alkylation of the $TiCl_3$ by $Al(Et)_3$ during catalyst preparation. It should be pointed out that not all the Ti^{3+} ions that are alkylated become active centers but only those that are in potentially stereospecific positions. These sites are on the edges of $TiCl_3$ crystals where there is sufficient room for an α-olefin to coordinate with the exposed titanium (Boor, 1963). A further appreciation of this point may be gained from consideration of the crystal structure of $TiCl_3$. Titanium trichloride can exist in α, β, or γ forms (Natta et al., 1961) with the α form giving the highest degree of stereoregulation in α-olefin polymerization. The structure consists of sheets in which a layer of titanium ions is sandwiched between two layers of chloride ions. The Ti^{3+} ions are located in the octahedral interstices of the chloride ion lattice (Fig. 2.13). If the bulk of the crystal is uniform, the requirement of electrical neutrality imposes certain boundary limitations in real crystals, viz., the existence of chloride ion vacancies at the surface.

Both experimental observations and theoretical calculations have shown that these vacancies exist at the edges of the elementary sheets and not in the basal planes of the crystal (Hargitay et al., 1959; Arlman, 1964). According to Arlman, the configuration of the stereospecific Ti^{3+} ions located on the $10\bar{1}0$ surface can be represented as shown in Fig. 2.14. The active site consists of a Ti^{3+} ion octahedrally coordinated with five chloride ions. The four chloride ions in the square base are not equivalent. Two (B) are sterically blocked while the other two protrude from the crystal. Of the latter, one (F) is attached to two Ti^{3+} ions and is considered fixed; the other chloride ion (L) is attached to only one Ti^{3+} ion and is considered loosely held. Further, L may interchange its position with the chloride ion vacancy as shown in Fig. 2.14. The structure may be represented schematically as

$$
\begin{array}{c}
\square \\
| \quad \diagup \text{Cl (B)} \\
\text{(B) Cl} - \text{Ti} - \text{Cl (L)} \\
\text{(F) Cl} \diagup \quad | \\
\quad \text{Cl}
\end{array}
$$

XII

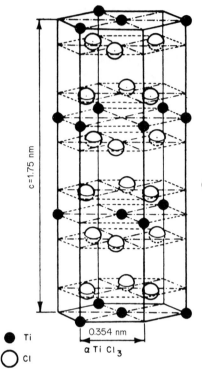

Ti

Cl

0.354 nm

α Ti Cl₃

c = 1.75 nm

FIG. 2.13.   α-TiCl₃ crystalline lattice structure (Natta and Pasquon, 1959).

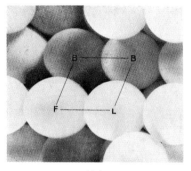

(b)

FIG. 2.14.   The configuration of the stereospecific Ti^{3+} ions on the $10\bar{1}0$ crystal surface of α-TiCl₃: (a) square base parallel to a axis; (b) L and chloride ion vacancy of (a) interchanged (Arlman, 1964).

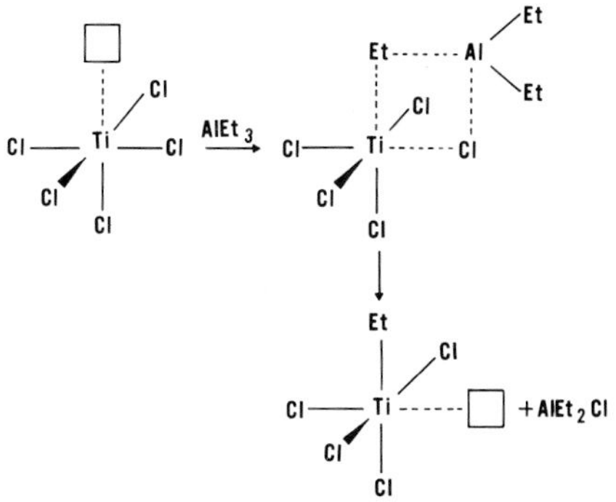

FIG. 2.15. Schematic representation of the proposed mechanism of alkylation of stereospecific Ti^{3+} ions.

in which the chloride ion vacancy is shown by a small square. In the initiation stage, this loosely held chloride ion is considered to undergo ligand exchange with $Al(C_2H_5)_3$ to give the active polymerization site (XI). This sequence of reactions is shown schematically in Fig. 2.15.

2.2.5.3. Propagation

According to the model first proposed by Cossee (1960, 1964), the olefin becomes coordinated to the transition metal at the vacant octahedral position via overlap of the π-electrons of the olefin with the vacant d orbitals of the metal. This results in a weakening of the metal–carbon bond, and propagation takes place by interposition of the coordinated olefin molecule between the transition metal and the alkyl group via a 4-membered ring transition state (Fig. 2.16). Cis opening of the double bond takes place with the unsubstituted carbon becoming attached to the metal center.

While the bulk of the evidence favors polymerization at the transition metal–carbon bond, there is still some question whether the active site is

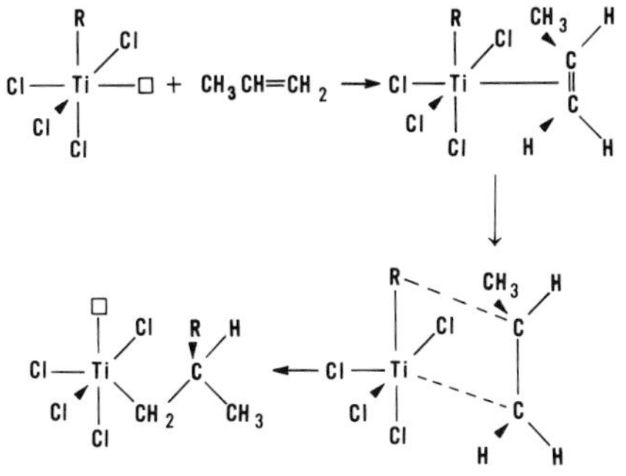

FIG. 2.16. Monometallic mechanism proposed by Cossee (1960, 1964).

exclusively monometallic or whether it is stabilized by the presence of the base metal derivative. For example, Rodrigues and Van Looy (1966) have proposed an active site (XIII) consisting of an alkylated octahedral transition metal complexed with the base metal alkyl.

$$
\begin{array}{c}
R \\
| \quad R \\
Al \\
R \quad Cl \\
| \quad / \\
Cl\!-\!M\!-\!\square \\
Cl \quad Cl
\end{array}
$$

XIII

While it may not be necessary to invoke such structures for heterogeneous catalysts because of the stabilizing effects of the crystal lattice, it seems likely that such structures do exist in solution where the active complex exists as individual molecules. For example, the active site in the case of the homogeneous catalyst dicyclopentadienyl titanium dichloride/AlEtCl$_2$ is believed to have the structure XIV in which the active octahedral site is stabilized by the

$$
\begin{array}{c}
R \quad Cl \quad C_p \quad R \\
\diagdown / \diagdown / \\
Al \quad Ti\!-\!\square \\
/ \diagdown / \quad | \\
Cl \quad Cl \quad C_p
\end{array}
$$

XIV

base metal derivative via halogen bridges (R = ethyl).

2.2.5.4.   Origin of Stereoregulation

A number of models have been proposed to explain stereoregulation in α-olefin and diene polymerization with heterogeneous and soluble catalysts. The models based on ligand–monomer and monomer–monomer interactions provide the most comprehensive and satisfactory explanation of the stereoregulating properties. The ligand–monomer interaction model was proposed by Arlman and Cossee (1964) for heterogeneous catalysts. The diagrammatic representation of the octahedral site and the schematic propagation mechanism (cf. Figs. 2.14 and 2.16) for propene have already been discussed. Consideration of the possible orientations of the propene molecule as it approaches an active site indicates that there is only one preferred orientation, all other possibilities being eliminated on steric grounds. This raises a problem with the polymerization model represented in Fig. 2.16, where it is seen

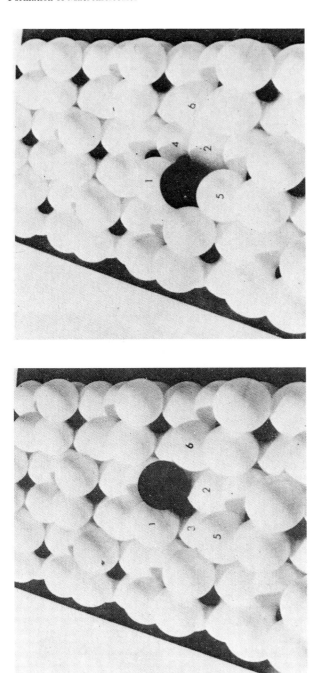

[a]

FIG. 2.17.   Configuration of the alkylated stereospecific Ti^{3+} sites. The structures shown in (a) and (b) differ in the geometry of the active titanium site. The positions of the propagating polymer chain (large black sphere) and the chloride ion vacancy are interchanged (chloride ions 1, 4, and 2 are equivalent to F, B, and B, respectively, in Fig. 2.14b) (Arlman and Cossee, 1964).

that propagation results in an interchange of the lattice vacancy and alkyl (polymer) substituent. These two positions are not structurally equivalent as may be appreciated from Fig. 2.17, which shows the configuration of the two nonequivalent positions of the alkylated stereospecific site. The two are mirror images and thus will present opposite steric possibilities.

In order to facilitate isotactic propagation, the alkyl group must flip back to its original position, as shown in Fig. 2.18. The driving force for the migration is provided by the lower energy of one of the configurations (most likely Fig. 2.17a since in this configuration the polymer chain protrudes much more from the polar TiCl₃ surface than it does in the other configuration). The overall extent of isotactic placement, therefore, depends on the rate of polymer chain migration relative to the rate of monomer insertion. If the latter is significantly high, the resultant polymer will contain sequences of syndiotactic placement. Whether this is the case or not depends on the particular monomer, catalyst system, and reaction temperature. Generally, under normal polymerization conditions involving temperatures of 50–100°C, the growing polymer chain is able to migrate to the most favored position before the next monomer molecule is inserted.

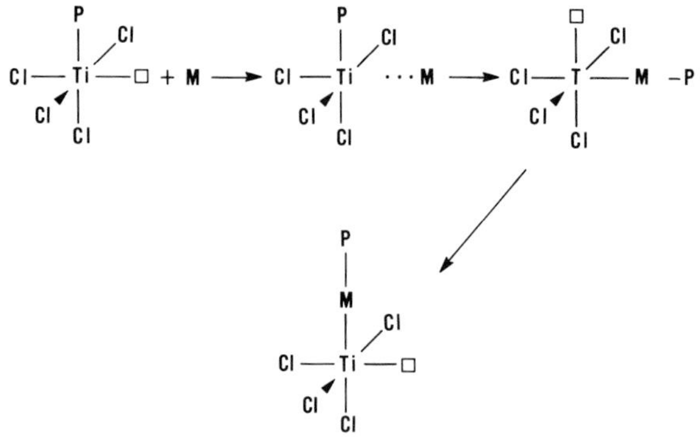

FIG. 2.18. Schematic illustration of the sequence of steps leading to an isotactic placement.

A monomer–monomer interaction model was proposed by Boor and Youngman (1966) to explain the syndiotactic influence on the polymerization of propene by the homogeneous catalyst system $Al(Et)_2Cl/VCl_4$ at $-78°C$. The model is similar to the isotactic model in that it is also based on an octahedral complex that has an alkyl ligand as the growth site and an open coordination site or vacancy through which the monomer is complexed. The active center in the homogeneous case is believed to be stabilized by complex formation with one or more aluminum alkyl molecules via halogen bridges as shown in structure XV. The sequence of steps leading to syndiotactic

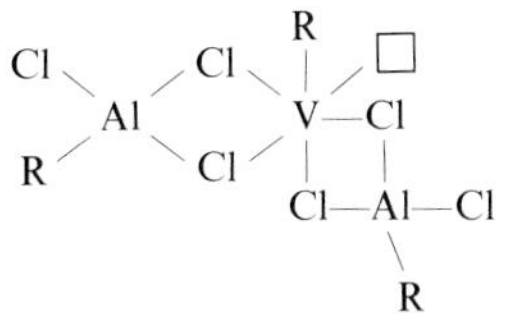

XV

placement is shown in Fig. 2.19. The main features of this model are

1.   The monomer complexing mode (structure B, Fig. 2.19) which leads to head to tail enchainment is energetically favored.

2.   Rotation about the R–V bond is hindered since it is difficult for the CH_3 group of the last added unit to pass over the adjacent ligand.

3.   One orientation of the monomer methyl group leading to syndiotactic placement is more favorable than the other due to methyl–methyl interaction.

4.   Whenever the rotational barrier indicated in (2) is overcome, e.g., at higher polymerization temperatures or when smaller ligands are attached to vanadium, a methylene unit faces the vacancy. As a result, the steric restrictions placed on the incoming monomer are the same for both syndiotactic and isotactic placements resulting in the production of atactic polymer.

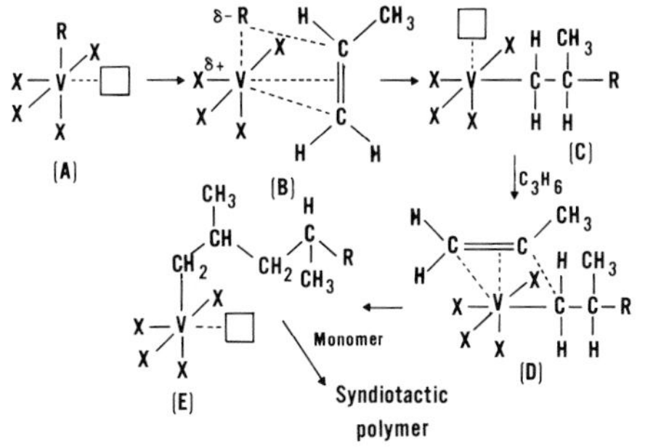

FIG. 2.19.   Schematic illustration of the sequence of steps leading to a syndiotactic configuration (Boor and Youngman, 1966).

Thus the major difference in the heterogeneous and homogeneous models is in the nature of the group interactions. Methyl–methyl interactions in the homogeneous model force the propene to be complexed in opposite configurations at each consecutive growth step. In the isotactic model, only one configuration is permitted at each growth step by the geometry of the chlorine vacancy of the active site on the crystal surface. It may also be noted that in the heterogeneous model of Arlman and Cossee, the R group must return to its original position between growth steps, whereas in the syndiotactic model, polymer is produced by growth at either octahedral position.

2.2.5.5. Polymerization of Conjugated Dienes

As was discussed in Section 2.2.2.2, 1,3 dienes such as butadiene, chloroprene, and isoprene can exist in several isomeric forms that may exhibit both geometric isomerism (when polymerization takes place by 1,4 addition) and stereoisomerism (when polymerization takes place by 1,2 or 3,4 addition).

Polymerization of 1,3 dienes via radical or ionic mechanisms generally yields a mixture of the various possible structural isomers. Only in the case of anionic polymerization in nonpolar solvents using lithium-based initiators was it possible to obtain a predominance of one structure, viz., the cis-1,4 isomer (Section 2.2.4.7). This predominance of cis-1,4 polymerization has been attributed to the formation of a π-complex between the diene and the lithium ion that locks the diene into a cis configuration before and after its addition to the propagating chain.

In contrast, most of the different stereoisomers of the 1,3 dienes have been obtained using Ziegler–Natta catalysts. In the case of butadiene, for example, all four possible polymer structures have been obtained. Table 2.10 summarizes the polymerization of butadiene with a number of selected catalysts. Not all catalysts show such a high degree of stereospecificity; most of the vast numbers that have been studied give mixtures of stereoisomers.

There is insufficient evidence to evaluate the different models proposed to explain stereoregulation in diene polymerization. The situation is much more complicated than that for α-olefin polymerization since diene polymerizations involve a larger number of complexing modes. Presumably, the active

TABLE 2.10

CATALYSTS FOR STEREOSPECIFIC POLYMERIZATION OF BUTADIENE[a]

Metal alkyl	Transition metal compound	Conditions	Structure of polymer (%)		
			1,2	cis 1,4	trans 1,4
Et_2AlCl, Et $AlCl_2$	Cobalt compounds	Al/Co > 100	—	95–98	—
Et_3Al	VCl_4, $VOCl_3$	Al/V = 2	—	—	95–100
Et_3Al	TiI_4	Al/Ti = 5	5	92	4
Et_3Al	Chromium acetylacetonate	Al/Cr = 10	Isotactic	—	—
Et_3Al	Chromium acetylacetonate	Al/Cr = 3	Syndiotactic	—	—
Et $AlCl_2$	β-$TiCl_3$	Al/Ti = 1–5	—	—	100
Et_3Al	$Ti(OBu)_4$	Al/Ti = 3	90	10	—
Et_3Al	$TiCl_4$	Al/Ti = 1	2	49	49

[a] From Boor (1967).

site of those catalysts, which possess a definite lattice structure, consists as before of an octahedral transition metal ion with a metal alkyl bond. When a conjugated diene inserts in a 1,4 manner into the organometallic bond, the product is an alkyl compound which may be a σ-covalent compound XVIa or a π-complex (XVIb)

$$RM_T + CH_2{=}CH{-}CH{=}CH_2 \Big\langle$$

$$R{-}CH_2{-}CH{=}CH{-}CH_2{-}M_T \qquad \text{XVIa}$$

$$RCH_2{-}CH{-}CH{-}CH_2$$

$$\downarrow$$

$$M_T$$

XVIb

Dynamic equilibria between σ-bonded and π-bonded alkyl metal compounds are well characterized and known to be affected by donor ligands. These possible variations in the structure and configuration of the propagating species coupled with different possible complexation modes of the monomer present difficulties for the understanding of the origins of stereoregulation in such systems. These difficulties are one reason for the interest in catalysts consisting of π-allyl derivatives of transition metals since these compounds constitute a good model of the active site structure. The reader is referred to reviews by Ledwith and Sherrington (1974) and Cooper (1970) for detailed discussion of possible mechanisms.

2.2.5.6. Mechanism of One-Component Catalyst Systems

The difference in polymerization mechanism between one-component and traditional Ziegler–Natta two-component catalysts seems to exist only in the initiation stage, while the mechanism of continued propagation of the polymer chain has many common features for all the catalytic systems based on transition metal compounds. Thus, most studies of the chromium oxide catalyst system, for example, deal either with the nature of the species on the catalyst surface or with the nature of the species responsible for polymerization. Such studies have shown that the formation of a surface chromate takes place by reaction of CrO_3 with silanol surface groups of the support (Karol, 1975).

$$
\begin{array}{l}
{-}Si{-}OH \\
\quad | \\
\quad O \qquad + CrO_3 \longrightarrow \\
\quad | \\
{-}Si{-}OH
\end{array}
\qquad
\begin{array}{l}
{-}Si{-}O \\
\qquad \diagdown \quad O \\
\quad O \qquad Cr \qquad + H_2O \\
\qquad \diagup \quad O \\
{-}Si{-}O
\end{array}
$$

Reduction of this surface chromate by ethylene or hydrogen or carbon monoxide results in the formation of a low valence chromium center.

The exact initiation mechanism is uncertain. There is little doubt that the reaction is a surface-catalyzed process requiring the monomer to be absorbed onto the catalyst surface. Initiation is believed to involve the formation of a metal–carbon σ-bond followed by coordination of incoming monomer molecules with subsequent insertion into the metal–carbon bond. The following mechanism has been proposed for the formation of the σ-bond between the metal and alkyl fragment (Yermakov and Zakharov, 1975).

The radicals formed in these reactions may also participate in the alkylation of ions of the transition metal.

Polymer growth then takes place by insertion into the metal–carbon bond as before. Transfer to monomer appears to be the main termination mode.

2.2.5.7. Kinetics

The kinetics of coordination addition polymerization are extremely complex. Homogeneous reactions can be handled in a manner similar to that employed for noncoordination polymerization reactions with reasonable success, although the kinetics are much more complicated. For example, Ballard (1975) has shown that the initial rate of polymerization of styrene by

Zr(benzyl)$_4$ can be expressed by

$$R_p = A[C_0][M_0]^2/(1 + B[M_0]) \quad (2.131)$$

where $[C_0]$ is the initial concentration of catalyst, $[M_0]$ the initial monomer concentration, and A and B are constants for a given temperature. Ballard proposed a mechanism involving rapid formation of a complex between the initiator and monomer followed by a slow reaction of the complex to form the reactive chain carrier. The detailed mechanism is shown below

$$CH_2{=}CHC_6H_5$$
$$+ \qquad\qquad\qquad\qquad\qquad CH_2{=}CHC_6H_5$$
$$(C_6H_5CH_2)_3Zr\ CH_2C_6H_5 \underset{\longleftarrow}{\overset{K_1}{\longrightarrow}} (C_6H_5CH_2)_3Zr\ CH_2C_6H_5$$

$$CH_2{=}CHC_6H_5$$
$$\downarrow$$
$$(C_6H_5CH_2)_3Zr\ CH_2C_6H_5 \xrightarrow[\text{slow}]{k_1} (C_6H_5CH_2)_3Zr(CHCH_2)\ CH_2C_6H_5$$
$$C_6H_5$$

$$CH_2{=}CHC_6H_5 \qquad\qquad\qquad\qquad\qquad CH_2{=}CHC_6H_5$$
$$+ \qquad\qquad\qquad\qquad\qquad\qquad\qquad \downarrow$$
$$(C_6H_5CH_2)_3Zr(CHCH_2)_xCH_2C_6H_5 \underset{\longleftarrow}{\overset{K_2}{\longrightarrow}} (C_6H_5CH_2)_3Zr(CHCH_2)_xCH_2C_6H_5$$
$$C_6H_5 \qquad\qquad\qquad\qquad\qquad\qquad\qquad\qquad C_6H_5$$

$$CH_2{=}CHC_6H_5$$
$$\downarrow$$
$$(C_6H_5CH_2)_3Zr(CHCH_2)_xCH_2C_6H_5 \xrightarrow[\text{fast}]{k_2} (C_6H_5CH_2)_3Zr(CHCH_2)_{x+1}CH_2C_6H_5$$
$$C_6H_5 \qquad\qquad\qquad\qquad\qquad\qquad\qquad C_6H_5$$

$$(C_6H_5CH_2)_3Zr(CHCH_2)_xCH_2C_6H_5 \xrightarrow{k_3} (C_6H_5CH_2)_3ZrH + C_6H_5CH_2(CHCH_2)_{x+1}\ CH{=}CH$$
$$C_6H_5 \qquad\qquad\qquad\qquad\qquad\qquad\qquad\qquad\qquad C_6H_5 \qquad C_6H_5$$

$$(C_6H_5CH_2)_3Zr(CHCH_2)_xCH_2C_6H_5 + CH_2{=}CH \xrightarrow{k_4} \text{inactive addition product}$$
$$C_6H_5 \qquad\qquad\qquad\qquad\qquad\qquad C_6H_5$$

from which it can be shown that the polymerization rate is given by

$$R_p = K_1K_2k_1k_2[C_0][M_0]^2/(k_3 + k_4[M_0]) \quad (2.132)$$

in agreement with the experimental rate equation.

The kinetics of heterogeneous catalysis are considerably more complicated than for homogeneous catalysis. It is generally recognized that equilibrium adsorption processes of both the monomer and metal alkyl on the surface of the transition metal halide crystal play an important role in determining the kinetic course of polymerization. The term adsorption is used to refer to the weak chemisorption of a species onto the active center. Tait (1975) has proposed an adsorption model for the polymerization of 4-methyl-1-pentene by the catalyst system $VCl_3/Al(i\text{-}Bu)_3$. The model assumes that the active site is the alkylated species VCl_2R and that chain propagation takes place between the active site and adsorbed monomer. Both monomer M and metal alkyl molecules A are able to compete for complex formation giving rise to the following equilibrium expressions

$$VCl_2P + M \xrightleftharpoons{K_M} VCl_2P \cdot M$$

$$VCl_2P + A \xrightleftharpoons{K_A} VCl_2P \cdot A$$

where K_M and K_A are the equilibrium constants for the adsorption of monomer and metal alkyl, respectively. The surface fraction θ_M covered by the monomer can be described by a Langmuir–Hinshelwood isotherm

$$\theta_M = \frac{K_M[M]}{1 + K_M[M] + K_A[A]} \tag{2.133}$$

where $[M]$ and $[A]$ are the bulk liquid equilibrium concentrations of monomer and metal alkyl, respectively.

The overall rate of polymerization is given by

$$R_p = k_p \theta_M C_0 \tag{2.134}$$

where C_0 is the active center concentration. Hence,

$$R_p = \frac{k_p K_M[M] C_0}{1 + k[M] + K[A]} \tag{2.135}$$

This model successfully predicts the observed kinetics and molecular weight dependencies of the $VCl_3/Al(i\text{-}Bu)_3$ system and appears to have fairly general application to many other Ziegler–Natta systems. There are major problems, however, in determining the actual number of active sites C_0. Radioactive quenching techniques are perhaps the most reliable, although a glance at the literature indicates a wide variation in experimental results (Yermakov and Zakharov, 1975). The quenching reaction with, for example, radioactive alcohol follows the scheme

$$L_xM_T\text{---}CH_2P + ROH^* \longrightarrow L_xM_TOR + H^*CH_2P$$

where L_x represents the ligands surrounding the transition metal M_T. Depending on the radioactivity of the polymer obtained, the number of propagation centers C_0 in the catalyst can then be calculated from

$$C_0 = A'G/aQ \tag{2.136}$$

where A' is the polymer activity (curies/gm), G is the polymer yield (gm), a is the quantity of catalyst (moles of transition metal), and Q is the specific radioactivity of a quenching agent (curies/mole). The units of C_0 are moles per mole of transition metal.

Much of the difficulty in reproducibility of kinetic measurements is probably a consequence of the heterogeneous nature of the catalyst surface, which varies with catalyst preparation and other experimental conditions. The kinetics can be further complicated by crystal fragmentation during polymerization. This process exposes new surface areas that react to form new activated sites. A review edited by Chien (1975) gives a detailed discussion of kinetic approaches to the mechanism of heterogeneous catalysis.

2.2.5.8.   Stereoselective and Stereoelective Polymerization

The polymerization of most vinyl monomers in the isotactic mode does not result in polymers that are optically active. The structural unit in the polymer can exist in either of two configurations but, as has already been noted, the tertiary carbon atom C^* is not a true asymmetric center since there is a plane of symmetry at C^* because of the equivalence of the two chain segments.

Optical activity is potentially observable in a polymer molecule when there is a true asymmetric center either in the main chain, as occurs with poly(propylene oxide) (XVII) or in the side chain, as in poly(4-methyl-1-hexene) (XVIII)

$$\text{\wavy}CH_2\text{—}CH\text{\wavy}$$

$$
\begin{array}{cc}
\text{H} & \text{CH}_2 \\
| & | \\
\text{\wavy}CH_2\text{—}C^*\text{—}O\text{\wavy} \quad\quad CH_3\text{—}{}^*C\text{—}H \\
| & | \\
\text{CH}_3 & \text{CH}_2\text{CH}_3 \\
\text{XVII} & \text{XVIII}
\end{array}
$$

Polymerization of the optically active enantiomer ((R) or (S)) of these monomers will result in an optically active polymer even using a nonstereospecific catalyst. Polymerization of either enantiomer of propylene oxide produces an optically active polymer that must be isotactic. The reason for this is that the S_N2 displacement at the less-hindered carbon atom does not disturb the chiral center. However, when the chiral center is in the side chain, polymerization of one or the other of the enantiomers produces an optically active polymer that can be either isotactic, syndiotactic, or atactic. The observed optical rotation may depend on the tacticity because of enhancement of the rotation for particular conformations, which in turn is determined by the configuration of the structural units.

Polymerization of racemic monomers will generally result in optically inactive polymers due to a mixture of both (R) and (S) enantiomers in the main chain. It has been found, however, that ordinary optically inactive Ziegler–Natta catalysts can polymerize racemic mixtures of certain α-olefins, such as 4-methyl-1-hexene, to an optically inactive isotactic polymer that can be separated into optically active fractions containing predominantly (R) or (S) units (Pino *et al.*, 1963). This process is called *stereoselective polymerization* and refers to the unique ability of the catalyst to choose one or the other enantiomer from the racemic monomer mixture with the result that there occurs a prevalence of one of the enantiomers in each single polymer chain.

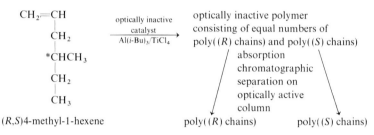

A similar result is obtained if an optically active (chiral) catalyst such as TiCl₄ plus tris-[(S)-2-Me-butyl]-aluminum or bis[(S)-2-Me-butyl]-zinc is used. However, when the asymmetric center is alpha to the double bond of the olefin (e.g., 3-Me-1-pentene), these chiral catalysts can actually exclusively convert one enantiomer to an optically active polymer whose chirality is opposite that of the unreacted monomer.

$$CH_2{=}CH$$
$$|$$
$$*CHCH_3 \xrightarrow[\text{catalyst}]{\text{optically active}} \text{optically active poly ((R) chain)}$$
$$| \qquad\qquad \text{(or (S)) + optically active}$$
$$CH_2 \qquad\qquad \text{monomer (S)} \qquad \text{(or (R))}$$
$$|$$
$$CH_3$$

(R,S)3-methyl-1-pentene

This process of obtaining an optically active polymer from a racemic monomer using an optically active catalyst is referred to as *stereoelective* polymerization. A further restriction on the stereoelective process is that the asymmetric center in the catalyst must be close to the metal atom. For example, the ZnX_2^S–TiCl₄ catalyst system $[X^S = (S)$-2-methylbutyl] polymerizes (R, S)3,7-dimethyl-1-octene to an optically active polymer (stereoelection) whereas the $LiAlX_2^R$–TiCl₄ catalyst system $[X = (R)(CH_2)_2-$ $*CHCH_3(CH_2)_3$–CH(CH₃)₂] results in an optically inactive polymer (stereoselection) (Tsuruta, 1972).

This phenomenon is well explained by Cossee's monometallic mechanism of propagation (Boor, 1967; Tsuruta, 1972). It is proposed that half the available sites on the heterogeneous catalyst surface are nonsuperimposable mirror images of the other half, designated as M^R–X and M^S–X, where X represents the catalyst alkyl group and M–X represents the active metal–

carbon growth bond site (Fig. 2.20). Stereoselection takes places because M^R–X sites complex and preferentially polymerize the (R) enantiomer while an equal number of M^S–X sites complex and preferentially polymerize the (S) enantiomer at an equal rate. When the X substituent is optically active, e.g., X^S, the active sites are designated M^S–X^S and M^R–X^S. Under these circumstances, it appears that when the asymmetric carbon atom of the α-olefin is in the α-position and that of the optically active catalyst alkyl group is close to the metal atom, the insertion of the first (R) enantiomer into the M^R–X^S bond is much more difficult than the insertion of the first (S) enantiomer into the M^S–X^S bond so that stereoelection results.

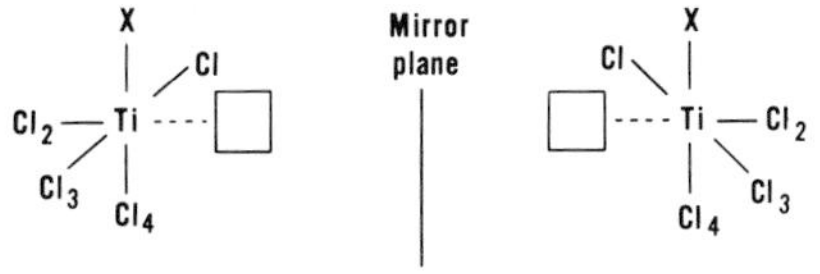

FIG. 2.20. Mirror image active polymerization sites X = alkyl group, e.g., X = Et (stereoselection); or X = (S)–2–Me–1–butyl (stereoelection).

2.2.6. COPOLYMERIZATION

The polymerization of the mixture of two (or more) monomers is referred to as *copolymerization* as distinct from *homopolymerization*, which refers to the polymerization of a single monomer. Provided both monomers can be polymerized by the same mechanism, it is found that the resulting *copolymer* contains units of both monomers in each chain. Similarly, *terpolymers* can be obtained from mixtures containing three monomers. The kinetics of the polymerization process become extremely complex when more than three monomers are present and such *multicomponent copolymerizations* have received little attention.

A wide variety of monomers will copolymerize with one another including some monomers that do not polymerize alone. For example, sulfur dioxide copolymerizes readily with a variety of olefins by a free radical mechanism to form poly(olefin sulfones). Neither sulfur dioxide nor the olefins (apart from ethylene) homopolymerize by a free radical mechanism yet the copolymerization of the two is extremely rapid.

Copolymerization may be employed in a variety of ways, allowing synthesis of polymers with a wide range of physical and mechanical properties, which depend on the nature and relative amounts of the two comonomers and their distribution within the polymer chain. Amorphous homopolymers, being single component systems, must consist of a single phase. In contrast, the phase structure of a copolymer system depends on the distribution of monomer units along the chain. When the copolymer has a random or alternating distribution of monomer units, it is generally a homogeneous single phase system in the solid state with properties intermediate between those of the parent homopolymers. When there are long sequences of each monomer in the copolymer chain, as in block copolymers, or when long sequences of the second monomer are attached (grafted) to the backbone chain of the other polymer, as in graft copolymers, the resulting copolymers are heterogeneous and multiphase with properties directly attributable to the multiphase character. For example, block copolymers having the composition polystyrene–polybutadiene–polystyrene (the so-called ABA copolymers) consist of two phases and show elastomeric or glassy properties depending on whether polybutadiene or polystyrene forms the continuous phase. This multiphase character arises from the inherent incompatibility of two polymer sequences of different composition.

For the most part, this section will consider the formation of single-phase homogeneous random and alternating copolymers which, in principle, are made by adding a suitable initiator to a mixture of two monomers. Techniques for preparing block and graft copolymers will be briefly discussed at the end of the section.

2.2.6.1. Copolymer Composition

The composition of a copolymer in most instances is quite different from that of the monomer mixture from which it was produced. This results from the fact that the reactivity of the propagating chain (which may be a free radical, carbenium ion, or carbanion) is generally not equal toward each monomer but rather depends on the monomer unit at the growing chain end. Kinetic equations were developed in the early 1940s that took into account the effects of the last added unit on the rates of addition of the two monomers (Mayo and Lewis, 1944; Alfrey and Goldfinger, 1944). There are four propagation steps that must be considered for the copolymerization of two monomers M_1 and M_2. These are shown below with the corresponding rate expression for the loss of monomer.

Propagation reaction

$$\text{\small{\(\sim\!\sim\!\sim\)}}\ M_1^* + M_1 \xrightarrow{\ k_{11}\ } \text{\small{\(\sim\!\sim\!\sim\)}}\ M_1^*$$

$$-d[M_1]/dt = k_{11}[M_1^*][M_1] \tag{2.137}$$

$$\text{\small{\(\sim\!\sim\!\sim\)}}\ M_1^* + M_2 \xrightarrow{\ k_{12}\ } \text{\small{\(\sim\!\sim\!\sim\)}}\ M_2^*$$

$$-d[M_2]/dt = k_{12}[M_1^*][M_2] \tag{2.138}$$

$$\text{\small{\(\sim\!\sim\!\sim\)}}\ M_2^* + M_1 \xrightarrow{\ k_{21}\ } \text{\small{\(\sim\!\sim\!\sim\)}}\ M_1^*$$

$$-d[M_1]/dt = k_{21}[M_2^*][M_1] \tag{2.139}$$

$$\text{\small{\(\sim\!\sim\!\sim\)}}\ M_2^* + M_2 \xrightarrow{\ k_{22}\ } \text{\small{\(\sim\!\sim\!\sim\)}}\ M_2^*$$

$$-d[M_2]/dt = k_{22}[M_2^*][M_2] \tag{2.140}$$

Each propagation reaction has a characteristic rate constant k_{ab}, where the first subscript refers to the active center and the second subscript refers to the monomer. The total rate of disappearance of M_1 is the sum of the two individual rate equations ((2.137) and (2.139)).

$$-d[M_1]/dt = k_{11}[M_1^*][M_1] + k_{21}[M_2^*][M_1] \qquad (2.141)$$

Likewise, the rate of disappearance of M_2 is given by

$$-d[M_2]/dt = k_{22}[M_2^*][M_2] + k_{12}[M_1^*][M_2] \qquad (2.142)$$

Division of Eq. (2.141) by Eq. (2.142) yields the ratio of the rates at which the two monomers enter the copolymer, i.e., the instantaneous copolymer composition, as

$$\frac{d[M_1]}{d[M_2]} = \frac{[M_1]}{[M_2]}\left\{\frac{k_{11}[M_1^*] + k_{21}[M_2^*]}{k_{22}[M_2^*] + k_{12}[M_1^*]}\right\} \qquad (2.143)$$

The terms in M_1^* and M_2^* can be removed by assuming a stationary-state condition for both M_1^* and M_2^*. This implies that the rate of disappearance of a given active center, say M_1^*, must be equal to its rate of appearance, i.e.,

$$k_{21}[M_2^*][M_1] = k_{12}[M_1^*][M_2] \qquad (2.144)$$

so that

$$[M_2^*] = k_{12}[M_2][M_1^*]/k_{21}[M_1] \qquad (2.145)$$

Substitution for $[M_2^*]$ in Eq. (2.143) yields

$$\frac{d[M_1]}{d[M_2]} = \frac{[M_1]}{[M_2]}\left\{\frac{k_{11}[M_1^*] + k_{12}[M_1^*]([M_2]/[M_1])}{(k_{22}k_{12}/k_{21})([M_2][M_1^*]/[M_1]) + k_{12}[M_1^*]}\right\} \qquad (2.146)$$

By dividing the nominator and denominator of Eq. (2.146) by $k_{12}[M_1^*]$, the instantaneous copolymer composition becomes

$$\frac{d[M_1]}{d[M_2]} = \frac{[M_1]}{[M_2]}\left\{\frac{(k_{11}/k_{12}) + ([M_2]/[M_1])}{(k_{22}/k_{21})([M_2]/[M_1]) + 1}\right\} \qquad (2.147)$$

$$= \frac{[M_1]}{[M_2]}\left\{\frac{(k_{11}/k_{12})[M_1] + [M_2]}{(k_{22}/k_{21})[M_2] + [M_1]}\right\} \qquad (2.148)$$

Defining the propagation rate constant ratios k_{11}/k_{12} and k_{22}/k_{21} as r_1 and r_2 respectively, one finally obtains

$$\frac{d[M_1]}{d[M_2]} = \frac{[M_1]}{[M_2]}\left\{\frac{r_1[M_1] + [M_2]}{r_2[M_2] + [M_1]}\right\} \qquad (2.149)$$

Equation (2.149) is known as the *copolymer equation* and it gives the composition of the copolymer being formed at any instant. It is seen to depend on the monomer concentrations in the feed and on the parameters r_1 and r_2, which are called *reactivity ratios.*

It is frequently helpful to discuss the copolymer equation in terms of mole fractions instead of concentrations. If F_1 and F_2 are the mole fractions of the monomers M_1 and M_2 in the polymer being formed at any instant, then

$$F_1 = 1 - F_2 = d[M_1]/(d[M_1] + d[M_2]) \tag{2.150}$$

Further, if f_1 and f_2 are the mole fractions of the monomers in the comonomer mixture, then

$$f_1 = 1 - f_2 = [M_1]/([M_1] + [M_2]) \tag{2.151}$$

By combining Eqs. (2.149), (2.150), and (2.151), the copolymer equation becomes

$$F_1 = (r_1 f_1^2 + f_1 f_2)/(r_1 f_1^2 + 2f_1 f_2 + r_2 f_2^2) \tag{2.152}$$

Thus F_1 can be calculated from Eq. (2.152) for any monomer feed ratio corresponding to a given pair of reactivity ratios.

2.2.6.2. Reactivity Ratios

The reactivity ratio is the ratio of the rate constant for a propagating species adding to its own monomer to the rate constant for its addition to the other monomer. Thus, $r_1 > 1$ means that M_1^* prefers to add M_1, whereas $r_1 < 1$ means that it prefers to add M_2. For example, in the radical copolymerization of styrene (M_1) with methyl methacrylate (M_2), $r_1 = 0.52$ and $r_2 = 0.46$, so that each radical adds the other monomer about twice as fast as its own. The overall rate of addition will of course also depend on the relative concentration of the two monomers.

Copolymerizations may be divided into three basic categories depending on the product of the two reactivity ratios:

1. If the relative reactivity of M_1^* and M_2^* toward both monomers is independent of the structure of the active centers, i.e., if the rate of reaction of $M_1^* + M_1$ compared to $M_1^* + M_2$, represented by k_{11}/k_{12}, is the same as that of $M_2^* + M_1$ compared to $M_2^* + M_2$, represented by k_{21}/k_{22}, then

$$k_{11}/k_{12} = k_{21}/k_{22} \tag{2.153}$$

i.e.,

$$r_1 = 1/r_2 \tag{2.154a}$$

or

$$r_1 r_2 = 1 \qquad (2.154b)$$

This criterion defines an *ideal* or *random copolymerization*. The end group has no influence on the relative rates of addition so that random copolymers are generated in which the relative amounts of the two units are determined by the feed composition and the relative reactivity of the two monomers. In this case, the copolymerization equation reduces to

$$d[M_1]/d[M_2] = r_1 [M_1]/[M_2] \qquad (2.155)$$

or

$$F_1 = r_1 f_1 / (r_1 f_1 + f_2) \qquad (2.156)$$

Plots of the dependence of instantaneous copolymer composition F_1 on the comonomer feed composition f_1 for varying values of r_1 for the condition $r_1 r_2 = 1$ are shown in Fig. 2.21. Obviously, when $r_1 = r_2 = 1$, the two monomers show equal reactivity toward both propagating species and the copolymer composition is the same as the comonomer composition. For the case where the two reactivities are different ($r_1 > 1$, $r_2 < 1$, or $r_1 < 1$, $r_2 > 1$), one of the monomers is more reactive and the copolymer will contain a larger proportion of the more reactive monomer in random placement.

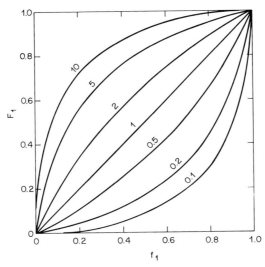

FIG. 2.21. Instantaneous composition of copolymer (mole fraction F_1) as a function of monomer composition (mole fraction f_1) for the indicated values of r_1 where $r_1 r_2 = 1$ (Billmeyer, 1971, after Mayo and Walling, 1950).

2. If each reactive species prefers to react exclusively with the other monomer, $r_1 = r_2 = 0$ and the copolymer equation reduces to

$$d[M_1]/d[M_2] = 1 \qquad (2.157)$$

or

$$F_1 = 0.5 \qquad (2.158)$$

Thus M_1^* can only react with M_2 and M_2^* only with M_1, so that M_1 and M_2 are compelled to enter into regular alternation. This type of copolymerization is referred to as *alternating copolymerization*.

3. If each reactive species prefers to react with its own monomer, i.e., $r_1 > 1$ and $r_2 > 1$, there will be a tendency to form block copolymer. In this case, M_1^* prefers to add M_1 and many units of M_1 would add until an M_2 unit happened to add, at which stage many M_2 units would successively add to the growing end until an M_1 happened to add, and so on. In the limit with $r_1 \gg 1$ and $r_2 \gg 1$, i.e., with the reactive species preferring to add their own monomers exclusively, each monomer will independently homopolymerize. This situation is extremely rare.

Most monomer pairs copolymerize with reactivity ratios given by $0 < r_1 r_2 < 1$. Curves for several nonideal cases are shown in Fig. 2.22. These curves show the effect of increasing tendency toward alternation as the product $r_1 r_2$ approaches zero, where a wider range of monomer compositions

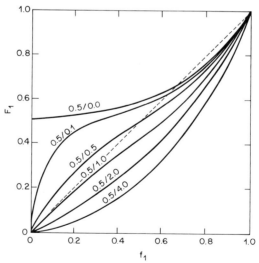

FIG. 2.22. Instantaneous composition of copolymer (mole fraction F_1) as a function of monomer composition (mole fraction f_1) for the values of the reactivity ratios r_1/r_2 indicated (Billmeyer, 1971, after Mayo and Walling, 1950).

yields copolymers containing significant amounts of each monomer. One interesting feature of these curves is that for values of r_1 and r_2 both less than unity, the F_1 versus f_1 curve crosses the diagonal representing $F_1 = f_1$. *Azeotropic copolymerization* is said to occur at this point since the copolymer and feed compositions are identical. Solution of the copolymer equation with $d[M_1]/d[M_2] = [M_1]/[M_2]$ gives the critical composition $(f_1)_c$ for azeotropic polymerization

$$[M_1]/[M_2] = (1 - r_2)/(1 - r_1) \qquad (2.159)$$

and

$$(f_1)_c = (1 - r_2)/(2 - r_1 - r_2) \qquad (2.160)$$

The expressions developed so far are equally applicable to radical, cationic, and anionic polymerization mechanisms. However the values of the reactivity ratios for a given pair of comonomers markedly depend on the polymerization mechanism. It will be recalled, for example, that r_1 and r_2 for the radical copolymerization of styrene (M_1) with methyl methacrylate (M_2) are 0.52 and 0.46, respectively. For cationic polymerization, $r_1 = 10$ and $r_2 = 0.1$, and for anionic polymerization $r_1 = 0.1$ and $r_2 = 6$. Accordingly, the instantaneous copolymer composition will vary with the monomer feed composition in a manner determined by the particular polymerization mechanism, as shown in Fig. 2.23.

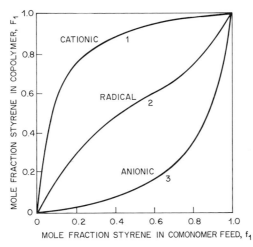

FIG. 2.23. Instantaneous composition of copolymer as a function of monomer composition for the polymerization of styrene–methyl methacrylate by cationic (SnCl$_4$) free radical (benzoyl peroxide) and anionic (Na) mechanisms (Odian, 1970, after Pepper, 1954; Landler, 1950).

2.2.6.3. Determination of Reactivity Ratios

The experimental measurement of r_1 and r_2 involves the determination of the composition of copolymers formed from several different feed compositions. Copolymerization is normally carried out to low conversion ($<5\%$) to ensure uniformity of copolymer composition; the composition is then determined by any of a number of analytical techniques. There are several methods of analyzing the data:

1. An F_1 versus f_1 plot is constructed and theoretical curves are then constructed corresponding to trial and error selections of r_1 and r_2. The values of r_1 and r_2 whose curve best fits the experimentally determined curve are taken as the values of the reactivity ratios. This direct curve-fitting procedure is tedious and can only give approximate values since the composition curve is rather insensitive to small changes in r_1 and r_2. A modification of this technique was proposed (Tidwell and Mortimer, 1965) that differs from the curve-fitting method in that the values of r_1 and r_2 satisfy the criterion that for the selected values of these ratios the sum of the squares of the differences between the observed and computed polymer compositions is minimized. The method employs initial approximations of r_1 and r_2 to define the copolymer composition relation. These are then refined by a series of nonlinear least-squares computations leading to values of r_1 and r_2 which minimize the sum of the mean-square deviations of the experimental points from the theoretical curve.

2. Rearrangement of the copolymer equation (Eq. (2.149)) to solve for one of the reactivity ratios results in the expression (Mayo and Lewis, 1944)

$$r_2 = \frac{[M_1]}{[M_2]}\left\{\frac{d[M_2]}{d[M_1]}\left(1 + r_1\frac{[M_1]}{[M_2]}\right) - 1\right\} \qquad (2.161)$$

Thus, knowing the feed composition and copolymer composition for a particular experiment, these values can be substituted into Eq. (2.161), thereby generating values of r_2 for assumed values of r_1. Since Eq. (2.161) is that of a straight line, a plot of r_1 versus r_2 will also be a straight line. A similar procedure is adopted for different feed compositions, resulting in a series of straight lines which, in the absence of experimental error, should intersect at a unique point in the $r_1 r_2$ plane. In practice, the lines rarely intersect at a point but rather intersect over an area in the $r_1 r_2$ plane as a result of experimental error in determining copolymer composition. The problem then becomes one of selecting the best point in the intersection area. Joshi and Joshi (1971) report an absolute analytical procedure involving setting up a condition whereby the coordinates of the intersection point are such that the sum of the squares of its perpendicular distances from all experimental lines is a minimum.

3. A variety of linear methods have been developed (Joshi 1973; Kelen and Tüdos, 1975). These involve rearrangement of the copolymer equation into a linearized form that may be graphically or numerically evaluated. Fineman and Ross (1950) rearranged the differential copolymer equation in the following form

$$B(1 - 1/b) = -B^2/(br_1) + r_2 \qquad (2.162)$$

where $B = [M_1]/[M_2]$ and $b = d[M_1]/d[M_2]$. Graphical plotting gives r_1 as the slope and r_2 as the intercept. A major disadvantage of this equation, and indeed of most graphically evaluable linear equations, is that on inversion of the data (reindexing of the monomers and reactivity ratios) the equations are not invariant, i.e., the reactivity ratio of a monomer would be calculated from a slope in one instance and from an intercept on inversion. This would present no problem if the data were exact. However, there is a degree of experimental error in the determination of monomer and copolymer composition and, as pointed out by Tidwell and Mortimer (1965), the experimental data are unequally weighted by Eq. (2.162). Consequently, it is found in practice that two different solutions are obtained for the same experimental data when the monomers are reindexed.

A numerically evaluable equation that retains the same form on inversion of data was introduced by Yezrielev *et al.* (1969). It has the form

$$(B/b^{1/2}r_1) - (b^{1/2}/Br_2) + b^{-1/2} - b^{1/2} = 0 \qquad (2.163)$$

for which a unique solution exists for both normal and inverted data. Recently Kelen and Tüdos (1975) introduced a graphically evaluable linear equation that is also invariant to data inversion. It has the form

$$G/(\alpha'' + F) = r_1 + (r_2/\alpha'')(F/(\alpha'' + F)) - (r^2/\alpha'') \qquad (2.164)$$

where $G = B(b - 1)/b$, $F = B^2/b$, and α'' is an arbitrary constant ($\alpha'' > 0$). Graphical plotting results in the determination of r_1 and r_2 as intercepts. The problem involves choosing a value of α so as to minimize experimental error.

4. The various computational methods involved in reducing the uncertainty in a given value of the reactivity ratio obtained from both graphical and algebraic procedures are tedious and still not entirely satisfactory. It would be convenient if values of r_1 and r_2 could be obtained from a single experiment. Such an approach is possible using nuclear magnetic resonance (NMR) techniques (Chapter 3, Section 3.3). The statistical distribution of sequence lengths for each type of repeating unit in the copolymer readily allows the calculation of reactivity ratios. It may be shown that the unconditional probability of finding a closed sequence with x units of the same

type M_1 $(P_1(x))$ is given by

$$P_1(x) = P_{11}^{x-1}P_{12} \qquad (2.165)$$

where P_{11} is the conditional probability that an active chain M_2–M_1^* adds a monomer molecule M_1 (forming an M_1–M_1 sequence) and P_{12} is the conditional probability that it adds M_2. Similarly,

$$P_2(x) = P_{22}^{x-1}P_{21} \qquad (2.166)$$

These probabilities are related to the reactivity ratios via the expressions

$$P_{11} = \frac{r_1[M_1]}{r_1[M_1] + [M_2]}, \qquad P_{12} = \frac{[M_2]}{r_1[M_1] + [M_2]} \qquad (2.167)$$

$$P_{22} = \frac{r_2[M_2]}{r_2[M_2] + [M_1]}, \qquad P_{21} = \frac{[M_1]}{r_2[M_2] + M_1} \qquad (2.168)$$

The actual sequence distribution in a polymer chain can be determined in favorable circumstances by high resolution NMR spectroscopy. Comparison with theoretical distributions calculated for different values of r_1 and r_2 then allows an assignment of values to r_1 and r_2 to be made. The reader is referred to Chapter 3, Section 3.3.6, for a discussion of this technique.

2.2.6.4. Variation of Copolymer Composition with Conversion

For the majority of copolymerizations, the values of r_1 and r_2 are such that the instantaneous composition of the copolymer is different from that of the monomer feed. Consequently, the composition of the feed and hence the polymer will vary with conversion. At 100% conversion, the average composition must be that of the original comonomer mixture, but the polymer itself will be quite heterogeneous in composition. The change in composition can be determined by integrating the copolymerization equation or by numerical and graphical methods. Integration results in the following expression for variation of the feed composition with degree of conversion (Meyer and Chan, 1967).

$$1 - \frac{M}{M_0} = 1 - \left(\frac{f_1}{(f_1)_0}\right)^{\alpha}\left(\frac{f_2}{(f_2)_0}\right)^{\beta}\left(\frac{(f_1)_0 - \delta}{f_1 - \delta}\right)^{\gamma} \qquad (2.169)$$

where the zero subscripts indicate initial quantities and α, β, γ, and δ are defined by

$$\alpha = \frac{r_2}{1 - r_2}, \qquad \beta = \frac{r_1}{1 - r_1}, \qquad \gamma = \frac{1 - r_1 r_2}{(1 - r_1)(1 - r_2)}, \qquad \delta = \frac{1 - r_2}{2 - r_1 - r_2}$$

$$(2.170)$$

More important than a knowledge of how the copolymer composition changes with conversion is how to prevent that change. This is obviously important in industrial application where a uniform product is desired. Molecular uniformity is generally accomplished by carefully regulating the monomer input so that the overall monomer ratio in the polymerizing system remains unchanged.

2.2.6.5. Deviations from the Copolymer Composition Equation

The copolymer equation was derived on the assumption that the reactivity of the propagating species depends only on the nature of the active terminal unit. It has been observed in several systems that the reactivity of the propagating species is affected by the nature of the monomer unit that precedes the terminal unit. This is referred to as the *penultimate effect*, and is manifested by the observation of different values of the reactivity ratios for different comonomer feed compositions. For example, in the radical copolymerization of styrene with fumaronitrile, the reactivity of fumaronitrile for a growing chain ending in styrene is reduced as the concentration of fumaronitrile in the starting monomer mixture is increased (Fordyce and Ham, 1951). This has been attributed to steric and polar interactions between the incoming fumaronitrile and the penultimate fumaronitrile unit in the propagating chain. The effect can be treated mathematically and involves eight propagation reactions and four reactivity ratios (Ham, 1967). Each monomer is therefore characterized by two monomer reactivity ratios, one representing the propagating species in which the penultimate and terminal units are the same, and the other representing the propagating species in which they are different.

$$r_{11} = k_{111}/k_{112}, \qquad r_{21} = k_{211}/k_{212}$$
$$r_{22} = k_{222}/k_{221}, \qquad r_{12} = k_{122}/k_{121} \qquad (2.171)$$

Expressions for the instantaneous copolymer composition may be derived in the same manner as that used to derive Eq. (2.149) (Ham, 1967).

Deviations from the copolymerization equation are also observed in systems where one of the monomers tends to depropagate, as may happen when the monomer to be added (say, M_2) is in the region of its ceiling temperature (Ivin, 1974). If the conditions are such that M_2 is far above its ceiling temperature, the addition of M_2 to M_2^* may be rapidly reversed while the addition of M_2 to M_1^* is not reversed. Under these conditions, the value of k_{22} and hence of r_2 will be effectively zero and Eq. (2.149) becomes

$$d[M_1]/d[M_2] = 1 + r_1[M_1]/[M_2] \qquad (2.172)$$

Figure 2.24 shows an example of this type of behavior for the copolymerization of α-methoxystyrene, itself an unpolymerizable monomer because of its
low ceiling temperature, with methyl acrylate and methyl methacrylate
(Lüssi, 1967). At low values of $[M_1]/[M_2]$, the copolymer tends to the alternating 1:1 composition; the slope gives $r_1 = 0.17$ for methyl acrylate, and
$r_1 = 2.5$ for methyl methacrylate. Ivin (1974) has summarized most of the
known cases of monomers which do not themselves polymerize but which
readily copolymerize.

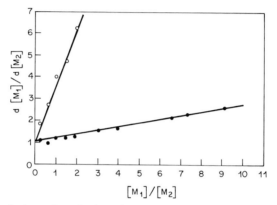

FIG. 2.24.   Radical copolymerization of α-methoxystyrene (M_2) with ($\bullet$) methyl acrylate
and ($\circ$) methyl methacrylate at $60°C$ (Ivin, 1974, after Lüssi, 1967).

It will be recognized that Eq. (2.172) represents a limiting case. It assumes
that a chain ending in $M_2M_2^*$ will shed the terminal unit prior to the addition
of M_1. There is, however, a finite probability that $M_2M_2^*$ will add M_1 before the depropagation reaction can take place, thereby locking the M_2
unit into the polymer chain. Yet another possibility is that M_2 may add to
$M_1M_2^*$ but not to $M_2M_2^*$, steric strain preventing the occurrence of the
propagation reaction only when it would lead to three consecutive M_2 units
in the chain. Further complications can occur if both monomers depropagate. General theoretical treatments of these possibilities that take into
account the reversal of one or more steps have been given (Kang et al.,
1972) and appear to adequately explain the copolymerization behavior of a
variety of such monomers (Ivin, 1974).

2.2.6.6.   Free Radical Copolymerization

As with homopolymerization, the chemistry of radical copolymerization
is the best understood. Reactivity ratios have been measured for a large
number of monomer pairs, some of which are listed in Table 2.11. Such

TABLE 2.11
REACTIVITY RATIOS FOR COPOLYMERIZATION AT $60°C$

M_1	M_2	r_1	r_2
Styrene	Acrylonitrile	0.4 ± 0.05	0.04 ± 0.04
Styrene	Methyl methacrylate	0.52 ± 0.02	0.46 ± 0.02
Styrene	Butadiene	0.78 ± 0.01	1.39 ± 0.03
Styrene	Vinyl acetate	55 ± 10	0.01 ± 0.01
Styrene	Maleic anhydride	0.02	0
Methyl methacrylate	Acrylonitrile	1.2 ± 0.14	0.15 ± 0.07
Methyl methacrylate	Vinyl acetate	20 ± 3	0.015 ± 0.015
Methyl methacrylate	Methyl acrylate	1.69	0.34
Vinyl acetate	Acrylonitrile	0.061 ± 0.013	4.05 ± 0.3
Vinyl acetate	Vinyl chloride	0.23 ± 0.02	1.68 ± 0.08
Vinylidene chloride	Isobutene	3.3	0.05

studies provide a powerful method for examination of the effect of structure on the reactivity of both radical and monomer.

2.2.6.6.1. Resonance Effects. The reactivity of a monomer toward a radical depends on the reactivities of both the monomer and the radical, which in turn depend on the nature of the substituents on the double bond. The inverse of the reactivity ratio gives the ratio of the rate of reaction of a radical with another monomer to its rate of reaction with its own monomer. Taking the latter as unity, relative monomer reactivities can be compared. Values of the relative reactivities of several monomers to reference radicals are listed in Table 2.12.

TABLE 2.12
RELATIVE REACTIVITIES $(1/r)$ OF MONOMERS WITH VARIOUS POLYMER RADICALS

Monomer	Polymer radical						
	Butadiene	Styrene	Vinyl acetate	Vinyl chloride	Methyl meth-acrylate	Methyl acrylate	Acrylo-nitrile
Butadiene		1.7		29	4	20	50
Styrene	0.7		100	50	2.2	6.7	25
Methyl methacrylate	1.3	1.9	67	10		2	6.7
Methyl vinyl ketone		3.4	20	10			1.7
Acrylonitrile	3.3	2.5	20	25	0.82	1.2	
Methyl acrylate	1.3	1.4	10	17	0.52		0.67
Vinylidene chloride		0.54	10		0.39		1.1
Vinyl chloride	0.11	0.059	4.4		0.10	0.25	0.37
Vinyl acetate		0.019		0.59	0.050	0.11	0.24

Comparison of the data within each column (values in different columns are not comparable since a different reference point is taken for each radical) shows that substituents enhance the reactivity of a monomer toward radical attack according to the following order

$$\text{\raisebox{-0.5em}{\includegraphics}} > -CH{=}CH_2 > -COCH_3 > -CN > -COOR > -Cl > -OCOCH_3 > -OR$$

i.e., in the order of resonance stabilization efficiency of a particular group. Thus styrene is found to be 30 to 50 times more reactive toward radicals than vinyl acetate because of resonance stabilization via the loosely held π-electrons of the benzene ring.

As might be expected, however, such substituents have an opposite effect on the reactivity of the subsequent radical, for the resonance stabilization that enhanced monomer reactivity will in turn suppress radical reactivity. Further, this effect of decreasing the radical reactivity is much greater than its effect in enhancing the activity of the monomer. Although styrene monomer is 30 to 50 times more reactive toward a given radical than vinyl acetate, the styrene radical is about 1000 times less reactive than the vinyl acetate radical. One would therefore expect that the propagation rate constant for the self-propagation of vinyl acetate would be greater than that of styrene and indeed, k_p for styrene is found to be 176 liter/mole sec compared to 3700 liter/mole sec for vinyl acetate.

Such wide differences in reactivity make the copolymerization of a resonance stabilized monomer with a nonresonance stabilized monomer extremely difficult. The order of the rate constants for the four possible propagation reactions is

$$R_s^. + M < R_s^. + M_s < R^. + M < R^. + M_s$$
$$\text{I} \qquad\qquad \text{II} \qquad\qquad \text{III} \qquad\qquad \text{IV}$$

where an s subscript indicates a resonance stabilized species. If two monomers are to copolymerize readily, there must be a combination of reactions I and IV. Usually reaction I is so slow that the $R_s^.$ generated in reaction IV is more likely to react with M_s, with the result that very little of the nonresonance stabilized monomer will be incorporated into the chain. This consideration explains the wide difference in reactivity ratios for styrene (M_s) for which $r = 55$ and vinyl acetate (M) for which $r = 0.01$. Copolymerization is not efficient since the styrene radical is too unreactive to add the unreactive vinyl acetate monomer.

It is not surprising then that most commercial polymers with molar compositions on the order of 50/50 consist either of two monomers both of which have stabilizing substituents (e.g. styrene–butadiene) or nonstabilizing substituents (vinyl chloride–vinyl acetate). Commercial polymers from monomers of mixed stability cannot be easily made at ~50:50 mole ratio for the reasons outlined above. Instead, one is limited to a ready incorporation of only about 10% of the unstabilized monomer and it is in these proportions that such pairs of monomers have found commercial application. Acrylonitrile, for example, is copolymerized with vinyl acetate to the extent of 8–12% to improve solubility of poly(vinyl acetate) in conventional hydrocarbon solvents.

2.2.6.6.2.   Steric Effects.   Steric factors also have a marked effect on reactivity. This is most easily seen in Table 2.13, which shows the rate constants (k_{12}) for the reactions of various chloroethylenes with vinyl acetate, styrene, and acrylonitrile radicals. The effect of a second substituent at the α-position enhances the monomer reactivity, e.g., vinylidene chloride is 2–10 times as reactive as vinyl chloride. However, when the substituent is present on the β-carbon atom, there is a marked decrease in reactivity due to steric hindrance between the monomer and radical to which it is adding. It is interesting to note that such α–β disubstituted monomers do in fact undergo copolymerization in spite of the fact that they exhibit a marked reluctance to homopolymerize. Their ability to copolymerize stems from the fact that the steric hindrance involved in the addition of another monomer to the α–β disubstituted monomer is not as great as in the addition of another α–β disubstituted monomer.

2.2.6.6.3.   Polarity Effects.   A third factor that influences reactivity is polarity. A substituent that lends a positive or negative character to a double bond has a similar effect on the radical and since opposite charges attract, copolymerization of two monomers with different electronegativities will tend toward alternation. The deviation of the $r_1 r_2$ product from unity and its approach to zero are a measure of this tendency toward alternation. By arranging monomers in the order of the electronegativity of their substi-

TABLE 2.13

RATE CONSTANTS (k_{12}) FOR RADICAL–MONOMER REACTIONS

Monomer	Polymer radical		
	Vinyl acetate	Styrene	Acrylonitrile
Vinyl chloride	10,100	8.7	720
Vinylidene chloride	23,000	78	2200
cis-1,2-Dichloroethylene	370	0.60	
trans-1,2-Dichloroethylene	2300	3.9	
Trichloroethylene	3450	8.6	29
Tetrachloroethylene	460	0.70	4.1

TABLE 2.14

Values of $r_1 r_2$ in Radical Copolymerization (e Values in Parentheses)

	Vinyl ethers (−1.3)[a]	Butadiene (−1.05)	Styrene (−0.80)	Vinyl acetate (−0.22)	Vinyl chloride (0.20)	Methyl methacrylate (0.40)	Vinylidene chloride (0.36)	Methyl vinyl ketone (0.68)	Acrylonitrile (1.20)	Diethyl fumarate (1.25)	Maleic anhydride (2.25)
Vinyl ethers (−1.3)[a]											
Butadiene (−1.05)											
Styrene (−0.80)		0.31									
Vinyl acetate (−0.22)			0.55								
Vinyl chloride (0.20)			0.34	0.39							
Methyl methacrylate (0.40)		0.19	0.24	0.30	1.0						
Vinylidene chloride (0.36)		<0.1	0.16	0.6	0.96	0.61					
Methyl vinyl ketone (0.68)		0.006	0.10	0.35	0.83	0.99					
Acrylonitrile (1.20)		0.0004	0.016	0.21	0.11	0.18	0.34	1.1			
Diethyl fumarate (1.25)		~0	0.021	0.0049	0.056	0.021	0.56				
Maleic anhydride (2.25)		~0.002	0.006	0.00017	0.0024	0.11					

[a] Ethyl, i-butyl, or dodecyl vinyl ethers.

tuents and examining the r_1r_2 product between monomers as shown in Table 2.14, it is apparent that the further apart the two monomers are in the table, the lower is the value of r_1r_2. Exceptions arise in instances where resonance or steric factors exert a dominating influence.

Highly alternating copolymers are prepared by copolymerizing strongly electron-accepting monomers such as maleic anhydride or sulfur dioxide with electron-donating monomers such as styrene, vinyl ethers, or isobutene. It has been suggested that alternation results from the homopolymerization of 1:1 complexes formed between the donor and acceptor monomers (Gaylord, 1970). Thus, in the familiar styrene–maleic anhydride system, the double bonds have opposite polarities, being electron rich and electron deficient, respectively, and complex formation results. Polymerization may then be envisaged in terms of a self-propagation of the equimolar monomer complex. An alternative explanation corresponds to the situation in which propagation proceeds through alternate monomer addition, implying that both r_1 and r_2 are effectively zero. Again, the electronic structures of the monomers are such as to encourage donor–acceptor interaction in the transition state of the radical–monomer additions.

In recent years, it has been found that the structure of a copolymer may be greatly modified by the incorporation of a Lewis acid in the reaction mixture (Bamford, 1975). Acrylic monomers, for example, which by themselves are thought to be too weak to act as acceptor molecules, become active in the presence of complexing agents such as $C_2H_5AlCl_2$ and other Lewis acids, thereby enabling them to form alternating copolymers with electron donating monomers. Monomers that give alternating copolymers in the presence of Lewis acids (at a sufficiently high concentration) can be divided into two groups: (a) the donor group contains electron-rich double bonds and has e-values less than 0.5 on the Alfrey–Price Q–e scheme (see below) and (b) the acceptor group generally contains nitrile or carbonyl groups conjugated with the double bonds and has higher e-values. A listing of donor and acceptor monomers is presented in Table 2.15.

2.2.6.6.4. Q–e Scheme. Various attempts have been made to quantify these effects of resonance stabilization, steric factors, and polarity on radical–monomer reactivity. The Q–e scheme of Alfrey and Price (1947) represents one such attempt that has found general utility. According to this scheme, two parameters characterize each participant radical and monomer, one representing its general reactivity (designated by P for radicals and Q for monomers) and a second representing its polarity (designated by e). The e values for a parent monomer and daughter radical are assumed to be identical. The rate constant for reaction between a radical $M_1\cdot$ and a monomer M_2 is then given by

$$k_{12} = P_1 Q_2 \exp(-e_1 e_2) \qquad (2.173)$$

TABLE 2.15

CLASSIFICATION OF MONOMERS

A group (donors)

Ethylene, propylene, isobutene, 1-hexene, 1-octadecene, styrene, α-methylstyrene, 2-butene, 2-methyl-2-butene, butadiene, isoprene, cyclopentene, 2-norbornene, β-methylstyrene, *trans*-stilbene, dihydronaphthalene, vinyl chloride, vinyl acetate, vinyl benzoate, vinylidene chloride, allyl chloride, allyl acetate.

B group (acceptors)

Methyl acrylate, methyl methacrylate, methyl α-chloroacrylate, *n*-butyl methacrylate, methyl crotonate, *n*-butyl crotonate, acrylonitrile, α-chloroacrylonitrile, methacrylonitrile, acrylic acid, *n*-octyl acrylate, acryloyl chloride, *N*-*n*-octylacrylamide, *N*,*N*-di-*n*-butylacrylamide, methyl vinyl ketone.

Similar expressions can be written for k_{11}, k_{21}, and k_{22} yielding the following expressions for the two reactivity ratios

$$r_1 = (Q_1/Q_2)\exp[-e_1(e_1 - e_2)] \qquad (2.174a)$$

$$r_2 = (Q_2/Q_1)\exp[-e_2(e_2 - e_1)] \qquad (2.174b)$$

The assignment of Q and e values to a monomer is quite empirical. In practice, it is necessary to assign an arbitrary reference value to one monomer (usually styrene for which $Q = 1$ and $e = 0.8$), and from this base to assign Q and e values to other monomers in order to obtain r values that agree with experimentally determined values. Values of Q and e for selected monomers are listed in Table 2.16. Despite many weaknesses in this scheme (e.g., no account is taken of steric factors that can exert an overwhelming

TABLE 2.16

Q AND e VALUES FOR SELECTED MONOMERS

Monomer	e	Q
t-Butyl vinyl ether	−1.58	0.15
Ethyl vinyl ether	−1.17	0.032
Butadiene	−1.05	2.39
Styrene	−0.80	1.00
Vinyl acetate	−0.22	0.026
Vinyl chloride	0.20	0.044
Vinylidene chloride	0.36	0.22
Methyl methacrylate	0.40	0.74
Methyl acrylate	0.60	0.42
Methyl vinyl ketone	0.68	0.69
Acrylonitrile	1.20	0.60
Diethyl fumarate	1.25	0.61
Maleic anhydride	2.25	0.23

influence in some copolymerizations), it has proved of general use in predicting the reactivity ratios for many monomer pairs.

It has long been assumed that monomer reactivity ratios show little sensitivity to the nature of the propagation medium. However, studies in recent years have demonstrated that the reactivity ratios are affected to some extent (albeit generally small) by the nature of the reaction medium. This has been attributed to the fact that the effective value of e in the Q–e scheme depends on the dielectric constant of the medium (Cameron and Esslemont, 1972).

Temperature has very little effect. There is a tendency toward random copolymerization with increasing temperature, but the effect of temperature is so small that the change in reactivity ratio with a reasonable change in temperature is experimentally significant only for those reactivity ratios that are very large (> 1.0) or small (< 0.1) (O'Driscoll, 1969).

2.2.6.7 Rate of Copolymerization

The kinetics of free radical copolymerization are similar to those derived for free radical homopolymerization except that now four propagation reactions (Section 2.2.6.1) and three termination reactions (corresponding to the termination between like radicals and cross termination between unlike radicals) must be considered:

$$M_1{\cdot} + M_1{\cdot} \xrightarrow{k_{t11}} ; \quad M_1{\cdot} + M_2{\cdot} \xrightarrow{k_{t12}} ; \quad M_2{\cdot} + M_2{\cdot} \xrightarrow{k_{t22}}$$

The overall rate of copolymerization is then given by the sum of the four propagation reactions.

$$-\frac{d[M_1] + d[M_2]}{dt} = k_{11}[M_1{\cdot}][M_1] + k_{12}[M_1{\cdot}][M_2]$$
$$+ k_{21}[M_2{\cdot}][M_1{\cdot}] + k_{22}[M_1{\cdot}][M_2] \quad (2.175)$$

The stationary-state condition is applied both to $[M_1{\cdot}]$ and $[M_2{\cdot}]$ as was done in the derivation of the copolymer equation

$$k_{21}[M_2{\cdot}][M_1] = k_{12}[M_1{\cdot}][M_2] \quad (2.176)$$

and to the overall active species concentration

$$R_i = 2k_{t11}[M_1{\cdot}]^2 + 2k_{t12}[M_1{\cdot}][M_1{\cdot}] + 2k_{t22}[M_1{\cdot}]^2 \quad (2.177)$$

The overall rate of copolymerization R_p is then obtained by combining Eqs. (2.175), (2.176), and (2.177) to obtain

$$R_p = \frac{(r_1[M_1]^2 + 2[M_1][M_2] + r_2[M_2]^2)R_i^{1/2}}{\{r_1^2\delta_1^2[M_1]^2 + 2\varphi r_1 r_2 \delta_2 \delta_1 [M_1][M_2] + r_2^2\delta_2^2[M_2]^2\}^{1/2}} \quad (2.178)$$

where

$$\delta_1 = \left(\frac{2k_{t11}}{k_{11}^2}\right)^{1/2}, \qquad \delta_2 = \left(\frac{2k_{t22}}{k_{22}^2}\right)^{1/2}, \qquad \text{and} \qquad \varphi = \frac{k_{t12}}{2(k_{t11}k_{t22})^{1/2}} \qquad (2.179)$$

The δ terms are the reciprocals of the $k_p/(2k_t)^{1/2}$ ratios for the homopolymerization of the individual monomers and are thus readily obtainable. The φ term compares the cross-termination rate constant to the geometric mean of the termination rate constants for like pairs of radicals. Hence $\varphi > 1$ indicates cross-termination is favored, whereas $\varphi < 1$ indicates that cross-termination is not favored. Studies of the rate of copolymerization of monomer pairs for which r_1 and r_2 are known therefore allow determination of the value of φ. Some typical values are shown in Table 2.17. It is apparent that cross-termination is favored in most systems, a fact which is also paralleled by the tendency toward cross-propagation, as indicated by the $r_1 r_2$ product, which is also shown in Table 2.17. This suggests that polar effects also influence the termination mode.

TABLE 2.17

VALUES OF ϕ AND $r_1 r_2$ IN RADICAL COPOLYMERIZATION

Comonomer system	ϕ	$r_1 r_2$
Styrene–butyl acrylate	150	0.07
Styrene–methyl acrylate	50	0.14
Methyl methacrylate–p-methoxystyrene	24	0.09
Styrene–methyl methacrylate	13	0.24
Styrene–p-methoxystyrene	1	0.95

2.2.6.8. Ionic Copolymerization

In contrast to radical copolymerization, the reactivity ratios in ionic copolymerization depend strongly on the nature of the reaction medium. Both the solvating power and dielectric constant of the solvent and the nature of the counterion can markedly affect the reactivity ratios both for cationic and anionic polymerizations. This is clearly seen in Table 2.18, which shows the effect of solvent and nature of counterion on the cationic copolymerization of isobutene and p-chlorostyrene. Such effects, coupled with dependence on resonance, polar, and steric factors of the reacting monomers, introduce additional complexities in elucidating the mechanism of ionic copolymerization. Only in very few instances have studies been able to assess the influence of a particular factor independently of others. For example, Table 2.19 shows the effect of methyl substituents in the α and β position of styrene on their cationic copolymerization with p-chlorostyrene.

TABLE 2.18

EFFECT OF SOLVENT AND GEGENION ON MONOMER REACTIVITY
RATIOS FOR COPOLYMERIZATION OF ISOBUTENE WITH p-CHLOROSTYRENE

Solvent	Catalyst	r_1 Isobutene	r_2 p-Chlorostyrene
Hexane ($\varepsilon = 1.8$)	AlBr$_3$	1.0	1.0
Nitrobenzene ($\varepsilon = 36$)	AlBr$_3$	14.7	0.15
Nitrobenzene ($\varepsilon = 36$)	SnCl$_4$	8.6	1.2

TABLE 2.19

STERIC EFFECTS IN COPOLYMERIZATION
OF α- AND β-METHYLSTYRENES (M$_1$)
WITH p-CHLOROSTYRENE (M$_2$)

M$_1$	r_1	r_2
Styrene	2.50	0.30
α-Methylstyrene	15.0	0.35
trans-β-methylstyrene	0.32	0.74
cis-β-methylstyrene	0.32	1.0

Introduction of a methyl group at the α position enhances the reactivity from a value of $r_1 = 2.5$ for styrene to 15.0 for α-methylstyrene. The enhanced reactivity stems from the electron-donating properties of the methyl group. Reactivity is markedly reduced relative to styrene when the methyl group is introduced into the β position in either the cis or trans configuration indicating that the steric effect far outweighs the increase in electron density at the double bond.

One other feature that has contributed to a paucity of data for ionic copolymerization systems is the fact that the product of the reactivity ratios $(r_1 r_2)$ is frequently >1 with $r_1 > 1$ and $r_2 < 1$. Thus polymers covering a wide range of sequence distributions are difficult to obtain and as a consequence, ionic copolymerization is of little practical importance. This limitation is obviously not important if one only wants to incorporate a small amount of the second monomer, as for example in the commercial synthesis of butyl rubber, which is composed of 97% isobutene and 3% isoprene. This reaction is initiated at $-100°C$ using AlCl$_3$ as catalyst in a solution of methylene chloride. The product of the reactivity ratios is unity so that isoprene units are arranged randomly in the chain and give rise to random cross-linking reactions during later vulcanization.

Saegusa et al. (1975) have recently reported on a novel type of alternating polymerization that proceeds in the absence of added catalyst. It involves the polymerization of a zwitterion complex formed between a nucleophilic monomer M_N and an electrophilic monomer M_E

$$M_N + M_E \longrightarrow {}^+M_N\text{—}M_E^-$$
$${}^+M_n\text{—}M_E^- + {}^+M_N\text{—}M_E^- \longrightarrow {}^+M_n\text{—}M_E\text{—}M_N\text{—}M_E^-$$

The general propagation reaction may be written

$${}^+M_N(\text{—}M_E\text{—}M_N)_{\overline{x}}M_E^- + {}^+M_N\text{—}M_E^- \longrightarrow {}^+M_N(\text{—}M_E\text{—}M_N)_{\overline{x+1}}M_E^- \quad (2.180)$$

In addition to the propagation reaction, which involves the generic or "monomeric" zwitterion, the intermolecular and intramolecular reactions of the polymeric zwitterion must also be considered. The former lead to higher molecular weight, the latter to the formation of macrocyclic molecules.

A variety of M_N monomers have been investigated; all are cyclic. Those containing a nitrogen atom in the ring are converted into a cyclic onium species during initiation. A somewhat wider variety of M_E monomers have been investigated, including cyclic compounds and electron deficient olefins. The following examples are representative of the zwitterion formation reaction

2.2.6.9. Block and Graft Copolymerization

Simultaneous copolymerization of two monomers for which $r_1r_2 = 1$ results in a copolymer in which the probability of finding long sequences of one monomer is extremely small, except of course for the case where one monomer is present in a large excess. Techniques are available whereby such structures containing long sequences of both monomers can be synthesized.

The sequences may be present within the main polymer chain as in *block copolymers*

$$\text{\large\char`~\char`~\char`~ AAA \char`~\char`~\char`~ ABBB \char`~\char`~\char`~ BAAA \char`~\char`~\char`~}$$

Alternatively, sequences of one monomer may be attached to the backbone of the polymer of the second monomer as in *graft copolymers*

```
~~~~~ AAA ~~~~~ AAA ~~~~~ AAA ~~~~~
      B          B          B
      B          B          B
      B          B          B
      ⌇          ⌇          ⌇
```

Techniques for the preparation of block and graft copolymers generally involve prior reaction of one of the monomers to form a parent homopolymer followed by subsequent reaction with the other monomer to produce a block or graft.

Anionic polymerization utilizing the living polymer technique is particularly well suited to preparing block copolymers. Depending on whether monofunctional or difunctional initiators are used, one or both chain ends remain active after monomer A has completely reacted. Monomer B is then added, and its polymerization is initiated by the living polymeric carbanion of polymer A. This method of sequential monomer addition can be used to produce block copolymers of several different types. With monofunctional initiators, AB, ABA, and higher block sequences such as ABC can be formed. Difunctional initiators can be used to form ABA, ABCBA, etc. The ABA block copolymers, where A is styrene and B is butadiene or isoprene, are of commercial importance because they exhibit elastomeric properties without the need for vulcanization. They constitute an important class of materials known as thermoplastic elastomers. The success of this technique depends on the ability of the macroanion of the parent sequence to promote addition of the second monomer. In the case of styrene with either butadiene or isoprene, both polymeric carbanions will initiate polymerization of either monomer so that multiblock sequences of the type ABABAB, etc., can be formed. On the other hand, whereas polystyrene anions will readily initiate polymerization of methyl methacrylate, the converse is not true and the number of block sequences will therefore be limited to two for monofunctional initiation or three for difunctional initiation.

Block copolymers can also be formed by free radical processes. Initiation usually takes place by either thermal or photolytic decomposition of groups deliberately introduced at the end of the parent chain. For example, polymerization of styrene in the presence of CBr_4, an efficient chain transfer agent, results in chains containing bromo and tribromomethyl groups at the chain ends. Subsequent photolysis results in the removal of a bromine atom producing free radicals, which in the presence of a second monomer can initiate polymerization to form a block copolymer. Free radicals can also be produced by thermal decomposition of peroxy groups. These can be introduced as end groups via redox initiation by alkyl hydroperoxides in the presence of organometallic complexes, e.g., the acetylacetonate complexes of transition metals such as copper(II) or chromium(III):

$$(CH_3)_3COOH + Cu^{2+} \longrightarrow CH_3COO\cdot + H^+ + Cu^+$$

Other techniques rely on the splitting of bonds within the main chain to produce free radicals that can initiate polymerization of a second monomer. Such bonds may be weak links, such as peroxy groups, or the initiation method may be such as to rupture the normal main chain bonds, e.g., polymer chains may be broken by mechanical degradation, direct photolysis, or high-energy radiation.

Generally speaking, the synthesis of block copolymers by the free radical route is much more versatile than that based on anionic polymerization since there is a wider range of monomers that can be incorporated into the blocks and hence a wider range of potential physical and mechanical properties available. In practice, however, free radical synthesis has several limitations. The mean block lengths cannot be as accurately controlled as in anionic polymerization since they are determined by statistical parameters. The molecular weight is much broader and, in addition, many free radical syntheses result in the formation of some homopolymer.

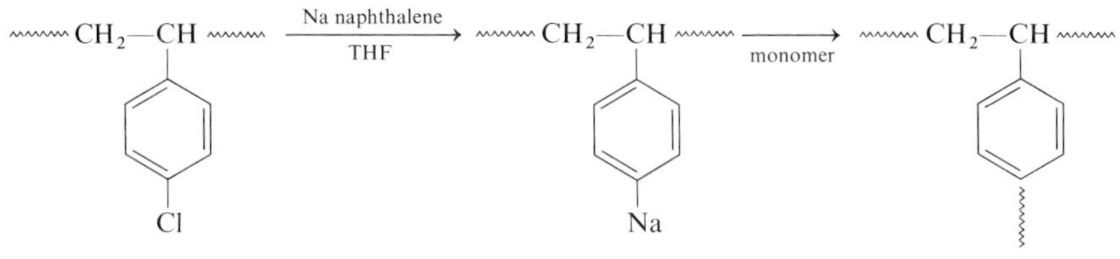

Graft copolymers can be prepared either by anionic, cationic, or free radical means. The scope of such reactions is enormous and only a few examples are given to illustrate the synthetic methods. Two basic methods have been used to form graft copolymers anionically. The first involves the production of anionic centers along the polymer backbone. This can be achieved by direct metallation or by metal–halogen interconversion on halogen-containing polymers. For example, by incorporating small amounts of p-chlorostyrene into a polystyrene backbone, the chloro group can be removed in the presence of sodium naphthalene in tetrahydrofuran (Greber and Tölle, 1962). The active polymer chain is then reacted with the desired monomer to produce side chains of the corresponding polymer. An alternative method of graft copolymer synthesis involves the coupling of side chains in the form of living polymers to a polymeric backbone containing a coupling site. For example, living polystyrene can be coupled to a partially chloromethylated polystyrene backbone:

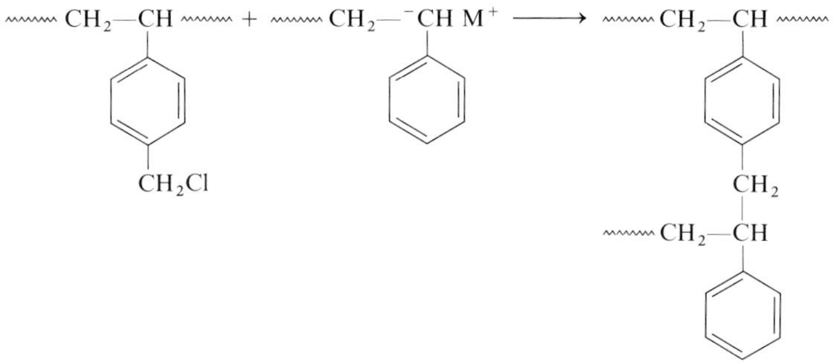

Other side reactions such as metal halogen exchange and elimination can occur, although they can be minimized by choice of appropriate experimental conditions. Such polymers are often referred to as "comb" polymers (if a monomeric multifunctional chlorine-containing compound such as silicon tetrachloride is used in place of a polymeric compound, the resultant graft is termed a "star" polymer). Similar comb polymers can be prepared by coupling of living polystyrene with the ester function of a methacrylate polymer (Finaz et al., 1962).

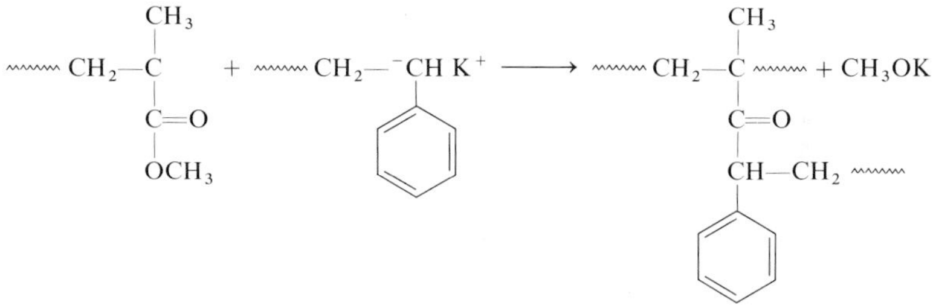

The application of cationic polymerization techniques to graft copolymer synthesis has until recently received little attention, primarily because of difficulties in controlling initiation and the occurrence of side reactions that lead to ill-defined product mixtures. Recent fundamental research by Kennedy and co-workers using di- and trialkylaluminum halides as initiators has provided a better understanding of the details of the initiation mechanism and has provided an efficient synthetic method that promises to be of far-reaching importance (Kennedy (1977)). These initiators require a coinitiator, usually an alkyl halide. By using a polymeric halide (PCl) such as a chlorinated ethylene–propylene copolymer (EPM), polymeric cations can be generated that can act as initiation sites for graft copolymerization (Kennedy and Smith, 1974).

$$AlEt_2Cl + PCl \rightleftharpoons P^+ \cdots AlEt_2Cl_2^- \xrightarrow{\ M\ } \text{graft copolymer}$$

Thus polystyrene may be readily grafted onto EPM by the following reaction sequence:

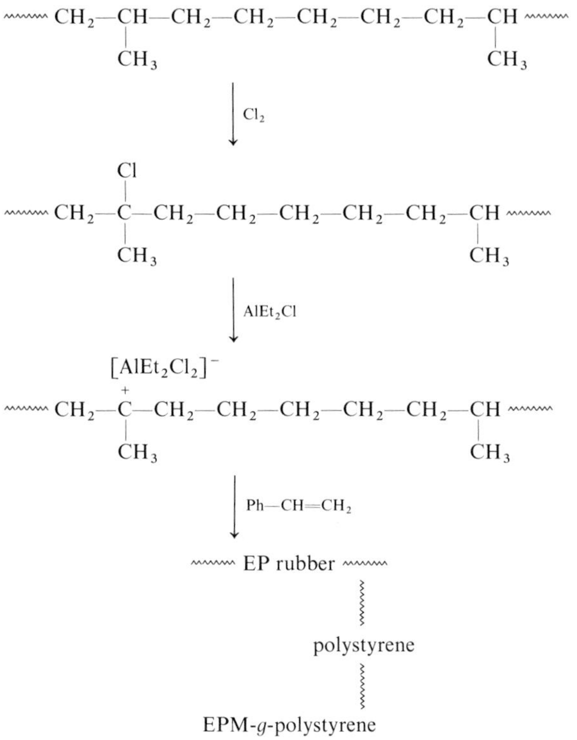

$$\sim\!\!\sim\!\!\sim CH_2\!-\!CH\!-\!CH_2\!-\!CH_2\!-\!CH_2\!-\!CH_2\!-\!CH_2\!-\!CH\sim\!\!\sim\!\!\sim$$
$$\qquad\qquad\qquad |\qquad\qquad\qquad\qquad\qquad\qquad\qquad\qquad |$$
$$\qquad\qquad\qquad CH_3\qquad\qquad\qquad\qquad\qquad\qquad\qquad CH_3$$

$\downarrow$ Cl_2

$$\qquad\qquad Cl$$
$$\qquad\qquad |$$
$$\sim\!\!\sim\!\!\sim CH_2\!-\!C\!-\!CH_2\!-\!CH_2\!-\!CH_2\!-\!CH_2\!-\!CH_2\!-\!CH\sim\!\!\sim\!\!\sim$$
$$\qquad\qquad |\qquad\qquad\qquad\qquad\qquad\qquad\qquad\qquad |$$
$$\qquad\qquad CH_3\qquad\qquad\qquad\qquad\qquad\qquad\qquad CH_3$$

$\downarrow$ $AlEt_2Cl$

$$[AlEt_2Cl_2]^-$$
$$\qquad\qquad +$$
$$\sim\!\!\sim\!\!\sim CH_2\!-\!C\!-\!CH_2\!-\!CH_2\!-\!CH_2\!-\!CH_2\!-\!CH_2\!-\!CH\sim\!\!\sim\!\!\sim$$
$$\qquad\qquad |\qquad\qquad\qquad\qquad\qquad\qquad\qquad\qquad |$$
$$\qquad\qquad CH_3\qquad\qquad\qquad\qquad\qquad\qquad\qquad CH_3$$

$\downarrow$ $Ph\!-\!CH\!=\!CH_2$

$$\sim\!\!\sim\!\!\sim \text{EP rubber} \sim\!\!\sim\!\!\sim$$
$$|$$
$$\text{polystyrene}$$
$$|$$
$$\text{EPM-}g\text{-polystyrene}$$

The preparation of graft copolymers via free radical methods is of commercial importance and has received considerable attention. Three main approaches have been used: (1) chain transfer to both saturated and unsaturated backbone or pendant groups; (2) ultraviolet and ionizing radiation; and (3) activation of pendant reactive groups. Grafting via chain transfer usually occurs by the abstraction of a hydrogen atom from the preformed polymer backbone. When polybutadiene is dissolved in styrene, and polymerization of the latter is initiated by a suitable free radical catalyst, some of the polystyrene radicals are grafted to the polybutadiene via the reaction

$$\text{\raisebox{0pt}{$\sim\!\sim\!\sim$}} CH_2\!-\!CH\!=\!CH\!-\!CH_2 \text{\raisebox{0pt}{$\sim\!\sim\!\sim$}} + \text{\raisebox{0pt}{$\sim\!\sim\!\sim$}} CH_2\!-\!\overset{\centerdot}{C}H \longrightarrow$$

$$\text{\raisebox{0pt}{$\sim\!\sim\!\sim$}} \overset{\centerdot}{C}H\!-\!CH\!=\!CH\!-\!CH_2 \text{\raisebox{0pt}{$\sim\!\sim\!\sim$}} + \text{\raisebox{0pt}{$\sim\!\sim\!\sim$}} CH_2\!-\!CH$$

$$\text{\raisebox{0pt}{$\sim\!\sim\!\sim$}} \overset{\centerdot}{C}H\!-\!CH\!=\!CH\!-\!CH_2 \text{\raisebox{0pt}{$\sim\!\sim\!\sim$}} + CH_2\!=\!CH \longrightarrow \text{\raisebox{0pt}{$\sim\!\sim\!\sim$}} CH\!-\!CH\!=\!CH\!-\!CH_2 \text{\raisebox{0pt}{$\sim\!\sim\!\sim$}}$$
$$(CH_2\!-\!CH)_x \text{\raisebox{0pt}{$\sim\!\sim\!\sim$}}$$

Homopolystyrene is also formed so that the polymerizing mixture contains polystyrene, polybutadiene, and poly(butadiene-g-polystyrene) dissolved in styrene. Ordinarily, mixtures of polystyrene and polybutadiene are incompatible and will form separate phases. In the present case, the graft copolymer plays an important role in stabilizing the two-phase structure so that one ends up with a fine dispersion of rubber particles in a polystyrene matrix. The polybutadiene portion of the graft may be thought of as being "solubilized" in the rubber particle with the whole particle being "anchored" to the polystyrene matrix by the polystyrene portion of the graft. This process forms the basis for the preparation of high-impact polystyrene.

Ultraviolet light and high-energy radiation may be used to initiate graft copolymerization. Both the parent polymer and the monomer to be grafted can be irradiated simultaneously or alternatively, the polymer can be preirradiated to produce trapped reactive species that can subsequently initiate polymerization. The former process is referred to as *mutual irradiation* and gives rise to a certain amount of homopolymer of the grafting monomer. This is not a complication with the latter process, which is termed *pre-irradiation*, although success depends on the ability to generate and trap an adequate concentration of radicals. Irradiation methods are particularly important in those cases where the parent homopolymer is insoluble, e.g., poly(tetrafluoroethylene) or cellulose. The grafting reaction is heterogeneous, involving the solid polymer somewhat swollen by and immersed in the liquid monomer. The relative rates of grafting and diffusion determine whether reaction occurs only at the polymer surface or throughout its volume (Chapiro, 1962).

Chemical grafting involves activating preformed labile groups on the polymer chain. Groups such as hydroperoxide can be easily introduced either in the initial polymerization or through subsequent chemical reaction of the polymer. Polystyrene containing amine groups (prepared, for example, by reducing partially nitrated polystyrene) can be diazotized in the presence of ferrous salts to give polymeric phenyl radicals capable of initiating polymerization.

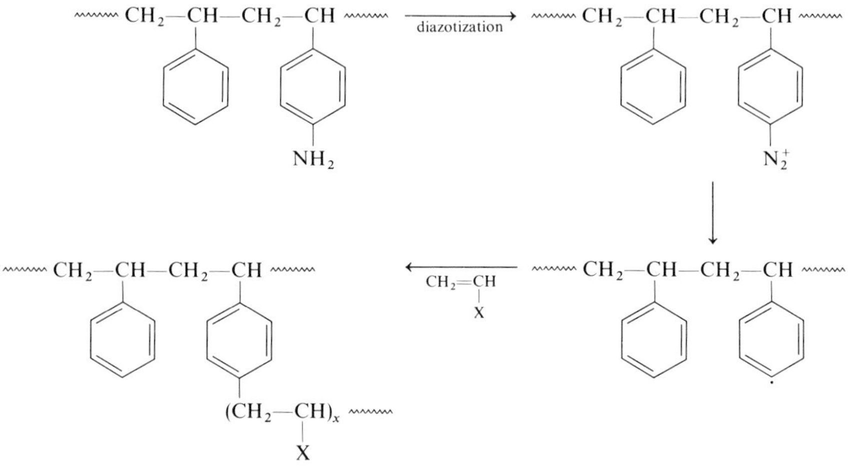

It is apparent from this brief discussion that the number of possible block and graft copolymers is immense and the reader is directed elsewhere for a more complete discussion of this aspect of copolymerization (Battaerd and Tregear, 1967; Allport and James, 1973).

2.3 Ring Opening Polymerization

Cyclic compounds constitute a potentially polymerizable class of monomers. The difunctionality criterion for polymerizability can be achieved by a ring opening process as shown below for ethylene oxide

Cyclic monomers should, therefore, be capable of being polymerized provided a suitable mechanism for opening the ring is available (kinetic criterion).

The principal factor determining the magnitude of the free energy change is the existence and extent of ring strain. Ring strain is a thermodynamic property caused by either forcing the bonds between ring atoms into angular distortion or by steric interaction of substituents on the ring atoms. It is the release of ring strain by polymerization that provides the principal driving force for the polymerization of cyclic monomers. Theoretical estimates of the free energy changes for the hypothetical polymerization of pure liquid cycloalkanes indicate that the polymerization of 3- and 4-membered ring compounds is accompanied by a substantial negative free energy change. The same applies to ring compounds containing greater than six atoms. These conclusions, which are based on the data for cycloalkanes (Fig. 2.25), are fairly general and can be applied to other cyclocompounds containing heteroatoms in the ring (Frisch and Reegen, 1969).

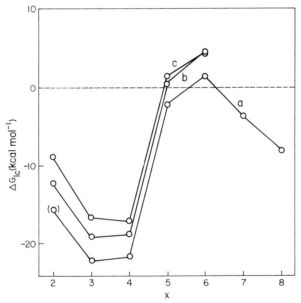

FIG. 2.25. Free energy of polymerization of cycloalkanes as a function of the number of atoms in the ring x: (a) unsubstituted; (b) methyl substituted; (c) 1,1-dimethyl substituted (Ivin, 1974).

The polymerizability of 5- and 6-membered rings depends very much on the nature of the groups in the ring. As can be seen from Fig. 2.25, the free energy changes are small because of the rather small strain energy. For this reason, minor perturbations in the physical conditions and chemical structure can have a marked effect in causing the sign of the free energy change to be in favor of or against polymerization. Thus, whereas 5-membered cyclic ethers such as tetrahydrofuran exhibit a negative free energy change and are consequently polymerizable, the 5-membered cyclic esters (γ-butyrolactone) exhibit a positive free energy change and are not. In contrast, 6-membered cyclic ethers do not polymerize while the corresponding cyclic esters do. With other cyclic systems such as imides and anhydrides, both 5- and 6-membered rings are polymerizable.

Substitution in the ring tends to make the free energy change more positive, thereby decreasing polymerizability (Ivin, 1974). Thus, whereas tetrahydrofuran is polymerizable, 2-methyltetrahydrofuran is not. However, 3-methyltetrahydrofuran has been polymerized to give low molecular weight polymer (Chiang and Rhodes, 1969). Certain bicyclic strained tetrahydrofurans are also known to form polymers very readily.

A wide variety of cyclic monomers have been successfully polymerized. They include homocyclic carbon monomers and heterocyclic monomers based on a variety of heteroatoms. These polymerization processes take place by an addition-type mechanism, i.e., only monomer molecules add to the growing chain in the propagation step. The actual chain propagation reaction is, however, a substitution reaction rather than an addition reaction as is the case with vinyl monomers. Consequently, transfer and propagation rate constants can have comparable values and the attainment of high molecular weight can prove difficult.

Ring opening polymerizations are primarily initiated by ionic initiators (including coordinate ionic) as well as initiators that are molecular species, e.g., water. This latter class of initiator is generally only effective for the more reactive cyclic monomers. Ionic initiators are usually more reactive and typically are the same as those used to initiate polymerization of vinyl monomers.

2.3.1. HOMOCYCLIC CARBON MONOMERS

The cycloalkanes and their derivatives have been successfully polymerized by Lewis acid catalysts and by certain Ziegler–Natta catalysts (Pinazzi et al., 1973). Cyclopropane for example can be polymerized with aluminum tribromide to ill-defined oils of low molecular weight.

$$x\text{CH}_2-\text{CH}_2 \atop \diagdown \diagup \atop \text{CH}_2 \xrightarrow{\text{AlBr}_3} -(\text{CH}_2-\text{CH}_2-\text{CH}_2)_x-$$

In contrast, 1,1-dimethylcyclopropane gives a uniform polymer which has the same structure as that obtained from 3-methyl-1-butene (Ketley, 1963).

$$x\text{CH}_2-\underset{\underset{\text{CH}_2}{\diagdown \diagup}}{\overset{\overset{\text{CH}_3}{|}}{\text{C}}}-\text{CH}_3 \xrightarrow{\text{AlBr}_3} +\text{CH}_2-\text{CH}_2-\underset{\underset{\text{CH}_3}{|}}{\overset{\overset{\text{CH}_3}{|}}{\text{C}}}+_x$$

Recent interest has centered on bicyclo[n.1.0] alkanes (XIX) and spiro[2.n] alkanes (XX). These compounds are polymerizable by Lewis acids with attack and subsequent ring opening occurring on the cyclopropane

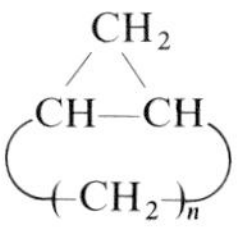

$$
\underset{\text{XIX}}{\overset{\text{CH}_2}{\overset{\displaystyle /\backslash}{\underset{\displaystyle \overset{\displaystyle \text{CH—CH}}{(\text{CH}_2)_n}}{}}}}
\qquad
\underset{\text{XX}}{\overset{\text{CH}_2\text{—CH}_2}{\overset{\displaystyle \backslash /}{\underset{\displaystyle \overset{\displaystyle \text{C}}{(\text{CH}_2)_n}}{}}}}
$$

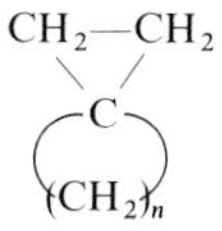

system. The products are usually oligomeric. For bicyclo[5.1.0]octane, the polymerization may be represented as

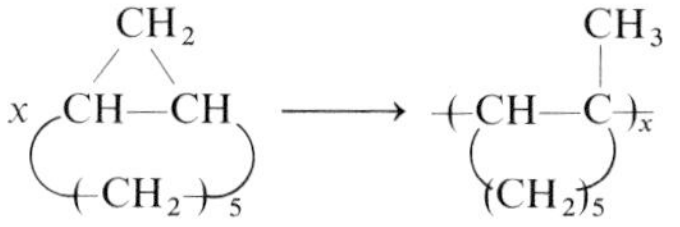

$$
x\ \overset{\text{CH}_2}{\overset{/\backslash}{\underset{(\text{CH}_2)_5}{\text{CH—CH}}}}
\quad\longrightarrow\quad
\overset{\text{CH}_3}{\underset{(\text{CH}_2)_5}{(\text{CH—C})_x}}
$$

Similarly, for spiro[2.6]nonane

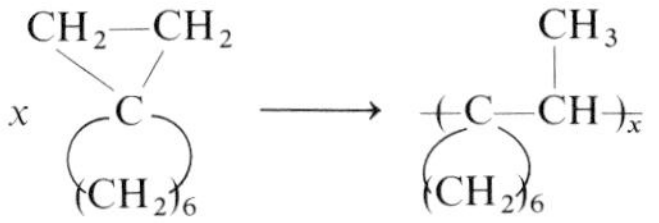

$$
x\ \overset{\text{CH}_2\text{—CH}_2}{\overset{\backslash /}{\underset{(\text{CH}_2)_6}{\text{C}}}}
\quad\longrightarrow\quad
\overset{\text{CH}_3}{\underset{(\text{CH}_2)_6}{(\text{C—CH})_x}}
$$

In both cases, the oligomeric product contains a methyl group in the side chain. These methyl groups stem from molecular rearrangement involving opening of the cyclopropyl group and hydride shift during the propagation step. With Ziegler–Natta catalysts, these monomers give oligomers whose structures are different from those observed with cationic catalysts. Spiro[2.n] alkanes, for example, when polymerized with catalysts such as $(C_2H_5)_3Al/TiCl_4$ give rise to the following structure

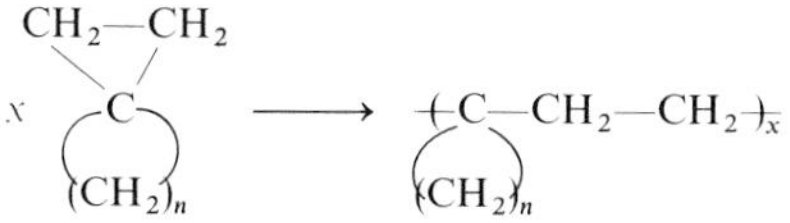

$$
x\ \overset{\text{CH}_2\text{—CH}_2}{\overset{\backslash /}{\underset{(\text{CH}_2)_n}{\text{C}}}}
\quad\longrightarrow\quad
\underset{(\text{CH}_2)_n}{(\text{C—CH}_2\text{—CH}_2)_x}
$$

The polymerization mechanism of these cyclic compounds becomes a good deal more complex when $n = 1$. Spiro[2.3] pentane, for example, polymerizes to complex cyclic structures similar to cyclized poly(1,4-isoprenes). Bicyclo[1.1.0] butane and certain derivatives polymerize to structures containing cyclobutane rings in the backbone.

2.3.2. HETEROCYCLIC MONOMERS

The polymerization of heterocyclic monomers has been extensively studied due in part to the fact that many of these systems are of industrial importance. Simple epoxy compounds such as ethylene and propylene oxides undergo ring opening polymerization to give polyethers, which find use as intermediates in the preparation of polyurethane elastomers

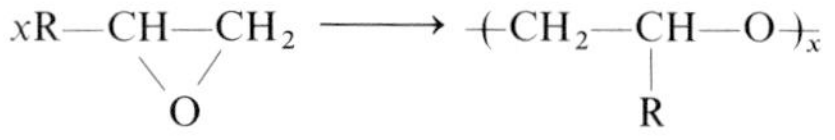

$$x\text{R}-\text{CH}\overset{\diagup}{\underset{\text{O}}{\diagdown}}\text{CH}_2 \longrightarrow +\text{CH}_2-\underset{\underset{\text{R}}{\mid}}{\text{CH}}-\text{O}+_x$$

The 7-membered cyclic amide caprolactam undergoes ring opening polymerization to yield the important polyamide nylon-6

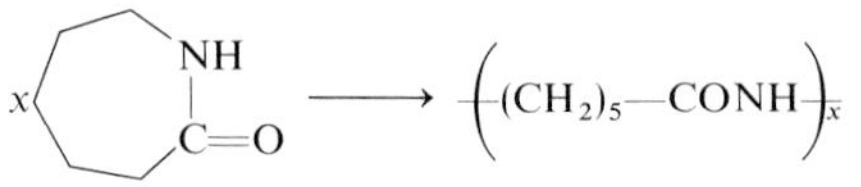

while 3,3-bischloromethyloxetane (BCMO) yields a tough thermoplastic known as Penton

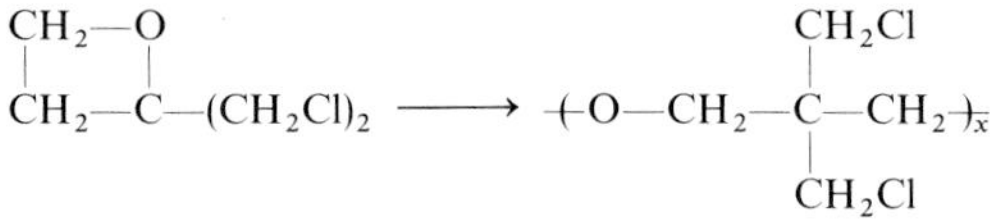

In addition to the cyclic ethers and cyclic amides, a variety of other heterocyclic compounds have been shown to undergo ring opening polymerization. They include cyclic acetals, amines, sulfides amides, esters, and anhydrides among others. Considerable work has also been done on the polymerization of heterocyclics containing more than one heteroatom in the ring.

2.3.2.1. Cyclic Ethers

This group of monomers has been widely studied. It includes the simple 1,2 epoxides (oxiranes) XXI, 1,3 epoxides (oxetanes or oxacyclobutanes) XXII, 1,4 epoxides (oxacyclopentanes) XIII, cyclic formals XXIV, and, of course, derivatives of these compounds. Examples of these generic types are shown below along with their common names.

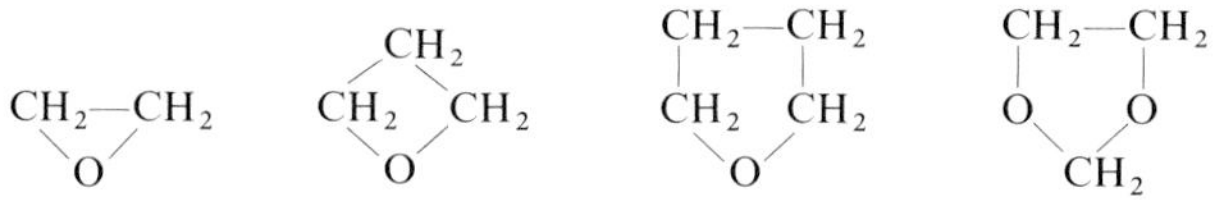

XXI	XXII	XXIII	XXIV
ethylene oxide	oxetane	tetrahydrofuran	1,3-dioxolane

The ether linkage is basic in the Lewis sense and consequently cyclic ethers can be readily polymerized by typical cationic initiators such as protonic or Lewis acid catalysts. In the case of proton donors, the mechanism appears quite straightforward and involves protonation of the epoxy group to form an oxonium ion

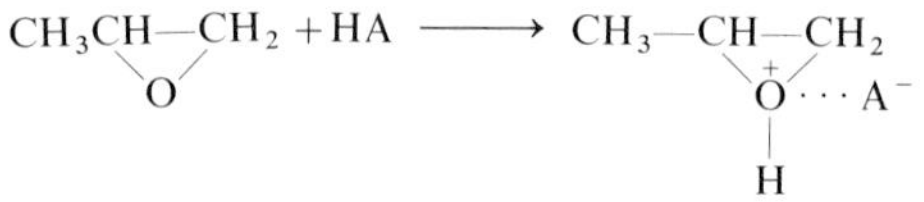

The oxonium ion then reacts with a second monomer molecule via nucleophilic attack by the oxygen of the monomer at the carbon bearing the more labile carbon–oxygen bond.

$$CH_3\!-\!CH\!-\!CH_2 \;+\; CH_3CH\!-\!CH_2 \;\longrightarrow\; HO\!-\!CH_2\!-\!CH\!-\!O^{+} \cdots \begin{array}{c} CH_3 \\ | \\ CH \\ \diagdown \\ CH_2 \end{array}$$

As with vinyl monomers, initiation efficiency is limited by the nucleophilicity of the acid anion. Sulfuric acid, for example, is not an initiator because the bisulfate anion is a stronger nucleophile than the monomer and competes successfully for the oxonium ion. On the other hand, acids such as $HClO_4$, HBF_4, and HSO_3Cl are effective initiators. Anions of low nucleophilicity are conveniently formed by complexing Lewis acid halides with a proton donor, although the detailed kinetics of Lewis acid initiator systems are frequently complex. In the polymerization of oxetane by BF_3/H_2O, for example, the rate of polymerization rises to a maximum and then declines with increasing concentration of water. When the initiator concentration is $0.1\ M$, a simple rate law is obeyed in which the polymerization rate is first order in monomer, initiator, and total concentration of added and adventitious water. This is consistent with the following initiation mechanism (Rose, 1956):

$$BF_3 + \begin{array}{c} CH_2 \\ \diagup \diagdown \\ CH_2 \quad CH_2 \\ \diagdown \diagup \\ O \end{array} \rightleftharpoons \text{complex}$$

$$\text{complex} + H_2O \xrightarrow{k_i} \begin{array}{c} CH_2 \\ \diagup \diagdown \\ CH_2 \quad \overset{+}{O}HBF_3OH^- \\ \diagdown \diagup \\ CH_2 \end{array}$$

XXV

$$\begin{array}{c} CH_2 \\ \diagup \diagdown \\ CH_2 \quad \overset{+}{O}HBF_3OH^- \\ \diagdown \diagup \\ CH_2 \end{array} + \begin{array}{c} CH_2 \\ \diagup \diagdown \\ CH_2 \quad CH_2 \\ \diagdown \diagup \\ O \end{array} \xrightarrow{\text{fast}} HO(CH_2)_3\ \overset{+}{O} \begin{array}{c} CH_2 \\ \diagup \diagdown \\ \quad CH_2 \\ \vdots \end{array}$$

$$BF_3OH^-$$

XXVI

The dependence on water content is explained by the two-step initiation process. Since the secondary oxonium ion **XXV** is much less reactive than the tertiary oxonium ion **XXVI**, its formation will be rate determining. With monomer present in excess, the equilibrium for complex formation reaction lies to the right so that the rate of initiation (R_i) may be written

$$R_i = k_i[\text{complex}][H_2O] = k_i[BF_3][H_2O] \qquad (2.181)$$

If a steady state is established with termination first order in propagating oxonium ions, then the steady-state rate of polymerization will be given by

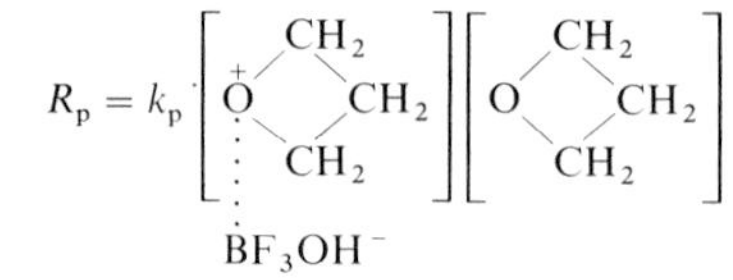

$$R_p = k_p \left[\begin{array}{c} CH_2 \\ \overset{+}{O} \quad CH_2 \\ CH_2 \\ \dot{B}F_3OH^- \end{array}\right] \left[\begin{array}{c} CH_2 \\ O \quad CH_2 \\ CH_2 \end{array}\right] \qquad (2.182)$$

$$= \frac{k_p k_i}{k_t} [BF_3][H_2O] \left[\begin{array}{c} CH_2 \\ O \quad CH_2 \\ CH_2 \end{array}\right] \qquad (2.183)$$

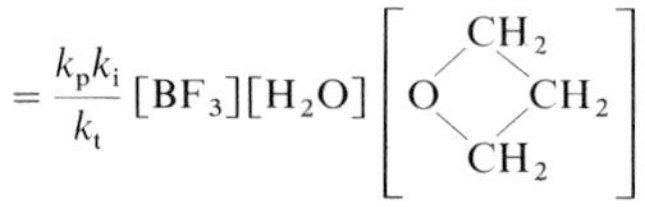

consistent with experimental observation.

In other systems, it appears that polymerization initiated by Lewis acids may not require a coinitiator. For example, in the polymerization of ethylene oxide initiated by $SnCl_4$, it is believed that the active species is a double inner oxonium ion salt that is formed by the following reaction

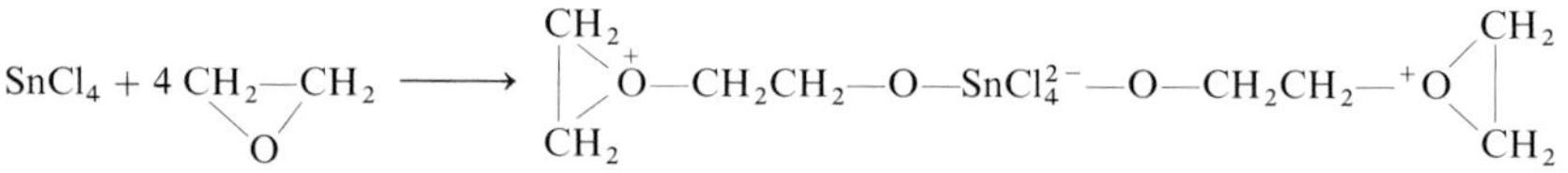

$$SnCl_4 + 4\, CH_2\!-\!CH_2 \ \overset{}{\longrightarrow}\ \begin{array}{c} CH_2 \\ \overset{+}{O}\!-\!CH_2CH_2\!-\!O\!-\!SnCl_4^{2-}\!-\!O\!-\!CH_2CH_2\!-\!{}^+O \\ CH_2 \end{array} \begin{array}{c} CH_2 \\ \\ CH_2 \end{array}$$

This mechanism provides a explanation for the fact that each stannic chloride molecule initiates two polymer chains. It suffers, however, from the disadvantage of requiring considerable charge separation in the growing chain unless, of course, chain coiling allows some ion–counterion interaction. The reaction is further complicated by the production of the cyclic dimer dioxane as a result of an intramolecular "back biting" ring expansion reaction followed by a displacement step

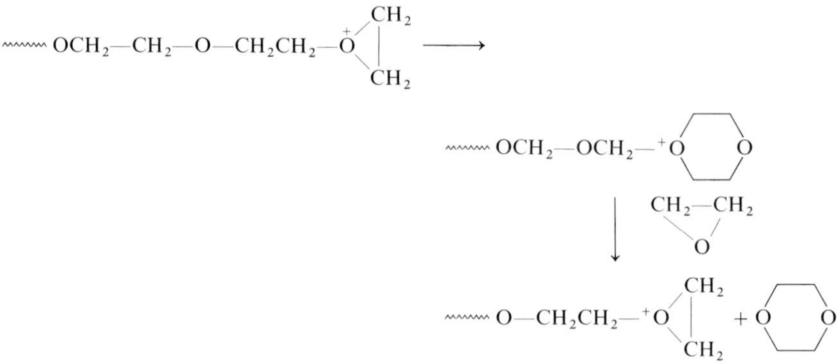

$$\text{\small wwww } OCH_2\!-\!CH_2\!-\!O\!-\!CH_2CH_2\!-\!\overset{+}{O}\begin{array}{c} CH_2 \\ \\ CH_2 \end{array} \longrightarrow$$

$$\text{\small wwww } OCH_2\!-\!OCH_2\!-\!{}^+O \bigcirc O$$

$$\Big\downarrow \quad \begin{array}{c} CH_2\!-\!CH_2 \\ O \end{array}$$

$$\text{\small wwww } O\!-\!CH_2CH_2\!-\!{}^+O\begin{array}{c} CH_2 \\ \\ CH_2 \end{array} + O \bigcirc O$$

Frequently, Lewis acid initiator/coinitiator systems are used in conjunction with a reactive cyclic ether termed a *promoter*. Initiation presumably occurs by the formation of the secondary and tertiary oxonium ions of the reactive ether, which then act as the actual initiating species. This technique has been used to initiate polymerization of tetrahydrofuran, itself a fairly unreactive monomer because of the small ring strain in the molecule as evidenced by its low heat of polymerization ($\sim 3.5\,kcal/mole$)

BF_3 alone in conjunction with epichlorohydrin will initiate the polymerization of tetrahydrofuran. Again, the following initiation mechanism, proposed by Dreyfus and Dreyfus (1969), is quite complex

It is readily apparent that initiation with Lewis acids is extremely complex and each monomer/initiator system must be examined individually. In common with all Lewis acid systems, the effects of adventitious impurities are frequently difficult to quantify.

A number of other initiating systems have been developed whose mechanisms are considerably less complex. Since the tertiary oxonium ion is the actual propagating species, polymerization can be conveniently initiated by trialkyl oxonium salts (Dreyfus, 1973)

Typical anions are tetrafluoroborate or hexachloroantimonate. Initiation is thus fairly straightforward. A similar unambiguous mode of initiation is provided by carboxonium ion salts and acylium ion salts. The latter, for example, are conveniently prepared by reaction of an active halide with a Lewis acid

$$CH_3COCl + SbCl_5 \longrightarrow CH_3CO^+ \ SbCl_6^-$$

Initiation then involves direct addition of the carbenium ion

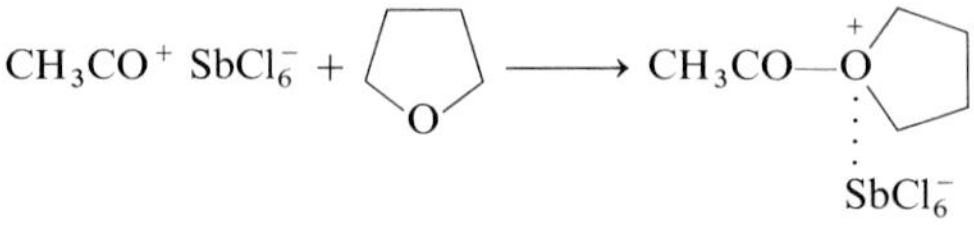

The stable carbenium ion salts, in particular salts of triphenylmethyl (Ph_3C^+) and cycloheptatrienyl ($C_7H_7^+$), have been extensively studied. In contrast to their initiation mechanism with olefins, where direct addition of the carbenium ion to the double bond occurs, initiation of cyclic ethers is effected by a hydride ion abstraction followed by proton loss resulting in dehydrogenation (Dreyfus, 1975):

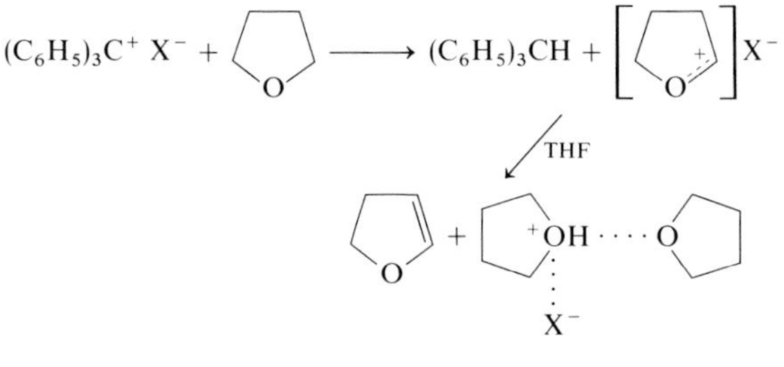

This reaction sequence results in the formation of the free acid of the counterion which is stabilized by complexing with the monomer. This free acid complex then reacts, probably quite slowly, with additional monomer to form the propagating cation. The acid is thus the true initiator.

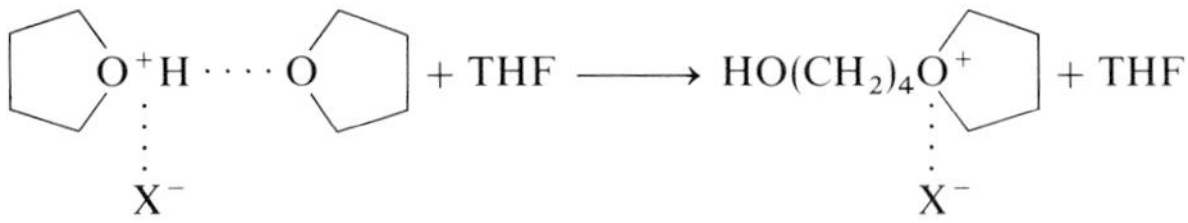

There are a variety of termination reactions that can take place. For example, with tetrahydrofuran initiated by BF_3/epichlorohydrin, transfer to the counterion can occur

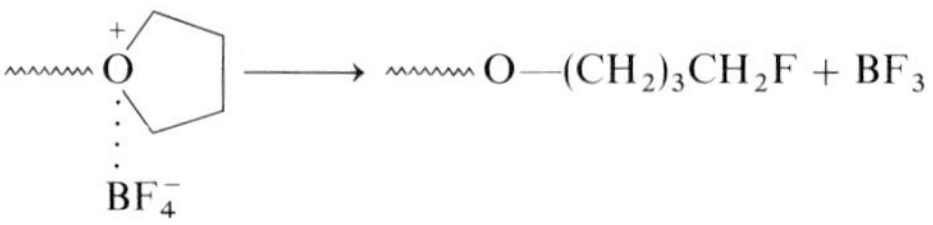

Transfer reactions are known to take place with acyclic ethers

The ether oxygens in the polymer can also interact in this way with the growing oxonium ion centers. Water and alcohols at low levels can also function as transfer agents.

Transfer and termination reactions are much more important in the case of 1,2 epoxides, where molecular weights on the order of only 1000 or less are usually obtained. Transfer reactions and reactions leading to cyclic oligomers appear to predominate although little is understood concerning the mechanism of the reactions.

Under certain conditions, characteristics of living polymerization are observed, indicating that termination is absent. This is particularly true of tetrahydrofuran initiated, for example, by oxonium tetrafluoroborate salts; such reactions can be described by kinetic expressions similar to Eq. (2.119). The ceiling temperature for polymerization is quite low (T_c for the bulk monomer is $\sim 80°C$) so that the polymerization of tetrahydrofuran approaches an equilibrium between monomer and polymer at normal polymerization temperatures. The rate of a living equilibrium polymerization is then given by

$$R_p = k_p C_{LE}([M] - [M]_{eq}) \tag{2.184}$$

where $[M]_{eq}$ is the monomer concentration for the system at equilibrium and C_{LE} is the total concentration of living ends. Integration yields

$$\ln\left(\frac{[M]_0 - [M]_{eq}}{[M] - [M]_{eq}}\right) = k_p C_{LE} t \tag{2.185}$$

Methods have been developed to determine the number of active ends (Saegusa and Matsumoto, 1968). These involve reaction of a phenoxy group with the growing center to form a phenyl ether group whose concentration can be determined by ultraviolet spectroscopy. More recently, Sangster and Worsfold (1972) have determined values of the rate constants for ion and ion pair propagation by techniques similar to those used by Szwarc[†] to determine the corresponding values in anionic polymerization of vinyl monomers. The calculated free ion rate is only a factor of seven greater than the ion pair rate compared with a factor of 10^3 in anionic polymerization of

[†] See Section 2.2.4.5.

vinyl monomers. This has been attributed to the fact that chain propagation in heterocyclics is a substitution reaction at a carbon atom adjacent to the charged atom, whereas ionic vinyl polymerization is an addition reaction involving attack on the charged atom.

So far, only initiation reactions involving the tertiary oxonium ion have been considered. Propagation takes place by nucleophilic attack on the α-carbon to the positive oxygen, as shown for tetrahydrofuran:

$$\sim\!\!\sim\!\!\sim O^+\!\!\big\rangle \longrightarrow \ \sim\!\!\sim\!\!\sim O\text{-}CH_2CH_2CH_2CH_2\text{-}O^+\!\!\big\rangle$$

(with attack from cyclic O below)

However, not all cyclic ethers propagate by an oxonium ion mechanism. Trioxane, the cyclic trimer of formaldehyde, polymerizes rapidly in the liquid or solid state with or without initiator and also by irradiation. The product is polyoxymethylene having the same structure as polyformaldehyde. In this case it is postulated that the propagating species is a carbenium ion (Furukawa and Tada, 1969):

$$HA + O\!\!\begin{array}{c}CH_2-O\\ \\ CH_2-O\end{array}\!\!CH_2 \longrightarrow H-O^+\!\!\begin{array}{c}CH_2-O\\ \\ CH_2-O\end{array}\!\!CH_2 \quad A^-$$

$$\downarrow$$

$$HO-CH_2OCH_2O\overset{+}{C}H_2 \cdots\cdots A^-$$

The carbenium ion is believed to be the propagating species (rather than the cyclic oxonium ion) because of the resonance stabilization from the adjacent oxygen atom.

The polymerization of trioxane is considerably more involved than the above equation indicates. Propagation is complicated by an equilibrium reaction between the propagating species and formaldehyde

$$\sim\!\!\sim\!\!\sim OCH_2-OCH_2-O\overset{+}{C}H_2 \ \rightleftharpoons\ \sim\!\!\sim\!\!\sim OCH_2-O\overset{+}{C}H_2 + CH_2O$$

and the long induction period associated with polymerization can be attributed to a build-up of an equilibrium formaldehyde concentration. Many of the experimental observations on this system are contradictory, and it is not possible at the present time to formulate a unified mechanism.

Of all the oxygen-containing heterocyclic compounds, the oxiranes (3-membered rings) are unique in that they can be polymerized by anionic initiators in addition to cationic initiators. Anionic polymerization can be induced by many Lewis bases such as metal oxides, hydroxides, alkoxides, carbonates, or organometallic compounds such as dialkyl zinc. The ring opening polymerization involving initiation and propagation may be formulated as follows:

$$CH_2—CH—R + B:^- \longrightarrow BCH_2—CH—O^-$$

$$BCH_2CH—O^- + CH_2—CH—R \longrightarrow BCH_2CH—CH_2CH—O^-$$

in which the propagating species is an alkoxide anion. The polymer molecular weights obtained in epoxide polymerizations are generally quite low as a result of several undesirable transfer reactions. Many reactions are carried out in the presence of an alcohol with which the propagating anion can undergo an exchange reaction

$$RO(CH_2CH—O)_x CH_2CH—O^- \; Na^+ + ROH \longrightarrow RO(CH_2CH—O)_x CH_2CH—OH$$
$$+ RO^-Na^+$$

The polymeric alcohol can itself undergo an exchange reaction with another propagating chain.

Transfer to monomer also occurs, especially in substituted ethylene oxides such as propylene oxide. The transfer reaction involves hydrogen extraction from the alkyl substituent on the epoxide ring followed by rapid ring cleavage to form an allyl ether anion

$$\text{~~~}CH_2—CH—O^- Na^+ + CH_3—CH—CH_2 \longrightarrow \text{~~~}CH_2—CH—OH$$
$$+ Na^+ \cdots \; ^-CH_2CH—CH_2$$

$$\downarrow$$

$$CH_2{=}CH—CH_2O^- \cdots Na^+$$

Generally speaking, there is no termination mechanism as such so that the kinetics are similar to those observed in living polymerizations.

The 1,2 epoxides may also be polymerized by a coordinate anionic mechanism. These initiators are believed capable of coordinating with the monomer and incorporating the monomer into the polymer in a highly stereoregular manner. They are generally of two types: (a) alkaline and alkaline earth compounds such as calcium amide ethoxide, and (b) organometallic compounds of the general structure M—Y where M is Li, Mg, Zn, Al, Sn, and Fe and Y is a ligand such as OH, OR, Cl, or R. The activity of this latter class stems from metal–oxygen bonds. Although organometallic compounds of the type M–R or M–Cl may be thought of as an exception, it appears that these catalysts undergo prior reaction with the cyclic ether or are used in conjunction with alcohols or water to form an M–OR structure. For example, the $FeCl_3$/propylene oxide system involves prior formation of an alkoxide-type catalyst of structure

$$
\begin{array}{c}
(OCH_2CH)_nCl \\
\underset{CH_3}{|} \\
\diagup \\
Cl\ Fe \\
\diagdown \\
(OCH_2CH)_mCl \\
\underset{CH_3}{|}
\end{array}
$$

where $m + n = 4$ or 5. These catalyst systems are generally heterogeneous and can exert a stereochemical influence over the polymerization, which in turn can lead to stereoselective and stereoelective polymerization. Polymerization of racemic propylene oxide by the $FeCl_3$/propylene oxide catalyst system above to a stereoregular isotactic polymer is an example of a stereoselective polymerization. The resulting polymer is not optically active but consists of a racemic mixture of poly((R)-propylene oxide) and poly((S)-propylene oxide), i.e., the catalyst itself has two asymmetric sites. Catalyst systems based on diethylzinc in conjunction with optically active alcohols, e.g., $Zn(Et)_2/d$-borneol are stereoelective and optically active poly(propylene oxide) is obtained from racemic monomer. As discussed in Section 2.2.5.9, this phenomenon is believed to arise from an interaction between the incoming monomer and the terminal or penultimate unit in the polymer.

2.3.2.2. Cyclic Amines

Cyclic amines or imines, as exemplified by ethylene imine, are polymerized to polyamines by Lewis acid initiators. The polymerization mechanism is similar to that for cyclic ethers with propagation proceeding via the immonium ion (Hauser, 1969):

$$
\begin{array}{c}
\wwww NHCH_2CH_2\overset{+}{N}\underset{H}{\diagup}{\diagdown}\overset{CH_2}{CH_2} + HN\diagup{\diagdown}\overset{CH_2}{CH_2} \longrightarrow \wwww NHCH_2CH_2NH\ CH_2\overset{+}{N}\underset{H}{\diagup}{\diagdown}\overset{CH_2}{CH_2}
\end{array}
$$

2.3.2.3. Cyclic Sulfides

Cyclic sulfides, like the cyclic ethers, are polymerized by a variety of cationic initiators such as strong mineral acids, Lewis acids, and strong alkylating agents. Initiation by triethyloxonium tetrafluoroborate is similar to the cyclic ethers and in the case of 3,3-dimethylthietane, proceeds as follows (Goethals *et al.*, 1973)

$$
\begin{array}{c}
CH_3 \quad CH_2 \\
\diagdown C \diagdown \\
CH_3 \quad CH_2 \diagup \\
\end{array}
S + (C_2H_5)_3 O BF_4^- \longrightarrow
\begin{array}{c}
CH_3 \quad CH_2 \\
\diagdown C \diagdown \\
CH_3 \quad CH_2 \diagup \\
\end{array}
\overset{+}{S}-C_2H_5 \; BF_4^- + (C_2H_5)_2O
$$

Again propagation occurs by nucleophilic attack of the monomer on the α-carbon of the cyclic sulfonium ion

$$
\text{~~~~}\overset{+}{S}
\begin{array}{c}
CH_2 \quad CH_3 \\
\diagdown C \diagdown \\
CH_2 \quad CH_3 \\
\end{array}
+ S
\begin{array}{c}
CH_2 \quad CH_3 \\
\diagdown C \diagdown \\
CH_2 \quad CH_3 \\
\end{array}
BF_4^-
\longrightarrow
\text{~~~~}S
\begin{array}{c}
CH_2 \quad CH_2 \quad CH_2 \quad CH_3 \\
\diagdown C \diagdown \overset{+}{S} \diagdown C \diagdown \\
CH_3 \quad CH_3 \quad CH_2 \quad CH_3 \\
\end{array}
BF_4^-
$$

Termination occurs by reaction of a sulfur atom on the already formed polymer with the cyclic sulfonium ion. This reaction forms a nonstrained sulfonium ion that cannot reinitaite polymerization.

$$
\text{~~~~}\overset{+}{S}
\begin{array}{c}
CH_2 \quad CH_3 \\
\diagdown C \diagdown \\
CH_2 \quad CH_3 \\
\end{array}
BF_4^-
+ S
\begin{array}{c}
CH_2 \\
\diagdown \\
CH_2 \\
\end{array}
\longrightarrow
\text{~~~~}S
\begin{array}{c}
CH_2 \quad CH_2 \quad CH_2\text{~~~} \\
\diagdown C \diagdown \overset{+}{S} \diagdown \\
CH_3 \quad CH_3 \quad CH_2\text{~~~} \\
\end{array}
BF_4^-
$$

A similar mechanism is believed to occur with propylene sulfide with the exception that termination is first order involving the formation of a 12-membered ring sulfonium salt

$$
\text{~~~~}CH_2-S-\underset{\underset{CH_3}{|}}{CH}-CH_2-S-\underset{\underset{CH_3}{|}}{CH}-CH_2-S-\underset{\underset{CH_3}{|}}{CH}-CH_2-\overset{+}{S}\diagup\underset{\underset{CH_2}{|}}{CH-CH_3} \xrightarrow{\text{termination}}
$$

$$
\begin{array}{c}
\quad\quad CH_3 \diagdown \quad CH_2-S-CH \diagdown CH_3 \\
\quad\quad\quad\quad CH \quad\quad\quad CH_2 \\
\text{~~~~}CH_2-\overset{+}{S} \quad\quad\quad\quad S \\
\quad\quad\quad\quad CH_2 \quad\quad\quad CH \\
\quad\quad CH-S-CH_2 \diagup \diagdown CH_3 \\
\quad CH_3 \\
\end{array}
$$

This terminated polymer is able to reinitiate polymerization by reaction of a monomer molecule at the exocyclic carbon atom of the sulfonium salt function. A cyclic tetramer of propylene sulfide is formed in this reaction. Trithiane, the cyclic equivalent of trioxane, is also polymerizable with cationic catalysts. The polymerization mechanism is believed to be similar to that of trioxane in that propagation proceeds via the formation of a sulfonium complex that opens up to give a resonance-stabilized carbenium ion.

By analogy with cyclic ethers, the 3-membered cyclic sulfides (thiiranes or episulfides) are also polymerizable by anionic initiation and various co-ordination catalysts. In the case of anionic polymerization, the propagating species is the sulfide (thiolate) anion $\sim\sim\sim S^-$. The anionic polymerization of propylene sulfide, for example, proceeds through perfectly stable thiolate living ends and gives living polymers, the molecular weight of which can be predicted from monomer and initiator concentrations (Boileau *et al.*, 1963).

Contrary to the behavior of the 4-membered cyclic ether oxetane, the 4-membered cyclic sulfide, thietane, can also be polymerized by an anionic mechanism (Morton *et al.*, 1971). However, it was found that the propagating species is a carbanion rather than a thiolate ion and could in fact be used to initiate polymerization of vinyl monomers.

The cyclic sulfides have also been polymerized by a variety of coordination catalysts. The field has been reviewed recently by Sigwalt (1969).

2.3.2.4. Cyclic Amides

The polymerization of cyclic amides has received considerable attention in view of their ability to form nylonlike polyamides. They may be polymerized by acidic or basic initiators as well as molecular initiators, viz., water (Sebenda, 1972). Polymerization initiated by water is termed hydrolytic polymerization and, as the name suggests, involves hydrolytic cleavage of the lactam to the corresponding amino acid:

$$HN-CO + H_2O \rightleftharpoons H_2N \qquad COOH$$

The reaction is carried out at about $200°C$ and the predominating propagation reaction consists of stepwise addition of lactam molecules to the end groups via the following ring opening reaction:

$$\text{wwww}NH_2 + CO\!-\!NH \rightleftharpoons \text{wwww}NHCO \quad NH_2$$

Since polymerization cannot be initiated by water-free carboxylic acids or water-free amines, it would appear that the zwitterion of the amino acid, $^-OOC(CH_2)_5NH_3^+$, is the real active species.

Lactams polymerize much more rapidly in the presence of strong bases such as alkali metals, metal hydrides, metal amides, and organometallic compounds. Initiation involves nucleophilic attack by the base to produce the lactam anion which reacts with the monomer in the second step of the initiation process by a ring opening transamidation.

$$HN\!-\!CO + B^- \rightleftharpoons {}^-N\!-\!CO + BH$$

$$HN\!-\!CO + {}^-N\!-\!CO \underset{\text{fast}}{\overset{\text{slow}}{\rightleftharpoons}} HN^- \quad CO\!-\!N\!-\!CO$$

Unlike the lactam anion, this primary amine anion is not resonance-stabilized and once formed it undergoes a rapid proton transfer with another monomer molecule to form an imide dimer (amino acyllactam) with the regeneration of the lactam anion.

$$HN\!-\!CO + {}^-NH \quad CO\!-\!N\!-\!CO \rightleftharpoons {}^-N\!-\!CO + NH_2 \quad CO\!-\!N\!-\!CO$$

The actual propagating center is the cyclic amide linkage of the N-acylated terminal lactam ring. Acylation of the amide nitrogen increases the electron deficiency of the amide linkage, and this increases the reactivity of the ring amide carbonyl toward attack by the nucleophilic lactam anion. This is followed by a fast proton exchange with the monomer and leads to an equilibrium between lactam and polymer–amide anions.

$$\text{wwww}CO\!-\!N\!-\!CO + {}^-N\!-\!CO \rightleftharpoons \text{wwww}CO\!-\!N^- \quad CO\!-\!N\!-\!CO$$

$$[-CON\!-]^- + HN\!-\!CO \rightleftharpoons -CONH\!- + {}^-N\!-\!CO$$

This is an unusual polymerization mechanism since the propagating center is not an anion but the acylated lactam. Further, it is not the monomer that adds to the propagating chain but rather it is the monomer anion. These reactions are characterized by long induction periods that are due to the very slow initiation reaction involving attack of the lactam anion on the amide linkage. Propagation involves nucleophilic attack by the lactam anion on the much more reactive imide linkage.

Instead of forming the propagation centers in the slow initiation reactions, the electronegatively substituted lactam can be added at the beginning of the reaction, e.g., as N-acyllactam, or formed *in situ* by fast reactions with suitable precursors such as acid chlorides, anhydrides, and isocyanates. Such precursors are termed activators or cocatalysts.

$$RCOCl + {}^-N\!\!-\!\!CO \longrightarrow RCO\!\!-\!\!N\!\!-\!\!CO + Cl^-$$

$$R\!\!-\!\!N\!\!=\!\!C\!\!=\!\!O + HN\!\!-\!\!CO \rightleftharpoons RNHCO\!\!-\!\!N\!\!-\!\!CO$$

Initiation involves reaction of the N-acyllactam with the lactam anion formed by the base initiator

$$RCON\!\!-\!\!CO + {}^-N\!\!-\!\!CO \longrightarrow RCO\!\!-\!\!{}^-N \quad CO\!\!-\!\!N\!\!-\!\!CO$$

followed by proton transfer and continued propagation as before. The rate of addition of the first lactam anion depends on the structure of the activator residue in the initiating center. Bulky acyl groups, such as in pivalocaprolactam, decrease the rate mainly by steric hindrance. Highly electronegative substituents like the benzoyl group increase the rate of addition of the first lactam anion.

The exact mechanism of the ring opening reaction is the subject of some debate. The most plausible explanation involves a nucleophilic attack of the lactam anion at the cyclic carbonyl group of an N-acylated lactam, giving the intermediate symmetric mesomeric anion, which then rearranges with ring opening. In this process, the terminal lactam molecule becomes incorporated into the polymer and the former lactam anion becomes the end N-acylated lactam.

$$RCO\!\!-\!\!N\!\!-\!\!CO + {}^-N\!\!-\!\!-\!\!CO \rightleftharpoons RCO\!\!-\!\!N\!\!-\!\!\overset{\overset{\displaystyle O^-}{|}}{C}\!\!-\!\!N\!\!-\!\!-\!\!CO \rightleftharpoons$$

$$RCO\!\!-\!\!N^- \quad CO\!\!-\!\!N\!\!-\!\!-\!\!CO$$

Lactams may also be polymerized by cationic initiators such as protonic acids, as well as salts of such acids with amines or amides. Initiation involves protonation of the lactam to give a protonated amide. Protonation occurs preferentially at the oxygen atom but a small fraction of the N-protonated amide is assumed to be present in the tautomeric equilibrium

$$\underset{\displaystyle}{HO}\!\!\diagdown\!\!\underset{+}{C}\!\!=\!\!\overset{+}{N}H \longleftrightarrow \overset{+}{HO}\!\!\diagdown\!\!C\!\!-\!\!NH$$

$$O\!\!=\!\!C\!\!-\!\!\overset{+}{N}H_2 \text{ (A)}$$

The amidium cation (A) is not resonance-stabilized and is considered to be the reactive species that undergoes nucleophilic attack by the monomer

$$CO—NH + CO—\overset{+}{N}H_2 \rightleftharpoons CO—N—CO \quad \overset{+}{N}H_3$$

The protonated lactam is regenerated by the equilibrium reaction

$$\text{wwww}\,\overset{+}{N}H_3 + CO—NH \rightleftharpoons \text{wwww}\,NH_2 + CO—\overset{+}{N}H_2$$

Now the strongest nucleophile is represented by the primary amine group and acylation of the latter by the protonated lactam results in the incorporation of one monomer unit into the polymer

$$\text{wwww}\,NH_2 + CO—\overset{+}{N}H_2 \rightleftharpoons \text{wwww}\,NHCO \quad \overset{+}{N}H_3$$

2.3.2.5. Cyclic Esters

The ring opening polymerization of lactones results in polyester formation. It represents a fairly facile route to polyester synthesis in that problems associated with melt condensation such as the necessity of exact stoichiometry, high temperatures, and long reaction cycles do not apply. However, this process has not found significant commercial utilization and detailed studies on the various polymerization mechanisms are limited. The synthesis and polymerization mechanism of a number of lactone monomers have been reviewed (Lunberg and Fox, 1969).

High molecular weight polyesters can be prepared using cationic or anionic catalysts. Cationic polymerization proceeds under the influence of various acid catalysts and involves attack of the initiating carbenium ion on the lactone molecule with cleavage of the acyl oxygen bond and generation of an acyl carbenium ion

$$R^+ + O—CO \longrightarrow RO^+—CO \longrightarrow R—O \quad \overset{C^+}{\underset{O}{\|}}$$

$$R—O \quad \overset{C^+}{\underset{O}{\|}} + O—CO \longrightarrow R—O \quad CO \quad \overset{C,}{\underset{O}{\|}} \quad \text{etc.}$$

An apparent exception to this is β,β-dimethyl-β-propiolactone, which undergoes ring opening to give the preferred tertiary carbenium ion intermediate

The anionic polymerization of lactones involves acyl–oxygen scission with subsequent propagation through the alkoxide anion

$$O\!-\!CO + B^- \longrightarrow O\!-\!\underset{\overset{|}{B}}{C}\!-\!O^- \longrightarrow B\!-\!\underset{\overset{\|}{O}}{C}\quad O^-$$

In certain instances, the terminal unit has been postulated to be a carboxylate anion formed as a result of alkyl oxygen cleavage

$$\text{\textasciitilde}\!C\!-\!O^- + O\!-\!CO \longrightarrow \text{\textasciitilde}\!C\!-\!O \qquad CO^-$$

Lactones may also be polymerized with conventional active hydrogen initiators such as alcohols, amines, and carboxylic acids. These reactions are generally slow, although the polymerization of pivalolactone by strained cyclic amines is reported to be rapid and to lead to high molecular weight polymer and high conversion (Brode and Koleske, 1972).

2.4. Step-Reaction Polymerization

Difunctional monomers with reactive end groups are also capable of coupling together. Many simple organic reactions are known in which two molecules are linked together, typical examples being the reaction of an isocyanate with an alcohol to yield a urethane

$$R\!-\!N\!=\!C\!=\!O + HOR' \longrightarrow R\!-\!NH\!-\!\underset{\overset{\|}{O}}{C}\!-\!OR'$$

and the reaction of an acid with an alcohol to yield an ester

$$R\!-\!\underset{\overset{\|}{O}}{C}\!-\!OH + HOR' \longrightarrow R\!-\!\underset{\overset{\|}{O}}{C}\!-\!O\!-\!R' + H_2O$$

In the latter case, a small molecule (H_2O) is eliminated and such reactions are frequently called condensation reactions.

Since the above reactants are monofunctional, only one coupling reaction can take place. If the initial reactants are difunctional, then the addition product will also have reactive groups at either end that can couple with the reactive group on another molecule. In other words, when the monomers have a functionality of two or greater, the adduct formed by the coupling reaction will also have the same functionality and hence coupling reactions can continue to take place indefinitely.

Thus *poly*urethanes can be formed by the reaction of a *di*isocyanate with a *di*alcohol

$$x\text{OCN}—\text{R}—\text{NCO} + x\text{HO}—\text{R}'—\text{OH} \longrightarrow \underset{\substack{\| \\ \text{O}}}{\text{[CNH}}—\text{R}—\underset{\substack{\| \\ \text{O}}}{\text{NHC}}—\text{O}—\text{R}'—\text{O]}_x$$

and polyesters can be formed by the reaction of *di*acids with *di*alcohols via a polycondensation reaction

$$x\text{HOOC}—\text{R}—\text{COOH} + x\text{HO}—\text{R}'—\text{OH} \longrightarrow \text{H[O}—\underset{\substack{\| \\ \text{O}}}{\text{C}}—\text{R}—\underset{\substack{\| \\ \text{O}}}{\text{C}}—\text{O}—\text{R}']_x\text{OH}$$

$$+ (2x - 1)\text{H}_2\text{O}$$

Alternatively, the same types of reaction can take place using a single monomer bearing two different functional groups. Thus, a polyurethane could also be formed by polymerization of an hydroxyisocyanate

$$x\text{OCN}—\text{R}—\text{OH} \longrightarrow \text{H[O}—\text{R}—\text{NH}—\underset{\substack{\| \\ \text{O}}}{\text{C}}]_{x-1}\text{O}—\text{R}—\text{NCO}$$

and a polyester by polymerization of a hydroxy acid

$$x\text{HO}—\text{R}—\text{COOH} \longrightarrow \text{H[O}—\text{R}—\underset{\substack{\| \\ \text{O}}}{\text{C}}]_x\text{OH} + (x - 1)\text{H}_2\text{O}$$

Consider the synthesis of a polyester from a diacid and a diol. The two monomers react in the initial step to form dimer with the elimination of a molecule of water

$$\text{HOOC}—\text{R}—\text{COOH} + \text{HO}—\text{R}'—\text{OH} \longrightarrow \text{HOOC}—\text{R}—\underset{\substack{\| \\ \text{O}}}{\text{C}}—\text{O}—\text{R}'—\text{OH} + \text{H}_2\text{O}$$

The dimer can then react with another dimer molecule or with unreacted monomer to form a tetramer or trimer, respectively. It can be seen then that all molecules in the system bear reactive groups and are functionally capable of coupling to give dimers, trimers, tetramers, and so on. Polymerization proceeds in this stepwise manner with the molecular weight continuously increasing. Hence this process is termed *step-reaction polymerization*. Since most step reactions take place with the elimination of a small molecule, the process is sometimes called *condensation polymerization*. However, this terminology has led to some confusion in the literature where it has often been used to encompass those step reactions in which no small molecule is eliminated, such as the formation of polyurethane; consequently the former terminology is preferred since it is more descriptive of the overall chain growth mechanism.

TABLE 2.20
COMPARISON OF CHARACTERISTICS OF ADDITION AND STEP-REACTION POLYMERIZATION

Addition polymerization	Step-reaction polymerization
Growth occurs by rapid addition of monomer to a small number of active centers	Growth occurs by coupling of any two species (monomer or polymer)
Monomer concentration decreases slowly during reaction	Monomer concentration decreases rapidly before any high polymer is formed
High polymer is present at low conversion	Polymer molecular weight increases continuously during polymerization; high polymer is present only at very high conversions
Mean degree of polymerization may be very high	Mean degree of polymerization is usually fairly low
Rate of polymerization is zero initially, quickly rises to a maximum as active centers are formed from the initiator, and remains more or less constant during the reaction before falling off when the initiator is consumed	Rate of polymerization is a maximum at the start and decreases continuously during the reaction as the concentration of functional groups decreases

This mechanism is in complete contrast to that of addition polymerization. In the latter case, high polymer is present at the lowest conversions as each initiation step leads to the rapid formation of high molecular weight polymer. At all stages of conversion, the polymerizing system contains high molecular weight polymer, monomer, and propagating active centers. The concentration of the latter is very small and usually remains constant throughout polymerization, while the monomer concentration decreases slowly, becoming zero at complete conversion. In contrast, step reaction polymerization is characterized by the disappearance of monomer very early in the reaction, e.g., in most step reactions, there is less than 1% of the original monomer molecules remaining at a point when the average polymer chain contains only 10 monomer units. In order to achieve a number average degree of polymerization of 1000 it is necessary to push the reaction to 99.9% conversion. These differences are summarized in Table 2.20.

2.4.1. REACTIVITY OF FUNCTIONAL GROUPS

At the beginning of a step polymerization, the primary reaction is coupling between monomer species to form dimer since the probability of a monomer colliding with another monomer molecule is higher than the probability of collision of any other species. However, as the reaction proceeds, a whole range of oligomers are produced, all of which are capable of reacting with each other. The growth reaction can therefore be represented by the general reaction

$$x\text{-mer} + y\text{-mer} \rightarrow (x + y)\text{-mer}$$

Such reactions would be extremely difficult to analyze kinetically if the rate constant for the coupling reaction depended on the degree of polymerization of both species. Fortunately, kinetic studies of simple esterification reactions involving acids of increasing chain length have shown that the rate constant is effectively independent of the chain length. These results lead to Flory's equal reactivity principle for step-reaction polymerization, which states that the intrinsic reactivity of all functional groups is constant and is independent of molecular size (Flory, 1953). The theoretical justification for this assumption is based on the fact that the collision frequency (which determines the reactivity of a functional group) is independent of molecular mobility, i.e., although the rate of diffusion of the larger molecules does depend on their size, the collision frequency of a functional group attached to the end of those chains does not. This results from the appreciable mobility of the terminal group due to conformational change.

One complication that can arise in step-reaction polymerization is the possibility of intramolecular reactions between end groups to form cyclic structures. This situation arises with bifunctional linear monomers such as hydroxy acids $HO\!\!-\!\!(CH_2)_n\!\!-\!\!COOH$ when n is 3, 4, or 5. The main factor affecting ring formation is the size of the ring that may be formed. Ring sizes corresponding to values of $n > 3$ are strain-free structures as opposed to the structures with $n < 3$. However, with $n > 5$, the probability of ring closure is negligible because of the distance of separation of the reactive groups.

2.4.2. KINETICS OF STEP-REACTION POLYMERIZATION

Most discussions of the kinetics of step-reaction polymerization deal with polyesterification reactions because there is extensive published data from which conclusions may be drawn. Information regarding the polymerization of other systems such as polyamides is somewhat sparse. Nevertheless, the conclusions derived from polyesterification studies appear to be quite general and will be presented here. In particular, the polyesterification of an hydroxy acid to a linear polyester will be considered

$$x\text{HO}\!\!-\!\!R\!\!-\!\!COOH \rightarrow H[\!\!-\!\!O\!\!-\!\!R\!\!-\!\!\underset{\overset{\|}{O}}{C}\!\!-\!\!]_x OH + (x - 1)H_2O$$

The reaction of an alcohol with a carboxylic acid is catalyzed by acids and several mechanisms have been proposed (Solomon, 1972). The mechanism commonly accepted involves protonation of the carboxylic acid either by an added acid catalyst or by another molecule of the carboxylic acid. The protonated acid (XXVII) then reacts with the alcohol to yield the ester (XXIX).

$$\text{\textasciitilde\textasciitilde\textasciitilde}\overset{\overset{O}{\|}}{C}\text{—OH} + H^+ \underset{k_{-1}}{\overset{k_1}{\rightleftharpoons}} \text{\textasciitilde\textasciitilde\textasciitilde}\overset{\overset{OH}{|}}{\underset{+}{C}}\text{—OH}$$

XXVII

$$\text{\textasciitilde\textasciitilde\textasciitilde}\overset{\overset{OH}{|}}{\underset{+}{C}}\text{—OH} + \text{\textasciitilde\textasciitilde\textasciitilde OH} \xrightarrow{k_2} \text{\textasciitilde\textasciitilde\textasciitilde}\overset{\overset{OH}{|}}{\underset{\underset{+}{\overset{|}{\text{\textasciitilde\textasciitilde\textasciitilde OH}}}}{C}}\text{—OH}$$

XXVIII

$$\text{\textasciitilde\textasciitilde\textasciitilde}\overset{\overset{OH}{|}}{\underset{\underset{+}{\overset{|}{\text{\textasciitilde\textasciitilde\textasciitilde OH}}}}{C}}\text{—OH} \xrightarrow{k_3} \text{\textasciitilde\textasciitilde\textasciitilde}\overset{}{\underset{\underset{O}{\|}}{C}}\text{—O\textasciitilde\textasciitilde\textasciitilde} + H_2O + H^+$$

XXIX

Normally, these reactions are reversible, but since the water is continuously removed in a polyesterification reaction, the equilibrium is shifted in the direction of polymer formation. The formation of XXVIII is the rate-determining step, hence the rate of polymerization R_p expressed as the rate of loss of carboxyl groups becomes

$$R_p = -d[\text{COOH}]/dt = k_2[\text{OH}][\text{COOH}_2^+] \qquad (2.186)$$

One can substitute for $[\text{COOH}_2^+]$ by means of the equilibrium expression

$$K = k_1/k_{-1} = [\text{COOH}_2^+]/[\text{COOH}][\text{H}^+] \qquad (2.187)$$

Combination of Eqs. (2.186) and (2.187) yields

$$-d[\text{COOH}]/dt = k_2[\text{OH}][\text{COOH}][\text{H}^+]K \qquad (2.188)$$
$$= k_2'[\text{OH}][\text{COOH}][\text{H}^+] \qquad (2.189)$$

where $k_2' = k_2 K$. Two distinct kinetic situations arise depending on whether $[\text{H}^+]$ arises from a strong acid added as an external catalyst or from the monomer itself acting as its own catalyst.

2.4.2.1. Externally Catalyzed Polymerization

Under these conditions, $[\text{H}^+]$ remains constant throughout the polyesterification so that Eq. (2.189) reduces to

$$-d[\text{COOH}]/dt = k_2''[\text{OH}][\text{COOH}] \qquad (2.190)$$

where $k_2'' = k_2'[\text{H}^+]$.

At any time t, the concentration of carboxyl and hydroxyl groups is equal

$$[COOH] = [OH] = C \qquad (2.191)$$

so that Eq. (2.190) becomes

$$-dC/dt = k_2'' C^2 \qquad (2.192)$$

which on integration yields the familiar second-order expression

$$(1/C) - (1/C_0) = k_2'' t \qquad (2.193)$$

where C_0 = concentration of carboxyl or hydroxyl groups when time $t = 0$.

By defining the extent of reaction p as the fraction of functional groups which has reacted in time t, $(p = (C_0 - C)/C_0)$, Eq. (2.193) expressed in terms of p becomes

$$1/(1 - p) = k_2'' C_0 t + 1 \qquad (2.194)$$

2.4.2.2. Self-Catalyzed Polymerization

In the absence of an external catalyst, a molecule of the acid acts as its own catalyst and, hence, the concentration of catalyst decreases with conversion. In this case, $[H^+]$ will be proportional to the concentration of acid in the system and the kinetic rate expression becomes

$$-d[COOH]/dt = k_2''' [OH][COOH]^2 \qquad (2.195)$$

As before, at any time t, $[OH] = [COOH] = C$, and Eq. (2.195) reduces to

$$-dC/dt = k_2''' C^3 \qquad (2.196)$$

which on integration yields

$$(1/C^2) - (1/C_0^2) = 2k_2''' t \qquad (2.197)$$

In terms of the extent of reaction p, Eq. (2.197) becomes

$$1/(1 - p)^2 = 2k_2''' C_0^2 t + 1 \qquad (2.198)$$

These expressions have been applied to a number of polyesterification reactions and have been found to fit the experimental data reasonably well, particularly at high conversion where high molecular weight polymer is being formed. Figure 2.26 shows a third-order plot for the self-catalyzed polyesterification of adipic acid with diethylene glycol. It is apparent that third-order kinetics are obeyed between 80 and 93% conversion. The deviation from third-order kinetics in the low-conversion region (0–80%) has been attributed to polarity changes in the reaction medium (Solomon, 1972). The reactions are generally carried out in the absence of added solvent and consequently the reactants themselves function as the solvent. Initially the medium is an equimolar mixture of alcohol and acid, whereas at complete reaction it is an ester. Thus there is a large change in "solvent" polarity

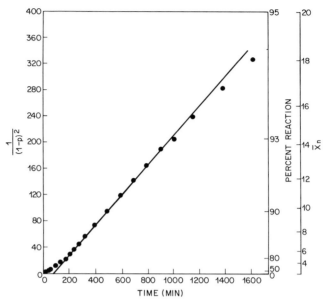

FIG. 2.26. Third-order plot of the self-catalyzed polyesterification of adipic acid with ethylene glycol (Odian, 1970, after Solomon, 1967, and Flory, 1939).

with conversion. In addition, because the initial mixture is very concentrated, it is likely to be thermodynamically nonideal. Under these circumstances, the thermodynamic activity may not be proportional to concentration and, consequently, a kinetic analysis based on concentration may be misleading. Deviation in the high-conversion region is probably attributable to an increase in the rate of the reverse reaction. As discussed earlier, polyesterification reaction are reversible and in order to drive the reaction to completion, it is necessary to totally remove the water byproduct. This becomes increasingly difficult as the viscosity increases with conversion.

2.4.3. MOLECULAR WEIGHT

We have seen that a major difference between addition polymerization and step-reaction polymerization is that in the latter the molecular weight or degree of polymerization increases steadily throughout the reaction, i.e., $\bar{X}_n$ is time dependent. In particular,

$$(\bar{X}_n)_t = \frac{\text{number of monomer molecules in the starting mixture}}{\text{total number of molecules present at time } t}$$

Since each molecule contains one acid end group and one hydroxyl end group, the initial number of molecules per unit volume is $C_0 N_{Avo}$ (N_{Avo} is Avogadro's number). The number of molecules present at time t is $C N_{Avo}$, whence

$$(\bar{X}_n)_t = C_0 N_{Avo}/C N_{Avo} = C_0/C = 1/(1-p) \qquad (2.199)$$

Table 2.21 shows the increase in $\bar{X}_n$ as $p \to 1$. It can be seen that in order to achieve a reasonably high molecular weight, the extent of reaction must exceed 99%.

TABLE 2.21

VARIATION OF NUMBER AVERAGE DEGREE OF POLYMERIZATION $\bar{X}_n$
WITH EXTENT OF REACTION p

p	0	0.5	0.8	0.9	0.95	0.99	0.999	1.0
$\bar{X}_n$	1	2	5	10	20	100	1000	∞

This expression for $\bar{X}_n$ was derived for the simple case of a single monomer containing two different functional groups. A similar argument applies to the situation in which there are two separate difunctional monomers. In this case, $\bar{X}_n$ refers to the number of structural units, where each diol or diacid group separately (not in pairs) constitutes a structural unit. There will therefore be two structural units in a repeating unit and since each structural unit is derived from a difunctional monomer, the total number of structural units present equals the total number of carboxyl groups (or hydroxyl groups) $C_0 N_{Avo}$ initially present. Similarly, the total number of molecules will be $C N_{Avo}$ so that $\bar{X}_n$ is again given by Eq. (2.199).

It has been shown that in the case of a step-reaction polymerization catalyzed by a strong acid, $1/(1-p)$ is directly proportional to the reaction time t. Consequently, $\bar{X}_n$ is also directly proportional to t. Likewise in the absence of catalyst, $1/(1-p)^2$ is directly proportional to t so that $\bar{X}_n$ is proportional to $t^{1/2}$. It is apparent then that molecular weight will increase much more slowly for self-catalyzed reactions than for externally catalyzed reactions and indeed the former are of limited practical importance.

2.4.4. MOLECULAR WEIGHT DISTRIBUTION

Expressions for the molecular weight distribution can be derived from kinetic or statistical considerations (Flory, 1953). As with addition polymerization, the statistical argument is simpler and will again be developed here. It will be recalled that probability can be expressed as number fraction, and hence the probability of finding a functional group reacted at time t is equivalent to the extent of reaction p at time t. Conversely, the probability of finding a functional group unreacted is simply $(1 - p)$. Consider a polymer prepared from a difunctional monomer A–B. The following question may be posed: What is the probability P of finding a molecule with degree of polymerization x? The molecule will contain $(x - 1)$ A groups reacted and one A group unreacted. Since the probability of finding one A group reacted is p, the probability of finding $(x - 1)$ A groups reacted is p^{x-1}. The xth A group has not reacted and the probability of finding this group is $(1 - p)$. Hence, P is given by

$$P = p^{x-1}(1 - p) \tag{2.200}$$

and therefore the number fraction n_x of molecules of degree of polymerization x has the same form, viz.,

$$n_x = p^{x-1}(1 - p) \tag{2.201}$$

The total number of molecules that are x-mers, N_x, is then given by

$$N_x = N_T p^{x-1}(1 - p) \tag{2.202}$$

where N_T is the total number of molecules. If N_0 is the total number of structural units,

$$N_T = N_0(1 - p) \tag{2.203}$$

and Eq. (2.202) becomes

$$N_x = N_0(1 - p)^2 p^{x-1} \tag{2.204}$$

Equation (2.202) is identical to that derived for the case of addition polymerization with termination by transfer or disproportionation (cf. Eq. (2.51)). The distribution is thus the most probable or random distribution for which $\bar{X}_n = 1/(1 - p)$ and $\bar{X}_w = (1 + p)(1 - p)$ with a polydispersity characterized by $\bar{X}_w/\bar{X}_n$, equal to $(1 + p)$. Thus as $p \to 1$, i.e., as the reaction approaches complete conversion, $\bar{X}_w/\bar{X}_n \to 2$.

The weight fraction w_x of molecules of degree of polymerization x is given by

$$w_x = W_x/\sum W_x = N_x x/\sum N_x x \tag{2.205}$$

where W_x is the total weight of all molecules having a degree of polymerization equal to x.

Substituting for N_x in Eq. (2.205) using Eq. (2.202) yields

$$w_x = x(1 - p)^2 p^{x-1} \tag{2.206}$$

Plots of the two distribution functions n_x and w_x for several values of p are

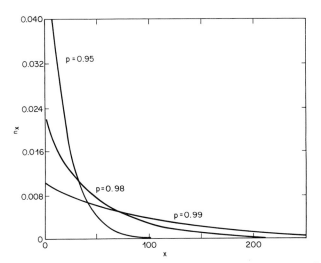

Fɪɢ. 2.27. Mole fraction distribution of chain molecules in a linear condensation polymer for several extents of reaction p (Flory, 1936).

shown in Figs. 2.27 and 2.28, respectively. These figures demonstrate that on a number basis, the smaller species are always more abundant but on a weight basis the proportion of low molecular weight species is very small and decreases as $p \to 1$. The weight fraction w_x passes through a maximum corresponding to $\bar{X}_n$.

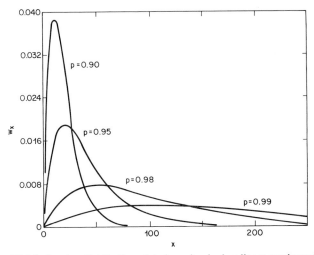

Fɪɢ. 2.28. Weight fraction distribution of chain molecules in a linear condensation polymer for several extents of reaction p (Flory, 1936).

2.4.5. MOLECULAR WEIGHT CONTROL

As in vinyl polymerization, it is frequently necessary to limit the molecular weight to a particular value in order to facilitate subsequent processing. Since the value of $\bar{X}_n$ is determined by the extent of conversion, it should be possible to stop the reaction at an extent of conversion corresponding to the desired molecular weight. However, the polymer still contains reactive ends that could further react at elevated temperature (in an absolutely pure stoichiometrically balanced system, it is theoretically conceivable for the reaction to proceed indefinitely to produce polymer of infinite molecular weight). One method of overcoming this difficulty is to adjust the concentrations of the two monomers so that they are slightly nonstoichiometric. This effect may be illustrated as follows.

Consider a system in which two bifunctional monomers A and B are present with the number of A and B groups being given by N_A and N_B, respectively. The number of A and B molecules is equal to $N_A/2$ and $N_B/2$, respectively. Suppose that $N_B > N_A$, the extent of stoichiometric imbalance r being given by $r = N_A/N_B$. Values of r fall in the range $0 \leq r \leq 1$. The total number of monomer molecules is given by $(N_A + N_B)/2$ or $N_A(1 + 1/r)/2$. If polymerization is carried out to an extent of reaction p (defined for A groups), then the number of A groups that have reacted is pN_A, which must equal the number of B groups reacted. Consequently, the extent of reaction with respect to B groups, p_B is given by $pN_A/N_B = rp$. This leads to the following relationships:

$$\text{total number of unreacted A groups} = N_A - pN_A = N_A(1 - p) \qquad (2.207)$$

$$\text{total number of unreacted B groups} = N_B - rpN_B = N_B(1 - rp) \qquad (2.208)$$

$$\begin{aligned}\text{total number of chain ends} &= N_A(1 - p) + N_B(1 - rp) \\ &= N_A(1 - p) + (N_A/r)(1 - rp)\end{aligned} \qquad (2.209)$$

Since each polymer chain has two chain ends, the number of polymer molecules will equal one-half the total number of chain ends, i.e.,

$$\text{total number of polymer molecules} = \tfrac{1}{2}N_A[1 - p + (1 - rp)/r] \qquad (2.210)$$

Hence the number average degree of polymerization at time t is given by

$$(\bar{X}_n)_t = \frac{N_A(1 + 1/r)/2}{N_A[1 - p + (1 - rp)/r]/2} \qquad (2.211)$$

$$= (1 + r)/(1 + r - 2rp) \qquad (2.212)$$

If the two reactants are present in stoichiometric amounts, i.e., $r = 1$, Eq.

(2.212) reduces to

$$(\bar{X}_n)_t = 1/(1 - p) \qquad (2.213)$$

which was derived in Section 2.4.3.

Further, if polymerization is taken to 100% conversion, i.e., until all functional groups of type A have reacted ($p = 1$), Eq. (2.212) reduces to

$$\bar{X}_n = (1 + r)/(1 - r) \qquad (2.214)$$

Consideration of the following example will illustrate how drastic an effect stoichiometric imbalance has on the molecular weight. If B is present in excess to the amount of 0.1 and 1 mole% ($r = 1000/1001$ and $100/101$, respectively), then at 100% conversion Eq. (2.214) indicates that the values of $\bar{X}_n$ will be 2001 and 201, respectively. It has been seen (Table 2.21) that in order to obtain high molecular weight, conversion must be nearly complete. Consequently it is important to control accurately the stoichiometry of the reactants if high molecular weight polymer is to be obtained. The exact combination of r and p necessary to obtain any particular degree of polymerization is obtained from Eq. (2.212). In practice, one frequently does not have complete freedom of control of these variables because of economic and practical considerations.

Molecular weight stabilization can also be achieved by the addition of monofunctional reactants. This is particularly important for those monomers such as hydroxy acids which possess internally supplied stoichiometry. The same is also true, for example, in the polymerization of a diamine with a diacid to form a polyamide. Internal stoichiometry is assured by the preparation of a balanced (1:1) salt

$$H_2N(CH_2)_nNH_2 + HOOC(CH_2)_mCOOH \rightarrow [H_3N(CH_2)_nNH_3]^{2+}[OOC(CH_2)_mCOO]^{2-}$$

which can be purified. Since the system is internally stoichiometrically balanced, it is not possible to add an excess of one reagent and in such cases monofunctional stabilizing reagents are often used, as in the molecular weight stabilization of nylons by acetic acid.

2.4.6. POLYFUNCTIONAL MONOMERS

When both reactants are difunctional, or if a single difunctional monomer is used, step-reaction polymerization can give rise only to a linear polymer. When one or more monomers with more than two functional groups are present, the resulting polymer will be branched, and under certain conditions will give rise to a three-dimensional network or cross-linked structure. Consider for example the polymerization of monomer A—B in the presence of a small amount of monomer A_f containing f functional groups per molecule. The structure arising from this polymerization will be branched with f chains attached to a central branch point. A functionality of 3 will give rise to structure XXX. There can be only one A_f molecule per chain. Further, the chains cannot link together since the growing ends of the chain only possess A-type functional groups. This type of polymerization is referred to as *multichain polymerization*.

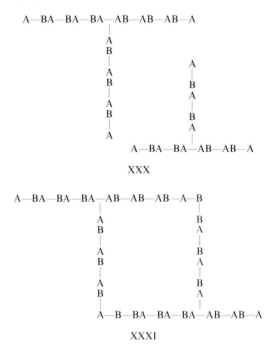

XXX

XXXI

The point in a molecule from which three or more chains emanate is called a *branch point*. Thus when chains growing from two separate branch points on two different chains are able to couple together, the two chains become linked with the branch joining the two chains being referred to as a *cross-link*. In structure XXX, cross-linking is impossible because all growing chains possess A-type functional groups at their ends. If B–B molecules are added to the system, polymerization can lead to a cross-linked polymer structure as depicted in structure XXXI. As a more general example, consider the polymerization of two difunctional monomers A–A and B–B in the presence of a trifunctional monomer

The presence of the trifunctional monomer introduces branch points into the chain so that at some intermediate stage of conversion the structure may be represented as

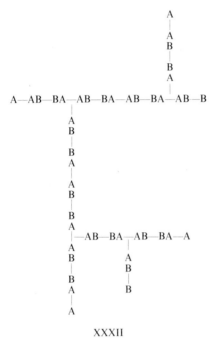

XXXII

The polymerizing system therefore consists of a mixture of highly branched molecules. As conversion increases, the probability of two such molecules coupling together also increases, producing a vast increase in the degree of polymerization and leading ultimately to the formation of an infinite network whose formation is characterized by a sudden enormous rise in viscosity of the mixture. This process is termed *gelation*. The point of transition from a soluble branched polymer to a cross-linked insoluble polymer or gel is termed the *gel point*. Gelation takes place well before all the reactants are bound together, so that reaction continues after gelation has occurred. Consequently, at the gel point a considerable portion of the mixture still exists as soluble polymer chains, termed *sol*, which can be extracted from the *gel* phase. As conversion increases beyond the gel point, these chains become incorporated into the infinite network so that the amount of gel increases at the expense of the sol until eventually the product may be considered to be a single infinite three-dimensional polymer molecule. In contrast to linear polymers (sometimes termed *thermoplastics*), cross-linked polymers do not flow when heated and are termed *thermosetting* polymers or *thermosets*. The cross-links impart dimensional stability, making these materials excellent engineering plastics.

2.4.7. STATISTICAL ANALYSIS OF POLYFUNCTIONAL
STEP-REACTION POLYMERIZATION

Flory (1953) and also Stockmayer (1953) developed a statistical approach to derive a useful expression for predicting the extent of reaction at the gel point. Using the previous example of two difunctional monomers A–A and B–B in the presence of a trifunctional monomer

$$A \mathbf{-} A$$
$$|$$
$$A$$

a portion of the cross-linked network can be depicted as shown in Fig. 2.29. It is assumed that a chain section selected at random from the resulting gelled polymeric structure lies in the first envelope of Fig. 2.29. The immediate problem is to determine under what conditions there is a finite probability that this element of polymer structure occurs as part of an infinite network. In Fig. 2.29 this chain section gives rise to branch units, one at each end of the chain. The four new chains that result lead to three new branch points (on envelope 2) and one terminal bifunctional group. The resulting six new chains lead to two new branch units and four terminal bifunctional groups on envelope 3 and so on.

Consider the ith envelope from the randomly selected chain section in Fig. 2.29. Suppose there are N_i branch units on the ith envelope. If all the chain sections emanating from these N_i branch units ended in branch units

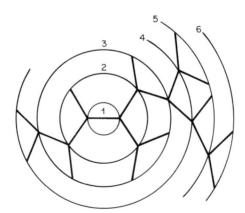

FIG. 2.29. Schematic representation of a trifunctionally branched network polymer (Flory, 1941).

on the $(i + 1)$th envelope, there would be $2N_i$ of them. However, there is only a certain probability that chains beginning at a branch unit will end at a branch unit. This probability is termed the branching coefficient α. (The probability that a functional group of a branch unit leads to an unreacted difunctional terminal group is $1 - \alpha$.) Thus the actual number of branch units on the $(i + 1)$th envelope will be $2N_i\alpha$. The criterion for continuous expansion of the network is that the number of chains emanating from the $(i + 1)$th envelope, viz., $2N_i\alpha$, be greater than the number of chains emanating from the ith envelope. This requires that $2N_i\alpha > N_i$ in which case α must be >0.5. When $\alpha < 0.5$, an infinite network will not be generated. This value defines the critical value of α for gelation, which is given by the general relationship

$$\alpha_c = 1/(f - 1) \tag{2.215}$$

where f is the functionality of the branch unit. Simply stated, the criterion for gel formation is that at least one of the $f - 1$ segments emanating from the branch unit must be connected to another branch unit.

It is possible to relate the value of α to the extent of reaction for a specific blend of reactants. The structures resulting from a mixture of A–A, B–B, and A_f consist of chain segments of the type

$$A_{(f-1)}\text{—}A(B\text{—}BA\text{—}A)_x B\text{—}BA\text{—}A_{(f-1)}$$

XXXIII

A parameter ρ is defined as the ratio of A groups, both reacted and unreacted, which are part of the branch units, to the total number of all A groups in the mixture. If p_A and p_B are the extents of reaction for A groups and B groups. respectively, then α may be related to the extent of reaction by determining the probability of obtaining the chain segment shown in structure XXXIII. The two equalities necessary for this calculation are given by Eqs. (2.216) and (2.217).

Probability that a B group has reacted with a branch unit $= p_B\rho$ (2.216)

Probability that a B group has
reacted with a nonbranch unit $= p_B(1 - \rho)$ (2.217)

The probability of finding structure XXXIII may be calculated as follows:

Probability that A on left-hand branch
unit has reacted with a difunctional B $= p_A$ (2.218)

Probability that B in next
position along the chain has reacted $= p_B$ (2.219)

Probability that this B reacted with a difunctional A $= p_B(1 - \rho)$ (2.220)

Probability that A has reacted with a B $= p_A$ (2.221)

Probability that there are x
such A–AB–B sections $= (p_A p_B(1 - \rho))^x$ (2.222)

Probability that final B has reacted with a trifunctional A $= p_B \rho$ (2.223)

Thus the probability p of obtaining the chain segment XXXIII is given by

$$p = p_A [p_A p_B(1 - \rho)]^x p_B \rho \qquad (2.224)$$

One is interested, however, in obtaining a total probability that is independent of the length of the chain section, i.e.,

$$\alpha = \sum_{x=0}^{\infty} p_A [p_A(1 - \rho)]^x p_B \rho \qquad (2.225)$$

which gives on summation

$$\alpha = p_A p_B \rho / [1 - p_A p_B(1 - \rho)] \qquad (2.226)$$

Either p_A or p_B can be eliminated from this expression by defining $r = N_A / N_B$ as before, in which case $p_B = r p_A$ and Eq. (2.226) becomes

$$\alpha = r p_A^2 \rho / [1 - r p_A^2(1 - \rho)] = p_B^2 \rho / [r - p_B^2(1 - \rho)] \qquad (2.227)$$

This equation gives the value of α at any degree of conversion for a step-reaction polymerization involving a polyfunctional moiety. There are a number of special cases. When the two functional groups are present in equal numbers, $r = 1$ and $p_A = p_B = p$ and Eq. (2.227) becomes

$$\alpha = p^2 \rho / [1 - p^2(1 - \rho)] \qquad (2.228)$$

When there are no A–A molecules in the system, $\rho = 1$ and Eq. (2.227) becomes

$$\alpha = r p_A^2 = p_B^2 / r \qquad (2.229)$$

When both conditions are met, $r = p = 1$ and Eq. (2.227) becomes

$$\alpha = p^2 \qquad (2.230)$$

For the special case with only polyfunctional monomers present, the probability that a branch unit leads to another branch unit is simply the probability that it has reacted, i.e.,

$$\alpha = p \qquad (2.231)$$

2.4.8. Gel Point Observations

The gel point is usually determined experimentally as the time when the reaction mixture suddenly loses fluidity or when bubbles cease to rise through it. Since gelation should occur at a critical value of α, which is 0.5 for a system containing trifunctional cross-linking agents, this should correspond to a theoretical critical extent of reaction p_c for infinite network formation.

Flory (1941) made several measurements of the gel points in reactions of diethylene glycol with succinic or adipic acid with varying proportions of the tribasic acid 1,2,3-propanetricarboxylic acid. His results are summarized in Table 2.22.

TABLE 2.22

Gel Point Determinations for Mixtures of
1,2,3-Propanetricarboxylic Acid, Diethylene Glycol, and
Either Adipic or Succinic Acid

$r = \dfrac{[CO_2H]}{[OH]}$	ρ	Extent of reaction at gel point(p_c)		α Observed at gel point
		Calculated from Eqs. (2.215) and (2.227)	Observed	
1.000	0.293	0.879	0.911	0.59
1.000	0.194	0.916	0.939	0.59
1.002	0.404	0.843	0.894	0.62
0.800	0.375	0.955	0.991	0.58

It is apparent from these results that the value of α observed at the gel point is higher than the value of α_c, 0.5, calculated from statistics. This is generally attributed to the failure of the simple statistical approach to take account of intramolecular cyclization reactions. Such intramolecular reactions make no contribution to establishing the infinite network and, consequently, polymerization has to be carried out to a slightly greater extent of reaction to reach the gel point.

It is further assumed in the statistical derivation that all functional groups have equal reactivity. This may not necessarily be the case. For example, in a series of experiments on the reactions of glycerol with several diacids, gelation occurred at $p_c = 0.765$ compared with a theoretical value of $p_c = 0.707$. Correction for the lower reactivity of the secondary hydroxyl of glycerol decreases the discrepancy but does not entirely eliminate it. Effects such as nonequality of functional group reactivity and the occurrence of intramolecular reactions can be incorporated into more sophisticated statistical treatments.

2.4.9. SIZE DISTRIBUTION IN POLYFUNCTIONAL SYSTEMS

2.4.9.1. Pre-Gel Region

Distribution functions for three-dimensional polymers can also be derived. They depend on the functionality and relative amounts of all the units involved. Only the simplest case is discussed here, viz., the step reaction involving self-condensation of a molecule such as A_f. This gives rise to structure XXXIV for $f = 3$.

XXXIV

The number average degree of polymerization can be easily calculated from stoichiometric considerations. The number of molecules present at time t corresponding to an extent of reaction p is given by the number of monomer units present initially minus the number of linkages formed

$$N_t = (N_A/f) - \tfrac{1}{2}pN_A = N_A(1 - \tfrac{1}{2}fp)/f \tag{2.232}$$

It will be recalled that for systems with only polyfunctional monomers present, the probability that a branch unit leads to another branch unit is simply the probability that it has reacted, i.e., $\alpha = p$. Hence Eq. (2.232) may be rewritten as

$$N_t = N_A(1 - \tfrac{1}{2}\alpha f)/f \tag{2.233}$$

and since $\bar{X}_n = N_0/N_t$, the expression for $\bar{X}_n$ becomes

$$\bar{X}_n = 1/(1 - \tfrac{1}{2}\alpha f) \tag{2.234}$$

The statistical derivation of the distribution is analogous to that for linear polymers, but is somewhat more difficult and only the results will be considered here. It can be shown that the number of molecules N_x of degree of polymerization x is given by the equation

$$N_x = N_0 \frac{(fx - x)!f}{x!(fx - 2x + 2)!} \alpha^{x-1}(1 - \alpha)^{(fx - 2x + 2)} \tag{2.235}$$

for which the weight fraction w_x and number fraction n_x are given by

$$w_x = \frac{(fx - x)!f}{(x - 1)!(fx - 2x + 2)!} \alpha^{x-1}(1 - \alpha)^{(fx - 2x + 2)} \tag{2.236}$$

and

$$n_x = \frac{(fx - x)!f}{x!(fx - 2x + 2)!(1 - \alpha f/2)} \alpha^{x-1}(1 - \alpha)^{fx - 2x + 2} \tag{2.237}$$

The various molecular weight averages may be computed by standard mathematical procedures. It is found that

$$\bar{X}_n = \frac{1}{1 - \frac{1}{2}\alpha f} \tag{2.238}$$

$$\bar{X}_w = \frac{1 + \alpha}{1 - \alpha(f - 1)} \tag{2.239}$$

and

$$\frac{\bar{X}_w}{\bar{X}_n} = \frac{(1 + \alpha)(1 - \alpha f/2)}{1 - \alpha(f - 1)} \tag{2.240}$$

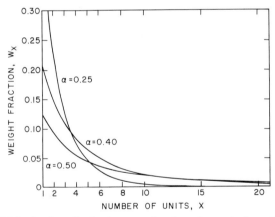

FIG. 2.30. Weight fraction distribution as a function of extent of reaction for a simple trifunctional condensation where $\alpha = p$ (Flory, 1946).

Comparison of Eq. (2.236) with Eq. (2.206) shows that the weight distribution of branched polymers is broader than that of linear polymers at an equivalent extent of reaction. The distribution of the branched polymer becomes broader as conversion increases. This is seen in Fig. 2.30, which shows the weight fraction of x-mers as a function of degree of polymerization for different values of α for a polymerization involving only trifunctional reactants. Comparison of Fig. 2.30 with Fig. 2.28 shows an additional contrast to the linear case in that the weight fraction of monomer in the trifunctional case is always greater than the amount of any one of the outer species. This is not true for the linear case, where it was seen that the weight fraction passed through a maximum at $\bar{X}_n$.

2.4.9.2.   Post-Gel Region

An examination of Fig. 2.30 shows that for the three values of α considered, the weight fraction of the monomer is always greater than that of any other species in the reaction provided $\alpha < \alpha_c$. This is further demonstrated in Fig. 2.31, which shows a plot of the weight fractions of the various finite species for the trifunctional polymerization. Also plotted in Fig. 2.31 is the weight fraction of the gel. It is seen that the extent of reaction at which the weight fraction of any species reaches a maximum shifts continuously to higher values for higher degrees of polymerization but in no case does the maximum occur beyond the gel point ($\alpha = 0.5$).

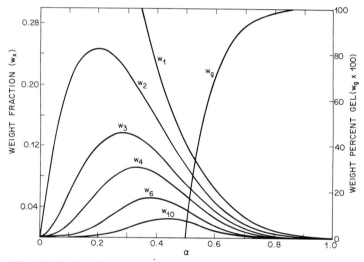

FIG. 2.31.   Weight functions of various finite species and of gel in a simple trifunctional condensation where $\alpha = p$ (Flory, 1946).

The weight fraction of the gel w_g can be simply calculated when it is realized that the sum of the weight fractions of the sol and the gel must be unity. Equation (2.236) applies to the sol portion during all stages of reaction although it is apparent that with $\alpha > \alpha_c, \sum w_x$ for all finite species ($\sum w_{xfs}) \neq 1$ and with $\alpha < \alpha_c, \sum w_x = 1$. It may be shown that for $\alpha > \alpha_c$

$$\sum w_{xfs} = w_{sol} = \frac{(1 - \alpha)^2 \alpha'}{(1 - \alpha')\alpha^2} \qquad (2.241)$$

where α' is the lowest root of the expression $\alpha(1 - \alpha)^{f-2}$ and the expression for w_g becomes

$$w_g = 1 - \frac{(1 - \alpha)^2\alpha'}{(1 - \alpha')^2\alpha} \tag{2.242}$$

The infinite network is first formed at the gel point and its weight fraction increases rapidly. The species in the sol decrease in average size as the larger species are preferentially tied into the infinite network. Thus the size distribution of the soluble portion and the averages characterizing it undergo retroversion over the same course they following up to the gel point as shown in Fig. 2.32.

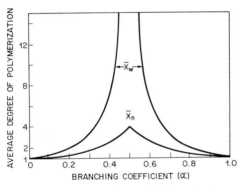

FIG. 2.32. Number and weight average degrees of polymerization as a function of α for a trifunctional polymerization. The portions of the curves after the gel point ($\alpha = 0.5$) are for the sol fraction only (Flory, 1946).

2.4.10. PRACTICAL METHODS

Several factors must be taken into consideration to successfully prepare polymers of high molecular weight by step-reaction polymerization:

(1) the starting materials must be as pure as possible in order to avoid premature termination of the growth reaction by monofunctional impurities;

(2) the reactants must be present in stoichiometric amounts; the molecular weight may be regulated by the presence of a controlled excess of one of the bifunctional reagents or a controlled amount of a monofunctional reagent;

(3) the reaction must be taken to as high a conversion as possible;

(4) since many step-reaction polymerizations are equilibrium reactions, e.g., polycondensation reactions, some means must be provided to shift the equilibrium in the direction of the polymer. This involves removing the

small molecule eliminated in the condensation reaction either by vacuum or chemical reaction.

Synthetic methods fall into two basic categories, viz., bulk or mass polymerization and solution polymerization. Of the two processes, bulk polymerization is the oldest and best established and is the preferred method for the preparation of many important polymers. The method involves mixing the required amounts of catalyst and reactants and heating this mixture to form a melt in which polymerization can take place. The reactions are not generally exothermic and consequently thermal control presents no great problem. High molecular weight polymer is not produced till the latter stages of the reaction so that mixing and by-product removal are easily accomplished during the early stages. At high conversion the accompanying high viscosity causes considerable difficulty in removing volatile by-products. Several examples will serve to illustrate the variety of approaches that can be taken.

Polyamides are usually prepared by the melt polycondensation of a diacid and diamine, e.g., nylon 6.6 is prepared by the direct amidation of hexamethylene diamine with adipic acid. As mentioned earlier, stoichiometric balance is obtained via formation of the 1:1 nylon salt. Polymerization is then carried out by heating an aqueous slurry containing about 60–80% of the salt at about $200°C$ in a closed autoclave under a pressure of about 15 atm.

$$H_2N(CH_2)_6NH_2 + HOOC(CH_2)_4COOH \rightarrow [H_3N(CH_2)_6NH_3]^{2+}[OOC(CH_2)_4COO]^{2-}$$

$$\downarrow$$

$$H\{NH(CH_2)_6NHCO(CH_2)_4CO\}_xOH + (2x - 1)H_2O$$

Unlike polyamidation reactions under melt conditions, the preparation of polyesters in the melt from diols and the esters of diacids does not always require exact reactant stoichiometry at the start. An excess of the diol leads to a low molecular weight hydroxy-ended polymer, which is then converted to high polymer by an ester exchange reaction with evolution of excess diol. This technique is used in the production of poly(ethylene terephthalate) (Mylar) from dimethyl terephthalate and ethylene glycol in the presence of a weakly basic catalyst such as calcium acetate or antimony trioxide. The first ester interchange reaction takes place at $200°C$ with methanol being continuously distilled off

The reaction temperature is then raised to $280°C$ under reduced pressure and ethylene glycol is removed in the second ester interchange reaction

$$HOCH_2CH_2OC-\bigcirc-C-OCH_2CH_2OH \longrightarrow$$
$$\qquad\quad\; O \qquad\qquad O$$

$$HOCH_2CH_2O-\left[C-\bigcirc-COCH_2CH_2O\right]_x H + (x-1)HOCH_2CH_2OH$$
$$\qquad\qquad\quad O \qquad\qquad O$$

The high temperatures frequently employed in bulk step-reaction polymerization stem from the low reactivity of the reactants. This necessitates that the reactants and polymer be thermally stable at the reaction temperature, thereby limiting the choice of monomer and hence the variety of polymers.

One method of overcoming this problem is to carry out polymerization reactions in solution using fairly reactive functional groups. The Schotten–Baumann reaction of an organic acid halide with a compound containing active hydrogens can be readily modified to produce polymer. Several examples are listed below:

$$\sim\sim NH_2 + \sim\sim C-Cl \longrightarrow \sim\sim C-NH\sim\sim \quad \text{polyamide}$$
$$\qquad\qquad\quad O \qquad\qquad\qquad O$$

$$\sim\sim OH + \sim\sim C-Cl \longrightarrow \sim\sim C-O\sim\sim \quad \text{polyester}$$
$$\qquad\qquad\quad O \qquad\qquad\qquad O$$

$$\sim\sim OH + Cl-C-Cl \longrightarrow \sim\sim O-C-O\sim\sim \text{polycarbonate}$$
$$\qquad\qquad\quad O \qquad\qquad\qquad O$$

Similarly, the reaction of isocyanates with diamines or diols is rapid at room temperature, particularly in the presence of catalysts such as dibutyl tin dilaurate. In contrast to high temperature polymerization, these reactions are irreversible. The acidic by-product (HCl) from these polycondensation reactions is usually removed chemically by an acid acceptor such as sodium hydroxide.

There are two methods for carrying out step-reaction polymerizations at low temperature: solution and interfacial polymerization. The former involves combining the reactants in a suitable inert solvent and providing for the removal of any acid intermediates by means of basic additives such as tertiary amines. Reaction may start with all reactants fully dissolved and the polymer may remain in solution or precipitate as the reaction progresses.

Interfacial polymerization involves dissolution of the reactants in a pair of immiscible liquids, one of which is usually water and the other an organic solvent such as hexane or carbon tetrachloride. The aqueous phase contains the diamine or diol and the alkali, while the organic solvent contains the diacid halide. If the two phases are now brought into contact without stirring and if the organic solvent is a non-solvent for the polymer, a thin

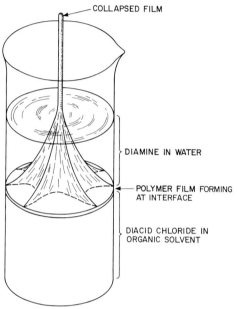

FIG. 2.33. Formation of a polyamide by interfacial polymerization ("nylon rope trick") (Morgan and Kwolek, 1959).

film of polymer forms at the interface (hence the term interfacial polymerization), which can be withdrawn (Fig. 2.33). Polymer continues to form at the interface until one of the reactants is used up. Rapid stirring of the system results in increased polymerization rates because of the increase in interfacial (reacting) surface area.

The molecular weight of the polymer produced by interfacial polymerization is generally high since monomer molecules diffusing to the interface are more likely to react with a growing chain before they can penetrate the polymer film and start a new chain.

These solution polymerization methods are very advantageous for laboratory preparations. This is particularly true of the interfacial method where factors such as purity and stoichiometry are not nearly as critical as in high-temperature melt processes. The molecular weight does not depend on the overall extent of conversion and is generally high. Consequently, this method has been applied to the synthesis of a wide range of polymers such as polyamides, polyesters, polyurethanes, polycarbonates, and many others. There are, however, a number of disadvantages in large-scale operation. These include (1) the formation of large quantities of by-product salt that must be removed from the polymer; (2) the need to handle and recover large quantities of organic solvent; and (3) the high cost of intermediates, particularly the diacid halides. Nevertheless, some reactions are of commercial importance, e.g., the production of polycarbonates from bisphenol A and phosgene.

polycarbonate

The same polymer can be produced by the ester interchange melt polymerization process similar to that discussed for poly(ethylene terephthalate).

The practical techniques for producing cross-linked or thermosetting polymers are quite different from the techniques discussed for linear or thermoplastic polymers. Whereas thermoplastic polymers may be processed by various techniques following preparation, e.g., extrusion or molding, no such latitude is afforded by thermosetting polymers. The latter are no longer capable of flow and consequently must be fabricated in the desired shape at the time of manufacture. Generally, these polymers are supplied to a fabricator in an already partially reacted form, which is termed a *prepolymer*. Prepolymers themselves may be classed as A-stage or B-stage depending on the extent of reaction p relative to the extent of reaction at gelation p_c. The A-stage prepolymer is soluble and fusible with $p \ll p_c$ while the B-stage prepolymer is still fusible but only barely soluble since p is close to p_c. Final cross-linking is then carried out by the fabricator to give the C-stage polymer.

The reaction of formaldehyde with phenol, urea, or melamine is an important industrial process that exemplifies the three stages of reaction. For example, base catalysis of phenol–formaldehyde mixtures containing an excess of formaldehyde over phenol yields mixtures of mononuclear methylol phenols (XXXV) and the corresponding dinuclear (XXXVI) and polynuclear compounds (XXXVII).

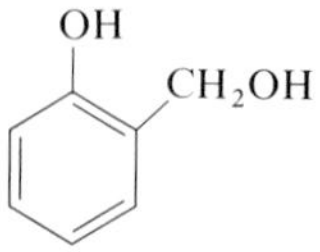

XXXVa 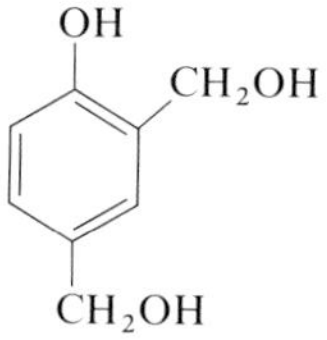   XXXVb 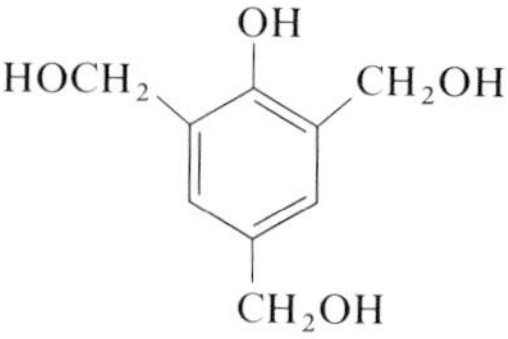   XXXVc

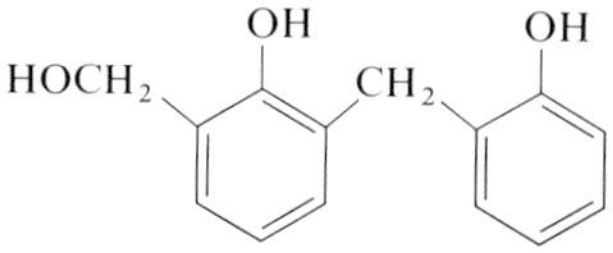

XXXVIa 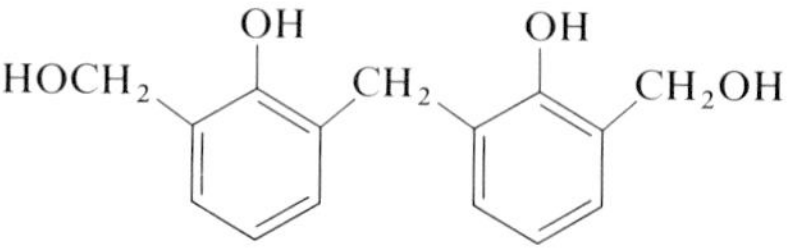   XXXVIb

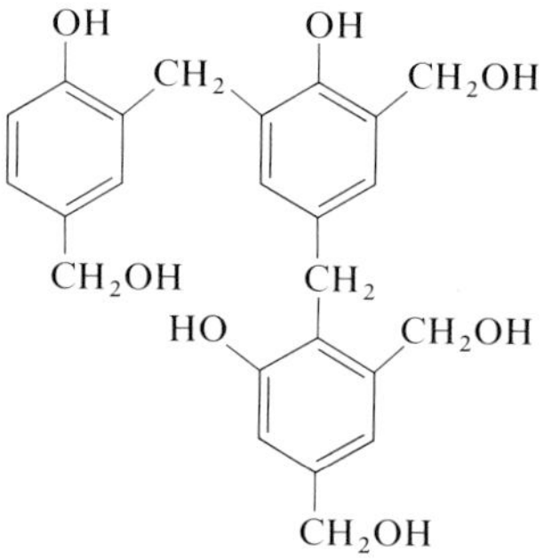

XXXVII

The reaction is carried out in aqueous solution and the products (A-stage) are of such low degree of polymerization that they remain soluble. The polynuclear compounds are formed by condensation of the lower molecular weight methylol phenols. As the condensation reactions continue, the products become insoluble and phase separation occurs. Vacuum is applied, water is removed, and the reaction progresses into the melt phase, where the degree of polymerization increases to a range of 100–200 to produce a branched B-stage polymer termed a *resole*. The resoles possess highly branched structures with methylene bridges linking the benzene rings. They are, however, still thermoplastic and may be removed from the reactor at this stage, quench-cooled, and compounded into a molding powder.

Subsequent polymerization to high molecular weight C-stage polymer can then be effected simply by heating. The resoles are generally neutralized or made slightly acid before the final condensation, which takes place by the formation of both methylene and dibenzyl ether linkages to form the cross-linked structure XXXVIII.

The cross-linking reactions that occur in these systems are quite random. Polymerization during the C-stage is usually a continuation of the reactions

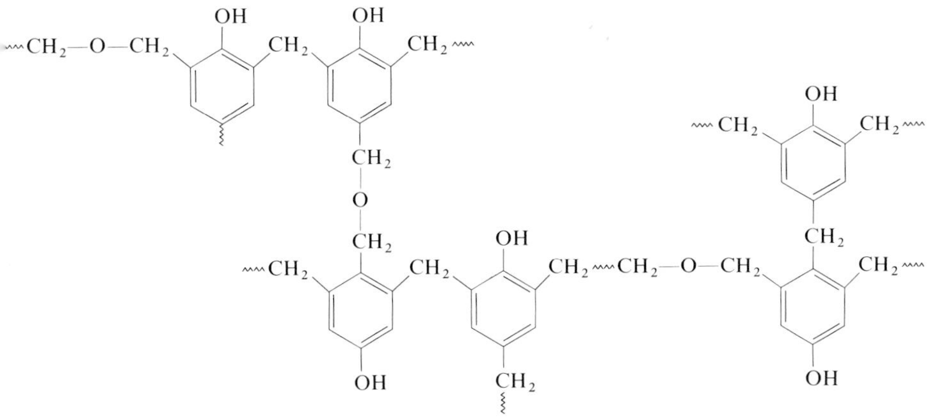

XXXVIII

that occurred in the formation of the B-stage. There are a number of thermo-set systems, however, which involve the use of specially designed prepolymers of a more well-defined structure. Such systems are usually cross-linked by the addition of a catalyst or some other reagent. Further, the cross-linking mechanism is usually quite different from that for prepolymer formation.

For example, epoxy resins are prepared by step-reaction polymerization between an epoxide such as epichlorohydrin and a dihydroxy compound such as 2,2'-bis (hydroxyphenyl) propane (bisphenol A)

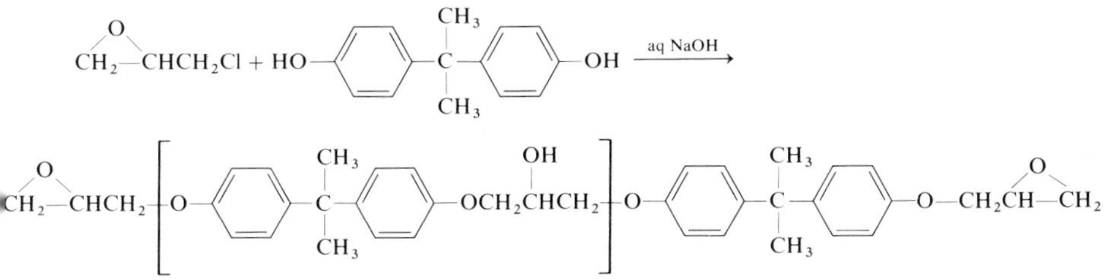

XXXIX

Depending on the number of repeating units, the prepolymer XXXIX varies from a viscous liquid to a solid.

Cross-linking of this prepolymer may be achieved either by opening the terminal epoxy group, as occurs when polyamines are used as curing agents, or through the independent hydroxy groups, as occurs, for example, when a polyanhydride is used as the cross-linking agent.

There are a wide variety of polymers that are formed on this general principle. They include polyurethanes, polysiloxanes, and unsaturated poly-esters among others. All are of high commercial significance where they find use as elastomers, foams, and coatings.

REFERENCES

Alexander, A. E., and Napper, D. H. (1971). *Progr. Polym. Sci.* **3**, 145.
Alfrey, T., and Goldfinger, G. (1944). *J. Chem. Phys.* **12**, 205.
Alfrey, T., and Price, C. C. (1947). *J. Polym. Sci.* **2**, 101.
Allen, P. E. M., and Patrick, C. R. (1963). *Trans. Faraday Soc.* **59**, 1819.
Allen, P. E. M., and Patrick, C. R. (1974). *In* "Kinetics and Mechanisms of Polymerization Reactions." Halsted Press, New York.
Allport, D. C., and James, W. H. (1973). *In* "Block copolymers" (D. C. Allport and W. H. James, eds.). Halsted Press, New York.
Altares, T., Wyman, D. P., Allen V. R., and Meyersen, K. (1965). *J. Polym. Sci. Part A* **3**, 4131.
Arlman, E. J. (1964). *J. Catal.* **3**, 89.
Arlman, E. J., and Cossee, P. (1964). *J. Catal.* **3**, 99.
Ballard, D. G. H. (1975). *In* "Molecular Behaviour and the Development of Polymeric Materials" (A. Ledwith and A. M. North, eds.). Halsted Press, New York.
Bamford, C. H. (1974). *In* "Reactivity Mechanism and Structure" (A. D. Jenkins and R. Ledwith, eds.), Chapter 3. Wiley, New York.
Bamford, C. H. (1975). *In* "Molecular Behaviour and the Development of Polymeric Materials" (A. Ledwith and A. M. North, eds.). Halsted Press, New York.
Bamford, C. H., and Denyer, R. (1966). *Trans. Faraday Soc.* **62**, 1567.
Bamford, C. H., Jenkins, A. D., and Johnston, R. (1959). *Trans. Faraday Soc.* **55**, 179.
Barnes, C. E., Elofson, R. M., and Jones, G. D. (1950). *J. Am. Chem. Soc.* **72**, 210.
Barney, A. L., Bruce, J. M., Coker, J. N., Jacobson, H. W., and Sharkey, W. H. (1966). *J. Polym. Sci. Part A-1*, **4**, 2617.
Battaerd, H. A., and Tregear, G. W. (1967). *In* "Graft Copolymers," Polymer Reviews, Vol. 16. Wiley (Interscience), New York.
Bawn, C. F. H., Carruthers, R., and Ledwith, A. (1965). *Chem. Commun.* 522.
Baysal, B., and Tobolsky, A. V. (1952). *J. Polym. Sci.* **8**, 529.
Berger, M. N., Boocock, G., and Harward, R. N. (1969). *Adv. Catal.* **19**, 211.
Bevington, J. C. (1961). *In* "Radical Polymerization." Academic Press, New York.
Bhattacharyya, D. N., Lee, C. L., Smid, J., and Szwarc, M. (1964). *Polymer* **5**, 54.
Bhattacharyya, D. N., Lee, C. L., Smid, J., and Szwarc, M. (1965). *J. Phys. Chem.* **69**, 612.
Billingham, N. C., and Jenkins, A. D. (1972). *In* "Polymer Science" (A. D. Jenkins, ed.), Vol. 1, p. 1. Elsevier, New York.
Billmeyer, F. (1971). *In* "Textbook of Polymer Science," 2nd ed. Wiley, New York.
Boileau, S., Champetier, G., and Sigwalt, P. (1963). *Makromol. Chem.* **69**, 180.
Boor, J. Jr. (1963). *J. Polym. Sci. Part C*, **1**, 257.
Boor, J. Jr. (1967). *Macromol. Rev.* **2**, 115.
Boor, J. Jr. (1974). *Am. Chem. Soc. Div. Polym. Chem. Preprints* **15(1)**, 359.
Boor, J. Jr., and Youngman, E. A. (1966). *J. Polym. Sci. Part A-1*, **4**, 1861.
Bovey, F. A., and Kolthoff, I. M. (1947). *J. Am. Chem. Soc.* **69**, 2143.
Bovey, F. A., Kolthoff, I. M., Medalia, A. I., and Meehan, E. J. (1955). *In* "Emulsion Polymerization," High Polymers, Vol. IX. Wiley (Interscience), New York.
Brandrup, J., and Immergut, E. H. (1975). "Polymer Handbook," 2nd ed. Wiley, New York.
Briers, F., Chapman, D. L., and Walters, E. (1926). *J. Chem. Soc.* 562.
Brode, G. L., and Koleske, J. V. (1972). *J. Macromol. Sci.-Chem.* **A6(6)**, 1109.
Brown, T. L. (1970). *Pure Appl. Chem.* **23(4)**, 447.
Burnett, G. M., and Melville, H. W. (1947). *Proc. Roy. Soc. London Ser. A* **189**, 486.
Bywater, S. (1975). *Prog. Polym. Sci.* **4**, 27.
Cameron, G. G., and Esslemont, G. F. (1972). *Polymer* **13**, 435.

Chapiro, A. (1962). *In* "Radiation Chemistry of Polymeric Systems," High Polymers, Vol. XV. Wiley (Interscience), New York.

Chiang, R., and Rhodes, J. H. (1969). *J. Polym. Sci. Part B,* **7(9)**, 643.

Chien, J. C. W. (1975). *In* "Coordination Polymerization" (J. C. W. Chien, ed.). Academic Press, New York.

Clark, A., Hogan, J. P., Banks, R., and Lanning, W. C. (1956). *Ind. Eng. Chem.* **48**, 1152.

Conley, R. T. (1970). *In* "Thermal Stability of Polymers" (R. T. Conley, ed.). Dekker, New York.

Cooper, W. (1970). *Ind. Eng. Chem. Prod. Res. Dev.* **9**, 457.

Cossee, P. (1960). *Tetrahedron Lett.* **12**, No. 17, 17.

Cossee, P. (1964). *J. Catal.* **3**, 80.

Cram, D. J. (1965). *In* "Fundamentals of Carbanion Chemistry." Academic Press, New York.

Dainton, F. S., and Ivin, K. J. (1958). *Quart. Rev. (London)* **12**, 61.

Dreyfuss, M. P. (1975). *J. Macromol. Sci.* **A9(5)**, 729.

Dreyfuss, P. (1973). *J. Macromol. Sci.* **A7(7)**, 1361.

Dreyfuss, P., and Dreyfuss, M. P. (1969). "Addition and Condensation Polymerization Processes," p. 335. American Chemical Society, Washington, D.C.

Finaz, G., Gallot, Y., Parrod, J., and Rempp, P. (1962). *J. Polym. Sci.* **58**, 1363.

Fineman, M., and Ross, S. D. (1950). *J. Polym. Sci.* **5**, 259.

Flory, P. J. (1936). *J. Am. Chem. Soc.* **58**, 1877.

Flory, P. J. (1939). *J. Am. Chem. Soc.* **61**, 3334.

Flory, P. (1940). *J. Am. Chem. Soc.* **62**, 1561.

Flory, P. (1941). *J. Am. Chem. Soc.* **63**, 3083.

Flory, P. J. (1946). *Chem. Rev.* **39**, 137.

Flory, P. J. (1953). *In* "Principles of Polymer Chemistry." Cornell Univ. Press, Ithaca, New York.

Fordham, J. W. L. (1959). *J. Polym. Sci.* **39**, 321.

Fordyce, R. G., and Ham, G. E. (1951). *J. Am. Chem. Soc.* **73**, 1186.

Frisch, K. C., and Reegen, S. L. (1969). *In* "Kinetics and Mechanisms of Polymerization" (K. C. Frisch and S. L. Reegen, eds.), Vol. 2, Ring-Opening Polymerization, Chapter 1, Dekker, New York.

Funt, B. L., Richardson, D., and Bhadani, S. N. (1966). *Can. J. Chem.* **44**, 711.

Furukawa, J., and Tada, K. (1969). *In* "Kinetics and Mechanisms of Polymerization" (K. C. Frisch and S. L. Reegen, eds.), Vol. 2, Ring-Opening Polymerization, Chapter 4. Dekker, New York.

Gardon, J. (1968). *J. Polym. Sci. Part A-1,* **6**, 2859.

Gardon, J. (1975). *In* "Applied Polymer Science" (J. K. Craver and R. W. Tess, eds.), p. 138, Org. Coat. and Plastics Chem. American Chemical Society, Washington, D.C.

Gaylord, N. G. (1970). *J. Polym. Sci. Part D,* **4**, 183.

Geacintov, C., Smid, J., and Szwarc, M. (1962). *J. Am. Chem. Soc.* **84**, 2508.

George, M. H. (1967). *In* "Kinetics and Mechanisms of Polymerization," Vol. 1, Vinyl Polymerization, Pt. 1 (G. Ham, ed.), Dekker, New York.

Gerrens, H. (1959). *Fortsch. Hochpolym. Forsch.* **1**, 234.

Goethals, E. J., Drijvers, W., Van Ooteghem, D., and Boyle, A. M. (1973). *J. Macromol. Sci.-Chem.* **A7(7)**, 1375.

Goldfinger, G., Yeě, W., and Gilbert, R. D. (1967). *Encycl. Polym. Sci.* **7**, 644.

Gons, J., Vorenkamp, E. J., and Challa, G. (1975). *J. Polym. Sci. Polym. Chem. Ed.* **13**, 1699.

Goode, W. E., Owens, F. H., Fellman, R. P., Snyder, W. H., and Moore, J. E. (1960). *J. Polym. Sci.* **46**, 317.

Greber, G., and Tölle, J. (1962). *Makromol. Chem.* **53**, 208.

Gregg, R. A., and Mayo, F. R. (1947). *Disc. Faraday Soc.* **2**, 328.

Ham, G. (1967). *In* "Kinetics and Mechanisms of Polymerization," Vol. 1, Vinyl Polymerization, Pt. 1 (G. Ham, ed.), Dekker, New York.

Hargitay, B., Rodriguez, L., and Miotto, M. (1959). *J. Polym. Sci.* **35**, 559.
Harkins, W. D. (1947). *J. Am. Chem. Soc.* **69**, 1428.
Hauser, M. (1969). *In* "Kinetics and Mechanisms of Polymerization" (K. C. Frisch and S. L. Reegen, eds.), Vol. 2, Ring-Opening Polymerization, Chapter 6. Dekker, New York.
Higginson, W. C. E., and Wooding, N. S. (1952). *J. Chem. Soc.* 760.
Hostalka, H., and Schulz, G. V. (1965). *J. Polym. Sci. Part B*, **3**, 175.
Inoe, S. (1969). *In* "Structure an Mechanism in Vinyl Polymerization" (T. Tsuruta and K. F. O'Driscoll, eds.), Chapter 5. Dekker, New York.
Irie, M., Yamamoto, Y., and Hayashi, K. (1975). *J. Macromol. Sci.-Chem.* **A9(5)**, 817.
Ivin, K. J. (1974). *In* "Reactivity, Mechanism and Structure in Polymer Chemistry" (A. D. Jenkins and A. Ledwith, eds.). Wiley, New York.
Jenkins, A. D. (1967). *In* "Kinetics and Mechanisms of Polymerization," Vol. 1, Vinyl Polymerization, Pt. 1 (G. Ham, ed.), Dekker, New York.
Joshi, R. M. (1973). *J. Macromol. Sci.-Chem.* **A7**, 1231.
Joshi, R. M., and Joshi, S. G. (1971). *J. Macromol. Sci.-Chem.* **A5(8)**, 1329.
Joshi, R. M., and Zwolinski, B. J. (1967). *In* "Kinetics and Mechanisms of Polymerization," Vol. 1, Vinyl Polymerization, Pt. 1 (G. Ham, ed.), Dekker, New York.
Kamienski, C. W., and Eastham, J. E. (1973a). U.S. patent 3,725,368, June 26.
Kamienski, C. W., and Merkley, J. H. (1973b). U.S. patent 3,751,501, August 7.
Kang, B. K., O'Driscoll, K. F., and Howell, J. A. (1972). *J. Polym. Sci. Part A-1*, **10**, 2349.
Karol, F. J. (1975). *In* "Applied Polymer Science" (J. K. Craver and R. W. Tess, eds.), p. 176, Org. Coat. and Plastics Chem. American Chemical Society, Washington, D.C.
Karol, F. J., and Carrick W. L. (1961). *J. Am. Chem. Soc.* **83**, 2654.
Kelen, T., and Tüdos, F. (1975). *J. Macromol. Sci.-Chem.* **A9(1)**, 1.
Kennedy, J. P. (1972). *J. Macromol. Sci.-Chem.* **A6(2)**, 329.
Kennedy, J. P. (1975). *In* "Cationic Polymerization of Olefins: a Critical Inventory." Wiley, New York.
Kennedy, J. P. (1977). *J. Appl. Polym. Sci. Appl. Polym. Symp.* **30**, 1.
Kennedy, J. P., and Gilham, J. K. (1972). *Fortsch. Hochpolym. Forsch.* **10**, 1.
Kennedy, J. P., and Smith, R. R. (1974). *Polym. Sci. Technol.* **4**, 303.
Kennedy, J. P., and Squires, R. G. (1967). *J. Macromol. Sci.* **A1(5)**, 831.
Kennedy, J. P., Kirshenbaum, I., Thomas, R. M., and Murray, D. C. (1963). *J. Polym. Sci. Part A*, **1**, 331.
Kennedy, J. P., Elliot, J. J., and Hudson, B. E. (1964a). *Makromol. Chem.* **79**, 109.
Kennedy, J. P., Minkler, L. S. Jr., Wanless, G. C., and Thomas, R. M. (1964b). *J. Polym. Sci. Part A-2*, **3**, 1441.
Kennedy, J. P., Borzel, P., Naegele, W., and Squires, R. G. (1966). *Makromol. Chem.* **93**, 191.
Ketley, A. D. (1963). *J. Polym. Sci., Part B* **1**, 313.
Landler, Y. (1950). *Compt. Rend.* **230**, 539.
Ledwith, A. (1969). "Addition and Condensation Polymerization Processes," p. 317. American Chemical Society, Washington, D.C.
Ledwith, A., and Sherington, D. C. (1974). *In* "Reactivity Mechanism and Structure in Polymer Chemistry" (A. Jenkins and A. Ledwith, eds.). Wiley, New York.
Lundberg, R. D., and Cox, E. F. (1969). *In* "Kinetics and Mechanisms of Polymerization" (K. C. Frisch and S. L. Reegen, eds.), Vol. 2, Ring-Opening Polymerization, Chapter 7. Dekker, New York.
Lüssi, H. (1967). *Makromol. Chem.* **103**, 68.
Mayo, F. R. (1959). *J. Chem. Educ.* **36**, 157.
Mayo, F. R., and Lewis, F. M. (1944). *J. Am. Chem. Soc.* **66**, 1594.
Mayo, F. R., and Walling, C. (1950). *Chem. Rev.* **46**, 191.
McCormick, C. L., and Butler, G. B. (1972). *J. Macromol. Sci. Rev. Macromol. Chem.* **C8(2)**, 201.

Meyer, V. E., and Chan, R. K. S. (1967). *Am. Chem. Soc.Div. Polym. Chem. Preprints* **8(1)**, 209.
Minoura, Y. (1969). *In* "Structure and Mechanism in Vinyl Polymerization" (T. Tsuruta and K. F. O'Driscoll, eds.). Dekker, New York.
Morgan, P. W., and Kwolek, S. L. (1959). *J. Chem. Educ.* **36**, 182.
Morton, M., and Fetters, L. J. (1975). *Rubber Chem. Technol.* **48(3)**, 359.
Morton, M., Bostick, E. E., and Livigni, R. A. (1961). *Rubber Plast. Age* **42**, 397.
Morton, M., Kammereck, R. F., and Fetters, L. J. (1971), *Br. Polym. J.* **3**, 120.
Nakane, R., Watanabe, T., and Kurihara, O. (1962). *Bull. Chem. Soc. (Jpn.)* **35**, 1747.
Natta, G., and Mazzanti, G. (1960). *Tetrahedron* **8**, 86.
Natta, G., and Pasquon, I. (1959). *Adv. Catalysis* **9**, 1.
Natta, G., Corradini, P., and Allegra, G. (1961). *J. Polym. Sci.* **51**, 399.
Norrish, R. G. W., and Smith, R. R. (1942). *Nature (London)* **150**, 336.
North, A. M. (1966). *In* "The Kinetics of Free Radical Polymerization." Pergamon, Oxford.
North, A. M., and Postlethwaite, D. (1969). *In* "Structure and Mechanism of Vinyl Polymerization" (T. Tsuruta and K. F. O'Driscoll, eds.). Dekker, New York.
Odian, G. (1970). *In* "Principles of Polymerization." McGraw-Hill, New York.
O'Driscoll, K. F. (1969). *J. Macromol. Sci.-Chem.* A3(2), 307.
O'Driscoll, K. F., and Ghosh, P. (1969). *In* "Structure and Mechanism in Vinyl Polymerization" (T. Tsuruta and K. F. O'Driscoll, eds.), Chapter 3. Dekker, New York.
Olah, G. (1972). *J. Am. Chem. Soc.* **94**, 808.
Ottolenghi, A., and Zilkha, A. (1963). *J. Polymer Sci. Part A*, **1**, 687.
Overberger, C. G., Yuki, H., and Urakawa, W. (1960). *J. Polym. Sci.* **45**, 127.
Parry, A. (1974). *In* "Reactivity, Mechanism and Structure in Polymer Chemistry" (A. D. Jenkins and A. Ledwith, eds.). Wiley, New York.
Pepper, D. C. (1949). *Trans. Faraday Soc.* **45**, 397.
Pepper, D. C. (1952). *Quart. Rev. (London)* **8**, 88.
Pepper, D. C., and Reilly, P. J. (1962). *J. Polym. Sci.* **58**, 639.
Pinazzi, C. P., Brosse, J. C., Pleurdeau, A., Brossas, J., Legeay, G., and Cattiaux, J. (1973). *In* "Polymerization Reactions and New Polymers" (N. A. J. Platzer, ed.). American Chemistry Society, Washington, D.C.
Pino, P., Ciardelli, F., and Lorenzi, G. P. (1963). *J. Polym. Sci. Part C*, **4**, 21.
Plesch, P. H. (ed.) (1963). *In* "The Chemistry of Cationic Polymerization." Pergamon, Oxford.
Price, C. C. (1946). *In* "Mechanism of Reactions at Carbon-Carbon Double Bonds." Wiley (Interscience), New York.
Reich, L., and Schlindler, A. (1966). *In* "Polymerization by Organometallic Compounds." Wiley (Interscience), New York.
Rodriguez, L. A. M., and Van Looy, H. M. (1966). *J. Polym. Sci. Part A-1*, **4**, 1951, 1971.
Rose, J. B. (1956). *J. Chem. Soc.* 542.
Saegusa, T., and Matsumoto, S. (1968). *J. Polym. Sci. Part A-1*, **6**, 1559.
Saegusa, T., Kobayashi, S., Kimura, Y., and Ikeda, H. (1975). *J. Macromol. Sci.* A9(5), 641.
Sangster, J. M., and Worsfold, D. J. (1972). *Macromolecules* **5(2)**, 229.
Sartori, G., Lammens, H., Stiffert, J., and Bernard, A. (1971). *J. Polym. Sci. Part B*, **9(8)**, 599.
Sawada, H. (1969). *J. Macromol. Sci. Rev. Macromol. Chem.* C3(2), 313.
Schultz, G. V., and Haborth, G. (1948). *Makromol. Chem.* **1**, 106.
Scott, G. E., and Senogles, E. (1973). *J. Macromol. Sci. Rev. Macromol. Chem.* C9(11), 49.
Sebenda, J. (1972). *J. Macromol. Sci.-Chem.* A6(6), 1145.
Shimomura, T., Tölle, K. J., Smid, J., and Szwarc, M. (1967a). *J. Am. Chem. Soc.* **89**, 796.
Shimomura, T., Smid, J., and Szwarc, M. (1967b). *J. Am. Chem. Soc.* **89**, 5743.
Sigwalt, P. (1969). *In* "Kinetics and Mechanisms of Polymerization" (K. C. Frisch and S. L. Reegen, eds.), Vol. 2, Ring-Opening Polymerization, Chapter 4. Dekker, New York.
Smid, J. (1969). *In* "Structure and Mechanism in Vinyl Polymerization" (T. Tsuruta and K. O'Driscoll, eds.). Dekker, New York.

Smith, D. G. (1967). *J. Appl. Chem.* **17**, 339.
Smith, W. V., and Ewart, R. H. (1948). *J. Chem. Phys.* **16**, 592.
Solomon, D. H. (1967). *J. Macromol. Sci.-Revs. Macromol. Chem.* **C1(1)**, 179.
Solomon, D. H. (1972). *In* "Kinetics and Mechanisms of Polymerization" (D. H. Solomon, ed.), Vol. 3, Step-Growth Polymerization, Chapter 1. Dekker, New York.
Solomon, D. H. (1975). *J. Macromol. Sci.-Chem.* **A9(1)**, 97.
Spach, G., Levy, M., and Szwarc, M. (1962). *J. Chem. Soc.* 355.
Stearns, R. S., and Forman, L. E. (1959). *J. Polym. Sci.* **41**, 381.
Stockmayer, W. H. (1953). *J. Polym. Sci.* **11**, 424.
Stockmayer, W. H. (1957). *J. Polym. Sci.* **24**, 314.
Szwarc, M. (1956). *Nature (London)* **178**, 1168.
Szwarc, M. (1968). *In* "Carbanions, Living Polymers and Electron Transfer Processes." Wiley (Interscience), New York.
Szwarc, M. (1975). *In* "Molecular Behavior and the Development of Polymeric Materials" (A. Ledwith and A. M. North, eds.), Chapter 1. Chapman and Hall, London.
Szwarc, M., Levy, M., and Milkovich, R. (1956). *J. Am. Chem. Soc.* **78**, 2656.
Tabata, Y. (1968). *Adv. Macromol. Chem.* **1**, 283.
Tabata, Y. (1969). *In* "Kinetics and Mechanisms of Polymerization," Vol. 1, Vinyl Polymerization, Pt. 2 (G. Ham, ed.), Chapter 7, Dekker, New York.
Tait, P. J. T. (1975). *Chemtech.* 688.
Tanaka, T., and Vogl, O. (1974). *J. Macromol. Sci.-Chem.* **A8(6)**, 1059.
Thompson, L. F. (1974). *Solid State Technol.* **17(7)**, 27.
Tidwell, P. W., and Mortimer, G. A. (1965). *J. Polym. Sci. Part A*, **3**, 369.
Trommsdorf, E., Köhle, H., and Lagally, P. (1948). *Makromol. Chem.* **1**, 169.
Tsukamoto, A., and Vogl, O. (1971). *Prog. Polym. Sci.* **3**.
Tsuruta, T. (1972). *J. Polym. Sci. D. Macromol. Rev.* **6**, 179.
Vanderhoff, J. W. (1969). *In* "Kinetics and Mechanisms of Polymerization" (G. Ham, ed.), Vol. 1, Vinyl Polymerization, Pt. 2, Chapter 1. Dekker, New York.
Vermal, C., Matheson, M., Lrach, S., and Muller, F. (1964). *J. Chem. Phys.* **61**, 596.
Vogl, O. (1975). *J. Macromol. Sci.-Chem.* **A9(5)**, 663.
Vollmert, B. (1973). *In* "Polymer Chemistry," Springer-Verlag, New York.
Walling, C. (1957). *In* "Free Radicals in Solution." Wiley, New York.
Weissermel, K. *et al.* (1975). *J. Polym. Sci. Polym. Symp.* **51**, 187.
White, D. M. (1960). *J. Am. Chem. Soc.* **82**, 5678.
Wilson, C. W., and Santee, E. R. (1965). *J. Polym. Sci. Part C*, **8**, 97.
Wilson, J. E. (1974). *In* "Radiation Chemistry of Monomers, Polymers and Plastics." Dekker, New York.
Winstein, S., and Robinson, G. C. (1958). *J. Am. Chem. Soc.* **80**, 169.
Witt, D. R. (1974). *In* "Reactivity Mechanism and Structure in Polymer Chemistry" (A. D. Jenkins and A. Ledwith, eds.). Wiley, New York.
Wood, D. W. (1975). *Plast. Eng.* **31(5)**, 51.
Worsfold, D. J., and Bywater, S. (1960). *Can. J. Chem.* **38**, 1891.
Worsfold, D. J., and Bywater, S. (1964). *Can. J. Chem.* **42**, 2884.
Yamazaki, N. (1969). *Fortsch. Hochpolym. Forsch.* **6(3)**, 377.
Yermakov, N., and Zakharov, V. (1975). *Adv. Catal.* **24**, 173.
Yezrielev, A. J., Brokhina, E. L., and Roskin, Y. S. (1969), *Vysokomol. Soedin.*, **A11**, 1670.
Ziegler, K., Holzkamp, E., Breil, H., and Martin, H. (1955). *Angew. Chem.* **67**, 541.
Zlamal, Z. (1969). *In* "Kinetics and Mechanisms of Polymerization" Vol. 1, Vinyl Polymerization, Pt. 2 (G. Ham, ed.), Chapter 6 Dekker, New York.

Chapter 3

MICROSTRUCTURE AND CHAIN CONFORMATION OF MACROMOLECULES

F. A. BOVEY AND T. K. KWEI

3.1. Introduction

We have seen in Chapter 1, Section 1.1, the types of structural isomerism of which polymer chains are capable—in particular the occurrence of (a) various types of stereochemical isomerism, (b) branching and cross-linking, and (c) monomer sequence isomerism in copolymers. We have also seen that polymer chains are usually capable of rotation about at least some of the main chain bonds and may thus assume a large number of different *conformations*. In this chapter we shall consider these matters in greater detail with respect to some aspects of their theoretical treatment and their experimental (principally spectroscopic) observation. Chain conformation is discussed further from a somewhat different viewpoint in Chapter 4.

3.2. Vibrational Spectroscopy

The spectroscopic method that has the longest history for the study of macromolecules is *infrared*. More recently applied and very closely related is *Raman* spectroscopy. (For more complete discussions, see General References at the end of the chapter). Both deal with relatively high frequency processes that involve variation of internuclear distances, i.e., molecular vibration. (Rotational and translational processes will not concern us in polymer spectra.) As a first approximation, let us imagine that these molecular vibrators can be considered as classical *harmonic oscillators*. For a diatomic molecule of unequal masses m_1 and m_2 connected by a a bond regarded as a spring with a force constant k, the frequency of vibration expressed in wavenumbers (i.e., cm^{-1} or reciprocal wavelengths) is given by

$$v = (1/2\pi c)(k/m_r)^{1/2} \tag{3.1}$$

where c is the velocity of light and m_r is the *reduced mass*, given by

$$m_r = m_1 m_2/(m_1 + m_2) \quad \text{or} \quad \simeq m_1 \quad \text{if} \quad m_2 \gg m_1 \tag{3.2}$$

Thus a small mass, such as a hydrogen or deuterium atom, vibrating against a larger one, such as a carbon or chlorine atom, will have essentially the frequency characteristic of the smaller mass. Most molecular vibrational frequencies of interest for polymer characterization will be in the range of 3000 to about $650\,cm^{-1}$, or in wavelength 2.5 to 15 μm.

Actual molecular vibrators differ from the classical oscillator in two respects. First, the total energy E, expressed in terms of the nuclear displacement x, cannot have any arbitrary value but is confined to discrete energy levels expressed in terms of integral quantum numbers n:

$$E = \left(n + \frac{1}{2}\right) \frac{h}{2\pi} \left(\frac{k}{m_r}\right)^{1/2} = (n + \tfrac{1}{2})hcv \tag{3.3}$$

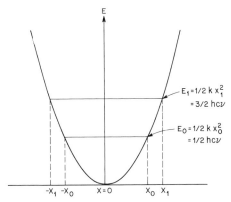

FIG. 3.1. The potential energy of a harmonic oscillator as a function of displacement from equilibrium.

the transition energies being given by $hc\nu$, as shown in Fig. 3.1. Only transitions between adjacent levels are allowed in a quantum mechanical harmonic oscillator. Second, the form of the vibrational potential energy is not parabolic, as in Fig. 3.1, for this would imply infinite bond strength; the true form is *anharmonic*, as shown in Fig. 3.2, the vibrational levels being more closely spaced as n increases. An important consequence of the departure from harmonicity is that the selection rules are relaxed, permitting transitions to levels higher than the next immediately higher one. Transitions from $n = 0$ to $n = 2$ correspond to the appearance of weak but observable *first overtone* bands having slightly less than twice the frequency of the *fundamental* band.

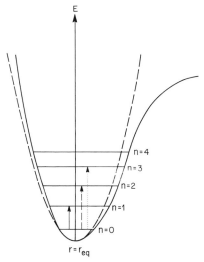

FIG. 3.2. The potential energy and energy levels of an anharmonic oscillator.

3.2.1. REQUIREMENTS FOR INFRARED AND RAMAN TRANSITIONS

The appearance of a vibrational absorption band in the infrared spectral region requires that the impinging radiation supply a quantum of energy ΔE just equal to that of the vibrational transition hcv. It is also necessary that the atomic vibration be accompanied by a *change in the electric dipole moment of the system*, thus producing an alternating electric field of the same frequency as the radiation field. This condition is often not met, as for example in the vibrations of homopolar bonds such as the carbon–carbon bonds in paraffinic polymers.

The Raman spectrum can give much the same information as the infrared spectrum, but they are in general not identical and can usefully complement each other. In Fig. 3.3 we see at the left the *Rayleigh scattering* process, in which the molecule momentarily absorbs a photon, usually of visible light, and then reradiates it to the ground vibrational state without loss of energy. (Such scattering can furnish a means of measuring the size of the polymer molecule in solution; this is discussed in Chapter 4, Section 4.5.3.) However, the excited molecule may also return to a higher vibrational state—the next highest in Fig. 3.3 (center)—and the reradiated photon will then be of lower frequency by $\Delta \bar{v}$. In a complex molecule there will be many such states, and so the Raman spectrum, like the infrared spectrum, will appear as a number of lines, called the *Stokes* lines, much weaker than the exciting radiation and appearing shifted to longer wavelength by a few hundred to 2–3 thousand wavenumbers.

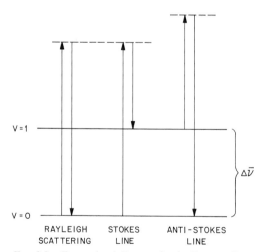

FIG. 3.3.   Energy level diagram for the Raman effect.

The irradiating light may also excite the molecule from a higher vibrational state ν_1, the emitted photon being then of *higher* frequency than the exciting radiation by $\Delta\bar{\nu}$, and appearing on the short wavelength side of the central line. Such *anti-Stokes* line will be weaker in intensity than the Stokes lines by the Boltzmann factor expressing the population of state ν_1. The Stokes lines are usually employed for evident practical reasons.

For Raman emission to occur, the *polarizability* of the bond must change during the vibration. For molecules that have a center of symmetry, an exclusion principle applies that states that transitions which are infrared active will not be Raman active, and *vice versa*. Polymer molecules generally lack the appropriate center of symmetry and most transitions appear in both types of spectra; an important exception are paraffinic carbon–carbon vibrations, which are inactive in the infrared spectrum, as we have seen, but are active in the Raman spectrum.

3.2.2. TYPES OF VIBRATIONAL BANDS

For polymer molecules (and nonlinear molecules in general) containing N atoms, there will be $3N - 6$ fundamental vibrations. In a complex polymer molecule, the number of transitions might be expected to be too great to deal with, but fortunately this does not happen because great numbers of them are degenerate, allowing us to recognize vibrational bands specific to particular types of bonds and functional groups. These appear in the high frequency region of the vibrational spectrum at similar positions regardless of the specific compound in which they occur. At the low frequency end of the spectrum, the vibrational bands are more characteristic of the molecule as a whole; this region is commonly called the "fingerprint" region, since detailed comparison here usually enables specific identification to be made.

In the region near 3000 cm^{-1} appear the *C–H bond stretching vibrations* (Fig. 3.4), which may be asymmetric or symmetric, as illustrated in Fig. 3.5. These occur in nearly all polymer spectra and so are not structurally diagnostic, although useful in a more fundamental sense. At lower frequencies, corresponding to smaller force constants, are the deformation vibrations involving valence angle bending or *scissoring*, giving a large band near 1500 cm^{-1}; *wagging* and *twisting* near 1300 cm^{-1}; and finally *rocking* deformations, appearing at the low energy end of the usual spectrum. (At still lower frequencies are torsion and skeletal as well as intermolecular and lattice vibrations which we shall not discuss here.)

In Fig. 3.4 are also shown a number of other characteristic vibrational bands and their frequency ranges. We may take particular note of the *carbonyl stretch* band near 1700 cm^{-1}, the *C=C stretch* band near 1600 cm^{-1}, and the *olefinic C–H bending* bands between 900 and 1000 cm^{-1}.

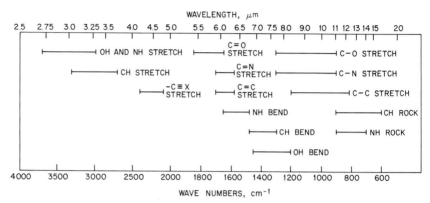

FIG. 3.4.   Infrared bands of interest in polymers.

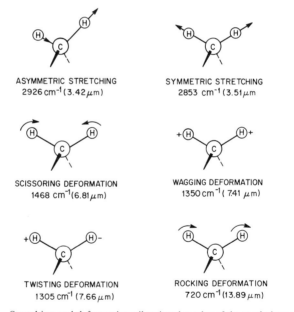

FIG. 3.5.   Stretching and deformation vibrational modes of the methylene group.

3.2.3.   INSTRUMENTATION AND MEASUREMENT

In Fig. 3.6 is a schematic diagram of a double beam infrared spectrophotometer. The light source S may be a *globar* (a silicon carbide rod), a *nichrome
wire*, or a *Nernst glower*, all electrically heated to at least 1100–1200°C in
order for the emitted blackbody radiation to cover the desired frequency

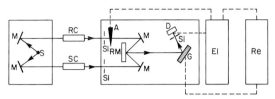

FIG. 3.6. Schematic diagram of a double-beam infrared spectrometer. S = source of radiation; M = mirror; SC = sample cell; RC = reference cell; A = attenuator; RM = rotating mirror; Sl = slit; G = grating or prism; D = detector; El = electronic amplifier; Re = recorder; —— = optical path; --- = electrical or mechanical connection (from Kossler, 1967).

range adequately. The *monochromator* G is a rock salt prism or a grating, the wavelength range being swept by rotating the mirror RM. Entrance and exit slits are provided to make the beam reaching the *detector* as monochromatic as possible. The latter may be a Golay cell or a bolometer. The double beam design permits one to blank out absorptions from solvents (if used), from moisture and CO_2 in the air, and in general to provide a flat, stable base line. The sample and reference beams are compared by moving the comblike attenuator in the reference beam, and recording its movement, until a null is obtained at the detector.

The requirements of a Raman spectrometer are in some respects similar to those of an infrared spectrophotometer, the principal differences being that one uses a monochromatic laser source (typically a helium–neon laser) and that instead of absorption one measures the light scattered at right angles to the sample. A schematic representation is shown in Fig. 3.7.

Infrared samples are commonly examined in the solid state either as films or as mulls in Nujol or fluorolube. They may also be ground up with KBr, which is transparent to infrared, and observed as pellets. Solutions

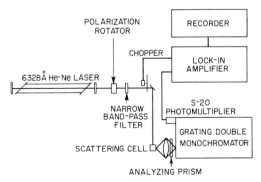

FIG. 3.7. Schematic diagram of a laser-excited Raman spectrometer (from R. F. Schaufele, Macromol. Rev. 4, 67 (1970)).

in CS_2 or CCl_4 are occasionally also used, but for polymers films are most commonly employed. The initial radiation intensity falling on the sample, I_0, will be attenuated in proportion to the path length b and, for solutions, to the concentration c; thus

$$I = I_0 e^{-a'bc} \tag{3.3a}$$

where a' is the *extinction coefficient* or *absorbtivity* characteristic of the band observed. For polymer films, c, if expressed in gm cm^{-3}, will be approximately unity. We then have

$$2.303 \log_{10}(I_0/I) = -a'bc \tag{3.3b}$$

The quantity $\log_{10}(I_0/I)$ is the *absorbance* or *optical density*, and is the ordinate on the left side of typical infrared spectra.

In Fig. 3.8 are shown the infrared spectra of (a) linear and (b) branched polyethylenes (Luongo, 1975). The principal bands are labeled in accordance

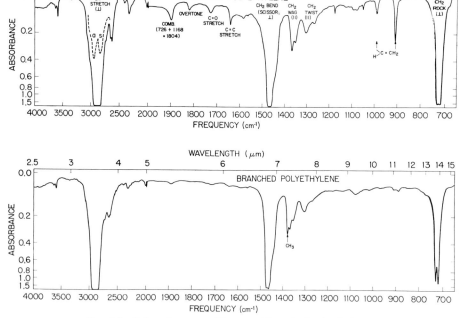

FIG. 3.8.   Infrared spectra of linear and branched polyethylenes.

with the vibrational assignments already discussed. The intense C–H stretch region is twofold owing to the splitting of symmetric and asymmetric bands, as shown in the inset dashed spectrum in (a). Bands for carbonyl groups ($1725\,cm^{-1}$) resulting from slight oxidation (see Chapter 7, Section 7.3) and terminal vinyl groups are observable in the spectrum of the linear polyethylene. If accurate values of a' (from model compounds) and of b are established, the content of these groups can be measured quantitatively. More commonly bands are reported qualitatively as vw (very weak), w (weak), m (medium), s (strong), and vs (very strong).

The measurement of the branch content of polyethylene from the infrared spectrum in Fig. 3.8b is discussed in Section 3.6.2.

The Raman spectra of polymers resemble the infrared spectra. In crystalline paraffinic hydrocarbons, bands appear below $800\,cm^{-1}$ corresponding to accordionlike vibrations of the chains, which are in a planar zigzag conformation. In polyethylene chains, similar bands may be observed for motions of chain folds (Chapter 5).

Vibrational bands that are Raman active but infrared inactive are those corresponding to C–C stretch; in crystalline n-$C_{44}H_{90}$ (and in polyethylene) these appear at 1140 and $1060\,cm^{-1}$, as shown in Fig. 3.9, which shows the Raman Stokes lines (Cheng and Ginsberg, 1975). In general, while infrared is more useful for identifying polar substituents on a polymer chain, Raman spectroscopy is more powerful in characterizing the homonuclear polymer backbone. It also can be used for very small samples, particularly those that cannot conveniently be made into films or mulls, since it is only necessary to fill the focused laser beam, which may be only $10\,\mu m$ in diameter.

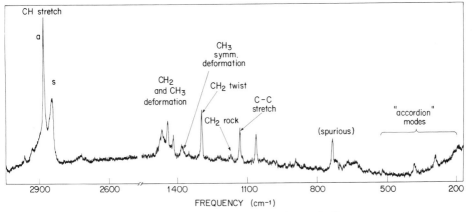

FIG. 3.9. Raman spectrum of crystalline n–$C_{44}H_{90}$.

More advanced techniques are available for such special problems as the identification and measurement of very thin films, as for example in studying oxidized surface groups, surface contaminants, and laminated structures. Among these are *attenuated total reflection* (ATR) and *Fourier transform* infrared spectroscopy. The latter, because it employs no grating or prism to disperse the light, allows the full strength of the source to reach the detector, the spectrum being generated by interferometry. The result is far greater sensitivity than that of conventional instruments.

An important technique, long used for the study of solid polymer samples, is the measurement of *infrared dichroism*. The radiation is polarized by a small rotatable grid of fine parallel wires placed in the beam. If the transition moment of the infrared band, i.e., the movement of the electrons accompanying the vibration, is parallel (or has a component parallel) to the electric vector of the advancing wave, absorption will be strong; if perpendicular, absorption will be weak. In Fig. 3.10 is shown in a schematic fashion the absorption of a polarized wave train when the planar zigzag chain of polyethylene is oriented in such a manner that the methylene rocking vibration of the planar zigzag chain is (a) perpendicular and (b) parallel to the polarization of the beam (Luongo, 1975). In crystalline polyethylene the rocking band at 725 cm^{-1} is actually split into two bands corresponding to the two

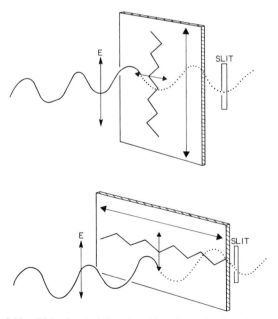

FIG. 3.10.   Dichroism in infrared rocking absorption in polyethylene.

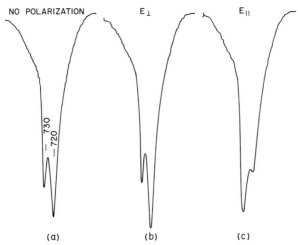

FIG. 3.11. Dichroism in the $720/730 \, \text{cm}^{-1}$ rocking doublet of oriented polyethylene (Luongo, 1975).

possible phase relationships of its motion to that of the neighboring chains. When the beam is unpolarized, the rocking band for an extruded sample of polyethylene is as shown in (a) in Fig. 3.11. The *dichroic ratio* is defined as $I_\perp/I_{||}$, the ratio of the absorption when the direction of orientation of the sample is perpendicular to the beam polarization to that when it is parallel. When polyethylene is drawn or extruded the chains tend to become oriented in the draw direction. The dichroism of the rocking band does not show the idealized behavior suggested by Fig. 3.10, but the $730 \, \text{cm}^{-1}$ band does show a dichroic ratio *less* than one, and is termed a parallel or π band; whereas the $720 \, \text{cm}^{-1}$ band shows a dichroic ratio markedly *greater* than one, as expected from Fig. 3.10, and is termed a perpendicular or σ band. Dichroism is observed in nearly all the bands of oriented polymer samples and gives very useful information concerning chain conformation and morphology.

3.3. Nuclear Magnetic Resonance Spectroscopy

3.3.1. THE BASIC PHENOMENON

The phenomenon of nuclear magnetic resonance depends on the fact that some nuclei possess *spin* or angular momentum. (See General References under Nuclear Magnetic Resonance Spectroscopy at the end of the chapter.) Such nuclei are described by spin quantum numbers I (usually referred to simply as "the spin") having integral or half-integral values. When placed in a magnetic field of strength H_0, such nuclei occupy quantized magnetic energy levels the number of which is equal to $2I + 1$ and the relative populations of which are normally given by a Boltzmann distribution. We shall deal only with nuclei for which $I = 1/2$: the proton (^{1}H) and the ^{13}C nucleus. These nuclei have two magnetic energy states separated by

$$\Delta E = h v_0 = 2\mu H_0 \tag{3.4}$$

in which μ is the magnetic moment of the nucleus. Transitions between these energy levels can be made to occur by means of a resonant radio frequency (rf) field of frequency v_0. In Fig. 3.12 Eq. (3.4) is plotted for the proton; the resonant rf frequency is shown for six different magnetic fields employed in current spectometers. (The fields from 52 to 85 kG require superconducting solenoid magnets.) For the ^{13}C nucleus, which has a magnetic moment one-fourth that of the proton, the resonant frequencies will be one-fourth of those indicated.

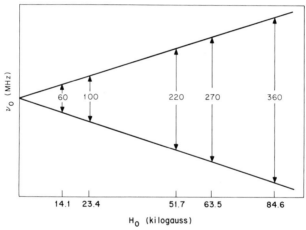

FIG. 3.12. The splitting of magnetic energy levels of protons, expressed as resonance frequency v_0 in varied magnetic field H_0, expressed in kilogauss (kG).

In Fig. 3.13 is shown a basic schematic diagram of a conventional continuous wave (cw) spectrometer. The sample is contained in the tube A to which the rf field is applied by means of coil B wrapped about the sample. A magnetic field is applied in a direction perpendicular to the axis of the coil and, by means of the sweep coils C, is slowly increased ("swept") until resonance occurs. At this point energy is absorbed from the rf field and this absorption may be detected and recorded. Alternatively, the nuclear moments as they are turned over at resonance induce a voltage in a second

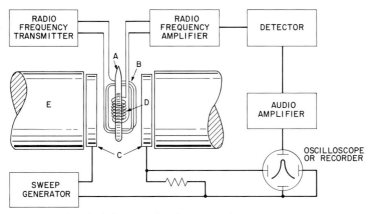

F<small>IG</small>. 3.13.   Block diagram of nuclear magnetic resonance spectrometer.

coil D, and this voltage is amplified and recorded. (It is also equally feasible to hold H_0 constant and develop the spectrum by sweeping v_0.)

3.3.2.   T<small>HE</small> C<small>HEMICAL</small> S<small>HIFT</small>

The value of NMR to the chemist lies in the fact that at any particular value of v_0, all nuclei of a given species—say, all protons—do not resonate at exactly the same value of H_0 (or *vice versa*). Equation (3.4) requires a modification, which is small in terms of relative magnitudes but very significant, because resonance actually occurs at slightly different values of H_0 for each type of proton, depending upon its chemical binding and position in the molecule. The cause of this variation is the cloud of electrons about each nucleus, which tends to shield the nucleus against the magnetic field, thus requiring a slightly higher H_0 to achieve resonance than for a bare proton. Protons attached to or near electronegative groups such as OR, OH, OCOR, CO_2R, and halogens experience a lower density of shielding electrons and resonate at lower H_0 (or at a higher value of v_0 if H_0 is held constant.) Protons removed from such groups, as in hydrocarbon chains, resonate at higher H_0. Similar structural relationships are observed for ^{13}C nuclei. These variations are termed *chemical shifts*, and are commonly expressed in relation to tetramethylsilane (TMS) as the zero of reference. The total range of variation of proton chemical shifts in organic compounds is only of the order of 10 ppm. For ^{13}C nuclei (in common with all other magnetic nuclei), it is much greater—over 600 ppm; this is a principal reason for the intense interest in the study of polymers by ^{13}C NMR despite certain inherent observing difficulties (*vid. inf.*). For any nucleus, the separation of chemically shifted resonances, expressed in hertz, is proportional to

H_0. An important advantage of high field magnets is the greater resolution of peaks and finer discrimination of structural features of polymer chains that they make possible.

3.3.3. Nuclear Coupling

Another important parameter in NMR spectra is *nuclear coupling*. Magnetic nuclei may transmit information to each other concerning their spin states through the intervening covalent bonds. If a nucleus has n sufficiently close, equivalently coupled neighbors, its resonance will be split into $n + 1$ peaks, corresponding to the $n + 1$ spin states of the neighboring group of spins. Intensities are given by simple statistical considerations and are therefore proportional to the coefficients of the binomial expansion. Thus, one neighboring spin splits the observed resonance to a doublet, two produce a $1:3:1$ triplet, three a $1:3:3:1$ quartet, and so on. The strength of the coupling is denoted by J and is expressed in hertz. The coupling of protons on adjacent saturated carbon atoms, termed *vicinal* coupling, varies with the dihedral angle Φ, (a) *trans* couplings, where $\Phi = 180°$, being substantially larger than (b) *gauche* couplings, where $\Phi = 60°$.

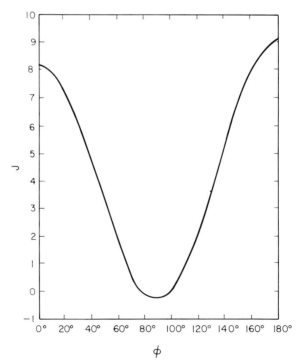

FIG. 3.14. J coupling of two protons, expressed in hertz, as a function of dihedral angle ϕ.

(a)   (b)

J_{gauche} is typically 2–4 Hz and J_{trans} ranges from 8 to 13 Hz. These considerations are of great importance in studying the conformations of polymer chains (Section 3.7). The dependence of J on Φ is generally well described by the Karplus relationship, shown in Fig. 3.14, although the magnitude of the coupling depends somewhat on the nature of the substituents on the bonded carbon atoms.

The chemical shifts and J couplings of nuclei in polymers do not essentially differ from those of small molecules of analagous structure. The spectrum of ethyl orthoformate, $CH(OCH_2CH_3)_3$, Fig. 3.15a, shows a single peak for the formyl proton and a quartet and triplet for the CH_2 and CH_3 protons, respectively, of the ethyl group, in accordance with the rules just described. The spectrum of poly(vinyl ethyl ether) (Fig. 3.15b),

$$\mathrm{-\!\!\left[CH_\alpha CH_{\beta 2}\right]_{\!n}}$$
$$\mathrm{OCH_2CH_3}$$

shows corresponding resonances and, in addition, those of the β protons appearing at 1.5 and of the α protons under the CH_2 quartet. Chemical shifts are expressed in ppm on the δ scale, on which tetramethylsilane appears at 0.0. The broadening of the lines in (b) is in part due to the slower motions of the large molecule in solution. [In the solid state, where motion is very slow, NMR lines are commonly so broad as to obscure all structural information (see Chapter 6, Section 6.2)]. The broadening is also in part due to structural complexities to be dealt with in Section 3.4.

The observation of carbon spectra presents special problems because the magnetic isotope ^{13}C has an abundance of only 1.1% and also, as we have seen, a relatively small magnetic moment—about $\frac{1}{4}$ that of the proton. The resulting decrease in observing sensitivity can be compensated by use of the *pulsed Fourier transform* technique, combined with spectrum accumulation. In this method, the rf field is supplied as periodic pulse of a few microseconds duration and of much greater power than used in conventional continuous wave spectroscopy. The pulse sets all the carbon nuclei over the usual chemical shift range into resonance at the same instant and obviates the need for sweeping the field or frequency, thus saving a factor of at least 100 in observing time. The time-domain spectra, appearing as interferograms following each pulse, are added in a computer; after enough are

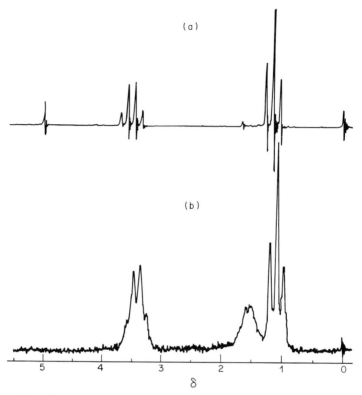

FIG. 3.15. The NMR spectrum of (a) ethyl orthoformate $HC(OCH_2CH_3)_3$ and (b) poly (vinyl ethyl ether), both observed at 60 MHz in approximately 15% solutions in carbon tetrachloride.

accumulated for an acceptable signal-to-noise ratio, the summed spectra are transformed to the frequency domain, i.e., to the usual form.

Nuclear coupling between ^{13}C nuclei and directly bonded protons is strong (125–250 Hz). The resulting multiplicity is often helpful in carbon assignments but is usually abolished by *double resonance*, i.e., by providing a second rf field tuned to the proton resonance frequency. The resulting multiplet collapse and accompanying *nuclear Overhauser enhancement* (see General References, II: Levy and Nelson, 1972, and Stothers, 1972) give a striking improvement in signal-to-noise ratio.

3.4. Stereochemical Configuration

In Fig. 3.16 are shown the proton spectra of poly(methyl methacrylate) prepared (a) with a free radical initiator and (b) with phenyl magnesium bromide in toluene, an anionic initiator (Bovey and Tiers, 1960). We have

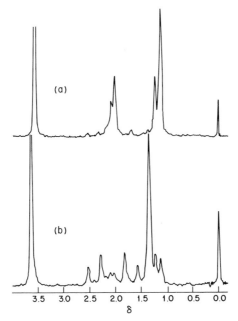

FIG. 3.16. 60 MHz NMR spectra of (a) predominantly syndiotactic and (b) predominantly isotactic samples of poly(methyl methacrylate).

seen (Chapter 2, Sections 2.2.2 and 2.2.4) that the nature of the initiator may profoundly influence the stereochemical configuration of many vinyl polymer chains. This is qualitatively evident from the marked differences between spectra (a) and (b). These are more than "fingerprint" differences and may be interpreted in detail. Let us consider the chain in terms of sequences of two monomer units or *dyads*. It is apparent that there are two possible types of dyads and that these have different symmetry properties. The syndiotactic or racemic[†] dyad (a), abbreviated *r*, has a twofold axis of symmetry, and consequently the two methylene protons (or β protons) are in equivalent

environments. They therefore have the same chemical shift and appear as

_____

[†] The terms *meso* and *racemic* are based on the standard nomenclature of analogous small molecules. They express the *relative* handedness of adjacent centers, the only experimentally observable stereochemical property of long vinyl polymer chains.

a singlet resonance despite the fairly strong (~ 15 Hz) *geminal* (this term refers to two protons on the same carbon) coupling between them. (For fundamental quantum mechanical reasons, the coupling of equivalent nuclei does not cause a splitting of their resonance.) The isotactic or *meso* dyad (b), abbreviated *m*, has no symmetry axis and so the two methylene protons are nonequivalent and should in general give different chemical shifts. When there is no vicinal coupling to α protons, as is the case in poly(methyl methacrylate), the methylene resonance of a syndiotactic chain is therefore expected to appear as a singlet, whereas that of an isotactic chain should appear as a quartet, i.e., doublets for H_A and H_B, the spacing of which equals the geminal coupling, J_{gem}. We see in Fig. 3.16b that the methylene spectrum is indeed such a quartet, while in Fig. 3.16a it is mainly a singlet. Polymer (a) is predominantly syndiotactic, while polymer (b) is predominantly isotactic. The additional resonances observable in the methylene region in each case arise from the fact that neither polymer (as is very commonly the case) is entirely stereoregular.

We thus see that proton NMR can provide absolute stereochemical information concerning vinyl polymer chains, without recourse to x ray or other methods. The information provided with regard to poly(methyl methacrylate) is entirely in accord with what has already been discussed in Chapter 2 (Section 2.2) concerning the influence of initiator type on chain stereoregularity. Somewhat more detailed, but not absolute, information can be gained from the α-methyl resonances, near 1.2δ. (The ester methyl resonance at 3.6δ is not informative.) In both spectra, we note three peaks appearing in the same positions but with greatly different intensities. These correspond to the α-methyl groups in the central monomer unit of the three possible *triad* sequences: (a) isotactic, (b) syndiotactic, and (c) *heterotactic*. These may also be appropriately and more simply designated in the *m* and *r* terminology, as indicated:

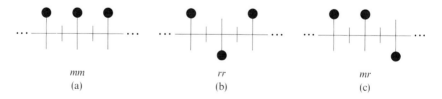

| mm | rr | mr |
| (a) | (b) | (c) |

Measurement of the relative intensities of the *mm*, *mr*, and *rr* α-methyl peaks, which, from what has already been said, must appear from left to right in both spectra in this order, gives a valid statistical representation of the structure of each polymer.

From the triad data, we may also gain considerable insight into the mechanism of polymerization. Let us designate by P_m the probability that the polymer chain will add a monomer unit to give the same configuration as that of the last unit as its growing end, i.e., that an m dyad will be generated. We assume that P_m is independent of the configuration of the growing polymer chain. In these terms, the generation of the chain is a Bernoulli-trial process. It is like reaching into a large jar of balls marked m and r and withdrawing a ball at random. The proportion of m balls in the jar is P_m. The probability of an r dyad is $1 - P_m$. Since two monomer additions are required to form a triad sequence, it can be readily seen that the probabilities of forming mm, mr, and rr triads are given by

$$[mm] = P_m^2 \tag{3.5}$$

$$[mr] = 2P_m(1 - P_m) \tag{3.6}$$

$$[rr] = (1 - P_m)^2 \tag{3.7}$$

A plot of these relationships is shown in Fig. 3.17. It will be noted that the proportion of mr (heterotactic) units rises to a maximum at $P_m = 0.5$, corresponding to a strictly random or atactic configuration, for which the proportion $[mm]:[mr]:[rr]$ will be $1:2:1$. For any given polymer, if Bernoullian, the mm, mr, and rr sequence frequencies, as estimated from the relative areas of the α-methyl resonances, would lie on a single vertical line in Fig. 3.17, corresponding to a single value of P_m. Spectrum (a) in Fig. 3.17 corresponds to these simple statistics, P_m being 0.24 ± 0.01. The polymer corresponding to (b) does not. The propagation statistics are in this case more complex and can be interpreted to indicate that the probability of isotactic propagation is not independent of the stereochemical configuration of the propagating chain. Free radical and cationic propagations always give predominantly syndiotactic chains; anionic initiators may also do so if strongly complexing ether solvents such as dioxane or glycol dimethyl ether are employed rather than hydrocarbon solvents as in Fig. 3.16b.

Vibrational spectra also reveal stereochemical differences. In Fig. 3.18 are shown infrared spectra of films of the same predominantly syndiotactic (top) and isotactic (bottom) methyl methacrylate polymers as in Fig. 3.17. It is evident that, in addition to other smaller differences, there is a conspicuous band at 1060 cm^{-1} in the syndiotactic polymer spectrum that is entirely absent in that of the isotactic polymer. This band can serve as a quick measure of the chain stereochemistry, but in general infrared is not as discriminating nor as quantitative as NMR.

Longer configurational sequences can be observed by NMR, particularly using larger magnetic fields or ^{13}C spectroscopy or both. With respect to β-methylene groups, one may expect to resolve tetrad sequences of monomer units, appearing as a fine structure on the m and r dyad resonances; α-carbons or their substituents may reveal pentad sequences as a fine structure on the triad peaks. In Table 3.1 are shown longer sequences up through pentad,

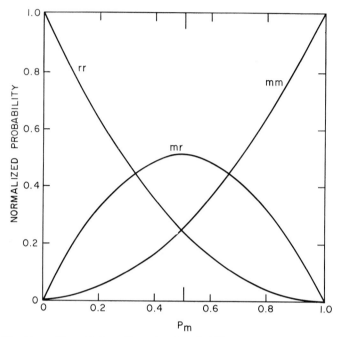

Fig. 3.17. Bernoullian triad configurational probability vs. P_m, the probability of isotactic placement.

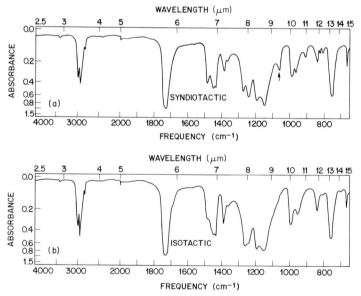

Fig. 3.18. Infrared spectra of predominantly (a) syndiotactic and (b) isotactic films of
polymethyl methacrylate.

TABLE 3.1

		α Substituent				β-CH$_2$		
	Designation	Projection	Bernoullian probability		Designation	Projection	Bernoullian probability
Triad	Isotactic, *mm* (*i*)		P_m^2	Dyad	*meso, m*		P_m
	Heterotactic, *mr* (*h*)		$2P_m(1 - P_m)$		*racemic, r*		$(1 - P_m)$
	Syndiotactic, *rr* (*s*)		$(1 - P_m)^2$				
Pentad	*mmmm* (isotactic)		P_m^4	Tetrad	*mmmm*		P_m^3
	mmmr		$2P_m^3(1 - P_m)$		*mmr*		$2P_m^2(1 - P_m)$
	rmmr		$P_m^2(1 - P_m)^2$		*rmr*		$P_m(1 - P_m)^2$
	mmrm		$2P_m^3(1 - P_m)$		*mrm*		$P_m^2(1 - P_m)$
	mmrr		$2P_m^2(1 - P_m)^2$		*rrm*		$2P_m(1 - P_m)^2$
	rmrm		$2P_m^2(1 - P_m)^2$		*rrr*		$(1 - P_m)^3$
	rmrr (heterotactic)		$2P_m(1 - P_m)^3$				
	rrmr		$P_m^2(1 - P_m)^2$				
	mrrm		$2P_m^2(1 - P_m)^2$				
	rrrm		$2P_m(1 - P_m)^3$				
	rrrr (syndiotactic)		$(1 - P_m)^4$				

together with their designations in terms of m and r and their Bernoullian probabilities. It should be pointed out that the written *direction* of the sequences is entirely arbitrary and therefore immaterial. Thus, for example, *mmr* cannot be observationally distinguished from *rmm*, nor *rmrr* from *rrmr* (It is also necessary to give such unsymmetrical sequences double statistical weights, as we have already seen in the case of heterotactic triad sequences.) Figure 3.19 shows the 68 MHz ^{13}C spectrum of polyvinyl chloride, prepared with a free radical initiator. The α-carbon pentad and β-carbon tetrad resonance are found to have Bernoullian intensities, corresponding to a P_m of 0.43. This is characteristic of this polymer. Both poly(vinyl chloride) and poly(methyl methacrylate) become somewhat more syndiotactic as the temperature of free radical polymerization is decreased, although the effect is not large. For many other monomers, e.g., styrene, vinyl acetate, acrylonitrile, and acrylate esters, the temperature coefficient of the stereochemical configuration is essentially zero.

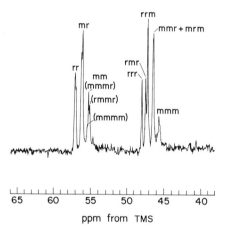

FIG. 3.19.   25 MHz ^{13}C spectrum of poly(vinyl chloride), 20% (w/v) in 1,2,4-trichlorobenzene at 140° (Schilling, 1974).

In Fig. 3.20 are shown the 25 MHz ^{13}C spectra of polypropylene prepared with (a) an isospecific Ziegler–Natta catalyst (TiCl$_3$·Al(C$_2$H$_5$)$_2$Cl); (c) a syndiospecific catalyst (VCl$_4$ · Al(C$_2$H$_5$)$_2$Cl); (b) is the spectrum of an atactic material. Polymer (b) was extracted with low boiling hydrocarbon solvents from the product of the isospecific polymerization leading to (a), the major fraction of which is the isotactic polymer. All carbon resonances are sensitive to configuration. This is particularly clear in spectrum (b), in which the methyl resonance is split into subpeaks corresponding to

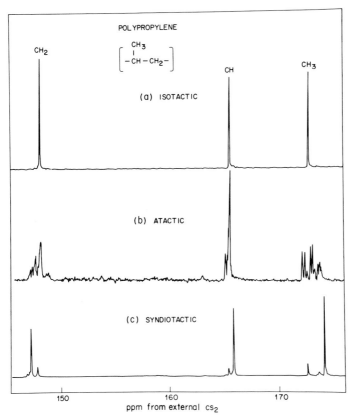

FIG. 3.20.   25 MHz ^{13}C spectra of three preparations of polypropylene: isotactic, atactic, and syndiotactic, observed as 20% (w/v) solutions in 1,2,4-trichlorobenzene at 140°.

nine of the ten possible (Table 3.1) pentad sequences. All of the polymers exhibit configurational statistics more complex than Bernoullian; this is generally true of chains generated by coordination catalysts.

3.5.   Geometrical Isomerism

Isomerism in diene polymer chains (Chapter 1, Section 1.3) can be readily detected and measured by both vibrational and NMR spectroscopy. In Fig. 3.21 (Pasteur, 1975) are shown the infrared spectra of poly(1,4-*cis*-butadiene), poly(1,4-*trans*-butadiene), and poly(1,2-butadiene), prepared with specific catalysts (Chapter 2, Section 2.2.5.6). These spectra cover only the 700–1200 cm^{-1} CH out-of-plane bending vibration region, which is best

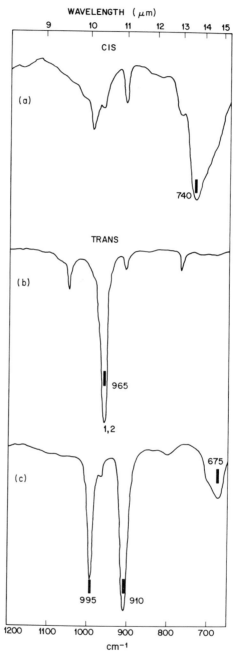

FIG. 3.21. Out-of-plane bending bands of (a) poly(*cis*-1,4-butadiene), (b) poly(*trans*-1,4-butadiene), and (c) poly(1,2-butadiene) (Pasteur, 1975).

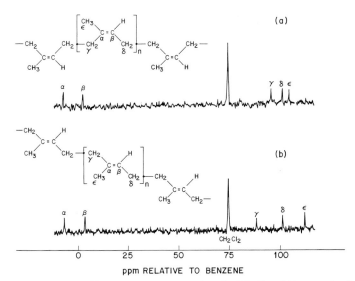

FIG. 3.22.   15.08 MHz ^{13}C spectra of 10% solutions in CH_2Cl_2 of (a) natural poly(*cis*-1,4-isoprene) and (b) natural poly(*trans*-1,4-isoprene) (balata). Benzene serves as the zero of reference (Duch and Grant, 1970).

adapted for the analysis of these polymers. The cis polymer shows a broad, strong band at $740 \, cm^{-1}$, whereas in the trans polymer the corresponding band appears at $965 \, cm^{-1}$ and is narrower. Poly(1,2-butadiene) shows the vinyl CH ($995 \, cm^{-1}$) and vinyl CH_2 ($910 \, cm^{-1}$) bending vibrations that we have already seen as a minor feature of the linear polyethylene spectrum (Fig. 3.8a). The spectra show that none of the three polymers is entirely regular; the trans polymer contains some 1,2 structures and the cis polymer contains both 1,2 and trans, while the 1,2 polymer exhibits a cis band at $\sim 670 \, cm^{-1}$. The spectra can yield quantitative analysis of the polymers by careful intercomparison even if extinction coefficients are not established.

In Fig. 3.22 (Duch, and Grant, 1970) are shown the ^{13}C spectra of (a) natural rubber, poly(*cis*-1,4-isoprene); and (b) balata, poly(*trans*-1,4-isoprene). The C_1 and methyl carbons are very sensitive to isomerism at the double bonds. Proton NMR may also be employed analytically to distinguish cis and trans structures and both proton and ^{13}C NMR can distinguish 1,2 and 3,4 units as well.

3.6.   Copolymer Sequences

We have seen (Chapter 2, Section 2.2.6) that the reactivity ratios r_1 and r_2 in the random copolymerization of two monomers may be obtained by determining the monomer ratio in the copolymer as a function of the monomer ratio in the feed and applying the copolymerization equation (Eq. (2.149)). The same treatment that successfully predicts overall copolymer composition also predicts monomer sequence lengths (Alfrey *et al.*, 1952). Comonomer sequences, ignoring stereochemical configuration, may be expressed as follows:

Dyads: m_1m_1 m_1m_2 (or m_2m_1) m_2m_2

Triads: $m_1m_1m_1$ $m_2m_2m_2$
        $m_1m_1m_2$(or $m_2m_1m_1$) $m_1m_2m_2$(or $m_2m_2m_1$)
        $m_2m_1m_2$ $m_1m_2m_1$

Tetrads: $m_1m_1m_1m_1$ $m_1m_1m_2m_1(m_1m_2m_1m_1)$ $m_2m_2m_2m_2$
       $m_1m_1m_1m_2(m_2m_2m_1m_1)$ $m_1m_1m_2m_2(m_2m_2m_1m_1)$ $m_2m_2m_2m_1(m_1m_2m_2m_2)$
                         $m_2m_1m_2m_1(m_1m_2m_1m_2)$
       $m_2m_1m_1m_2$ $m_2m_1m_2m_2(m_2m_2m_1m_2)$ $m_1m_2m_2m_1$

The dyad probabilities (i.e., frequencies of occurrence) are given by

$$[m_1m_1] = F_1P_{11} \tag{3.8}$$

$$[m_1m_2] \quad (\text{or} \quad [m_2m_1]) = 2F_1(1 - P_{11}) \tag{3.9}$$

$$[m_2m_2] = F_2P_{22} \tag{3.10}$$

Triad probabilities are given by

$$[m_1m_1m_1] = F_1P_{11}^2 \tag{3.11}$$

$$[m_1m_1m_2] \quad (\text{or} \quad [m_2m_1m_1]) = 2F_1P_{11}(1 - P_{11}) \tag{3.12}$$

$$[m_2m_1m_2] = F_2(1 - P_{22})(1 - P_{11}) \tag{3.13}$$

In these relationships, F_1 and F_2 are the overall mole fractions of m_1 and m_2, respectively, in the polymer (Chapter 2, Section 2.2.6), and:

$$P_{11} = \frac{r_1 f_1}{1 - f_1(1 - r_1)} \tag{3.14}$$

$$P_{22} = \frac{r_2 f_2}{1 - f_2(1 - r_2)} \tag{3.15}$$

P_{11} expresses the probability that a chain ending in m_1 will add another m_1, P_2 being the corresponding probability for m_2. As we have seen (Chapter 2, Section 2.2.6) f_1 and f_2 are the monomer mole fractions in the feed. Thus, from a single polymer sample, r_1 and r_2 can be calculated if the dyad probabilities are known.

Although the vibrational spectra of copolymers generally differ appreciably from a summation of the spectra of the homopolymers, it is not possible to deduce quantitative sequence information from such data. NMR spectra can be used effectively for this purpose provided the resonances can be correctly assigned. Comonomers of the type $CH_2{=}CX_2$, which do not give rise to asymmetric centers, are amenable to the above analysis. An example is provided by copolymers of vinylidene chloride and isobutylene (Kinsinger *et al.*, 1966, 1967; Hellwege *et al.*, 1966). In Fig. 3.23 are shown proton spectra of the homopolymers (a) and (b), and a copolymer (c) containing 70 mol% vinylidine chloride (m_1). The resonances are grouped in three chemical ranges: m_1 CH_2 at low field, m_2 CH_2 and CH_3 at high field, and peaks near 3δ that occur only in the copolymer spectra and must correspond to m_1m_2-centered units. If only dyad sequences were distinguished, there would be only a single CH_2 resonance at low field for m_1m_1, a single CH_2 resonance at intermediate field for m_1m_2 and a single resonance at high field for m_2m_2. It is evident that tetrad sequences are being clearly resolved, the assignments being as indicated in the caption to Fig. 3.23. The upfield isobutene peaks show considerable overlap and the assignments are somewhat less certain. At the dyad level (i.e., lumping all m_1m_1-centered sequences together, and all m_2m_2-centered sequences together), r_1 and r_2 values can be

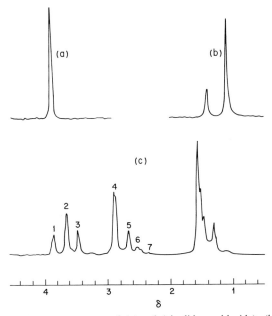

FIG. 3.23. 60 MHz proton spectra of (a) poly(vinylidene chloride); (b) polyisobutene; (c) a vinylidene chloride (m_1): isobutene (m_2) copolymer containing 70 mole% m_1. Peaks are identified with monomer tetrad sequences as follows: (1) $m_1m_1m_1m_1$; (2) $m_1m_1m_1m_2$; (3) $m_2m_1m_1m_2$; (4) $m_1m_1m_2m_1$; (5) $m_2m_1m_2m_1$; (6) $m_1m_1m_2m_2$; (7) $m_2m_1m_2m_2$ (from Fischer *et al.*, 1966).

calculated that agree quite well with those obtained by the conventional method (Chapter 2, Table 2.11). However, analysis of the tetrad intensities shows that they do not obey simple first-order Markov statistics (Hellwege *et al.*, 1966). In this system $r_2 \simeq 0$, since in a free radical copolymerization isobutene has only a very small probability of adding to a growing chain ending in an isobutene unit. Tetrad analysis shows that vinylidene chloride is about twice as likely to add to a growing chain ending in vinylidene chloride if the penultimate unit is isobutene.

If stereochemical configuration is to be considered, the problem of spectral analysis and assignment may be greatly complicated. We must consider the structures shown in Table 3.2 if both monomers are of the $CAB=CH_2$

TABLE 3.2
CONFIGURATIONAL SEQUENCES IN COPOLYMERS[a]

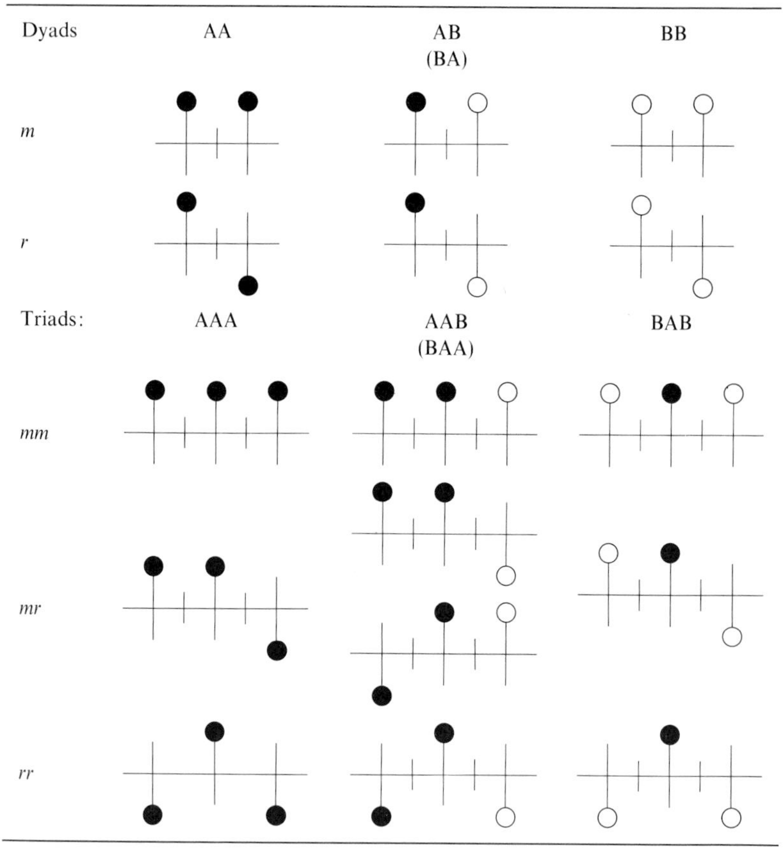

Dyads	AA	AB (BA)	BB
m			
r			
Triads:	AAA	AAB (BAA)	BAB
mm			
mr			
rr			

[a] +10 others with ● and ○ reversed.

type. The styrene–methyl methacrylate system has received particular attention (Harwood, 1965). The ester methoxyl proton resonance, normally insensitive to stereochemical configuration (Section 3.4), is found to be markedly shifted upfield if flanked by (a) two styrene units that are "co-meso" to it, while if only (b) one is "co-meso", the shielding effect is smaller, and is nil if the neighboring styrenes are (c) "co-racemic". Analysis in these

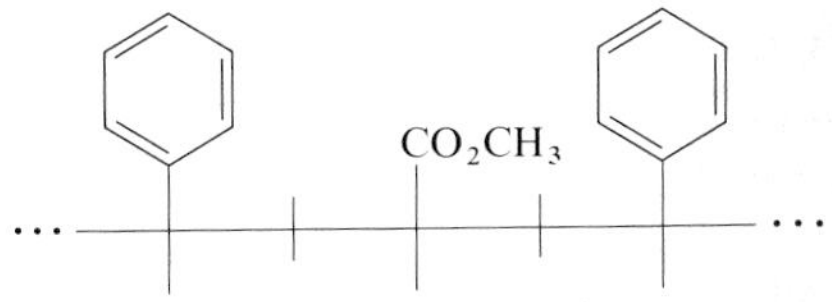

(a) 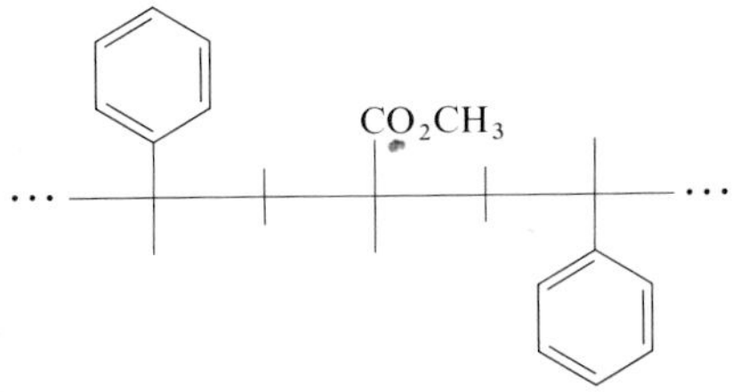　　　　　　　　　　　　　　　　　　(b)

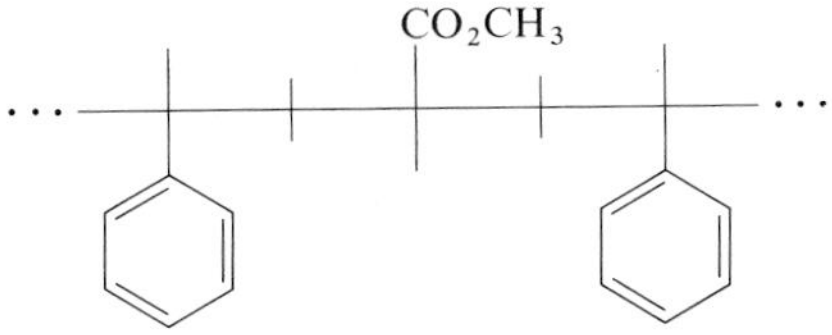

(c)

terms provides information not only concerning relative monomer and radical reactivities but also concerning steric preferences in monomer placement (Ramey and Brey, 1967; Willis and Cudby, 1968; Woodbrey, 1968).

It is evident that NMR can be employed for the study of copolymers that do not conform to the random statistics normally considered in copolymerization theory. Thus, anionic copolymers of styrene and methyl methacrylate show little or no complexity in the methoxyl region of the proton spectra, indicating a blocklike structure (Overberger and Yamamoto, 1965; Ito and Yamshita, 1965); this would not be revealed by the usual chemical analysis.

3.7. Ring-Opening Polymers

3.7.1. POLYPROPYLENE OXIDE

Most ring opening homopolymers (Chapter 2, Section 2.3) do not present stereochemical problems and the microstructure of the chain is evident from the nature of the polymerization. Among the exceptions to this statement are polymers of propylene oxide. The monomer and polymer contain true asymmetric centers. The chains have a sense of direction so that, for example, *ddl* is not equivalent to *ldd*. There are therefore four types of triad sequences, represented only in terms of relative configuration[†]:

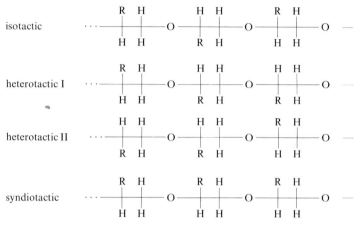

The isotactic chains are optically active, being either levo- or dextro-rotatory depending on the chirality of the active centers, as we have seen (Chapter 2, Section 2.3.2.1). The syndiotactic chain is internally compensating, i.e., meso, and would not be expected to be optically active if it could be prepared. The analysis of stereoirregular polypropylene oxide chains is very difficult; even ^{13}C and high field proton NMR have not fully resolved the problem, particularly as both head-to-tail and head-to-head:tail-to-tail propagation may occur, and the latter isomerism affects the NMR chemical shifts at least as much as the stereochemical isomerism. Highly crystalline polymer (m.p. 75°) has been shown to be ~99% head-to-tail isotactic; a polymer with m.p. 60° was found to have a head-to-tail isotactic dyad concentration of 90% (Schaefer, 1974).

3.7.2. POLYPROPYLENE SULFIDE

The polymerization of propylene sulfide with activated $CdCO_3$ yields a hard, predominantly isotactic polymer, while $ZnCO_3$ yields a softer, atactic material (Ivin and Navratil, 1970). The proton spectrum of polymers prepared from the α-d_1-monomer showed two AB quartets for the methylene

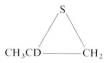

[†] Note that in the planar zigzag projection the R groups now appear alternately on *opposite* sides of the zigzag plane in the isotactic form and on the *same* side in the syndiotactic.

protons with no indication of heterotactic units, presumably because these protons are sensitive only to dyad configurations. It was thus not possible to tell whether the configurations are distributed in blocks or randomly. That the latter (which is more probable mechanistically) is the case was shown by oxidation to the polysulfone, in which four quartets could be resolved, corresponding to triad sequences of the type illustrated. These

$$\left[\begin{array}{c} CH_3 \quad\; O \\ | \qquad\; \| \\ C{-}CH_2{-}S \\ | \qquad\; \| \\ D \qquad\; O \end{array}\right]_n$$

polymers are thus statistical rather than block in configurational character.

3.7.3. N-CARBOXYANHYDRIDES

The chain structure and conformation of polypeptides, the products of the ring-opening polymerization of N-carboxy anhydrides (Chapter 1, Section 1.5), are discussed in Chapter 8 (Section 8.2).

3.8. Branches and Defect Structures

3.8.1. HEAD-TO-HEAD: TAIL-TO-TAIL STRUCTURES

The content of head-to-head: tail-to-tail units in vinyl polymers is usually very small because of resonance stabilization of the secondary radical or ionic end of the growing chain. In unusual cases, head-to-head units can be determined by chemical means. Flory and Leutner (Flory and Leutner, 1948; 1950) used periodic acid to cut the chains of polyvinyl alcohol (from hydrolysis of polyvinyl acetate) at the occasional head-to-head 1,2 glycol

$$\cdots CH_2CHCH_2CHCHCH_2CH_2CH \cdots \;\xrightarrow{\;HIO_4\;}\; \cdots CH_2CHO + HCOCH_2 \cdots \quad (3.16)$$
$$\underset{OH}{|}\quad \underset{OH}{|}\underset{OH}{|}\qquad \underset{OH}{|}$$

units, and determined the resulting decrease in molecular weight by intrinsic viscosity measurements. These were found to the extent of 1–2%, the fraction increasing with polymerization temperature with a coefficient corresponding to an additional 1.3 kcal of activation energy for head-to-head units.

Polymers of vinyl fluoride and vinylidene fluoride show a substantial content of head-to-head units, higher than for any other known vinyl polymers. This has been clearly demonstrated by ^{19}F NMR, which is very sensitive to structural details. In Fig. 3.24 is shown the 56.4 MHz ^{19}F spectrum of polyvinyl fluoride (Wilson and Santee, 1965). At least seven peaks can be

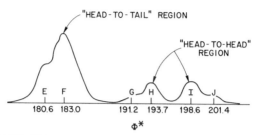

FIG. 3.24.   56.4 MHz ^{19}F spectrum of poly(vinyl fluoride) in a solvent composed of equal volumes of N,N'-dimethylacetamide and hexamethyl phosphoramide, observed at 89° (from Wilson and Santee, 1965).

seen. Some of the multiplicity may be due to configurational sequences, but the upfield group of resonances is due to head-to-head units:

$$\cdots CFHCH_2 \cdot CH_2CFH \cdot CFHCH_2 \cdots$$

About 32% of the ^{19}F intensity is in this upfield region, corresponding to one monomer out of six being inserted "backward." Poly(vinylidene fluoride) shows similar spectral features (Wilson and Santee, 1965; Wilson, 1963) (Fig. 3.24). Complete assignments were possible by measurements on model compounds. These are indicated in the figure and in the representative chain segment below:

```
   →    →    →    →    →    ←    →    →    →    →    →
  H  F  H  F  H  F  H  F  H  F  F  H  H  F  H  F  H  F  H  F  H  F  H  F
—C—C—C—C—C—C—C—C—C—C—C—C—C—C—C—C—C—C—C—C—C—C—C—C—
  H  F  H  F  H  F  H  F  H  F  F  H  H  F  H  F  H  F  H  F  H  F  H  F

        A     A     A     C  D     B     A     A     A     A
```

The letters refer to the corresponding peaks in Fig. 3.25. A single reversed unit leads to three distinguishable fluorine resonances upfield from the "normal" peak. Since peaks B, C, and D have equal intensities, it is evident that the reversed units do not propagate further in this sense. About 5–6% of the monomer units are head-to-head:tail-to-tail.

Although reversed units have been reported for several other polymers, none of these has been as clearly demonstrated as in these cases. Their possible presence in poly(vinyl chloride) has been the object of many studies, but there is no spectroscopic evidence for them and the chemical evidence lacks conviction.

3.8.2.   BRANCHING

Next to molecular weight and its distribution, branching is the most important chain variable (assuming a given basic type of structural unit) influencing the properties of polymers and polymer solutions. We have seen

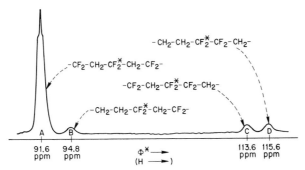

FIG. 3.25.   56.4 MHz ^{19}F spectrum of polyvinylidene fluoride in 25% solution in N,N'-dimethylacetamide, observed at room temperature (from Wilson and Santee, 1965; Wilson 1963).

in Chapter 1 (Section 1.2) and Chapter 2 (Section 2.4.6) that branches may be introduced deliberately by use of polyfunctional monomers. We speak rather here of branching introduced by processes that are under less specific control, and involve chain transfer reactions of various types. Such reactions are particularly to be expected for highly reactive polymer radicals that are not stabilized by resonance (Chapter 2, Section 2.2), such as those from ethylene, vinyl chloride, and vinyl acetate. Chain transfer of growing polyethylene chains to polymer already formed has been shown to lead to the formation of *long branches* in high-pressure polyethylene. Of still greater significance in polyethylene technology is the corresponding intramolecular process, leading to *short branches* with a probability approximately an order of magnitude greater, and having a profound effect on the crystal morphology and physical properties (Chapter 5).

That there are short branches in high-pressure polyethylene has long been realized but the measurement of their frequency and type has engendered some disagreement and controversy. The traditional method for their observation is by measurement of the CH_3 distortion band at 1375 cm^{-1} in the infrared spectrum in Fig. 3.8b, which measures the total content of branch ends. This band strongly overlaps the CH_2 wagging bands at 1350–1360 cm^{-1}. Earlier measurements did not correct properly for this overlap and tended to overestimate the branch frequency. Willbourn (1959) showed that by using the linear material as a blank, correct quantitative results could be obtained. However, their length, type (whether trifunctional or tetrafunctional), and distribution are not revealed by vibrational spectroscopy. It has been shown (Dorman et al., 1972; Randall, 1973; Bovey et al., 1976) that the ^{13}C chemical shifts of paraffinic hydrocarbons are not only highly sensitive to sterochemical configuration, as we have already seen (Section 3.2), but also to the type, length, and distribution of branches. In

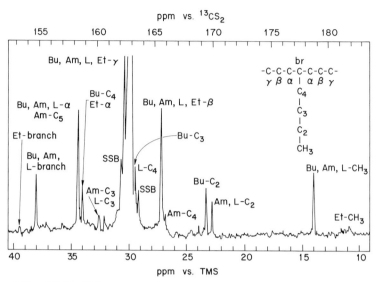

FIG. 3.26.   25 MHz ^{13}C spectrum of low density polyethylene observed as 20% solution in 1,2,4-trichlorobenzene at 110°. The diagram at the upper right shows the nomenclature employed for the carbons associated with a branch. The end carbon (i.e., C_1) is designated as CH_3; Et = ethyl, Bu = *n*-butyl, Am = *n*-amyl, and L = "long" in the sense described in the text; SSB refers to spinning sidebands (from Bovey *et al.*, 1976).

Fig. 3.26 is shown the 25 MHz spectrum of a branched polyethylene; the resonances are labeled according to the scheme indicated in the upper right-hand portion of the figure. The quantitative results are summarized in Table 3.3. All branches are trifunctional, even though the enhanced reactivity of branch-point hydrogens toward free radical attack might lead one to expect some tetrafunctional branches. These results are fairly typical, although the total branch content and their ratios vary somewhat, probably

TABLE 3.3

BRANCHING IN HIGH-PRESSURE POLYETHYLENE

Type of branch	Number of branches/1000 backbone carbons
—CH_3	0.0
—CH_2CH_3	1.0
—$CH_2CH_2CH_3$	0.0
—$CH_2CH_2CH_2CH_3$	9.6
—$CH_2CH_2CH_2CH_2CH_3$	3.6
Hexyl or longer	5.6
	19.8

mainly because of variations in polymerization temperature. The "hexyl or longer" branches are believed from other independent evidence (Bovey *et al.*, 1976; see also Chapter 4, Section 4.4.4) to be actually of the truly long type formed by intermolecular transfer. It may be concluded that if the "backbiting" mechanism is correct, the reaction indicated by Eq. (3.17) is

$$-\mathrm{CH_2} \quad \cdot\mathrm{CH_2} \quad \longrightarrow \quad -\mathrm{CH}\cdot \quad \mathrm{CH_3} \qquad (3.17)$$
$$(\mathrm{CH_2})_n \qquad\qquad\qquad (\mathrm{CH_2})_n$$

most probable when $n = 3$ or 4, has a low but finite probability when $n = 1$, and zero probability when $n = 0$ or 2.

Branches have been demonstrated in poly(vinyl chloride) by reduction to polyethylene with lithium aluminum hydride and infrared measurements; they increase in frequency with the temperature of polymerization to a maximum of about 3 per 1000 backbone carbon atoms (George *et al.*, 1958, Rigo *et al.*, 1972). By $^{13}\mathrm{C}$ measurements (Abbas *et al.*, 1975; Bovey *et al.*, 1975), these have been demonstrated to be $-\mathrm{CH_2Cl}$ groups, with possibly as many as one long branch per 1000 backbone carbons.

Long branches in polyvinyl acetate are believed to be formed by chain transfer to the acetoxy methoxyl group, leading to

$$\cdots-\mathrm{CH_2CH}-\cdots \qquad \text{main chain}$$
$$|$$
$$\mathrm{O}$$
$$|$$
$$\mathrm{C}{=}\mathrm{O}$$
$$|$$
$$\mathrm{CH_2}-\mathrm{CH_2}-\mathrm{CH}-\cdots \qquad \text{branch}$$
$$|$$
$$\mathrm{O}$$
$$|$$
$$\mathrm{C}{=}\mathrm{O}$$
$$|$$
$$\mathrm{CH_3}$$

Such branching may be measured by hydrolysis of the ester links followed by molecular weight measurement, but this method is correct only if the mechanism is correct. Stein (1964) has made careful molecular weight measurements indicating a rapid increase in the latter stages of monomer conversion, as expected. Such conclusions may be correct, but spectroscopic or other similar direct structural evidence is lacking.

3.9. Chain Conformation

A wide variety of shapes can be assumed by flexible chains. In a long chain of connected segments, the spatial relationship between two neighboring segments is governed by bond length and bond angle and, where rotation about at least some of the bonds is possible, by the rotational state of these bonds. Such rotation is possible in most macromolecular chains and leads to an astronomically large number of permissible spatial arrangements of the segments. Since it is impossible to consider every permissible conformation individually, a statistical approach must be adopted. We deal with average quantities, such as the average size of the molecule, and distribution functions, such as the probability that a certain segment is located at a specified coordinate relative to another.

A quantity that provides a useful measure of the chain conformation is the *end-to-end distance*. A schematic diagram of a chain consisting of n segments is shown in Fig. 3.27. Each segment is represented by a vector $\mathbf{l}_i$ numbered from 1 to n. The end-to-end distance for a given conformation is simply the vector connecting the two ends:

$$\mathbf{r} = \sum_{i=1}^{n} \mathbf{l}_i \tag{3.18}$$

Ordinarily, only the scalar magnitude of r can be related conveniently to experimentally measured quantities. The square of the scalar quantity is

$$r^2 = \mathbf{r} \cdot \mathbf{r} = \sum_{i,j} \mathbf{l}_i \cdot \mathbf{l}_j \tag{3.19}$$

where the summation is carried out from 1 to n for i and j.

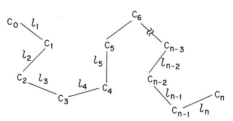

FIG. 3.27.   Schematic representation of a polymer chain backbone.

The summation in Eq. (3.19) can be written in an alternative form,

$$r^2 = \sum_{i} l_i^2 + 2 \sum_{i \neq j} \mathbf{l}_i \cdot \mathbf{l}_j \tag{3.20}$$

The first term is simply the product of $\mathbf{l}_i$ with itself and the second term is the scalar product of $\mathbf{l}_i$ with $\mathbf{l}_j$ when i is not equal to j. The factor of 2 results from the equivalence of $\mathbf{l}_i \cdot \mathbf{l}_j$ and $\mathbf{l}_j \cdot \mathbf{l}_i$.

Another parameter that is useful in the description of the size of the macromolecule is the radius of gyration s. Consider the chain as an assembly of mass elements, each located at a distance s_i from the center of gravity. The radius of gyration is defined as

$$s^2 = \sum_i ms_i^2 \Big/ \sum_i m = \sum_{i=1}^{n} s_i^2 \Big/ n \tag{3.21}$$

where m is the mass of each element or segment.

To obtain a statistical mechanical average of the above quantities, the averaging process is performed for all possible conformations. The average quantity is denoted by angle brackets, for example, $\langle r^2 \rangle$. The reference state for such a calculation is the one in which the molecule is free of external constraints, such as applied force or interaction with solvent. The *unperturbed state dimension* is denoted by the subscript 0 appended to the angle bracket, e.g., $\langle r^2 \rangle_0$ or $\langle s^2 \rangle_0$. For the purpose of illustration, the averaging process for a model chain is given below.

3.9.1. THE FREELY JOINTED CHAIN

The *freely jointed chain* is a hypothetical linear chain where all the segments are of equal length and the angle between two successive segments can assume any value. Bond rotation is also free. For such a hypothetical chain, where there is no correlation between the direction of neighboring bonds and the angle θ between two successive bonds can assume any value with equal probability, the average of $\mathbf{l}_i \cdot \mathbf{l}_j$, i.e., the average of $l_i l_j \cos \theta$ over all possible values of θ is zero when $i \neq j$. Summing over i and j values from 1 to n in Eq. (3.20), we obtain

$$\langle r^2 \rangle_0 = nl^2 \tag{3.22}$$

Although a real polymer chain is far from freely jointed, the correlation between bonds i and $i + j$ must vanish as j increases. Consequently, it is sometimes useful to represent the real chain as an ensemble of "statistical segments" forming an equivalent freely jointed chain, subject to the condition that the mean square end-to-end distance and the fully extended length, r_{max}, of this hypothetical chain equal those of the real chain,

$$\langle r^2 \rangle_0 = n'l'^2 \tag{3.23}$$

and

$$r_{max} = n'l' \tag{3.24}$$

Regardless of the exact nature of the conformational restrictions in a real chain, the proportionality between the unperturbed dimensions, as measured by $\langle r^2 \rangle_0$, and nl^2 is always maintained. The *characteristic ratio* $\langle r^2 \rangle_0 / nl^2$, sometimes designated as C_∞ (see Section 3.9.4), is greater than unity for all real chains, and is a measure of the departure from a freely jointed or freely rotating model. The theoretical calculation of C_∞ is discussed in Section 3.9.4 and its experimental measurement in Chapter 4, Section 4.4.

3.9.2. THE FREELY ROTATING CHAIN

If the angles between successive bonds in a chain are held fixed but the rotation of the bond is free, the resulting chain is termed *freely rotating*. Unlike the case of the freely jointed chain, the average of the scalar product of $\mathbf{l}_i$ with $\mathbf{l}_j$ is no longer zero. The projection of bond $i + 1$ on bond i is $l \cos \theta$ and

$$\langle \mathbf{l}_i \cdot \mathbf{l}_{i+j} \rangle = l^2 (\cos \theta)^j \tag{3.25}$$

(Note that the traverse projection of $\mathbf{l}_{i+1}$ on $\mathbf{l}_i$ averages to zero because of free rotation.) Substitution into Eq. (3.20) yields, for large n,

$$\langle r^2 \rangle_0 \cong \left(\frac{1 - \cos \theta}{1 + \cos \theta} \right) n l^2 \tag{3.26}$$

3.9.3. THE DISTRIBUTION OF CHAIN SEGMENTS

Having illustrated the concept of average quantities, we now describe briefly the spatial distribution of segments relative to each other. The distribution function is usually expressed as $W(\mathbf{r}_{ij})$ or $W(\mathbf{r})$. The latter is defined as the probability that the end-to-end distance is situated between $\mathbf{r}$ and $\mathbf{r} + d\mathbf{r}$, and $W(\mathbf{r}_{ij})$ is similarly defined as the probability that the distance between the ith and jth segments is located between $\mathbf{r}_{ij}$ and $\mathbf{r}_{ij} + d\mathbf{r}_{ij}$. If one chain end is placed at the origin of the coordinate system, the probability that the end-to-end vector will be within a spherical shell of thickness dr is $W(\mathbf{r})4\pi r^2\, dr$. Summed over all values of r, the distribution function must satisfy the condition

$$\sum_0^\infty W(\mathbf{r})4\pi \mathbf{r}^2\, d\mathbf{r} = 1 \tag{3.27}$$

The mean square end-to-end distance is given by

$$\sum_0^\infty W(r)4\pi \mathbf{r}^4\, d\mathbf{r} = \langle r^2 \rangle \tag{3.28}$$

Let us again use the freely jointed chain as an example. We consider first the one-dimensional case where n segments of length b are linked with equal probability in the positive and negative direction along the x axis. The probability that there are n_+ segments pointed toward the positive direction and n_- segments in the negative direction is

$$W(n_+) = \left(\frac{1}{2}\right)^n \frac{n!}{n_+! n_-!} \tag{3.29}$$

The end-to-end distance is $x = (n_+ - n_-)b = (2n_+ - n)b$. For large values of n, x is small compared to nb because the number of segments placed in the positive direction will not be greatly different from that in the negative direction. By the use of Stirling's approximation for factorials

$$\ln n! = n \ln n - n \tag{3.30}$$

and the series expansion

$$\ln\left(1 + \frac{x}{nb}\right) \cong \frac{x}{nb} \quad \text{for} \quad \frac{x}{nb} \ll 1 \tag{3.31}$$

we obtain

$$W(x)\,dx = (2\pi nb^2)^{1/2} \exp(-x^2/2nb^2)\,dx \tag{3.32}$$

For the three-dimensional case, the expression is

$$W(r)\,dr = \left(\frac{2\pi nl^2}{3}\right)^{-3/2} \exp\left(-\frac{3r^2}{2nl^2}\right)dr \tag{3.33}$$

The spherically symmetric distribution of end-to-end vectors is therefore of the Gaussian form. We also obtain $\langle r^2 \rangle = nl^2$, in view of Eq. (3.28). The most probable value of the end-to-end distance, denoted by r^* and obtained by setting $d \ln[4\pi r^2 W(\mathbf{r})]/d\mathbf{r} = 0$, is

$$(r^*)^2 = \tfrac{2}{3}nl^2 \tag{3.34}$$

For a chain of 10^4 segments, r^* equals $81.5l$, or only about 0.81% of the full contour length of the chain. The probability decreases rapidly at larger values of r, as shown in Fig. 3.28.

3.9.4. CHAINS WITH RESTRICTED ROTATION

For real polymer chains there are intrachain hindrances that restrict each bond to a small number of distinguishable rotational states. If the relative energies of these states are known, the conformational partition function can be computed by matrix methods and from it any conformation-dependent property of the chain can be calculated. These include the mean square end-to-end distance, the dipole moment, and a number of optical properties (see General References, III. Vol'kenshtein, 1963; Birshtein and Ptitsyn, 1966; Flory, 1969; and Tonelli, 1977). We now wish to examine these conformations and their energies in greater detail with respect to both computation and experimental observation.

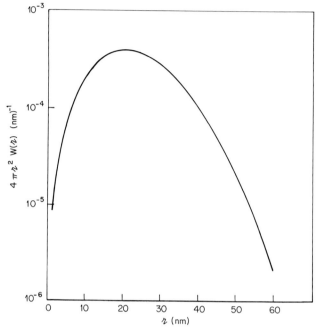

FIG. 3.28.   End-to-end distance probability.

Spectroscopic and electron diffraction studies of a variety of small molecules have demonstrated the nature of the hindering potentials that one may
expect to be present also in macromolecular chains. It is well known that
rotation about the C–C bond in ethane is characterized by a symmetric
three-fold potential with energy minima corresponding to staggering of the
C–H bonds, the depths of the wells being ~ 3 kcal. All the staggered conformers are equivalent. In n-butane, three staggered conformers are present
about the central (C_2–C_3) bond, corresponding also to a threefold potential
(Fig. 3.29), which in this case, however, is not symmetric, the *gauche* conformers g^+ and g^- having about 0.5 kcal higher energy than the *trans* t;

The g^+ and g^- conformers are mirror images, identical in energy, and are
formally generated by rotating about the C_2–C_3 bond through an angle ϕ

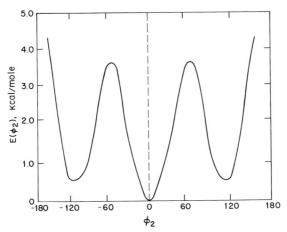

FIG. 3.29.   Torsional potentials about the central (C_2-C_3) bond in *n*-butane.

of $+$ and $-$ 120°, i.e., clockwise and counterclockwise, respectively, as indicated above (looking from C_3 toward C_2). The depths of the wells are such that, although small oscillations of ϕ of $\pm \sim 15-20°$ occur, the populations of the eclipsed conformers corresponding to the tops of the barriers are negligible; yet the barriers are low enough so that the conformers interchange with each other very rapidly, their lifetimes being of the order of 10^{-10} sec at ordinary temperatures.

When we lengthen the chain by one carbon, further complications develop. The conformations must now be specified by two rotational angles ϕ_1 and ϕ_2:

tt
$\phi_1 = \phi_2 = 0°$

$tg^+ (\equiv g^+ t)$
$\phi_1 = 0°;$
$\phi_2 = +120°$

$tg^- (\equiv g^- t)$
$\phi_1 = 0°;$
$\phi_2 = -120°$

$g^+ g^+ (\equiv g^- g^-)$
$\phi_1 = \phi_2 = +120°$

$g^+ g^- (\equiv g^- g^+)$
$\phi_1 = +120°;$
$\phi_2 = -120°$

The energy of a given rotational state of the C_2–C_3 bond depends on the rotational state of C_3–C_4 bond and *vice versa*. Relative to the *tt* state, the energies of the mirror-image tg^+ and tg^- states (and of the equivalent g^+t and g^-t states) are only slightly higher than for the g^+ and g^- states of butane. The g^+g^+ ($=g^-g^-$) and g^+g^- ($=g^-g^+$) states are very different in energy; in the g^+g^+ state, the methyl groups experience an approximately neutral (i.e., neither repulsive nor attractive) interaction, while in the g^+g^- state the methyls experience a severe repulsive interaction that strongly tends to exclude this state. Calculations (Abe *et al.*, 1966) based on semi-empirical interatomic potentials, both attractive and repulsive, and allowing considerable deviation ($\sim 40°$) of either ϕ_1 or ϕ_2 from strict staggering to relieve steric overlaps, indicate that the $g^{\pm}g^{\mp}$ conformations are about 3200 cal mole^{-1} above the *tt* state or about 2000 cal mole^{-1} above $g^{\pm}g^{\pm}$. (At exact staggering, the $g^{\pm}g^{\mp}$ conformations are of very high energy.)

Clearly, both three-bond or *first-order* interactions, like those in *n*-butane, and the additional four-bond or *second-order* interactions encountered in *n*-pentane and longer chains, must be taken account of even in the simplest of polymer chains, such as those of strictly linear polyethylene (often termed "polymethylene"). To a good degree of approximation, only the states of nearest-neighbor bonds need be considered, higher-order interactions being ordinarily negligible. Such a chain corresponds to the mathematical concept termed a one-dimensional Ising system.

If all bond lengths and valence angles are fixed, the conformation of a chain of n bonds can be specified by assigning a rotational state to each of the $n-2$ nonterminal bonds. This mode of treatment of chain statistics is called the *rotational isomeric state* (RIS) model, and clearly has the advantage of being much more realistic than the idealized models treated in Sections 3.9.1 and 3.9.2. If there are v rotational states about each bond, there will be v^{n-2} possible conformational states. For the common case that $v = 3$, even a chain of only 20 bonds will have over one billion conformational states; chains of polymeric length would seem to present insuperable computational obstacles. However, they can be readily treated by matrix methods, employing a digital computer.

Consider the segment of a *polymethylene chain* shown below:

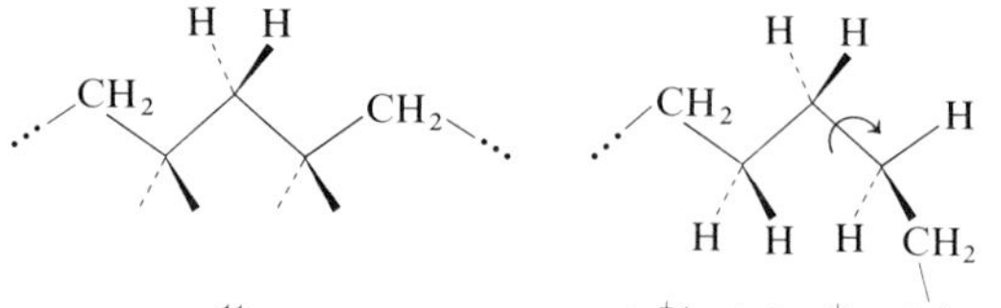

tt

$tg^+(\equiv tg^-, g^+t, g^-t)$

$u = \sigma$

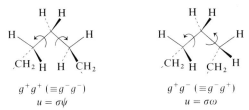

$$g^+g^+ (\equiv g^-g^-)$$
$$u = \sigma\psi$$

$$g^+g^- (\equiv g^-g^+)$$
$$u = \sigma\omega$$

We define *statistical weights* for these conformations as Boltzmann factors

$$u_i = \exp[-E_i(\phi_i - 1,\ \phi_i)/RT] \tag{3.35}$$

where E_i is the energy of the conformer corresponding to the rotational angles ϕ_{i-1} and ϕ_i for neighboring bonds. It is convenient to represent the statistical weights of specific conformations by symbols: tt is taken as having a statistical weight of 1; the tg^+ ($=g^+t = tg^- = g^-t$ for a polymethylene chain) conformation is assigned a statistical weight σ, the $g^{\pm\pm}$ conformation a statistical weight $\sigma\psi$, and the $g^{\pm}g^{\mp}$ conformation (almost excluded) a statistical weight $\sigma\omega$. The last two are compounded of two symbols because the conformations involve both the simple "butanelike" gauche energy and a new interaction. From what we have said before, however, concerning the methyl group interactions in the $g^{\pm}g^{\pm}$ conformation of pentane, i.e., that they are neither attractive nor repulsive, we may assume that $\psi \simeq 1$. The statistical weight matrix for two successive bonds, including both first- and second-order steric interactions, is then

$$U_i = i - 1 \begin{cases} (t) \\ (g^+) \\ (g^-) \end{cases} \begin{array}{c} \overbrace{\begin{array}{ccc} (t) & (g^+) & (g^-) \end{array}}^{i} \\ \begin{bmatrix} 1 & \sigma & \sigma \\ 1 & \sigma & \sigma\omega \\ 1 & \sigma\omega & \sigma \end{bmatrix} \end{array} \tag{3.36}$$

Here, the columns are indexed on the bond i and the rows on the preceding bond. Since $E_\sigma - E_t$ (from butane) $\simeq 500$ cal mole^{-1}, σ (at $300°$K) $\simeq 0.43$; $E_\omega \cong 3200 - 2E_\sigma \cong 2200$ cal mole^{-1}, and therefore $\omega \simeq 0.026$.

The statistical weight of a particular conformation of a chain of n bonds is

$$\Omega_{(\phi)} = \prod_{i=2}^{n-1} u_i \tag{3.37}$$

the first and last bonds being excluded from consideration since they are without effect. The conformational ("configurational") partition function is obtained by summing over all possible conformations:

$$Z = \sum_{(\phi)} \prod_{i-2}^{n-1} u_i \tag{3.38}$$

Application of matrix methods (see General References, III) leads to

$$Z = J^* \left[\prod_{i=2}^{n-1} u_i J \right] \qquad (3.39)$$

where J^* and J are the row and column vectors:

$$J^* = [1, \sigma, \sigma]; \quad J = \begin{bmatrix} 1 \\ 1 \\ 1 \end{bmatrix} \qquad (3.40)$$

Equations (3.39) and (3.40) allow one to calculate the conformational partition function provided the bond rotational states and their energies can be estimated.

By further matrix operations which we shall not detail here, any conformation-dependent property of a polymer chain can be calculated. Table 3.4 shows the experimental characteristic ratio C_∞ (Section 3.7.1; Chapter 4, Section 4.4) of the polymethylene chain calculated on various assumptions, including (a) the freely rotating chain; (b) a chain with a reasonable value for the energy of the gauche state but with the assumption of independence of the rotational states of neighboring bonds; and (c) the neighbor-dependent 3-state model with the values of E_σ and E_ω indicated above, as well as higher values (d) and (e). The energies in (c), deduced from small paraffins, give the correct value of C_∞, as well as its temperature coefficient; larger values increase C_∞ beyond its experimental value but decrease its temperature coefficient.

TABLE 3.4

CALCULATED AND EXPERIMENTAL VALUES OF THE CHARACTERISTIC RATIO $C_\infty = \langle r^2 \rangle_0 / nl^2$ AND ITS TEMPERATURE COEFFICIENT FOR THE POLYMETHYLENE CHAIN

Model	Characteristic ratio C_∞	Temperature coefficient $dC_\infty/dT \times 10^3$
Experimental	6.7	$-1.1 \pm 0.1 (\sim 120\text{--}150^\circ)$
(a) Freely rotating	2.1	0
(b) 3-State RIS, $E_\sigma = 500$ cal, $E_\omega = 0$ (independent rotations)	3.4	—
(c) 3-State RIS, $E_\sigma = 500$ cal, $E_\omega = 2200$ cal	6.7	-1.0
(d) 3-State RIS, $E_\sigma = 800$ cal, $E_\omega = 2200$ cal	8.3	-1.5
(e) 3-State RIS, $E_\sigma = 500$ cal, $E_\omega = 3500$ cal	7.1	-0.7

For the *polyoxymethylene* chain, which has a two-bond repeat unit, two statistical weight matrices are involved, one for the bonds to the O atom and one for the bonds to the CH_2 group:

$$U_a = \begin{bmatrix} 1 & \sigma & \sigma \\ 1 & \sigma & \sigma\omega_a \\ 1 & \sigma\omega_a & \sigma \end{bmatrix} \tag{3.41}$$

$$U_b = \begin{bmatrix} 1 & \sigma & \sigma \\ 1 & \sigma & \sigma\omega_b \\ 1 & \sigma\omega_b & \sigma \end{bmatrix} \tag{3.42}$$

The statistical weights refer to conformations analagous to those for the polymethylene chain. The gauche conformation is known from a variety of evidence, including the known crystal structures (p. 261), to be of considerably lower energy than the trans, in contrast to the behavior of the polymethylene chain, presumably because of favorable coulombic interactions. There are two $g^{\pm}g^{\mp}$ conformations, corresponding to (a) and (b) below. The first is

$(g^{\pm}g^{\mp})_a$ $(g^{\pm}g^{\mp})_b$

very rigorously excluded by severe $CH_2 \cdots CH_2$ overlap and ω_a may be taken as 0; the corresponding $O \cdots O$ interaction in (b) is more weakly excluded and ω_b is assigned a statistical weight of ~ 0.05. Assuming tetrahedral geometry, a gauche angle $\phi \simeq 120°$, and a value of ~ 12 for σ (i.e., $E_{\sigma} - E_t \simeq 1500\,\text{cal-mole}^{-1}$), one obtains a value of about 10 for C_{∞}, in agreement with the rather approximate experimental values for this polymer, and with the dipole moments for smaller model molecules. If the $g^{\pm}g^{\mp}$ conformers are allowed to contribute appreciably, only a positive value of E_{σ}, clearly excluded by other evidence, predicts the observed value of C_{∞}; a negative value would yield a very much smaller predicted C_{∞}.

Rotational isomeric state model calculations of *polyoxyethylene* chains (Mark and Flory, 1966a,b), which have a 3-atom repeat unit, require three statistical weight matrices, corresponding to c–a, a–b, and b–c bond pairs in the sense indicated:

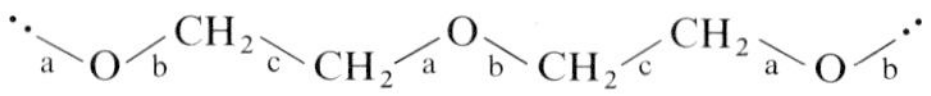

$$\overset{..}{\underset{a}{\diagdown}} O \underset{b}{\diagup} \overset{CH_2}{\diagdown} CH_2 \underset{c}{\diagup} \overset{O}{\underset{a}{\diagdown}} \underset{b}{\diagup} CH_2 \overset{CH_2}{\underset{c}{\diagdown}} \overset{..}{\underset{a}{\diagdown}} O \underset{b}{\diagup}$$

There are two gauche conformations with the estimated energies shown (referred to the trans conformation):

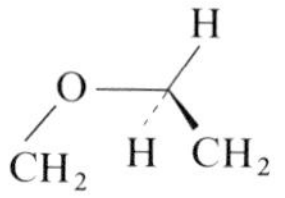

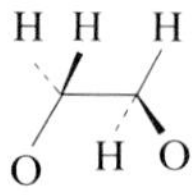

$E_\sigma = 900$ cal $E_{\sigma'} = -250$
$(\sigma = 0.26 \pm 0.03)$ $(\sigma' = 1.50 \pm 0.20)$

The second is favored over trans, although less so than the gauche conformation in polyoxymethylene, whereas the other is more unfavorable than the gauche polymethylene conformation because C–O bond lengths (0.143 nm) are shorter by 0.010 nm than the C–C bond lengths in the latter. Of the two $g^\pm g^\mp$ conformations, that involving four-bond $-CH_2 \cdots CH_2-$ interactions is given a statistical weight of zero, as in polyoxymethylene and polymethylene, while that involving $-O \cdots CH_2-$ interactions is assigned an energy E_ω of 800 (± 400) cal ($\omega = 0.30 \pm 0.20$). In Table 3.5 is shown a compilation of experimental properties compared to those predicted using these parameters. A number of other conformation-dependent properties, including optical properties, are also correctly predicted, and so the unperturbed conformation of this polymer appears to be well understood on the basis of a 3-state rotational isomeric model.

TABLE 3.5

CALCULATED AND EXPERIMENTAL CONFORMATION-DEPENDENT PROPERTIES
FOR POLYOXYETHYLENE[a]

Property	Temperature (°C)	Calculated	Observed
C_∞	40	4.8	4.8
$\left(\dfrac{d\ln C_\infty}{dT}\right) \times 10^3$	60	0.18	0.23
$\left(\dfrac{\langle\mu^2\rangle}{n\mu^2}\right)^{b}_{n\to\infty}$	25	0.58	0.58
$\left(\dfrac{d\ln\langle\mu^2\rangle}{dT}\right)^{b}_{n\to\infty} \times 10^3$	25	2.5	2.6

[a] See General References III, Tonelli (1977).
[b] μ = dipole moment of C—O bond.

The treatment of the conformations of vinyl homopolymer chains having asymmetric centers involves further complications in that the steric requirements of the side chains must be taken into account. There is also the formal complication that, because of the asymmetric centers, one must to some degree view the chain stereochemistry in an arbitrary manner and must generate chain conformations according to a consistent convention for bond rotations. For first-order interactions (adopting an arbitrary l chirality for the bonds about C_α), we have the following rotational states:

$$t, \quad u_1 = \sigma\eta \qquad\qquad g^+, \quad u_1 = \sigma \qquad\qquad g^-, \quad u_1 = \sigma\tau$$

The $C_\alpha H \cdots R$ interaction in the trans conformation is given the statistical weight $\sigma\eta$; if R is equivalent to CH_2, η will be unity and the interaction is essentially equivalent to the gauche interaction in a polymethylene chain. In general, this will not be true and η will be the ratio of the statistical weight of the first-order interaction involving R to the corresponding interaction involving CH_2. The g^+ conformations involve a $CH_2 \cdots CH$ interaction and are weighted by σ. The g^- conformations involve both $CH_2 \cdots CH$ and $CH \cdots R$ interactions and are assigned the statistical weight $\sigma\tau$. The relative statistical weights for t, g^+, and g^- are thus η, 1, and τ, respectively, and may be expressed in terms of the diagonal matrix:

$$\begin{array}{ccc} t & g^+ & g^- \end{array}$$
$$U_1(l) = \text{diag}(\eta \quad 1 \quad \tau); \tag{3.43}$$

it turns out that we must also take into account the d chirality:

$$\begin{array}{ccc} t & g^+ & g^- \end{array}$$
$$U_1(d) = \text{diag}(\eta \quad \tau \quad 1) \tag{3.44}$$

Second-order steric interactions about the bonds flanking the β-carbons are appropriately defined in terms of meso and racemic dyads. These conformations are indicated below together with their statistical weights. These are the ll dyads; the dd dyads (not shown) are their mirror images. The

$$tt, \quad u_2 = \omega'' \qquad\qquad g^+t, \quad u_2 = \omega' \qquad\qquad g^-t, \quad u_2 = 1$$
$$\qquad\qquad\qquad\qquad\qquad tg^- \qquad\qquad\qquad\qquad tg^+$$

$g^-g^-, \quad u_2 = \omega'$

g^+g^+

$g^+g^-, \quad u_2 = \omega\omega''$

$g^-g^+, \quad u_2 = \omega$

ll (meso)

statistical weight matrices incorporating both first- and second-order interactions are

$$
U_{m_{12}}(ll) = \begin{array}{c} \\ t \\ g^+ \\ g^- \end{array}
\overset{\displaystyle t \quad\quad g^+ \quad g^-}{
\begin{bmatrix}
\eta\omega'' & 1 & \tau\omega' \\
\eta\omega' & \omega' & 0 \\
\eta & \omega & \tau\omega'
\end{bmatrix}}
\tag{3.45a}
$$

$$
U_{m_{12}}(dd) = \begin{array}{c} \\ t \\ g^+ \\ g^- \end{array}
\overset{\displaystyle t \quad\quad g^+ \quad g^-}{
\begin{bmatrix}
\eta\omega'' & \tau\omega' & 1 \\
\eta & \tau\omega' & \omega \\
\eta\omega' & 0 & \omega'
\end{bmatrix}}
\tag{3.45b}
$$

The statistical weight of the g^+g^- conformation may be taken as 0.
    For racemic dyads:

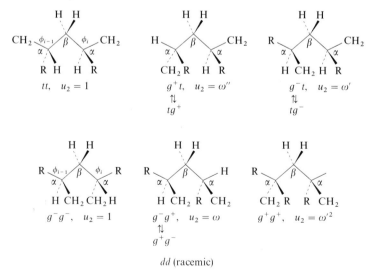

$tt, \quad u_2 = 1$

$g^+t, \quad u_2 = \omega''$

tg^+

$g^-t, \quad u_2 = \omega'$

tg^-

$g^-g^-, \quad u_2 = 1$

$g^-g^+, \quad u_2 = \omega$

g^+g^-

$g^+g^+, \quad u_2 = \omega'^2$

dd (racemic)

The complete statistical weight matrices are

$$
U_{r_{12}}(ld) = \begin{array}{c} \\ t \\ g^+ \\ g^- \end{array}
\begin{array}{c}
\quad t \qquad\ g^+ \qquad g^- \\
\left[\begin{array}{ccc}
\eta & \tau\omega'' & \omega' \\
\eta\omega'' & 0 & \omega \\
\eta\omega' & \tau\omega & 1
\end{array}\right]
\end{array}
\tag{3.46a}
$$

$$
U_{r_{12}}(dl) = \begin{array}{c} \\ t \\ g^+ \\ g^- \end{array}
\begin{array}{c}
\quad t \qquad\ g^+ \qquad g^- \\
\left[\begin{array}{ccc}
\eta & \omega' & \tau\omega \\
\eta\omega' & 1 & \tau\omega \\
\eta\omega'' & \omega & 0
\end{array}\right]
\end{array}
\tag{3.46b}
$$

It is evident that here the $g^+g^+(ld)$ (and $g^-g^-(dl)$) conformations will be very strongly excluded and that the corresponding statistical weight may be assumed to be negligible whatever the nature of R.

The rotational states about the bonds flanking the α-carbons must also be taken into account:

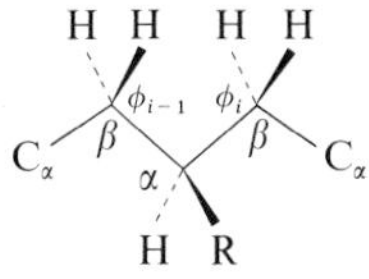

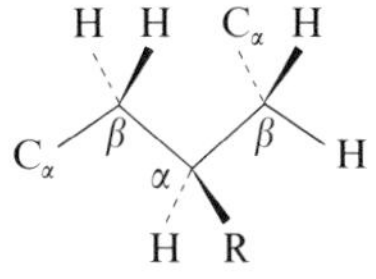

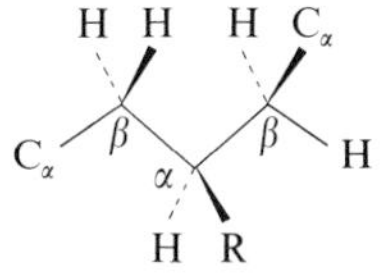

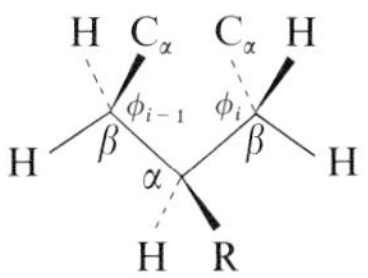

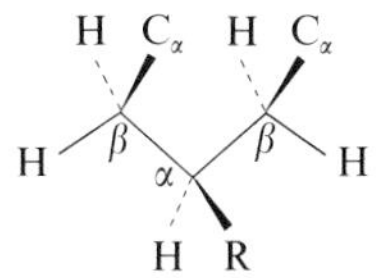

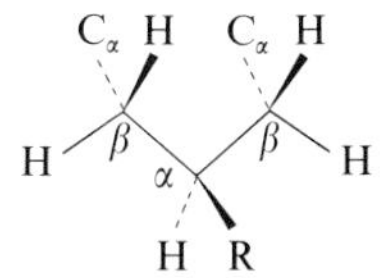

The $g^\pm g^\pm$ conformations involves no appreciable net interaction of the C_αs and their attached R groups, and are weighted as in the polymethylene chain (p. 249); but the g^+g^- and g^-g^+ conformations cause severe steric overlaps not only between the C_αs but also between their attached R groups. These conformations may be assigned zero statistical weight unless the first atom in the R group is relatively small (F, OR', OCOR') or involves special interactions, such as the $-O-H\cdots O-$ hydrogen bonding in polyvinyl alcohol. The complete statistical weight matrices for the conformational states about α-carbons are accordingly:

$$
U_{\alpha_{12}}(d) = \begin{matrix} & t & g^+ & g^- \\ t & \\ g^+ & \\ g^- & \end{matrix}\begin{bmatrix} \eta & 1 & \tau \\ \eta & 1 & \tau\omega \\ \eta & \omega & \tau \end{bmatrix} \tag{3.47a}
$$

$$
U_{\alpha_{12}}(1) = \begin{matrix} & t & g^+ & g^- \\ t & \\ g^+ & \\ g^- & \end{matrix}\begin{bmatrix} \eta & \tau & 1 \\ \eta & \tau & \omega \\ \eta & \tau\omega & 1 \end{bmatrix} \tag{3.47b}
$$

For calculation of the partition function and for other purposes, the matrices $U_{m_{12}}$ or $U_{r_{12}}$ are multiplied alternately in succession with the matrix $U_{\alpha_{12}}$. Chains may be generated corresponding to any value of P_m from isotactic ($P_m = 1$) to syndiotactic ($P_m = 0$), and their conformation-dependent properties can be calculated in the same manner as we have already described for simpler, symmetric chains.

The general conclusions from the consideration of the steric interactions in monosubstituted vinyl polymer chains are that *isotactic* chains tend to assume a conformation of alternating gauche (i.e., g^+, g^- being strongly excluded by $CH_2 \cdots C_\alpha H$ interactions) and trans bond rotational states. Such $(gt)(gt)$ sequences, regularly repeated, generate a 3_1 helix, i.e., a helix that makes one turn for each three monomer units and consequently exhibits threefold symmetry viewed along the helical axis (see Fig. 3.33a). In solution, such helical sequences will be right-handed and left-handed with equal probability at any instant, but will reverse their chirality at random and with great rapidity, for the conformational states have very short lifetimes, as we have seen (p. 247). The tendency of $(gt)(gt)$ sequences to propagate themselves depends on the energies of the tt and $g^\pm g^\mp$ conformations, which of course in turn depend upon the nature of the α-substituents. The junction between left- and right-handed helices represented by $\cdots (gt)(gt)(tg)(tg) \cdots$ is permitted to a substantial extent because the tt bond conformation at C_α (a) is energetically relatively favorable. However, the sequence $\cdots (tg)(tg)(gt)(gt) \cdots$ requires a $g^\pm g^\mp$ conformation (b), which involves $C_\alpha \cdots C_\alpha$ contacts, the energy of which, $E_{\omega''}$, is high. If the statistical weight of (b) is zero, only one junction of two helices of opposite sign is present; the chain cannot return again via conformation (a) to the other chirality. In fact, steric restrictions are not this stringent, and isotactic chains behave essentially as random coils.

For syndiotactic sequences, tt conformations are normally favored, with a substantial weight to g^+g^+, particularly in polyolefin chains, where dipole interactions are absent. In polyvinyl chloride, for example, such interactions

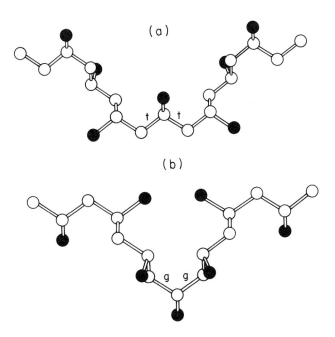

(a)

(b)

may be expected to favor *tt* conformations (Shimanouchi and Tasumi, 1961). For syndiotactic chains in solution, it is reasonable to expect substantial occurrence of helical conformations having sequences of the type $\cdots (gg)(tt)(gg)(tt) \cdots$. (As we shall see in Section 3.9.6, these expectations are generally borne out by the known crystalline conformations.)

—The successful calculation of conformation-dependent properties—unperturbed solution dimensions in particular—depends of course upon the proper choice of conformational states, i.e., rotational angles corresponding to energy minima, and upon the use of appropriate atom–atom or group–group potential energy functions, including electrostatic interactions in polar chains. Valence angles (not necessarily tetrahedral) and bond lengths must also be known. Many of these parameters may be fairly accurately estimated from x-ray and spectroscopic data on small molecules, but some remain adjustable within fairly wide limits; among these are E_η, E_ω, $E_{\omega'}$, and $E_{\omega''}$. In addition, the 3-state model is itself an oversimplified one, although often adequate. (This can be seen even for *n*-pentane, in which there are actually energy minima near the high energy $g^\pm g^\mp$ state achievable by appropriate adjustments of ϕ_2 and ϕ_3 to relieve the $CH_3 \cdots CH_3$ steric overlaps.) The result is that conformational calculations for most vinyl polymers—including α,α'-disubstituted chains, which we do not consider here—are semiempirical and must be accommodated to the experimental observations, usually by considerably complicating the relatively simple picture we have employed in this discussion.

As examples, let us briefly summarize the theoretical treatments of *polypropylene* and *polystyrene*. Calculations by Flory (Flory *et al.*, 1966), employing a 3-state model under the assumption that all the statistical weights ω, ω' and ω'' are zero, i.e., that conformations involving the four-bond CH_2 and side-chain interactions we have previously discussed are excluded, showed that the characteristic ratio C_∞, though somewhat sensitive to the allowed deviation of rotational angles, is relatively insensitive to the stereochemical configuration of the chain. It decreases somewhat from values of 10–12 when $P_m = 0$ (i.e., a syndiotactic chain), to ~ 9 as P_m increases from 0.2 to 0.9 and then rises very rapidly to values exceeding 30 as P_m exceeds 0.95. Actual measurements of C_∞ for a polypropylene believed from proton NMR measurements to have an isotactic dyad content of ~ 0.98 yielded a value of 4.7 (Heatley *et al.*, 1969). Such high values of isotactic content are confirmed by the ^{13}C NMR measurements discussed in Section 3.4. More recent calculations (Suter and Flory, 1975), however, using a 5-state model with values of E_ω of ~ 1200 cal predict a monotonic decrease of C_∞ with isotactic dyad content to a minimum of less than 5 for a purely isotactic chain.

Similarly, earlier calculations (Flory *et al.*, 1966) of the dimensions of isotactic polystyrene chains indicated exclusion of the *tt* conformation (see p. 253 *et seq*) owing to severe steric overlaps of the phenyl groups. More recent calculations (Yoon *et al.*, 1975) take into account the π–π attractive energy (~ 5 kcal) of the phenyl groups in this conformation. The absence of this attraction in other conformations is partially compensated for by solvent interactions; the net result is that the *tt* conformation in the meso configuration is believed to be permitted to an extent of ~ 8–12%. The calculated value of C_∞, excessively high in previous treatments, now agrees with the observed value of ~ 11.

3.9.5. EXPERIMENTAL OBSERVATION OF POLYMER CHAIN CONFORMATIONS IN SOLUTION

The detailed experimental observation of *local* polymer conformations in solution—as distinguished from end-to-end distances, dipole moments, and other similar "one-parameter" quantities—is not particularly easy, and has been accomplished principally by spectroscopic methods. Vibrational spectroscopy has had limited application, being mainly useful for biopolymers, which we shall discuss in Chapter 8. There is ultraviolet spectroscopic evidence (Longworth, 1966; Vala and Rice, 1963) for a helical conformation of isotactic polystyrene. Probably the most powerful method is NMR, chiefly through exploitation of the strong angular dependence of the vicinal proton–proton J coupling. We cannot discuss the method in detail here, as it involves considerable complexities when long chains are involved. The observations are somewhat hampered by dipolar broadening, as already discussed in Section 3.3.3. Interpretation of vinyl polymer spectra has been greatly aided by use of small model compounds: *meso* and *racemic* 2,4-disubstituted pentanes and 2,4,6-trisubstituted heptanes having "isotactic," "heterotactic," and "syndiotactic" configurations. (For a fuller discussion see Bovey, 1972.) The results of such studies in general support and confirm the steric influences and the results of the conformational calculations that we have just discussed.

3.9.6. EXPERIMENTAL OBSERVATION OF POLYMER CHAIN CONFORMATIONS IN THE CRYSTALLINE SOLID STATE

In the *solid state*, vibrational spectroscopy has proved to be of considerable power for the determination of chain conformations, particularly in crystallizable polymers. The splitting of the CH_2 rocking band reflects the interaction of the chains with their neighbors in the crystalline array and is consistent with a planar zigzag conformation. By the method of *normal coordinate analysis* (see Herzberg, 1945), it is possible to account for the positions and splittings of the vibrational lines in IR and Raman spectra and thereby to establish the intrachain and interchain force field. This has been particularly successful for polyethylene (Tasumi and Krimm, 1967) and for stereoregular vinyl polymers (Miyazawa, 1967).

The primary method for the determination of macromolecular conformations in the crystalline solid state is *x-ray diffraction*. Valuable contributions have also been made using *electron diffraction*. Space does not permit us to discuss the basis of the method here. For this there are many excellent sources; some of these are listed under "X-Ray Diffraction" in the General References. (Further discussion, with respect to crystal morphology, will be found in Chapter 5.) We shall deal only with the results.

The influences that determine the conformation of a polymer chain in the crystalline state are primarily those that we have already described as determining the solution conformation. In addition to these intramolecular requirements, one must also consider intermolecular energies associated with chain packing. These are generally relatively small (except in cases where interchain hydrogen bonding is involved, as in polyamides (p. 266) but may affect the choice between conformations having nearly equal internal energies and will play a part in first-order transitions between crystalline forms.

We have seen that for *polyethylene* the trans–trans or planar zigzag conformation is the form of lowest energy and that the virbational spectra are consistent with this finding. The crystal structure is shown in Fig. 3.30. The repeat distance along the chain axis is 0.253 nm, and represents the length of one monomer unit, i.e., twice the projection of the C–C bond

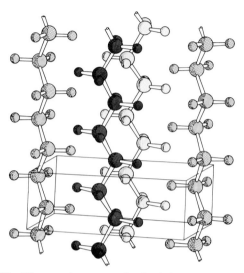

FIG. 3.30.   The crystal structure of polyethylene (from Bunn, 1946).

distance on the axis. The portion of the structure that reproduces the whole when repeated in three dimensions is the *unit cell* and is represented by the parallelopiped at the bottom of the figure. It contains segments of two chains (shown in the center; for clarity, neighboring chains are reproduced in lighter tones) and is the same as that of crystalline linear paraffins. The morphology of polyethylene is actually considerably more complex than Fig. 3.30 indicates, owing to the occurrence of amorphous regions, lattice defects, and folds. This is discussed in Chapter 5.

The substitution of hydrogen by the somewhat larger fluorine atom increases the internal energy of the trans–trans conformation enough so that the chains of *polytetrafluoroethylene* exhibit instead a twisted structure. The three-bond rotational angle is increased from $0°$ to $14°$ to minimize steric interactions, and this slow-turning ribbon consequently repeats every 13 carbon atoms (1.69 nm), constituting a 13_6 helix, corresponding to 2.17 monomers per turn. Unlike those of polyethylene, the chains of polytetrafluoroethylene are nearly cylindrical in cross section. The structure at a temperature below approximately $19°$ is shown in Fig. 3.31. Above $19°$, the conformation undergoes a first-order crystalline transformation in which the chains exhibit a 15_7 helical conformation with a torsional angle of about $12°$ and a repeat distance of 1.95 nm. Above $30°$, the structure becomes a torsionally oscillating one.

We have seen (p. 251) that in the chains of *polyoxymethylene*, unlike those of polyethylene, the gauche conformation is of lower energy than the trans.

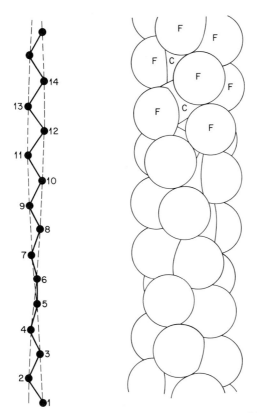

FIG. 3.31. Conformation of polytetrafluoroethylene (from Bunn and Howells, 1954).

The most stable crystalline chain conformation is in consequence a 9_5 helix (Fig. 3.32). The repeat distance is 1.73 nm. The COC bond angle is 112° and the OCO bond angle 111°. The helices interlock tightly and are all of one chirality in a single crystal or in large portions of it (Uchida and Tadokoro, 1967).

We have seen that in isotactic vinyl polymer chains the preferred conformation is often the 3_1 helix, generated from alternating gauche and trans conformations. This is true for polypropylene, polybutene-1, poly(5-methylhexene)

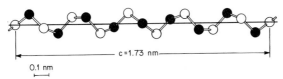

FIG. 3.32. The conformation of polyoxymethylene. Black circles represent carbon atoms and white circles oxygen atoms (from Uchida and Tadokoro, 1967).

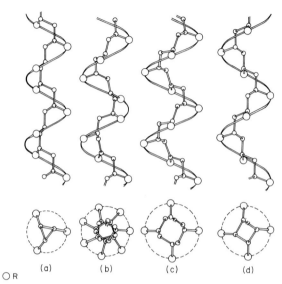

O R

(a)   (b)   (c)   (d)

FIG. 3.33.   Helical conformations of isotactic polymers with varying side-chains (from Natta and Corradini, 1960).

(a)   R = —CH$_3$,   —C$_2$H$_5$,   —CH=CH$_2$,
      —CH$_2$CH(CH$_3$)$_2$,   —OCH$_3$,
      —OCH$_2$CH(CH$_3$)$_2$

(b)   R = —CH$_2$CH(CH$_3$)CH$_2$CH$_3$,   —CH$_2$CH(CH$_3$)$_2$

(c)   R = —CH(CH$_3$)$_2$,   —C$_2$H$_5$

(d)   R =

(R=(CH$_3$)$_2$CHCH$_2$CH$_2$–), poly(vinyl methyl ether), poly(vinyl isobutyl ether) (R=—O–CH$_2$CH(CH$_3$)$_2$), and polystyrene (Fig. 3.33a). If the side chain is bulkier, particularly near its attachment to the α-carbon, the helix expands; thus poly(4-methylhexene-1) (R=CH$_2$CH(CH$_3$)CH$_2$CH$_3$) and poly(4-methylpentene-1) (R=CH$_2$CH(CH$_3$)$_2$) exhibit 7$_2$ helices (Fig. 3.33b), while poly(3-methylbutene-1) (R=i-propyl), poly(o-methylstyrene), poly(o-methyl-p-fluorostyrene) and poly (α-vinylnaphthalene) have 4$_1$ helices. Surprisingly, poly(o-fluorostyrene) crystallizes with a 3$_1$ helical conformation, whereas with the fluorine in the para position a 4$_1$ helix is preferred. This is an instance of chain conformation being dictated by crystalline packing requirements.

A number of isotactic vinyl polymers show *polymorphism*. Polypropylene has been reported to exist in four crystalline modifications, all 3$_1$ helices but differing in the details of the packing of the chains. Polybutene-1 initially crystallizes from the melt as an 11$_3$ helix, which then slowly transforms to a room-temperature stable 3_1 helical form; the equilibrium transition between them occurs at about 110° (Luongo, 1975).

The packing of vinyl chains having the same conformation may occur in several ways owing to the fact that the helices have a *sense of direction* (even though the basic chain structure does not) as well as chirality. This is illustrated in Fig. 3.34, which shows a side view of a right-handed 3_1 helix of isotactic polystyrene. In common with all such helices, the α-substituents make an angle with the helical axis. (In the case of polystyrene, the orientation of the plane of the phenyl group is such as to bisect the $C_\beta C_\alpha C_\beta$ angle.) Packing energies may differ depending upon whether each chain is surrounded by other chains—usually with a coordination of 3—of the same or of opposite direction, and of the same or of opposite chirality. In polystyrene and polypropylene, it appears that alternating chirality prevails, but it is not obvious whether the directions of the helices are (a) entirely random or (b) uniform over small domains or even whole single crystals. It is probable that in the most stable form of polypropylene, (b) is the case.

There is relatively little information concerning the conformations of syndiotactic vinyl polymer chains in the crystalline state owing to the scarcity of authentically established syndiotactic polymers. Syndiotactic *poly(1,2-butadiene)* (Natta and Corradini, 1956) and *poly(vinyl chloride)*

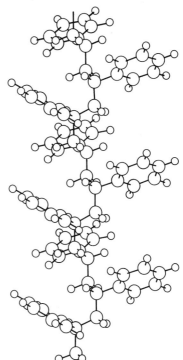

FIG. 3.34.   Side view of the 3_1 helix of isotactic polystyrene (from Natta, 1960).

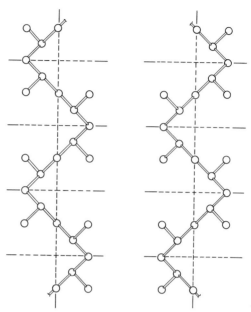

FIG. 3.35. Side view of enantiomorphous helices of syndiotatic polypropylene (from G. Natta, P. Corradini, and P. Ganis, *Makromol. Chem.* 39, 238 (1960)).

(Natta, 1961) crystallize in a planar zigzag conformation. Syndiotactic *polypropylene* forms a $\cdots (gg)(tt)(gg)(tt) \cdots$ helix. The side and end views, together with the crystal packing, are shown in Figs. 3.35 and 3.36. There are four monomer units per turn (Natta *et al.*, 1960).

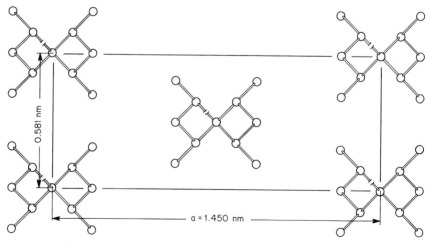

FIG. 3.36. Mode of packing of syndiotactic polypropylene helices (from Natta *et al.*, 1960).

Of the many α, α-disubstituted vinyl polymer chains, *polyisobutylene* is of particular interest. This polymer is rubbery in the normal solid state, but the structural regularity of the chains permits a considerable degree of crystallinity to develop in stretched fibers. Severe steric overlap of the methyl groups rules out a planar zigzag model. The conformation in best agreement with x-ray (Fuller *et al.*, 1940; Liquori, 1953, 1955; Bunn and Holmes, 1958) and energy calculations (Allegra *et al.*, 1970) appears to be an 8_3 helix (Fig. 3.37).

As we have already seen (Chapter 1, Section 1.3), natural rubber (poly(*cis*-1, 4-isoprene)) also crystallizes on stretching. This structure, one

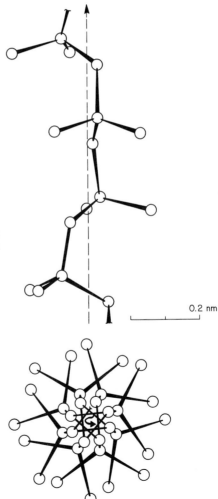

FIG. 3.37.   Crystalline conformation in stretched fibers of polyisobutylene (from Allegra *et al.*, 1970).

0.2 nm

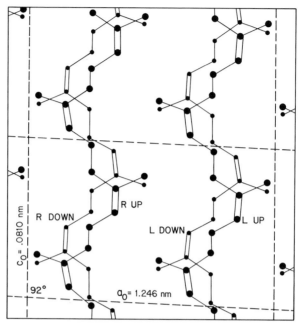

FIG. 3.38.   The chain conformation and crystal packing in stretched natural rubber (from Bunn, 1942).

of the earliest determined by x ray for a macromolecule, is shown in Fig. 3.38. The repeat distance along the chain is 0.81 nm, indicating two isoprene units per conformational repeat (Bunn, 1942).

The chain conformation and crystal morphology of *polyamides* (see Chapter 2, Section 2.4) are strongly influenced by the presence of hydrogen bonds between the amide carbonyl and NH groups of neighboring chains, resulting in the existence of sheets. The forces stabilizing the sheets are considerably greater than those between them. The chains are planar zig-zags and in nylon 66 are packed as shown in Fig. 3.39a (Bunn and Garner, 1947), which represents the crystalline form designated as α. There are also β and λ forms that have the same structure within each sheet but differ somewhat in the offset arrangement of the sheets with respect to each other.

In nylon 6, unlike nylon 66, the chains have a structural sense of direction and hence may in principle be arranged parallel or antiparallel in the hydrogen bonded sheets. Both models and x-ray data indicate an antiparallel arrangement, in which, as shown in Fig. 3.39b, all hydrogen bonds can be readily formed (Holmes *et al.*, 1955). The best defined form is again termed α; in this form, the arrangement of sheets is the same as in the α form of nylon 66.

(a)

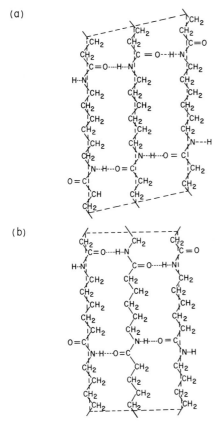

(b)

FIG. 3.39. Chain conformations of polyamides: (a) nylon 66; (b) nylon 6 (from Holmes et al. 1955).

The related but much more complex conformational problems presented by polypeptides and proteins are discussed in Chapter 8, Section 8.3.

3.9.7. EXPERIMENTAL OBSERVATION OF POLYMER CHAIN CONFORMATIONS IN THE AMORPHOUS SOLID STATE

For some years there has been a question concerning the conformations of macromolecules in the bulk amorphous state, both rubbery and glassy. One school of thought has maintained that the polymer chains coil upon themselves and do not mingle with their neighbors, whereas another school has maintained that the molecules do quite the opposite, responding to their neighbors by forming bundles of chains that are correlated to a substantial extent, though falling short of the degree of order detectable by x-ray diffraction (for reviews, see Tonelli, 1970; Flory, 1973). Flory postulated long ago (Flory, 1953) that in fact they do neither, but instead exhibit the unperturbed dimensions characteristic of their behavior in Θ solvents, i.e., the dimensions discussed in Section 3.9.4. The controversy has been kept alive by the absence of truly definitive experimental means for directly determining the mean square end-to-end distances or radii of the chains in amorphous polymers.

This lack has now been supplied by *small angle coherent neutron scattering* measurements (see Allen, 1976). In this technique, a monochromated and collimated beam of "cold" neutrons is made to impinge on a block of polymer and the angular dependence of the scattering determined by an array of counters. In order to detect individual molecules in the sample, a small fraction (1–2%) of normal polymer is randomly dispersed in pre-deuterated polymer. For example, normal poly(methyl methacrylate) may be dissolved in perdeuterated monomer, $CD_2{=}C(CD_3)COOCD_3$, and the latter then polymerized. The greater scattering power of the deuterons compared to the protons provides the equivalent of the difference in refractive indices of polymer and solvent in Rayleigh light scattering measurements Chapter 4, Section 4.4.2), and in fact the two experiments are exactly parallel. (One may also use a dilute deuterated polymer in a protonated matrix.) The results entirely confirm Flory's postulate. For example, Kirste *et al.* (1973) obtained the results shown in Table 3.6 for a nearly monodisperse poly(methyl methacrylate) of $\bar{M}_w = 2.5 \times 10^5$ at 25°: More extensive studies of Benoit *et al.* (1973), on polystyrene have confirmed these findings for several fractions of polystyrene covering a 50-fold span of molecular weight.

TABLE 3.6

Neutron Scattering Results for Poly(Methyl Methacrylate)

In the glass	$\langle r^2 \rangle_z^{1/2\,a} = 12.5$ nm; $A_2{}^b = 0$
In *n*-butyl chloride (a θ solvent)	$\langle r^2 \rangle_w^{1/2} = 11.0$ nm; $A_2 = 0$
In dioxane (a "good" solvent)	$\langle r^2 \rangle_w^{1/2} = 17.0$ nm; $A_2 = 5 \times 10^{-4}$ cm^3 g^{-2}

^{a} $\langle r^2 \rangle_z^{1/2}$ and $\langle r^2 \rangle_w^{1/2}$ are the mean square z-average and weight average chain dimensions, respectively.

^{b} A_2 is the first virial coefficient as obtained from light scattering experiments; see Chapter 4, Section 4.4.2.

REFERENCES

Abbås, K. B., Bovey, F. A., and Schilling, F. C. (1975). *Makromol. Chem. Suppl.* **1**, 227.
Abe, A., Jernigan, R. L., and Flory, P. J. (1966). *J. Am. Chem. Soc.* **88**, 631.
Alfrey, T. Jr., Bohrer, J. J., and Mark, H. F. (1952). "Copolymerization." Wiley (Interscience), New York.
Allegra, G., Benedetti, E., and Pedone C. (1970). *Macromolecules* **3**, 727.
Allen, G. (1976). "Structural Studies of Macromolecules by Spectroscopic Methods" (K. J. Ivin, ed.). Wiley, New York.

Benoit, H. *et al.* (1973). *Nature (London)* **245**, 13.
Bovey, F. A. (1972). "High Resolution NMR of Macromolecules," Chapter IX. Academic Press, New York.
Bovey, F. A., and Tiers, G. V. D. (1960). *J. Polym. Sci.* **44**, 173.
Bovey, F. A., Abbås, K. B., Schilling, F. C., and Starnes, W. H. (1975). *Macromolecules* **8**, 437.
Bovey, F. A., Schilling, F. C., McCrackin, F. L., and Wagner, H. L. (1976). *Macromolecules* **9**, 76.
Bunn, C. W. (1942). *Proc. R. Soc. London Ser A* **180**, 40.
Bunn, C. W. (1946). "Chemical Crystallography." Oxford Univ. Press, London and New York.
Bunn, C. W., and Garner, E. V. (1947). *Proc. R. Soc. London Ser. A* **189**, 39.
Bunn, C. W., and Holmes, D. R. (1958). *Discuss. Faraday Soc.* **25**, 95.
Bunn, C. W., and Howells, E. R. (1954). *Nature (London)* **174**, 549.
Cheng, H. N., and Ginsberg, A. P. (1975). Private communication.
Dorman, D. E., Otocka, E. P., and Bovey, F. A. (1972). *Macromolecules* **5**, 574.
Duch, M. W., and Grant, D. M. (1970). *Macromolecules* **3**, 165.
Fischer, T., Kinsinger, J. B., Wilson, C. W III (1966). *J. Polym. Sci. Part B* **4**, 379.
Flory, P. J., and Leutner, F. S. (1948). *J. Polym. Sci.* **3**, 880.
Flory, P. J., and Leutner, F. S. (1950). *J. Polym. Sci.* **5**, 267.
Flory, P. J., Mark, J. E., and Abe, A. (1966). *J. Am. Chem. Soc.* **88**, 639.
Fuller, C. S., Frusch, C. J., and Pape, N. R. (1940). *J. Am. Chem. Soc.* **62**, 1905.
George, M. H., Grisenthwaite, R. J., and Hunter, R. F. (1958). *Chem. Ind. (London)* 1114.
Harwood, H. J. (1965). *Angew. Chem. Int. Ed. Engl.* **4**, 1051.
Heatley, F., Salovey, R., and Bovey, F. A. (1969). *Macromolecules* **2**, 619.
Hellwege, K. H., Johnsen, U., and Kolbe, K. (1966). *Kolloid-Z* **214**, 45.
Herzberg, G. (1945). "Infrared and Raman Spectra," Chapter 3. Van Nostrand-Reinhold, Princeton, New Jersey.
Holmes, D. R., Bunn, C. W., and Smith, D. J. (1955). *J. Polym. Sci.* **17**, 159.
Ivin, K. J., and Navratil, M. (1970). *J. Polym. Sci. Part B* **8**, 51.
Ito, K., and Yamashita, Y. (1965), *J. Polym. Sci. Part B* **8**, 51.
Kinsinger, J. B., Fischer, T., and Wilson, C. W. III (1966). *J. Polym. Sci. Part B* **4**, 379.
Kinsinger, J. B., Fischer, T., and Wilson, C. W. III (1967). *J. Polym. Sci. Part B* **5**, 285.
Kirste, R. G., Fruse, W. A., and Schelten, J. (1973). *Makromol. Chem.* **162**, 299.
Kössler, I. (1967). *Encycl. Polym. Sci. Technol.* **7**, 620.
Liquori, A. M. (1953). *IUPAC Meeting, 13th, Stockholm.*
Liquori, A.M. (1955). *Acta Crystallogr.* **9**, 345.
Longworth, J. W. (1966). *Biopolymers* **4**, 1131.
Luongo, J. P. (1975). Private communication.
Mark, J. E., and Flory P. J. (1966a). *J. Am. Chem. Soc.* **87**, 1415.
Mark, J. E., and Flory, P. J. (1966b). *J. Am. Chem. Soc.* **88**, 3702.
Miyazawa, T. (1967). "The Stereochemistry of Macromolecules" (A. D. Ketley, ed.), Vol. 3, Chapter 3. Dekker, New York.
Natta, G. (1960). *Makromol. Chem.* **39**, 238.
Natta, G., and Corradini, P. (1956). *J. Polym. Sci.* **20**, 251.
Natta, G., and Corradini, P. (1960). *Nuovo Cimento Suppl.* **15 1**, 9.
Natta, G., Corradini, P., and Ganis, P. (1960). *Makromol. Chem.* **39**, 238.
Natta, G. *et al.* (1961). *Atti Accad. Naz. Lincei Rend., Classe Sci. Fis. Mat. Nat.* **31**, 17.
Overberger, C. G., and Yamamoto, N. (1965). *J. Polym. Sci. Part B* **3**, 569.
Pasteur, G. (1975). Private communication.
Ramey, K. C., and Brey, W. S. Jr. (1967). *Macromol. Sci. Rev. Macromol. Chem.* **1**, 263.
Randall, J. C. (1973). *J. Polym. Sci. Polym. Phys. Ed.* **11**, 275.

Rigo, A., Palma, G., and Talamini, G. (1972). *Makromol. Chem.* **153**, 219.
Schaefer, J. (1974). "Topics in ^{13}C NMR Spectroscopy" (G. C. Levy, ed.), Vol. 1, pp. 159–163. Wiley (Interscience), New York.
Schilling, F. C. (1974). Private communication.
Schilling, F. C., and Bovey, F. A. (1975). Unpublished observations.
Shimanouchi, T., and Tasumi, M., *Spectrochim. Acta*, **17**, 755 (1961).
Stein, D. J. (1964). *Makromol. Chem.* **76**, 170.
Suter, U. W., and Flory, P. J. (1975). *Macromolecules* **8**, 765.
Tasumi, M., and Krimm, S. (1967). *J. Chem. Phys.* **46**, 755.
Uchida, T., and Tadokoro, H. (1967). *J. Polym. Sci. Part A-2* **5**, 63.
Vala, M. T. Jr., and Rice S. A. (1963). *J. Chem. Phys.* **39**, 2348.
Willbourn, A. H. (1959). *J. Polym. Sci.* **34**, 569.
Willis, H. A., and Cudby, M. E. A. (1968). *Appl. Spectrosc. Rev.* **1**, 237.
Wilson, C. W. III (1963). *J. Polym. Sci. Part A* **1**, 1305.
Wilson, C. W. III, and Santee, E. R. Jr. (1963). *J. Polym. Sci. Part A* **1**, 1305.
Wilson, C. W. III, and Santee, E. R. Jr. (1965). *J. Polym. Sci. Part C* **8**, 97.
Woodbrey, J. C., (1968). "The Stereochemistry of Macromolecules" (A. D. Ketley, ed.), Vol. 3. Dekker, New York.
Yoon, D. Y., Sundararajan, P. R., and Flory, P. J. (1975). *Macromolecules* **8**, 776.
Zambelli, A., Dorman, D. E., Brewster, A. I. R., and Bovey, F. A. (1973). *Macromolecules* **6**, 925.

GENERAL REFERENCES

I Vibrational Spectroscopy
General

Colthup, N. B., Daly, L. H., and Wiberly, S. E. (1964). "Introduction to Infrared and Raman Spectroscopy." Academic Press, New York.
Potts, W. J. Jr. (1963). "Chemical Infrared Spectroscopy," Vol. I, Technique. Wiley, New York.

Polymers
Hendra, P. J. (1969). Laser-Raman spectra of polymers, *Adv. Polym. Sci.* **6**, 151.
Hummel, D. O. (1966). "Infrared Spectroscopy of Polymers." Wiley (Interscience), New York (Elementary).
Kössler, I. (1967). Infrared-absorption spectroscopy, *In* "Encyclopedia of Polymer Science and Technology" (N. M. Bikales, ed.). Wiley (Interscience), New York (Elementary).
Krimm, S. (1960). Infrared spectra of high polymers, *Adv. Polym. Sci.* **2**, 51 (Moderately Advanced).
Krimm, S. (1968). Infrared spectra and polymer structure, *Pure Appl. Chem.* **16**, 369 (Advanced).
Zbinden, R. (1964). "Infrared Spectroscopy of High Polymers." Academic Press, New York (Advanced).

II Nuclear Magnetic Resonance
General
Becker, E. D. (1969). "High Resolution NMR." Academic Press, New York.
Bovey, F.A. (1969). "Nuclear Magnetic Resonance Spectroscopy." Academic Press, New York.
Emsley, J. W., Feeney, J., and Sutcliffe, L. H. (1966). "High Resolution Nuclear Magnetic Resonance," 2 Vols. Pergaman, Oxford.

Levy, G. C., and Nelson, G. L. (1972). "Carbon-13 Nuclear Magnetic Resonance for Organic Chemists." Wiley (Interscience), New York.
Stothers, J. B. (1972). "Carbon-13 NMR Spectroscopy." Academic Press, New York.

Polymers
Bovey, F. A. (1972). "High Resolution NMR of Macromolecules." Academic Press, New York.
Schaefer, J. (1974). In "Topics in Carbon-13 NMR Spectroscopy" (G. C. Levy, ed.), Chapter 4. Wiley (Interscience), New York.

III   Conformations of Polymer Chains
Birshtein, T. M., and Ptitsyn, O. B. (1966). "Conformations of Macromolecules" (translated from the Russian by S. N. Timasheff and M. J. Timasheff). Wiley (Interscience), New York.
Flory, P. J. (1969). "Statistical Mechanics of Chain Molecules." Wiley (Interscience), New York.
Tonelli, A. E. (1977). Polymer conformation and configuration, In "Encyclopedia of Polymer Science and Technology" (N. Bikales, ed.), 2nd ed.
Vol'kenshtein, M. V. (1963). "Configurational Statistics of Polymer Chains" (translated from the Russian edition by S. N. Timasheff and M. J. Timasheff). Wiley (Interscience), New York.

IV   X-Ray Diffraction and Polymer Chain Conformations in the Crystalline State
Corradini, C. P. (1968). "The Stereochemistry of Macromolecules," Vol. 3, Chapter 1. Dekker, New York.
Geil, P. H. (1963). "Polymer Single Crystals." Wiley (Interscience), New York.
Tanford, C. (1961). "Physical Chemistry of Molecules," Chapter 2. Wiley, New York.
Wunderlich, B. (1973). "Macromolecules Physics," Vol. 1. Academic Press, New York,

Chapter 4

MACROMOLECULES IN SOLUTION

T. K. KWEI

4.1. Introduction

Macromolecules, by virtue of their high molecular weight, are soluble only in selected solvents. As a rule, solubility decreases when strong intermolecular forces are present. Thus crystalline polymers are often soluble only at elevated temperatures, and polymers containing polar groups and hydrogen bonds dissolve usually in solvents that interact strongly with them.

Although the basic principles of physical chemistry are applicable to molecules of all sizes, the adaptation of these principles to macromolecular

273

solutions deserves special consideration for several reasons. From the experimental viewpoint, many standard techniques of measuring solution properties, such as the lowering of vapor pressure of the solvent, become insensitive for solutions of macromolecules. On the other hand, a number of techniques that have found only limited use for solutions of small molecules, for example, solution viscosity, become important in the study of macromolecules. There are also conceptual differences on theoretical grounds. In classical physical chemistry dealing with small molecules, a limiting case that allows rigorous mathematical analysis is the "dilute" solution in which the solute molecules are considered to be separated effectively from each other. In macromolecular solutions, the interconnectivity of segments results in high local concentration of segments even though the total number of molecules in the solution is small. Therefore a dilute solution of macromolecules does not conform strictly to the classical definition of "dilute" solution for small molecules. A third point of departure is that experimental measurements yield "average" quantities for macromolecules. The concept of average size in the statistical-mechanical sense has already been elucidated in Section 3.9. There is still another kind of average that arises from the polydisperse nature of macromolecules. The measured quantities represent, with rare exceptions of some monodisperse biological macromolecules, average properties for species of different chain lengths. This is a feature not encountered with small molecules. In this chapter, we shall consider the theoretical framework that takes into account these characteristics and provides the basis for experimental measurement of molecular weight and size.

4.2. Thermodynamics of Macromolecules in Solution

Before we tackle the complex problem of macromolecules in solution, we shall first outline briefly the properties of binary solutions consisting of small molecules of equal or comparable sizes. The special features of solutions containing long chain solutes will then be discussed.

A summary of thermodynamic terms and relations is given below. The Gibbs free energy G of a system containing n_i moles of species i at absolute temperature T and pressure p is defined as

$$G = H - TS = E + pV - TS \tag{4.1}$$

where H is the enthalpy, E the internal energy, V the volume, and S the entropy. The total differential of G in the variables T, p, n_i is

$$dG = -S\,dT + V\,dp + \sum_i \mu_i\,dn_i \tag{4.2}$$

$$\partial G/\partial T = -S, \qquad \partial G/\partial p = V, \qquad \partial G/\partial n_i = \mu_i \qquad (4.3)$$

The partial molar quantities at constant temperature and pressure are defined by

$$Y_i = (\partial Y/\partial n_i)_{T,p} \qquad (4.4)$$

where Y is any extensive variable whose value depends on the amount of material in the system. For example, G, H, V, and S are extensive variables. The partial molar free energy of species i is called the chemical potential μ_i.

In the following sections, our interest concerns mixtures of two or more components. We shall always compare the properties of the mixture to those of pure components. The change in the thermodynamic function upon mixing will be denoted by a subscript m. For example, the free energy of mixing of a binary system is written as

$$\Delta G_m = G(n_1,n_2) - G^0(n_1) - G^0(n_2) = n_1(\mu_1 - \mu_1^0) - n_2(\mu_2 - \mu_2^0) \qquad (4.5)$$

with the superscript zero referring to pure components. The difference between μ_i and μ_i^0 is often expressed in terms of activity a_i of the species i, defined by

$$\mu_i - \mu_i^0 = RT \ln a_i \qquad (4.6)$$

The chemical potentials of the components in a mixture are related by the Gibbs–Duhem equation,

$$\sum n_i \, d\mu_i = 0 \quad \text{at constant } T \text{ and } p \qquad (4.7)$$

It is a common practice to compare the thermodynamic function of mixing of an actual system with the corresponding value in an ideal solution. The difference is called the thermodynamic excess function denoted by superscript E. The excess functions can also be expressed in partial molar quantities, e.g., h^E, s^E, and g^E. Their values for the system n-hexane and cyclohexane (Prigogine and Mathot, 1950; Mathot, 1950) are shown in Fig. 4.1. Bear in mind that in an ideal solution ΔH_m and ΔV_m are zero but ΔS_m and ΔG_m are not zero. This will be discussed in greater detail in Section 4.2.1.

4.2.1. SOLUTIONS OF SMALL MOLECULES

Ideal solutions are defined as those in which the activity of each component is equal to its mole fraction.

$$a_i = N_i \qquad (4.8)$$

Solutions of small molecules tend to approach ideal behavior when infinite dilution is neared. Let us consider two spherical molecules 1 and 2 that are identical in size and intermolecular forces. Molecules of one type may

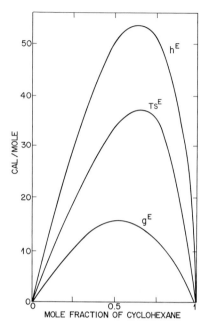

FIG. 4.1.   Thermodynamic excess function for mixtures of n-hexane (1) and cyclohexane (2) at $20°C$ (Prigogine and Mathot, 1950).

exchange neighbors with molecules of the other type without affecting the total energy of the system. The interchangeability of neighbors gives rise to a configurational entropy term because the solution may contain a large number of distinguishable arrangements of the two types of molecules. The configuration entropy contribution is the entropy of mixing ΔS_{m}.

The evaluation of configurational entropy is usually performed by placing the n_1 and n_2 molecules on a lattice of $(n_1 + n_2)$ sites (Fig. 4.2). The number

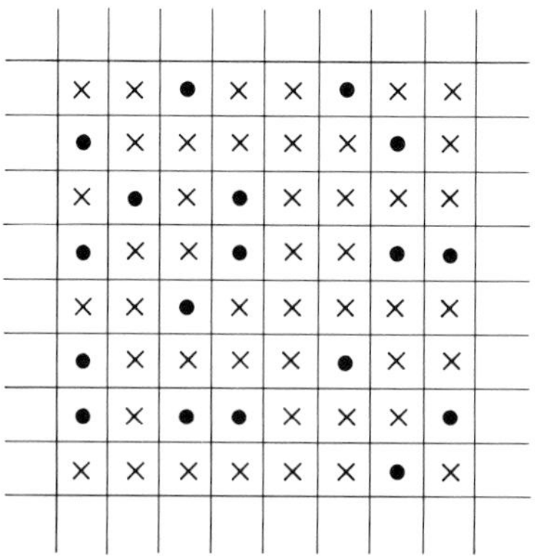

FIG. 4.2.   Schematic representation of monomeric solute molecules in a binary solution by the lattice model.

of distinguishable arrangements is

$$\Omega = (n_1 + n_2)!/n_1! n_2! \tag{4.9}$$

from which we obtain

$$\Delta S_m = k \ln \Omega = k(n_1 \ln N_1 + n_2 \ln N_2) \tag{4.10}$$

where N_1 and N_2 are mole fractions of molecules 1 and 2, respectively. The partial molar entropy $\partial \Delta S_m / \partial n_i$ is

$$S_1 - S_1^0 = -R \ln N_1 \quad \text{and} \quad S_2 - S_2^0 = -R \ln N_2 \tag{4.11}$$

Since the two types of molecules have the same force field, the heat of mixing is zero and the free energy of mixing ΔG_m equals $-T \Delta S_m$. The chemical potentials of the components in the solution are

$$\mu_1 - \mu_1^0 = RT \ln N_1 \quad \text{and} \quad \mu_2 - \mu_2^0 = RT \ln N_2 \tag{4.12}$$

or $a_i = N_i$, identical to Eq, (4.8). This gives a molecular description of the special conditions by which a solution behaves ideally.

Departure from ideality occurs when there is a finite heat of mixing that is related to the intermolecular forces operative between similar and dissimilar molecules. Let ε_{ij} be the energy required to separate molecules i and j. The energy of mixing that accompanies the formation of a contact between molecules 1 and 2 is

$$\Delta \varepsilon = \tfrac{1}{2} \varepsilon_{11} + \tfrac{1}{2} \varepsilon_{22} - \varepsilon_{12} \tag{4.13}$$

If ε_{12} happens to be the arithmetic mean of ε_{11} and ε_{12}, the heat of mixing remains zero, i.e., the solution is athermal. It is more likely, however, that the value of ε_{12} between dissimilar nonpolar molecules is approximated by the geometric mean of ε_{11} and ε_{12} (van Laar and Lorenz, 1925). Since the geometric mean is always smaller than the arithmetic mean, $\Delta \varepsilon$ is positive and mixing is generally endothermic for nonpolar molecules. (This conclusion does not hold true when strong interaction between dissimilar molecules is present.) With the use of the geometric mean approximation, Eq. (4.13) can be rewritten as follows:

$$\Delta \varepsilon = \tfrac{1}{2} (\varepsilon_{11}^{1/2} - \varepsilon_{22}^{1/2})^2 \tag{4.14}$$

Since ε_{ii} is proportional to the molar energy of vaporization ΔE_i^v, the reasoning outlined above leads to the following equation by Hildebrand and Scott (1950)

$$\Delta H_m = V \phi_1 \phi_2 [(\Delta E_1^v / V_1)^{1/2} - (\Delta E_2^v / V_2)^{1/2}]^2 \tag{4.15}$$

where V, V_1, V_2 are the molar volumes of the solution and the components, respectively, and ϕ_1, ϕ_2 are the volume fractions. The term $\Delta E_i^v / V_i$ is named "cohesive energy density" and its square root is commonly referred to as the solubility parameter δ. Typical values of δ for solvents are listed in Table 4.1 (Burrell and Immergut, 1967). They provide a convenient estimate of the magnitude of the enthalpy of mixing. Small (1953) has published a table of molar attraction constants that allows the estimation of the solubility parameter merely from the structural formula of the compound and its density.

TABLE 4.1
VALUES FOR δ FOR DIFFERENT SOLVENTS

Solvent	δ
Poorly hydrogen-bonded	
n-Pentane	7.0
n-Heptane	7.4
Apco thinner	7.8
Solvesso 150	8.5
Toluene	8.9
Tetrahydronaphthalene	9.5
o-Dichlorobenzene	10.0
1-Bromonaphthalene	10.6
Nitroethane	11.1
Acetonitrile	11.8
Nitromethane	12.7
Moderately hydrogen-bonded	
Diethyl ether	7.4
Diisobutyl ketone	7.8
n-Butyl acetate	8.5
Methyl propionate	8.9
Dibutyl phthalate	9.3
Dioxane	9.9
Dimethyl phthalate	10.7
2,3-Butylene carbonate	12.1
Propylene carbonate	13.3
Ethylene carbonate	14.7
Strongly hydrogen-bonded	
2-Ethylhexanol	9.5
Methyl isobutyl carbinol	10.0
2-Ethylbutanol	10.5
n-Pentanol	10.9
n-Butanol	11.4
n-Propanol	11.9
Ethanol	12.7
Methanol	14.5

It should be noted that Eq. (4.15) predicts only zero or positive values of ΔH_m but cannot account for exothermic mixing when there is a favorable interaction between unlike molecules, for example, through hydrogen bonding between chloroform and acetone.

When the heat of mixing is nonzero, reflecting a difference in the intermolecular forces between the two types of molecules, the entropy of mixing is expected to depart from the value calculated on the basis of a random distribution of molecules. A rigorous treatment is exceedingly difficult. But a good approximation is afforded by the concept of "regular solution" (Hildebrand and Scott, 1962) in which the entropy of mixing is considered to be the same as in ideal solution, and the deviation from ideal solution behavior is entirely accounted for by the heat of mixing. This assumption is reasonably sound when the ΔH_m is small and appears to represent experimental results adequately in many cases. The concept of a regular solution is also used in the formulation of the solution theory of macromolecules.

We have discussed the factors affecting the energetics and randomness of mixing. As a general rule, real solutions are seldom ideal except at the limit of infinite dilution. A convenient way to describe the nonideal behavior in a solution is to express the chemical potential of the solvent in terms of a power series of the mole fraction or concentration of the solute c_2,

$$\mu_1 - \mu_1^0 = RT\bar{V}_1 c_2((1/M_2) + A_2 c_2 + A_3 c_2^2 + \cdots) \qquad (4.16)$$

where $\bar{V}_1$ is the molar volume of the solvent and M_2 is the molecular weight of the solute. Note that the first term on the right-hand side is the limiting value for infinite dilution. The coefficients A_2 and A_3 are called the second and third virial coefficients, respectively, analogous to their usage in describing nonideal gases, for which the equation of state can be written in a power series of pressure p,

$$pV/RT = 1 + A_2 p + A_3 p^2 + \cdots \qquad (4.17)$$

The virial coefficients represent binary and higher-order interactions of solute molecules due to the volume excluded by a solute molecule and made unavailable for occupation by another solute molecule. The excluded volume arises from the fact that the distance between the centers of two rigid molecules cannot be less than the diameter of the molecule. The molecule therefore excludes a volume larger than its physically occupied space. When the total number of ways of placing the solute molecules in a dilute solution is evaluated, with due consideration of the "excluded volume" effect (Zimm, 1946), the virial coefficients can be shown to be related directly to the excluded volume U:

$$A_2 = U/M_2^2 \qquad (4.18)$$

The evaluation of the exact value of U, however, is possible only for simple geometry and the application of the excluded volume concept to dilute solution of macromolecules will be discussed in Section 4.2.3.

4.2.2. SOLUTIONS OF MACROMOLECULES—FLORY–HUGGINS THEORY

In a binary mixture of small molecules of the same size and shape, deviation from ideal solution behavior is considered to be associated with the heat of mixing. However, nonideality can also arise when there is a large difference in the sizes of the molecules even though the heat of solution is negligible. When one component of the solution is a macromolecule, the connectivity of chain segments imposes a special restriction on the arrangement of the two components in space. A quantitative theory was formulated by Flory (1942) and Huggins (1942a–c), who evaluated the number of distinguishable ways in which n_1 solvent molecules and n_2 macromolecular chains could be placed on a lattice (Fig. 4.3). Each macromolecule is represented by x segments, with the molar volume of the segment equal to that of the solvent molecule $\bar{V}_1$. There are a total of $(n_1 + xn_2)$ lattice sites, and the corresponding volume is $(n_1 + xn_2)\bar{V}_1$. (Note the implicit assumption of no volume change upon mixing.) The number of nearest neighbors of each lattice site is z, the coordination number. In the process of placing a chain segment on the lattice, a crucial assumption is made that the probability of occupancy of a given site may be approximated by the overall fraction of occupied sites, f, when i macromolecules are already present in the lattice.

$$f = xn_i/(n_i + n_2 x) \tag{4.19}$$

This assumption is invalid when the solution is very dilute and a few isolated macromolecules are separated from each other by large numbers of solvent

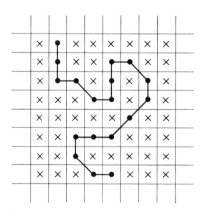

FIG. 4.3. Schematic representation of a macromolecule located in a liquid lattice.

molecules. Clearly, the segment density is nonuniform in a very dilute solution and f cannot be related to the probability of occupancy. But the approximation is a reasonable one in the concentration range in which the segment density is uniform as a result of intermeshing of the chains.

The first segment of the $(i + 1)$th chain can be placed on any vacant site. There are $(n_1 + xn_2)(1 - f_i)$ different ways for its placement. The second segment may be placed on any of the vacant adjacent sites; the number of available sites is $z(1 - f_i)$. The third segment has a choice of $(z - 1)(1 - f_i)$ sites, the factor $(z - 1)$ arising because the site already occupied by the first segment is unavailable. All the subsequent segments are placed in the same way as the third one. The number of distinguishable ways, v_{i+1}, to place the $(i + 1)$th chain is

$$v_{i+1} = \tfrac{1}{2}(1 - f_i)^x(n_1 + xn_2)z(z - 1)^{x-2} \qquad (4.20)$$

the factor $\tfrac{1}{2}$ resulting from the equivalence of two chain ends.

The total number of distinguishable ways of placing the n_2 chains on $(n_1 + xn_2)$ sites is

$$\Omega = (1/n_2!) \prod^{n_2 - 1} v_{i+1} \qquad (4.21)$$

Again, the factor $1/n_2!$ takes into account the indistinguishability of the n_2 chains. Once the chain segments are fixed, there is only one way to arrange the solvent molecules. Therefore Ω is also the total number of distinguishable ways to arrange the entire mixture. The number of ways Ω_2 by which chains can be placed on xn_2 sites is obtained by setting $n_1 = 0$ in Eq. (4.21).

The configurational entropy of mixing is

$$\Delta S_m = k \ln(\Omega/\Omega_2) = -k(n_1 \ln \phi_1 + n_2 \ln \phi_2) \qquad (4.22)$$

where ϕ_1 and ϕ_2 are volume fractions of the two components, respectively. It is important to note that in the above equation, the entropy of mixing is expressed in terms of volume fraction, unlike Eq. (4.10) for small molecules where mole fraction is the appropriate quantity.

The heat of mixing is obtained by multiplying the energy of formation of a segment–solvent contact, $\Delta\varepsilon$ in Eq. (4.13), by the number of such contacts.

$$\Delta H_m = z\Delta\varepsilon n_1 \phi_2 = kT\chi n_1 \phi_2 \qquad (4.23)$$

where

$$\chi = z\,\Delta\varepsilon/kT \qquad (4.24)$$

is commonly referred to as "Flory–Huggins interaction parameter." If the solvent molecule contains x_1 segments instead of only one, Eq. (4.24) should be amended by the factor x_1.

Applying the concept of a regular solution, the free energy of mixing is obtained by the combination of Eqs. (4.22) and (4.23),

$$\Delta G_{\mathrm{m}}/RT = n_1 \ln \phi_1 + n_2 \ln \phi_2 + \chi n_1 \phi_2 \qquad (4.25)$$

The partial molar quantities are

$$(\mu_1 - \mu_1^0)/RT = \ln a_1 = \ln \phi_1 + (1 - (1/x))\phi_2 + \chi \phi_2^2 \qquad (4.26)$$

and

$$(\mu_2 - \mu_2^0)/RT = \ln \phi_2 - (x - 1)\phi_1 + \chi x \phi_1^2 \qquad (4.27)$$

In a polydisperse system for which the number average value of x is x_n, the appropriate equations are

$$(\mu_1 - \mu_1^0)/RT = \ln \phi_1 + (1 - (1/x_n))\phi_2 + \chi \phi_2^2 \qquad (4.28)$$

$$(\mu_x - \mu_x^0)/RT = \ln \phi_x - (x - 1) + x(1 - (1/x_n))\phi_2 + \chi x \phi_1^2 \qquad (4.29)$$

An alternate form of Eq. (4.26) is obtained by the expansion of $\ln \phi_1$ in series form of ϕ_2,

$$(\mu_1 - \mu_1^0)/RT = 1 - (\phi_2/x) - (\tfrac{1}{2} - \chi)\phi_2^2 + \cdots \qquad (4.30)$$

Noting that $\phi_2/x v_1 = c_2/M_2$, where M_2 is the molecular weight of the chain molecule, we obtain

$$(\mu_1 - \mu_1^0)/RT = \bar{V}_1 c_2[(1/M_2) + A_2 c_2 + \cdots] \qquad (4.31)$$

where

$$A_2 = (\rho_1/M_1 \rho_2^2)(1/2 - \chi) \qquad (4.32)$$

and ρ is the density.

Equation (4.31) is of particular importance because the activity or the partial vapor pressure of the solvent can be readily measured. Examination of the experimental data soon reveals the following. First, the interaction parameter is often dependent on concentration (Bawn et al., 1950; Newing, 1950; Gee and Orr, 1946) (Fig. 4.4). This can be accounted for by the recently developed corresponding state theory to be described later. Second, the value of χ calculated from Eq. (4.31) is usually not identical to that obtained from the direct measurement of the heat of mixing, i.e., Eq. (4.23). The discrepancy has its origin in the assumption of a regular solution, which apportions all the deviation from ideality to the enthalpy term. As was pointed out previously, the difference in the molecular forces between like and unlike species that produces the nonzero heat of mixing should also

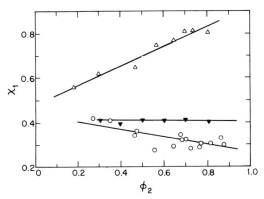

FIG. 4.4.   Interaction parameters for polystyrene in toluene, $\bigcirc$; rubber in benzene, $\blacktriangle$; and polydimethylsiloxane in benzene, $\triangle$ (Bawn *et al.*, 1950; Newing, 1950; Gee and Orr, 1946). Reprinted from Paul J. Flory: "Principles of Polymer Chemistry." Copyright © 1953 by Cornell University. Used by permission of the publisher, Cornell University Press.

lead to a nonrandom distribution of the molecules. It is reasonable therefore to consider χ obtained from the free energy expression, Eq. (4.25), to consist of both enthalpy and entropy deviations from ideality. A commonly used expression is

$$\chi = \chi_h + \chi_s \tag{4.33}$$

where χ_s, the entropy term, is a constant and χ_h, the enthalpy term, is inversely proportional to temperature and is equal to $-T(\partial\chi/\partial T)$.

Assuming that the van Laar heat of mixing is applicable to solvent–polymer mixtures, one can estimate the solubility parameter of a polymer, which is inaccessible to direct experimental measurement, by two methods. In the first method, a list of solvents in which the polymer is completely soluble is tabulated. Usually, the solvents are grouped within a certain range of δ. The solubility parameter of the polymer is taken as the midpoint of the solubility range. In the second method, a lightly cross-linked polymer is prepared, for example, cross-linked polystyrene from copolymerization of styrene with a small amount of divinylbenzene, or poly(methyl methacrylate) cross-linked by copolymerizing with ethylene glycol dimethacrylate. The cross-linked material is immersed in a series of solvents of varying δ. The solubility parameter of the polymer is assigned the same value as the solvent in which the highest degree of swelling is observed. By inference the uncross-linked polymer also has the same δ value. Typical values are listed in Table 4.2 (Burrell and Immergut, 1967).

The solubility parameter concept has also been extended to highly polar polymers and to solvents with hydrogen-bonding capability. Although the approach is empirical, practical utility has been demonstrated.

TABLE 4.2
SOLUBILITY PARAMETERS OF POLYMERS

Polymer	δ_2 (cal/cm^3)$^{1/2}$
Polyethylene	7.9
Polystyrene	8.6–9.1
Poly(methyl methacrylate)	9.1–9.5
Poly(vinyl chloride)	9.5–10
Poly(ethylene terephthalate)	10.7
66 Nylon	13.6
Polyacrylonitrile	15.4

4.2.3. DILUTE SOLUTIONS OF MACROMOLECULES

In dilute solutions, coils of flexible macromolecules, representing high local concentration of segments, are separated by large regions of pure solvent. The imhomogeneity in segment distribution in solution makes the Flory–Huggins theory inapplicable because the assumption that the probability of occupancy of a lattice site is equal to the overall fraction of occupied sites becomes clearly invalid. Instead, the problem has been dealt with via excluded volume considerations, given in Section 4.2.1. However, the evaluation of the excluded volume of a flexible macromolecular coil is a formidable task. The flexible coil, unlike a rigid sphere, does not have a uniform segment density in space. Furthermore, the size of the coil is dependent on the nature of the solvent. In a good solvent, the preference of segments for solvent molecules as neighbors increases the effective dimension of the coil and its excluded volume. Conversely, the excluded volume decreases in a poor solvent. Therefore, the excluded volume is related to the interaction parameter χ. In evaluating the excluded volume, Flory and Krigbaum (1950; Carpenter and Krigbaum, 1958) considered the centers of two coils approaching each other so that a volume element δV contains segments from both molecules. Within the volume element δV, Flory–Huggins treatment remains valid. The interpenetration of coils in δV is associated with an excess chemical potential of the solvent equal to $RT(\frac{1}{2} - \chi)\phi_2^2$, where ϕ_2 is the local segment volume fraction. The excluded volume is obtained by calculating the probability g of such interpenetration integrated from limits of no interpenetration to complete interpenetration. The final expressions are

$$U = 4V_e[1 - g(Y)] \qquad (4.34)$$

where

$$Y = (\tfrac{1}{2} - \chi)V_2^2/V_e V_1 \qquad (4.35)$$

Here, V_e is the effective molar volume of the coil and $\bar{V}_1, \bar{V}_2$ are the molar volumes of the solvent and macromolecule, respectively. The function $g(Y)$ decreases with increasing value of Y and approaches zero for $Y \to \infty$, i.e., very long chains. Interpenetration then becomes negligible. When $(\frac{1}{2} - \chi) = 0$, Y equals zero and $g(Y)$ has the value of unity; the excluded volume vanishes, in accordance with Eq. (4.34). The condition has special significance in that the second virial coefficient becomes zero and deviation from ideality vanishes. When x is small but finite, the excluded volume is approximated by

$$\lim_{Y \to 0} U = (\tfrac{1}{2} - \chi)\bar{V}_2^2/\bar{V}_1 \tag{4.36}$$

and the second virial coefficient is proportional to $(\frac{1}{2} - \chi)$.

We can write the excess enthalpy h^E and excess entropy s^E in alternate forms as follows:

$$h^E = RT\kappa\phi_2^2 \tag{4.37}$$

and

$$s^E = R\psi\phi_2^2 \tag{4.38}$$

so that

$$\psi - \kappa = \tfrac{1}{2} - \chi \tag{4.39}$$

The ratio of h^E and s^E, $\kappa T/\psi$, is defined by Flory as Θ, which has the dimension of temperature such that

$$\psi - \kappa = \psi(1 - (\Theta/T)) \tag{4.40}$$

At temperature T equal to Θ, the excess enthalpy and entropy contributions cancel each other. The Θ point can be reached by changing either the temperature of the solution or the nature of the solvent. The size of the macromolecule in a Θ solvent corresponds to the unperturbed dimensions.

4.2.4. CORRESPONDING STATE THEORY

One of the most powerful approaches developed in the last decade in the study of polymer mixtures is the utilization of the corresponding state theory. An extensive discussion of the subject is beyond the scope of this book. Only the basic principles will be outlined. Interested readers should consult Prigogine's book (General References, Prigogine et al., 1957).

The potential energy of a pair of molecules i and j is characterized by an interaction $\varepsilon(r_{ij})$ depending on the distance of separation r_{ij} between the two molecules. It can be represented by a universal function ϕ and two scale factors r^* and ε^* according to Eq. (4.41).

$$\varepsilon/\varepsilon^* = \phi(r/r^*) \tag{4.41}$$

The two scale factors allow us to introduce "reduced variables." The three dimensionless quantities, reduced temperature, volume, and pressure are defined as

$$\tilde{T} = kT/\varepsilon^*, \qquad \tilde{V} = V/r^{*3}, \qquad \tilde{p} = pr^{*3}/\varepsilon^* \tag{4.42}$$

The reduced pressure is a universal function of $\tilde{T}$ and $\tilde{V}$, the nature of this function being dependent on the potential energy function ϕ,

$$\tilde{p} = \tilde{p}(\tilde{T}, \tilde{V}) \tag{4.43}$$

Hence, Eq. (4.43) is known as the equation of state. An example of the equation of state is the one used by Flory (1965),

$$\frac{\tilde{p}\tilde{V}}{T} = \frac{\tilde{V}^{1/3}}{\tilde{V}^{1/3} - 1} - \frac{1}{\tilde{V}\tilde{T}} \tag{4.44}$$

but other equations have also proved successful (Nanda and Simha, 1964; Olabisi and Simha, 1975). The configurational entropy and free energy can be calculated through universal functions of $\tilde{T}$ and $\tilde{V}$ by statistical mechanical methods. The important point is that the reduced variables are defined in terms of molecular parameters and the concept can be applied to mixtures.

In a binary mixture we have six scale factors, namely, $\varepsilon_{11}^*, \varepsilon_{22}^*, \varepsilon_{12}^*, r_{11}^*, r_{22}^*$, and r_{12}^*. The same function ϕ is assumed to be valid for the description of interaction between dissimilar species. It follows that the second virial coefficient of the mixture and the excess volume of mixing depend on the relative magnitudes of the ε and r values for the three types of interactions. Thus, the interaction between unlike species can, in principle, be computed from experimental measurements. In practice, however, it is difficult to calculate ε^* and r^* directly from experimental data. Rather, simplifying assumptions are made in various models concerning the relation of ε_{12}^* with ε_{11}^* and ε_{22}^* as well as the potential energy function ε^* vs. r^*. The excess volume, excess free energy, and the equation of state are deduced explicitly from the reduced variables. Comparison with experimental results then allows judgment to be made about the success of the model.

One of the major triumphs of the corresponding state theory applied to mixtures is its ability to account for the excess volume of mixing, which is important when molecular sizes and interaction forces of the two components are significantly different. The excess volume is often negative. These considerations also give rise to additional contributions to the free energy of mixing, apart from Flory–Huggins combinatorial entropy and van Laar heat of mixing. When these contributions are taken into account, the variation of χ with concentration is predicted correctly (Eichinger and Flory, 1968).

4.3. Phase Equilibrium

The mutual solubility of a binary mixture can be best understood by referring to the free energy of the system as a function of composition. A schematic representation of the free energy of mixing is shown in Fig. 4.5a. Curve A is concave upward and has no inflection point. If two separate phases were present in the system, they will always correspond to a higher

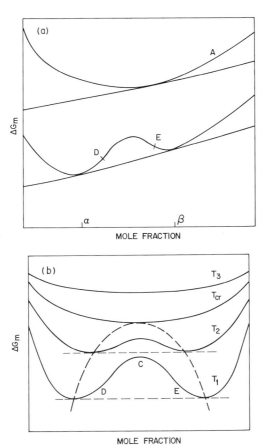

FIG. 4.5. (a) Free energy of mixing of binary mixtures; (b) transition from complete to partial miscibility.

free energy and, hence, are unstable with respect to the homogeneous phase. Therefore, curve A represents the condition where the binary mixture is miscible in all proportions. The case of partial miscibility is indicated by curve B. The two phases represented by α and β share a common tangent of the ΔG_m curve. They have identical values of μ_1 and μ_2, respectively, and are in equilibrium with each other. A homogeneous phase is stable if its composition is either smaller than α or larger than β. At compositions inter-mediate between α and β, any single phase is thermodynamically unstable. As the free energy curve passes from one minimum to another, there are two inflection points D and E at which $\partial^2 \Delta G_m/\partial N_2^2 = 0$. The third derivative of ΔG_m, $\partial^3 \Delta G_m/\partial N_2^3$, is negative at D but positive at E; therefore, the second derivative must go through a minimum at which the third derivative is zero.

Let us now consider the situation of enhanced mutual solubility at elevated temperature, depicted schematically by a set of ΔG_m curves in Fig. 4.5b. We define a critical temperature T_c below which phase separation occurs but above which the homogeneous phase is stable. The critical point corre-sponds to the coalescence of the two inflection points, characterized by the conditions:

$$\partial^2 \Delta G_m/\partial N_i^2 = 0 \quad \text{or} \quad \partial \mu_i/\partial N_i = 0 \tag{4.45}$$

$$\partial^3 \Delta G_m/\partial N_i^3 = 0 \quad \text{or} \quad \partial^2 \mu_i/\partial N_i^2 = 0 \tag{4.46}$$

4.3.1. BINARY SOLUTION OF MACROMOLECULES

In our examination of the stability limit of a solution of macromolecules, we assume that the free energy of mixing of the system is described by the Flory–Huggins theory. We further assume that the polymer is monodisperse. Application of the thermodynamic criteria of the critical point $\partial \mu_1/\partial \phi_1 = 0$ and $\partial^2 \mu_1/\partial \phi_1^2 = 0$ to Eq. (4.26) results in the following

$$\phi_{2c} = 1/(1 + x^{1/2}) \tag{4.47}$$

$$\chi_c = \tfrac{1}{2}(1 + x^{-1/2})^2 \tag{4.48}$$

It is seen that for very large values of x, the above expressions reduce to

$$\phi_{2c} = x^{-1/2} \tag{4.49}$$

$$\chi_c = \tfrac{1}{2} \tag{4.50}$$

A noteworthy feature of Eq. (4.49) is that the critical polymer concentration is very low for high molecular weight species. For $x = 10^3$, the critical con-centration is only about 3%. Phase diagrams for three polyisobutylene frac-tions in diisobutyl ketone (Shultz and Flory, 1952) are shown in Fig. 4.6. The critical points indeed occur at low polymer concentration although higher than the values predicted by Eq. (4.49). The second important con-clusion is that the critical value of χ approaches a limiting value of $\tfrac{1}{2}$ for

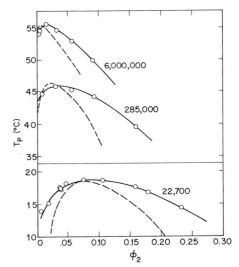

FIG. 4.6. Phase diagrams for three polyisobutylene fractions. Dashed curves were calculated from theory (Shultz and Flory, 1952).

infinite molecular weight (4.50). This corresponds to the Θ point in Eqs. (4.39) and (4.40), i.e., $T_c = \Theta$ for $x \to \infty$. Equation (4.40) can be rewritten as

$$\tfrac{1}{2} - \chi_c = \psi(1 - (\Theta/T_c))$$ (4.51)

Combination with Eq. (4.48) gives

$$\frac{1}{T_c} = \frac{1}{\Theta}\left[1 + \frac{1}{\psi}\left(\frac{1}{x^{1/2}} + \frac{1}{2x}\right)\right]$$ (4.52)

which predicts a linear relation between $1/T_c$ and $(1/x^{1/2} + 1/2x)$ and is verified by experimental results (Shultz and Flory, 1952) (Fig. 4.7). It can be seen from Eq. (4.52) that the theta temperature is also the critical miscibility temperature in the limit of infinite molecular weight. Representative values of Θ for several polymers are shown in Table 4.3.

TABLE 4.3

THERMODYNAMIC DATA FOR POLYMER–SOLVENT MIXTURES

Polymer	Solvent	Θ, °K	ψ	Ref.
Polyisobutene	Diisobutyl ketone	331.1	0.65	Elias et al. (1967)
Polystyrene	Cyclohexane	307.2	1.06	Krigbaum and Flory (1953)
Poly(methyl methacrylate)	4-Heptanone	305	0.61	Fox (1962)
Poly(dimethylsiloxane)	Butanone	298	0.43	Flory et al. (1952)

4.3.2. BINARY SYSTEMS CONTAINING TWO POLYMERS

The lattice model of Flory and Huggins can also be applied to mixtures of two polymers and generalized expressions have been obtained. For the

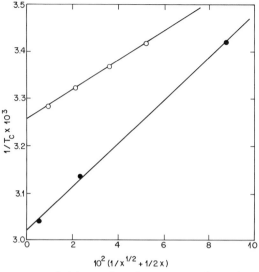

$10^2 (1/x^{1/2} + 1/2 x)$

Fig. 4.7. Relationship between critical temperature and chain length (Shultz and Flory, 1952).

purpose of this section, however, only a simple case is given to illustrate the important feature of miscibility of polymer mixtures. Let the chain lengths of the two polymers be equal, $x = x_2 = x_3$. The interaction parameter per molecule χ_{ij} is defined as

$$\chi_{ij} = x_i z \, \Delta \varepsilon_{ij} / kT \tag{4.53}$$

Obviously, $\chi_{23} = \chi_{32}$ for $x_2 = x_3$. The chemical potentials of the two components are

$$(\mu_2 - \mu_2^0)/RT = \ln \phi_2 + \chi_{23} \phi_3^2 \tag{4.54}$$

$$(\mu_3 - \mu_3^0)/RT = \ln \phi_3 + \chi_{23} \phi_2^2 \tag{4.55}$$

The critical composition is $\phi_2 = \phi_3 = \frac{1}{2}$ and the value of χ_{23} at the critical point is 2. The interaction per segment, χ_{23}/x, at the critical point is $2/x$, which is very small when x is large. The key to polymer miscibility therefore lies in the exceedingly small value of χ_c. Positive heat of mixing tends to negate the chance of complete miscibility while a negative heat of mixing favors compatibility. The underlying physical reason for this conclusion is that the gain in the entropy of mixing is small owing to the small numbers of molecules involved and is easily nullified by a positive heat of mixing. Indeed, very few chemically dissimilar polymers are compatible over the entire range of compositions. Examples of compatible polymer blends are poly(vinyl chloride)–poly(ethyl methacrylate) (Kern, 1958), polystyrene–poly(2,6-dimethyl-1,4-phenylene oxide) (Cizek, 1968), and polystyrene–poly(vinyl methyl ether) (Bank et al., 1971).

4.3.3. TERNARY SYSTEMS[†]

In ternary systems, the chemical potential of each component is characterized by three pair interaction parameters and the exact expressions become quite lengthy. Only a special case of a nonsolvent(1), a solvent(2), and a polymer(3) is discussed here because the situation is encountered in fractionation by the addition of a nonsolvent. Phase diagrams calculated by Tompa (1949) are shown in Fig. 4.8 for $x_3 = 10$, 100, and ∞. The χ values used in the computation are $\chi_{23} = 0$ and $\chi_{12} = \chi_{13} = 1.5$.

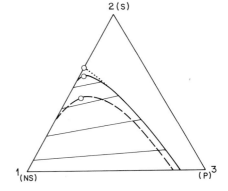

FIG. 4.8. Phase diagrams calculated by Tompa for three component systems consisting of nonsolvent (1), solvent (2), and polymer (3): (---) $\chi_3 = 10$; (——) $\chi_3 = 100$; and ($\cdots$) $\chi_3 = \infty$ (Tompa, 1949).

The critical points occur at very low polymer concentrations. An increase in chain length decreases the value of ϕ_{3c}, as expected. The composition of the solvent–nonsolvent mixture at the critical point for $x_3 = \infty$ corresponds to the Θ condition. However, this does not imply that the mixed solvent can be regarded as a single component and the fractionation phenomenon in a ternary system can be treated by the procedures given for binary solutions.

4.3.4. PRINCIPLES OF FRACTIONATION BY SOLUBILITY

When a solution of a heterogeneous polymer undergoes phase separation, the different molecular weight species redistribute themselves in the two phases. We wish to examine the extent of the molecular weight redistribution that forms the theory of polymer fractionation. For the two phases at equilibrium, the chemical potential of each component is identical in the concentrated or the gel phase, designated by a prime, and in the dilute phase, unprimed,

$$\mu_1 = \mu_1'$$
(4.56)

[†] Scott (1949).

for the solvent and

$$\mu_x = \mu_x' \tag{4.57}$$

for each macromolecule species. Using Eqs. (4.28) and (4.29), we obtain the distribution of species x in the two phases as

$$\ln(\phi_x'/\phi_x) = Ax \tag{4.58}$$

where

$$A = [1 - (1/x_n)]\phi_2 - [1 - (1/x_n')]\phi_2' + \chi(\phi_1^2 - \phi_1'^2) \tag{4.59}$$

and can be considered, to the first approximation, as a constant. Since the ratio ϕ_x'/ϕ_x increases exponentially with x_1, there is a strong tendency for the high molecular weight species to concentrate itself in the gel phase. This is the basis of fractionation by precipitation.

The fraction of species x remaining in the dilute phase is governed by the constant A and the volumes of the phases V and V',

$$f_x = V\phi_x/(V\phi_x + V'\phi_x') = 1/(1 + (V'/V)e^{Ax}) \tag{4.60}$$

If the volume of the dilute phase is made much larger than that of the gel phase, the value of f_x will be close to unity for small x. The low molecular weight species tend to remain in the dilute phase. The high molecular weight species will distribute preferentially in the gel phase because of the much larger value of e^{Ax}. Therefore, the rule for efficient fractionation is the use of a dilute solution coupled with a small volume of the gel phase in each of the successive steps.

The above discussion refers specifically to the fractionation of a binary solvent–polymer mixture by lowering of temperature. If fractionation is achieved by the addition of a nonsolvent at constant temperature, equations for ternary systems should be used, but the general considerations outlined in the foregoing discussion are still applicable.

4.3.5. Fractionation Methods

Two important methods of polymer fractionation based on solubility differences are fractional precipitation from solution by the addition of a nonsolvent and fractional dissolution of the polymer by gradual increase in solvent power.

Fractional precipitation is usually carried out by adding small quantities of nonsolvent into a large volume of dilute solution of polymer maintained at constant temperature until precipitation, indicated by turbidity, occurs. The precipitated phase that contains the high molecular weight species is

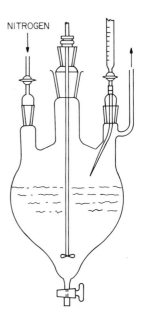

NITROGEN

Fig. 4.9.   Schematic diagram of an apparatus for fractional precipitation (Kotera, 1967).

allowed to settle to a compact layer and the supernatant solution is removed for further precipitation. A typical apparatus (Kotera, 1967) is shown in Fig. 4.9.

The precipitated phase is still polydisperse, as discussed in Section 4.3.4. To improve the efficiency of separation, the use of a very dilute solution is imperative and refractionation is often used.

In fractional dissolution, the polymer is first deposited from solution onto a column of finely divided inert particles such as sand or glass. A series of mixtures of solvents and nonsolvents of gradually increasing solvent power are introduced. The lowest molecular weight species dissolve in the least powerful mixture and are eluted first. The apparatus can be designed to regulate input and outflow volumes to produce a continuous gradient of solvent–nonsolvent composition in the column, hence, the name solvent gradient elution (Kokle and Billmeyer, 1965) (Fig. 4.10).

The results of fractionation are usually represented by a graph of cumulative weight fraction versus molecular weight, shown in Fig. 4.11.

4.3.6.   Lower Critical Solution Temperature

In the foregoing discussion, phase separation in a binary solution upon lowering the temperature was analyzed. It is taken for granted that the temperature coefficient of χ is negative in the region of our interest. At temperatures above the critical point, the value of χ decreases to less than

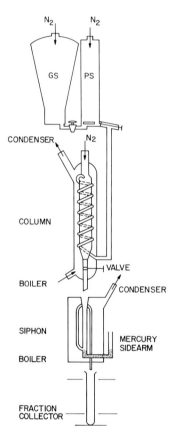

FIG. 4.10.   Schematic diagram of a column elution apparatus (Kokle and Billmeyer, 1965).

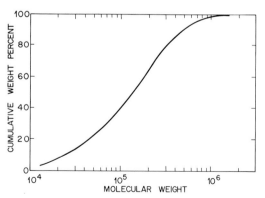

FIG. 4.11.   Typical cumulative molecular weight distribution.

the critical value of $\frac{1}{2}$ and the mixture is homogeneous. This is a natural consequence of the definition of χ, or more appropriately χ_h, which is inversely proportional to temperature. In the framework of a regular solution approximation with a van Laar type heat of mixing term, solubility always increases with temperature. The maximum temperature for phase separation is designated upper critical solution temperature (UCST).

In recent years, phase separation has also been observed for several polymer–solvent mixtures when the temperature of the solution is raised. Polyisobutylene–n-alkanes (Freeman and Rowlinson, 1960), polystyrene–benzene (Saeki et al., 1973), and polystyrene–2-butanone (Saeki et al., 1973; Patterson, 1969) are notable examples. The critical point associated with this phenomenon is called the lower critical solution temperature (LCST) (Patterson, 1969; Delmas and Patterson, 1962), although it is actually at a higher temperature than UCST. The LCST often lies above the normal boiling point of the solvent. Selected data for polystyrene–cyclopentane (Saeki et al., 1973) are shown in Fig. 4.12.

The increase in the value of χ at high temperatures to exceed the critical value for homogeneity calls for an explanation beyond the scope of random

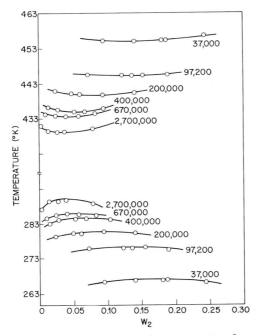

FIG. 4.12. Phase diagrams for polystyrene–cyclopentane system for samples of indicated molecular weight (Saeki et al., 1973).

mixing with no accompanying volume change. The details of the mathematical treatment based on corresponding state theory will not be given here. One contributing factor is the dissimilarity in the thermal expansion coefficients of the two components. At high temperature, the solvent molecules in the pure state occupy a larger molar volume than the polymer. Upon mixing, the solvent molecules are required to fit into a denser medium because the macromolecules, comprised of connected segments, are unable to expand easily. Thus the mixing process causes a negative volume change and a negative contribution to the entropy of mixing. The magnitude of χ therefore increases again at high temperatures.

The LCST phenomenon has also been observed in polymer–polymer mixtures, for example, polystyrene–poly(vinyl methyl ether) (Bank *et al.*, 1972) and polycaprolactone–(styrene–acrylonitrile) copolymer (McMaster, 1973).

4.4. Measurement of Molecular Weight and Size

Molecular weights of polymers may be determined by chemical analysis of functional groups or by physical measurements on very dilute solutions. End group analysis of condensation polymers is reliable for molecular weights below 3×10^4, while physical methods are generally suitable for molecular weights in excess of 5×10^3. In this section, molecular weight determination by measurement of colligative properties, light scattering, ultracentrifuge, solution viscosity, and gel permeation chromatography will be discussed.

Since synthetic polymers invariably contain species of different chain lengths, molecular weight measurements always yield average values. The number and weight average molecular weights have been defined in Chapter 2, Eqs. (2.52) and (2.53). Another average molecular weight, called the z-average, is given by

$$\bar{M}_z = \sum n_i M_i^3 / \sum n_i M_i^2 \tag{4.61}$$

The determination of molecular weights is not merely a matter of academic interest. In engineering applications, the tensile strength of a polymer has been correlated with $\bar{M}_n$ and melt viscosity with $\bar{M}_w$. A more precise description of the dependence of flow and ultimate properties of a polymer on molecular weight requires a detailed knowledge of the molecular weight distribution (MWD). The MWD can be obtained by fractionation or by gel permeation chromatography for linear polymers.

4.4.1. COLLIGATIVE PROPERTY MEASUREMENT

The measurement of colligative properties is based on the elementary consideration that at sufficiently high dilutions the activity a_1 of the solvent must be equal to the mole fraction N_1 (Raoult's law).

$$a_1 = N_1 = 1 - N_2 \qquad (4.62)$$

The activity of the solvent in a dilute binary solution therefore affords a direct measure of N_2. The molecular weight of the solute is then computed from its weight concentration. All colligative methods depend on this general principle. The boiling point elevation ΔT_b and freezing point depression ΔT_f are

$$\lim_{c \to 0} \Delta T_b/c = RT^2/\rho \, \Delta H_v M \qquad (4.63)$$

$$\lim_{c \to 0} \Delta T_f/c = RT^2/\rho \, \Delta H_f M \qquad (4.64)$$

Here, ρ is the density of the solvent and ΔH_v and ΔH_f are the latent heats of vaporization and fusion, respectively. With benzene as solvent, a polymer of $M = 5 \times 10^4$ will produce at $c = 1.0$ gm/100 ml a ΔT_b or ΔT_f of about $1 \times 10^{-3}\,°C$. While a temperature difference of $10^{-3}\,°C$ can be measured with precision with modern instrumentation, the use of alternate measurements is clearly desirable.

In the osmotic pressure method, the lowering of the activity of the solvent in the solution with respect to that of the pure solvent is compensated by applying a pressure π on the solution. According to Eq. (4.3)

$$RT d \ln a_1/dp = \bar{V}_1 \qquad (4.65)$$

where $\bar{V}_1$ is the partial molar volume of the solvent and may be replaced by molar volume V_1 in dilute solutions. At osmotic equilibrium,

$$(\mu_1 - \mu_0) = \int_0^\pi V_1 \, dp \qquad (4.66)$$

$$\mu_1 - \mu_0 = -\pi V_1 \qquad (4.67)$$

The osmotic pressure equation is obtained by subsituting Eq. (4.67) into (4.16)

$$\pi/c_2 = RT((1/M_2) + A_2 c_2 + A_3 c_2^2 + \cdots) \qquad (4.68)$$

and the limiting value at infinite dilution is

$$\lim_{c_2 \to 0} \pi/c_2 = RT/M \qquad (4.69)$$

The osmotic pressure for the same polymer solution used in the previous calculation of ΔT_f and ΔT_b is about 6 cm of solvent, a much larger effect than cryoscopic or ebulliometric differences.

All colligative methods measure $\bar{M}_n$ of a polydisperse sample. For example, we write the osmotic pressure π_i due to species i as

$$\lim_{c \to 0} \pi_i = RTc_i/M_i \tag{4.70}$$

The total osmotic pressure π is the sum of the π_i's:

$$\lim_{c_2 \to 0} \pi = \sum \pi_i = RT \sum c_i/M_i = RTc_2/\bar{M}_n \tag{4.71}$$

In membrane osmometry, the osmometer consists of two compartments separated by a semipermeable membrane that allows the passage of solvent molecules only. The most commonly used organic membrane materials are collodion (nitrocellulose) and gel cellophane. Among the many different osmometers, the Zimm–Myerson (1946) design is shown schematically in Fig. 4.13. Two membranes are clamped against a glass cell containing the polymer solution. The cell is placed in a large glass tube partially filled with solvent. The hydrostatic head, corrected for capillarity by the use of a

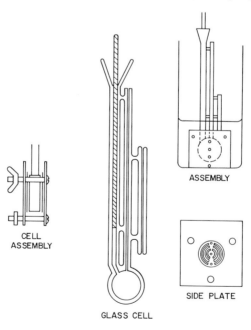

FIG. 4.13. Diagram of a Zimm–Myerson osmometer (Zimm and Myerson, 1946).

reference capillary of the same diameter for the solvent, gives a direct measure of the osmotic pressure. Osmotic equilibrium is attained typically in several hours to several days.

More recently, automatic osmometers have made use of pressure-sensing devices to monitor the height of the solvent compartment. A servo system adjusts the solvent level through a reservoir to balance the osmotic pressure. As a result of this rapid action, osmotic equilibrium can be reached in minutes (Coll, 1967).

In a typical osmotic experiment, measurements are made at several concentrations (Krigbaum and Flory, 1953a). A plot of π/RTc versus c is made (Fig. 4.14), the value of π/RTc at $c = 0$ is equal to $1/M_2$. The slope of the osmotic pressure plot gives the second virial coefficient A_2. It is equal to $\rho_1(\frac{1}{2} - \chi)/M_1\rho_2^2$, which can be seen readily by combining Eqs. (4.31) and (4.67). The interaction parameter χ is then computed from A_2.

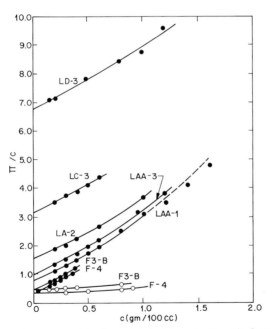

FIG. 4.14. Osmotic pressure data for polyisobutylene fractions in ● cyclohexane and ○ benzene (Krigbaum and Flory, 1953a).

The success of the osmotic experiment relies on perfect semipermeability of the membrane. In practice, existing membranes always allow the diffusion of low molecular weight species to take place. The measured molecular weight tends to give too high a value of $\bar{M}_n$ for unfractionated samples containing a substantial number fraction of low molecular weight materials.

4.4.2. LIGHT SCATTERING

Light scattering by molecules is caused by the interaction of the electro-magnetic wave with the induced oscillating dipole in the molecule, which then emits light in all directions (Rayleigh, 1871, 1881). In a dilute medium in which the molecules are randomly distributed at large distances from each other, the scattering contribution can be considered to be directly proportional to the number of molecules v. The scattered intensity i_0 observed at distance r and angle θ from the incident beam I_0 is characterized by Rayleigh's ratio R_θ defined by Eq. (4.72).

$$R_\theta = i_\theta r^2 / I_0 \tag{4.72}$$

It is proportional to the square of the molecular polarizability α and inversely to the fourth power of the wavelength.

$$R_\theta = 8\pi^4 v\alpha^2(1 + \cos^2\theta)/\lambda^4 \tag{4.73}$$

The molecular polarizability is related to the dielectric constant D or refractive index n of the medium. The total scattering, called turbidity τ, is given by $I/I_0 = e^{-\tau l}$ where l is the length of the sample, and is equal to the integration of i_θ over all angles.

In dense media such as liquids and solutions, destructive interference reduces the intensity of scattered light. The problem can be dealt with by the use of fluctuation theory, which evaluates the local fluctuation of the dielectric constant on a scale compared to the wavelength of light. The local fluctuation in dielectric constant may arise from density fluctuation in a liquid or from concentration fluctuation in a solution. According to the fluctuation theory (Debye, 1944), the mean square concentration fluctuation is

$$\langle(\Delta c_2)^2\rangle = \frac{kT}{(d^2 G/dc_2^2)} \tag{4.74}$$

where G is the free energy. The corresponding fluctuation in dielectric constant can be computed and the contribution to light scattering from concentration fluctuation is denoted by ΔR_θ. (Note that ΔR_θ is not the total light scattering.)

ΔR_θ is inversely proportional to $d^2 G/dc_2^2$ through Eq. (4.74). If we recall that the osmotic pressure is derived from dG/dc_2, the connection between osmotic pressure and light scattering becomes apparent. The underlying physical reason is that the local fluctuation in solute concentration that results in concentration higher than the equilibrium value is opposed by osmotic pressure. The equations relating ΔR_θ to osmotic pressure are as follows:

$$\Delta R_\theta = \frac{KRTc_2}{(d\pi/dc_2)} \tag{4.75}$$

where

$$K = \frac{2\pi^2 n^2 (dn/dc_2)^2}{v\lambda^4} \qquad (4.76)$$

Finally, differentiation of the osmotic pressure from Eq. (4.68) results in

$$Kc_2/\Delta R_\theta = (1/M_2) + 2A_2c_2 + 3A_2c_2^2 + \cdots \qquad (4.77)$$

A plot of $Kc_2/\Delta R_\theta$ versus c_2 yields $1/M_2$ as intercept, according to the above equation. The slope of the straight line is $2A_2$, twice the value of the osmotic pressure plot.

For a polydisperse system, additivity of ΔR_θ contributions from all species gives, in the limit of $c_2 \to 0$,

$$\lim \Delta R_\theta = \sum_i (\Delta R_\theta)_i = K\sum c_i M_i \qquad (4.78)$$

and

$$\lim Kc_2/\Delta R_\theta = \sum c_i / \sum c_i M_i = 1/\bar{M}_w \qquad (4.79)$$

Therefore, the light scattering experiments afford a measure of the weight average molecular weight. The method is suitable for molecular weight ranges of 5×10^3 to about 10^7. A schematic diagram of a light scattering photometer is depicted in Fig. 4.15. The use of the laser as a light source has greatly improved the accuracy of light scattering measurements.

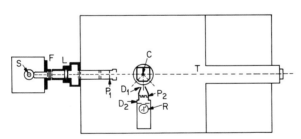

FIG. 4.15. Schematic diagram of a light scattering apparatus: S, light source; F, filter; L, lens; P, polarizer; C, cell containing polymer solution; D_1, D_2, slits; R, phototube (from Billmeyer, "Textbook of Polymer Science." Wiley, New York, 1971).

When the dimensions of the scattering particle are larger than $\frac{1}{20}\lambda$, the phase difference between scattering in different parts of the particle becomes large enough for destructive interference to occur. As shown schematically in Fig. 4.16, the difference in optical paths in the forward direction is smaller than in the backward direction. Destructive interference owing to phase difference is larger for larger θ. It vanishes as θ approaches zero. A correction factor for the angular dependence, $P(\theta)$, is computed by averaging over all possible relative positions of the scattering points to the incident beam.

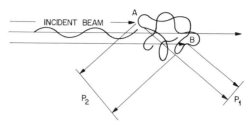

FIG. 4.16. Interference of light scattered by different parts of a macromolecule.

The values of $P(\theta)$ have been calculated for spheres (Rayleigh, 1910), Gaussian coils (Debye, 1947; Zimm, 1948), and rods (Neugebauer, 1942; Zimm *et al.*, 1945). The limiting value of $(Kc_2/\Delta R_\theta)$ at $c_2 = 0$ is given by

$$\left(\frac{Kc_2}{\Delta R_\theta}\right)_{c_2=0} = \frac{1}{M_2 P(\theta)} = \frac{1}{M_2}\left(1 + \frac{16\pi^2}{3\lambda^2}\langle s^2 \rangle \sin^2\frac{\theta}{2} + \cdots\right) \qquad (4.80)$$

where s is the radius of gyration. When the values of $Kc_2/\Delta R_\theta)_{c_2=0}$ obtained at different scattering angles are plotted against $\sin^2 \theta/2$, the mean square radius of gyration may be calculated from the slope.

A widely adopted procedure for the treatment of experimental scattering data consists of the double extrapolation of the measured $(Kc_2/\Delta R_\theta)$ values at various concentrations and angles to zero concentration and zero angle. The resulting diagram is called a Zimm plot (1948) shown in Fig. 4.17. The

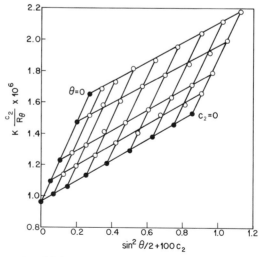

FIG. 4.17. Zimm plot of light scattering data of polystyrene in butanone (Zimm, 1948).

factor in front of c_2 is selected arbitrarily to provide a convenient scale for the abscissa. The lines corresponding to $c_2 = 0$ and $\theta = 0$ should intersect the ordinate at the same point, equal to $1/M_2$. The slope of the $\theta_2 = 0$ line yields the second virial coefficient and that of the $c_2 = 0$ line yields the radius of gyration.

We have already proved that the average molecular weight computed from light scattering data is a weight average. The average radius of gyration obtained here is a z-average according to the following consideration.

$$\frac{\sum(\Delta R_\theta)_i}{K c_i} = \frac{\sum c_i M_i P_i(\theta)}{c_i} = \bar{M}_w \bar{P}(\theta) \tag{4.81}$$

$$\bar{P}(\theta) = \frac{\sum c_i M_i P_i(\theta)}{\bar{M}_w \sum c_i} = \frac{\sum c_i M_i P_i(\theta)}{\sum c_i M_i} = \frac{\sum n_i M_i^2 P_i(\theta)}{\sum n_i M_i^2} \tag{4.82}$$

4.4.3. ULTRACENTRIFUGE[†]

When a solution is placed in a large gravitational field, say $10^4\, g$, the contribution of the gravitational force to the chemical potentials of the components becomes appreciable and causes a change in the equilibrium distribution of the system.

The chemical potential at constant temperature is now not only a function of pressure p and concentration c_2, but also depends on gravitational force F. For a solute of molecular weight M_2 in a centrifuge operating at an angular velocity ω at a distance r from the axis of rotation, the gravitational force is $M_2\omega^2 r$ and its contribution to the chemical potential is

$$-\int_0^r M_2\omega^2 r\, dr \qquad \text{or} \qquad -\tfrac{1}{2}M_2\omega^2 r^2 \tag{4.83}$$

The effect of the applied force is to make p and c_2 continuously varying functions of r. Therefore we write μ_2 as $\mu_2(p, c_2, r)$ which, at equilibrium, is uniform throughout the system, that is,

$$\frac{d\mu_2}{dr} = \left(\frac{\partial \mu_2}{\partial p}\right)_{c_2}\frac{dp}{dr} + \left(\frac{\partial \mu_2}{\partial c_2}\right)_p\frac{dc_2}{dr} - \frac{M_2\omega^2 r^2}{2} = 0 \tag{4.84}$$

The partial derivative of μ_2 with respect to p is $\bar{V}_2$ (Eq. (4.3)), or $M_2\bar{V}_2$, and the term dp/dr is simply $\rho\omega^2 r$, where $\bar{V}_2$ is the specific volume of component 2 and ρ is the density of the solution. Substitution of these relationships into Eq. (4.84) yields

$$M_2(1 - \bar{V}_2\rho)\omega^2 r = (\partial \mu_2/\partial c_2)(dc_2/dr) \tag{4.85}$$

[†] Williams et al. (1958).

The term $\partial\mu_2/\partial c_2$ can be evaluated from $\partial\mu_1/\partial c_2$ by the use of Gibbs–Duhem relation (Eq. (4.7)); hence ultracentrifuge equilibrium is again related to osmotic virial coefficients,

$$\frac{(1 - \bar{V}_2\rho)\omega^2 r c_2}{(dc_2/dr)} = \frac{RT}{M_2}(1 + 2A_2M_2c_2 + \cdots) \tag{4.86}$$

All the quantities on the left-hand side of Eq. (4.86) can be measured easily with the exception of dc_2/dr which can be obtained with precision only by elaborate optical methods. In the majority of ultracentrifuge equilibrium experiments, the schlieren optical system measures the refractive index gradient, which is proportional to dc_2/dr. The treatment of experimental data is complicated by the fact that both c_2 and dc_2/dr are dependent on r. The molecular weight of the specimen must be averaged over all values of r. The averaging process can be performed in two ways. On a section-shape ultracentrifuge cell (Fig. 4.18), the amount of solute in a layer between r and $r + dr$ is proportional to c_2r. The total amount of solute in the cell is

$$\int_{r_b}^{r_t} c_2 r\, dr = \tfrac{1}{2}c_2^0(r_b^2 - r_t^2) \tag{4.87}$$

where c_2^0 is the concentration at the beginning of the experiment and r_b and r_t are distances from the bottom and the top of the cell to the center of rotation. In the limit of infinite dilution, Eq. (4.86) becomes

$$\frac{RT}{M_2(1 - \bar{V}_2\rho)\omega^2} = \frac{\int_t^b c_2 r\, dr}{\int_t^b dc_2} = \frac{c_2^0(r_b^2 - r_t^2)}{2(c_{2b} - c_{2t})} \tag{4.88}$$

For a polydisperse mixture Eq. (4.88) is applied to each species,

$$\frac{RT}{(1 - \bar{V}_2\rho)\omega^2} = \frac{c_i^0 M_i(r_b^2 - r_t^2)}{2(c_{ib} - c_{it})} \tag{4.89}$$

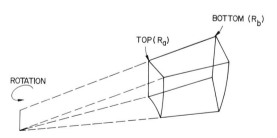

FIG. 4.18. Section-shaped ultracentrifuge cell.

Summing over all species and dividing both sides by $\sum c_i^0$, we have

$$\frac{RT}{(1 - \bar{V}_2\rho)\omega^2} \frac{\sum(c_{ib} - c_{it})}{c_2^0} = \frac{\sum c_i M_i}{\sum c_i^0}(r_b^2 - r_t^2) = \bar{M}_w(r_b^2 - r_t^2) \qquad (4.90)$$

The average molecular weight thus obtained is the weight average $\bar{M}_w$.

An alternative procedure is to start from Eq. (4.85) which, in the limit of infinite dilution, simplifies to

$$\frac{RT}{r} \sum \frac{dc_i}{dr} = (1 - \bar{V}_2\rho)\omega^2 \sum c_i M_i \qquad (4.91)$$

Evaluation of the concentration gradients (dc_i/dr) at r_b and r_t results in Eq. (4.92).

$$\left[\frac{1}{r_b} \sum\left(\frac{dc_i}{dr}\right)_{r_b} - \frac{1}{r_t} \sum\left(\frac{dc_i}{dr}\right)_{r_t}\right] \Big/ \left[\sum c_{ib} - \sum c_{it}\right] = \frac{(1 - \bar{V}_2\rho)\omega^2}{RT} \frac{\sum c_i^0 M_i^2}{\sum c_i M_i} \qquad (4.92)$$

which yields a z-average molecular weight.

The fact that both $\bar{M}_w$ and $\bar{M}_z$ can be obtained from the same experiment is of great importance. Under properly chosen conditions, $\bar{M}_n$ and $\bar{M}_{z+1}$ can also be computed. Studies of sedimentation velocity and approach to equilibrium are employed with success. A new refinement that appears to have great potential is equilibrium sedimentation in a density gradient (Hermans and Ende, 1963) produced by a mixed solvent, a method that is highly sensitive to small differences in the densities of solute species due to inhomogeneity in the composition of the solvent.

4.4.4. Viscosity

In addition to osmometry, light scattering, and ultracentrifugation, which afford absolute determination of molecular weight, there are two indirect methods that, after proper calibration, serve as convenient measures of molecular weight. The first method is solution viscosity, the importance of which was recognized more than 40 years ago.

When a liquid is subject to a shear force F, the frictional resistance per unit area A between two neighboring layers of the fluid is governed by the viscosity η_0 of the liquid and the velocity gradient dv/dx between layers. At steady state of flow, the definition of Newtonian viscosity is

$$F/A = \eta_0(dv/dx) \qquad (4.93)$$

The rate of energy dissipation per unit volume is

$$q = \eta_0(dv/dx)^2 \qquad (4.94)$$

If a rigid sphere is suspended in the fluid, the flow perturbation due to the presence of the sphere causes additional energy dissipation Δq and

$$\Delta q/q = (\eta - \eta_0)/\eta_0 \tag{4.95}$$

The term $(\eta - \eta_0)/\eta_0$ is called the *specific viscosity* η_{sp}. For a dilute solution of suspended spheres in which interaction among spheres is negligible, the specific viscosity was obtained by Einstein (1906)

$$\eta_{sp} \cong \tfrac{5}{2}\phi_2 \tag{4.96}$$

where ϕ_2 is the volume fraction of the spheres. In this limiting case, η_{sp} depends only on the total volume of the spheres but not on the size of the individual sphere.

For more concentrated solutions, higher powers of ϕ_2 come into play. Regardless of the shape of the solute particle or whether the solute particle is completely impermeable to solvent, η_{sp} can be expressed as a power series of ϕ_2.

$$\eta_{sp} = a_1\phi_2 + a_2\phi_2^2 + \cdots \tag{4.97}$$

The volume fraction ϕ_2 is equal to $(V_e/M_2)c_2$, where V_e is the effective hydrodynamic volume of the solute particle, and we have,

$$\eta_{sp} = a_1\left(\frac{V_e}{M}\right)c_2 + a_2\left(\frac{V_e}{M}\right)^2 c_2^2 + \cdots = [\eta]c_2 + k'[\eta]^2 c_2^2 + \cdots \tag{4.98}$$

where $[\eta]$, the *intrinsic viscosity*, is the limiting value of (η_{sp}/c_2) as c_2 approaches zero and is a measure of the effective hydrodynamic volume of the solute divided by molecular weight. The coefficient k' is called the Huggins constant. A common practice in the treatment of experimental data is to plot η_{sp}/c_2 versus c_2 and obtain $[\eta]$ as intercept and k' from the slope divided by $[\eta]^2$.

If the solute particle is a flexible macromolecular chain, we have to consider the fact that the flexible coil, rather than being a rigid sphere, is swollen by the solvent. In the limit, the swollen coil in a flow field may be considered as free-draining. Many theories were developed to evaluate $[\eta]$ for different models. A completely free draining model appears to overestimate the permeability effect because solvent molecules in the center of a macromolecular coil move with a velocity more nearly that of the macromolecule than the solvent. Therefore solvent molecules in the interior of the coil tend to move together with the macromolecule as a unit. This leads to the concept of an equivalent hydrodynamic sphere, impenetrable to solvent, which has the same viscosity as the actual coil. The volume of the equivalent sphere is proportional to $\langle r^2 \rangle$ (or $\langle s^2 \rangle$) and we have, from Einstein's equation

$$[\eta] = \tfrac{5}{2}V_e/M = \Phi\langle r^2 \rangle^{3/2}/M = \Phi'\langle s^2 \rangle^{3/2}/M \qquad (4.99)$$

where Φ and Φ' are universal constants. Different model calculations predict slightly different values of Φ; experimental results point to a Φ value of $(2.1 \text{ to } 2.5) \times 10^{21}$ ($[\eta]$ in deciliters per gram, r in centimeters, M_2 in grams per mole).

The mean square end-to-end distance $\langle r^2 \rangle$ in solution is larger than the unperturbed dimension $\langle r_0^2 \rangle$. The multiplication factor α is called expansion factor defined as

$$\langle r^2 \rangle = \alpha^2 \langle r^2 \rangle_0 \qquad (4.100)$$

Equation (4.99) may then be rewritten as

$$[\eta] = \Phi \frac{\langle r^2 \rangle_0^{3/2}}{M} M^{1/2}\alpha^3 \qquad (4.101)$$

Since $\langle r^2 \rangle_0$ for a linear polymer is proportional to the number of segments, the ratio $\langle r^2 \rangle_0/M$ is independent of M, and we have

$$[\eta] = KM^{1/2}\alpha^3 \qquad (4.102)$$

where $K = \Phi(\langle r^2 \rangle_0/M)^{3/2}$ is a constant independent of molecular weight and the nature of the solvent.

Several important conclusions follow from Eq. (4.102). The effect of solvent on intrinsic viscosity enters into play only through the expansion factor α. In a Θ solvent, $\alpha = 1$ and

$$[\eta]_\Theta = KM^{1/2} \qquad (4.103)$$

In a good solvent α is greater than unity and increases in magnitude as M increases. Therefore the dependence of intrinsic viscosity on molecular weight is often written as

$$[\eta] = K'M^a \qquad (4.104)$$

where a is greater than $\tfrac{1}{2}$.

The dependence of α on M is elucidated from a consideration of the free energies of solvent–segment mixing and coil deformation (Flory and Fox, 1951). The resulting equation is

$$\alpha^5 - \alpha^3 \propto (\tfrac{1}{2} - \chi)M^{1/2} \qquad (4.105)$$

or

$$\alpha^5 - \alpha^3 \propto \psi(1 - (\theta/T))M^{1/2} \qquad (4.106)$$

For sufficiently large values of the right-hand side of Eq. (4.105), α^3 can be neglected compared with α^5 and we have $\alpha \approx M^{0.1}$. The exponent a in Eq. (4.104) would then be 0.80, an upper limit.

The effect of temperature on intrinsic viscosity involves changes in both K and α. The expansion factor α decreases with increases in temperature in a good solvent, is independent of temperature in an athermal solvent, and increases with temperature in a poor solvent. The temperature dependence of K reflects the alteration of $\langle r^2 \rangle_0$.

An important contribution of Eq. (4.105) is the establishment of a link between a hydrodynamic property $[\eta]$ and a thermodynamic property χ through the expansion factor. The combination of Eqs. (4.101) and (4.105) allows the calculation of the unperturbed dimension from molecular weight, second virial coefficient, and intrinsic viscosity without recourse to a θ solvent (Orofino and Flory, 1957). The unperturbed dimension thus obtained is then used to compute the characteristic ratio discussed in Chapter 3.

As can be seen from the above analysis, $[\eta]$ is not an absolute measurement of molecular weight owing to the various proportionality constants appearing in the equation. But once K' and a are obtained for a given polymer–solvent pair by using carefully fractionated samples, the viscosity–molecular weight correlation becomes a valuable tool for molecular characterization because solution viscosity can be easily measured. Extensive tabulations of K' and a are available (Kurata et al., 1967).

Since intrinsic viscosity measures the effective hydrodynamic radius of the macromolecular coil, it does not differentiate between branched and linear molecules. A branched chain has a smaller radius of gyration than a linear one of the same molecular weight. The reduction in size is usually expressed by a factor g (Zimm and Stockmayer, 1949; Zimm and Kilb, 1959) defined as

$$g = \overline{s^2}(\text{branched})/\overline{s^2}(\text{linear}) \qquad (4.107)$$

The values of g have been calculated for various types of branching; for example, one random branch reduces $\overline{s^2}$ (linear) by about 10%. Recent advances in the synthesis of star-shaped macromolecules of controlled chain length (Roovers and Bywater, 1974) have confirmed theoretical predictions.

Measurements of solution viscosity are usually made by comparing the flow time t of the polymer solution through a capillary with the flow time t_0 of the solvent. The construction of the viscometer entails only simple designs. A popular version, the Ubbelohde viscometer, is shown schematically in Fig. 4.19. The specific viscosity η_{sp} is computed from $(t - t_0)/t_0$. The correction for capillary end effect is minimized by using long flow times in excess of at least 100 sec. Typical η_{sp}/c versus c plots are shown in Fig. 4.20. Double logarithmic plots of $[\eta]$ versus M are given in Fig. 4.21 and 4.22.

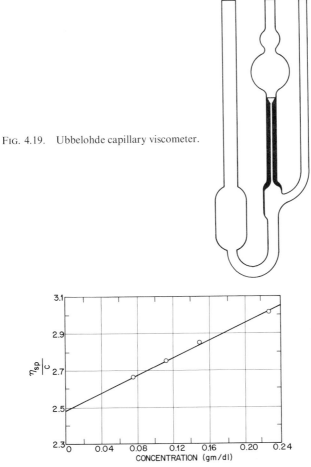

FIG. 4.19.   Ubbelohde capillary viscometer.

FIG. 4.20.   Extrapolation of viscosity data to infinite dilution.

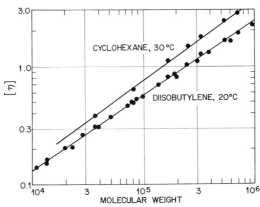

FIG. 4.21.   Intrinsic viscosity–molecular weight relationship for polyisobutylene in two solvents (Krigbaum and Flory, 1953a).

309

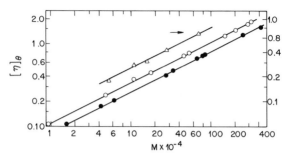

FIG. 4.22. Intrinsic viscosity–molecular weight relationship in Θ solvents; △, poly-(dimethylsiloxane) in 2-butanone at 20°C; ○, polyisobutylene in benzene at 24°C; ●, polystyrene in cyclohexane at 34°C (Flory *et al.*, 1952; Krigbaum and Flory, 1953b; Fox and Flory, 1951; replotted by Flory).

The slopes of the $[\eta]_\Theta$ versus M lines are indeed one-half, as predicted by theory.

Intrinsic viscosity measurements yield the viscosity average molecular weight defined as

$$\bar{M}_v = \left[\sum w_i M_i^a\right]^{1/a} = \left[\sum n_i M_i^{1+a} / \sum n_i M_i\right]^{1/a} \tag{4.108}$$

4.4.5. GEL PERMEATION CHROMATOGRAPHY

Another indirect method of molecular weight determination that has gained wide acceptance in recent years is gel permeation chromatography (GPC).

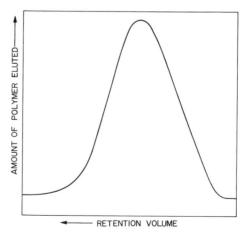

FIG. 4.23. Typical gel permeation chromatogram (Cazes, 1970).

The method utilizes a chromatographic column packed with porous beads usually made of glass or cross-linked polystyrene. The pore size is selected to be of the same order as the dimensions of the polymer coil. When a polymer solution is introduced into the column, the molecules can diffuse into the porous structures of the beads. The diffusion of large polymer molecules into pores, however, is restricted by the available pore size. Some will be completely excluded from the pores and eluted from the column first. Smaller molecules have longer residence times owing to their ability to penetrate the interior pore structure. The retention times of the different species vary inversely with molecular sizes. The amount of polymer eluted can be determined by optical or spectrophotometric methods. A typical gel permeation chromatogram is shown in Fig. 4.23 (Cazes, 1970).

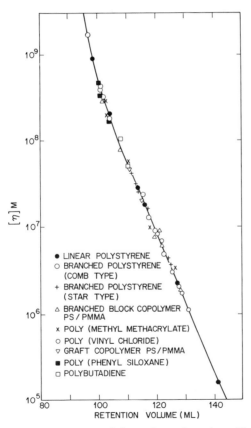

FIG. 4.24.   Universal correlation between $[\eta]M$ and retention volume (Grubisic *et al.*, 1967).

Since gel permeation chromatography differentiates size rather than molecular weight directly, empirical calibration of the retention volume–molecular weight relationship is necessary. This is usually accomplished by the use of narrow-distribution polystyrenes. The product $[\eta]M$, which is proportional to $\langle r^2 \rangle^{3/2}$ (Eq. (4.99)), is plotted against retention volume to obtain a universal curve (Grubisic *et al.*, 1967) (Fig. 4.24) from which the molecular weight distribution of an "unknown" sample is obtained. Various molecular weight averages can be computed from the MWD curve. The GPC method has acquired increasing popularity because the operation lends itself readily to automation. It is possible to complete the measurement and calculation in less than one hour. Results from GPC determination agree well with other methods.

4.5. Polyelectrolytes

In water or other strong polar solvents, polyelectrolytes ionize to form macroions. The solution properties of polyelectrolytes differ from those of simple electrolytes in one important respect. The electrostatic forces between charges on an individual macroion obviously do not vanish at infinite dilution. Furthermore, the counterions are attracted by the high charge density of the macroion and strong electrostatic interaction between them persists at low concentrations. If the macroion has a well-defined shape, e.g., a rigid sphere as in globular proteins, the electrostatic forces can be calculated by the use of Debye–Hückel theory, which is used with great success in the study of simple electrolytes. The case of a flexible polyelectrolyte chain, however, has presented a special challenge to theoretical analysis on more basic and conceptual grounds. The electrostatic repulsion of the fixed charges on a flexible macroion (Fig. 4.25) results in an expansion of the macromolecular coil but the expansion is also countered by the retractive force due to the rubberlike elasticity of the chain (Chapter 6). If added salt is present, it alters the ionic strength of the medium and has a shielding effect for the fixed charges. The interaction of counterions or added ions with the macroion modifies the electrostatic forces between fixed charges but the charge density and the shape of the macroion in turn determine its interaction with simple ions. This interdependence leads to complexities beyond the scope of this chapter. Pertinent experimental evidences for macroion expansion are given below for the purpose of illustration. Extensive discussions can be found in the books by Morawetz (General References, 1975) and by Tanford (General References, 1961).

Experimental evidence for the expansion of macroions is found in light scattering and viscosity studies. Figure 4.26 shows the observation of Oth

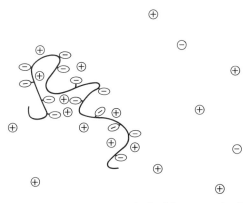

FIG. 4.25.   Schematic representation of a flexible macroanion in solution.

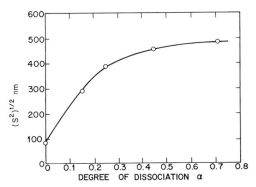

FIG. 4.26.   The radius of gyration of a sample of polymethacrylic acid as a function of the degree of dissociation of the carboxyl groups (data of Oth and Doty, 1952, replotted by Tanford, 1961).

and Doty (1952) of an increase in the radius of gyration of poly(methacrylic acid) from ~ 8 to ~ 48 nm as the degree of dissociation of the carboxyl groups is increased. The effect of added salt is demonstrated in sodium poly(styrene sulfonate) solutions in which the expansion of macroion increases as the concentration of added salt decreases (Fig. 4.27) (Takahashi *et al.*, 1967). The same effect is manifested in viscosity measurements Poly(vinyl butyl pyridinium bromide) is an often-quoted sample. In Fig. 4.28, there is a drastic increase in the value of η_{sp}/c with decreasing polyelectrolyte concentration when the amount of added salt is low. Therefore the conventional viscosity plot does not always permit a valid extrapolation to zero concentration (Fuoss, 1951).

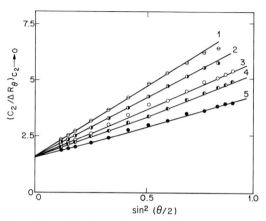

FIG. 4.27.   Light scattering of sodium polystyrenesulfonate ($M_w = 1.55 \times 10^6$) in solutions containing 0.005 M (1), 0.01 M (2), 0.05 M (3), 0.1 M (4) and 0.5 M (5) NaCl. The radius of gyration is calculated from the slope of the linear plot (Takahashi *et al.*, 1967).

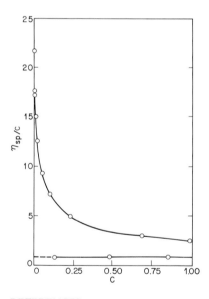

FIG. 4.28.   Reduced viscosity of poly(*n*-butyl-4-vinylpyridinium bromide) in aqueous solutions. Curve 1, in pure water; curve 2 in 0.0335 N KBr (Fuoss, 1951).

REFERENCES

Bank, M., Leffingwell, J., and Thies, C. (1971). *Macromolecules* **4**, 43.
Bank, M., Leffingwell, J., and Thies C. (1972). *J. Polym. Sci. Part A-2*, **10**, 1097.
Bawn, C. E. H., Freeman, R. F. J., and Kamaliddin, A. R. (1950). *Trans. Faraday Soc.* **46**, 677.
Burrell, H., and Immergut, B. (1967). "Polymer Handbook" (J. Brandrup and E. M. Immergut eds.). Wiley, New York.

Carpenter, D. K., and Krigbaum, W. R. (1958). *J. Chem. Phys.* **28**, 513.
Cazes, J. (1970). *J. Chem. Educ.* **47**, A461-471, A505-514.
Cizek, E. P. (1968). U.S. Patent No. 3,383,435.
Coll, H. (1967). *Makromol. Chem.* **109**, 38.
Debye, P. (1944). *J. Appl. Phys.* **15**, 338.
Debye, P. (1947). *J. Phys. Colloid Chem.* **51**, 18.
Delmas, G., and Patterson, D. (1962). *Official Digest* **1**.
Eichinger, B. E., and Flory, P. J. (1968). *Trans. Faraday Soc.* **64**, 2035.
Einstein, A. (1906). *Ann. Phys.* [4], **19**, 289.
Elias, H-G., Adank, G., Dietschy, Hj., Etter, O., Gruber, U., and Ibrahim, F. W. (1967). *In* "Polymer Handbook" (J. Brandup and E. H. Immergut, eds.). Wiley, New York.
Flory, P. J. (1942). *J. Chem. Phys.* **10**, 51.
Flory P. J. (1965). *J. Am. Chem. Soc.* **87**, 1833.
Flory, P. J., and Fox, T. G. (1951). *J. Am. Chem. Soc.* **73**, 1909.
Flory, P. J., Krigbaum, W. R. (1950). *J. Chem. Phys.* **18**, 1086.
Flory, P. J., Mandelkern, L., Kinsinger, J. B., and Shultz, W. B. (1952). *J. Am. Chem. Soc.* **74**, 3364.
Fox, T. G. (1962). *Polymer* **3**, 111.
Fox, T. G. Jr., and Flory, P. J. (1951). *J. Am. Chem. Soc.* **73**, 1915.
Freeman, P. I., and Rowlinson, J. S. (1960). *Polymer* **1**, 20.
Fuoss, R. M. (1951). Disc. Faraday Soc. **11**, 125.
Gee, G., and Orr, W. J. C. (1946). *Trans. Faraday Soc.* **42**, 507.
Grubisic, Z., Rempp, P., and Benoit, H. (1967). *J. Polym. Sci. Part B* **5**, 753.
Hermans, J. J., and Ende, H. A. (1963). *In* "Newer Methods of Polymer Characterization" (B. Ke, ed.), Wiley (Interscience), New York.
Hildebrand, J. H., and Scott, R. L. (1950). "The Solubility of Nonelectrolytes." Van Nostrand-Reinhold, Princeton, New Jersey.
Hildebrand, J. H., and Scott, R. L. (1962). "Regular Solutions." Prentice Hall, Englewood Cliffs, New Jersey. ·
Huggins, M. L. (1942a). *Ann. N.Y. Acad. Sci.* **43**, 1.
Huggins, M. L. (1942b). *J. Am. Chem. Soc.* **64**, 2716.
Huggins, M. L. (1942c). *J. Phys. Chem.* **46**, 151.
Kern, R. J. (1958). *J. Polym. Sci.* **33**, 524.
Kokle, V., and Billmeyer, F. W. Jr. (1965). *J. Polym. Sci. Part C* **8**, 217.
Kotera, A., (1967). *In* "Polymer Fractionation" (M. J. R. Canton, ed.). Academic Press, New York.
Krigbaum, W. R., and Flory, P. J. (1953a). *J. Am. Chem. Soc.* **75**, 1775, 5254.
Krigbaum, W. R., and Flory, P. J. (1953b). *J. Polym. Sci.* **11**, 37.
Kurata, M., Iwama, M., and Kamada, K. (1967). *In* "Polymer Handbook" (J. Brandrup and E. H. Immergut, eds.). Wiley, New York.
Mathot, V. (1950). *Bull. Soc. Chim. Belg.* **57**, 111.
McMaster, L. (1973). *Macromolecules* **6**, 760.
Nanda, V. S., and Simha, (1964). *J. Chem. Phys.* **41**, 1884.
Neugebauer, T. (1942). *Ann. Phys.* [5], **42**, 509.
Newing, M. J. (1950). *Trans. Faraday Soc.* **46**, 613.
Newman, S., Krigbaum, W. R., Laugier, C., and Flory, P. J. (1954). *J. Polym. Sci.* **14**, 451.
Olabisi, O., and Simha, R. (1975). *Macromolecules* **8**, 211.
Orofino, T. A., and Flory, P. J. (1957). *J. Chem. Phys.* **26**, 1067.
Oth, A., and Doty, P. (1952). *J. Phys. Chem.* **56**, 43.

Patterson, D. (1969). *Macromolecules* **2**, 672.
Prigogine, I., and Mathot, V. (1950). *J. Chem. Phys.* **18**, 765.
Lord Rayleigh (1871). *Phil. Mag.* [4], **41**, 447.
Lord Rayleigh (1881). *Phil. Mag.* [5], **12**, 81.
Lord Rayleigh (1910). *Proc. R. Soc. London* **A84**, 25.
Roovers, J. E. L., and Bywater, S. (1974). *Macromolecules* **7**, 443.
Saeki, S., Kawahara, N., Konno, S., and Kaneko, M. (1973). *Macromolecules* **6**, 589.
Scott, R. L. (1949). *J. Chem. Phys.* **17**, 268.
Shultz, A. R., and Flory, P. J. (1952). *J. Am. Chem. Soc.* **74**, 4760.
Small, P. A. (1953). *J. Appl. Chem.* **3**, 71.
Takahashi, A., Kato, N., and Nagasawa, M. (1967). *J. Phys. Chem.* **71**, 2001.
Tompa, H. (1949). *Trans. Faraday Soc.* **45**, 1142.
VanLaar, J. J. and Lorenz, R. (1925). *Z. Anorg. Gllgem. Chem.* **146**, 42.
Williams, J. W., VanHolde, K. E., Baldwin, R. L., and Fujita, H. (1958). *Chem. Rev.* **58**, 715.
Zimm, B. H., and Kilb, R. W. (1959). *J. Polym. Sci.* **37**, 19.
Zimm, B. H. (1946). *J. Chem. Phys.* **14**, 164.
Zimm, B. H. (1948). *J. Chem. Phys.* **16**, 1099.
Zimm, B. H., and Myerson, I. (1946). *J. Am. Chem. Soc.* **68**, 911.
Zimm, B. H., and Stockmayer, W. H. (1949). *J. Chem. Phys.* **17**, 1301.
Zimm, B. H., Stein, R. S., and Doty, P. (1945). *Polym. Bull.* **1**, 90.

GENERAL REFERENCES

Allen, P. W. (1959). "Techniques of Polymer Characterization." Butterworths, London.
Billmeyer, F. W. Jr. (1971). "Textbook of Polymer Science." Wiley, New York.
Flory, P. J. (1953). "Principles of Polymer Chemistry." Cornell Univ. Press, Ithaca, New York.
Hildebrand, J. H. and Scott, R. L. (1957). "The Solubility of Non-Electrolytes." Van Nostrand-Reinhold, Princeton, New Jersey.
Morawetz, H. (1975). "Macromolecules in Solution." Wiley, New York.
Prigogine, I., Bellemans, A., and Mathot, V. (1957). "The Molecular Theory of Solutions." Wiley (Interscience), New York.
Stacey K. A. (1956). "Light Scattering in Physical Chemistry." Academic Press, New York.
Svedberg, T., and Pederson, K. O. (1940). "The Ultracentrifuge." Oxford Univ. Press, London and New York.
Tanford, C. (1961). "Physical Chemistry of Macromolecules." Wiley, New York.
Tompa, H. (1956). "Polymer Solutions." Butterworths, London.

Chapter 5

MACROMOLECULES IN THE SOLID STATE: MORPHOLOGY

F. A. Bovey

5.1. Introduction

The crystalline state of organic macromolecules has engaged the attention of x-ray crystallographers and polymer scientists for well over 50 years (see the discussion of early work in Astbury (1933) and Hermans (1949) in the General References) and continues to do so today with no noticeable diminution of interest. One reason is the great complexity of polymer solid state morphology and the wide diversity of effects and phenomena that can be observed. Another more practical reason is that the solid state morphology of crystalline polymers is directly related to their mechanical properties.

The term *morphology* refers to the organization of crystals on a scale larger than that which would be the normal domain of a crystallographer. It concerns the size and shape of individual crystallites and the manner of their aggregation. Our principal tools in this field are the optical and electron microscopes rather than the x-ray diffractometer. In this chapter we shall

317

describe some of the commoner forms that crystallites take when formed from polymer solutions and from polymer melts, and we shall also deal with the kinetics of formation of these structures, although in a simplified and abbreviated way. We shall deal only with synthetic polymers, and will omit consideration of naturally occurring polymers (Chapter 8).

5.2. Polymer Crystallinity

5.2.1. REQUIREMENTS FOR CRYSTALLINITY

In order for crystallinity to occur, the polymer chains must be capable of packing closely together in a regular, parallel array. As will be seen in Chapter 8, some natural polymers—notably globular proteins—may crystallize even though their chains are folded into complex spheroidal shapes. But the distinctive feature of such polymer molecules is that, for a given type, they are all identical in chain length and manner of folding and can be stacked like tennis balls. For synthetic polymers this is not the case, and crystallinity requires packing side by side in extended form either as planar zigzags or helices.

In order to pack in this way, the chains must be at least fairly regular in structure. We have seen in Chapter 3 that in vinyl chains irregularity may be introduced in a variety of ways. We may have head-to-head:tail-to-tail monomer units; these are seldom present to a substantial extent. Much more common is stereochemical irregularity (Chapter 3, Section 3.4). Unless a chain is predominantly isotactic or (much more rarely) syndiotactic, it usually cannot be fitted sufficiently well to its neighbors to crystallize. Exceptions are poly(chlorotrifluoroethylene), $+CFClCF_2+_n$, in which both chlorine and fluorine atoms can be fitted into the lattice regardless of their orientation, and poly(vinyl alcohol), where hydrogen bonding plays a role. Both of these polymers are atactic, but it is the general rule that such chains will not crystallize. Isotactic and syndiotactic chains in general do crystallize and give quite different crystal structures (Chapter 3, Section 3.9.6).

Another form of irregularity that can inhibit crystallization is represented by copolymers. If these are of the usual random sort, crystallinity does not occur. Block copolymers form domain structures with a fairly high degree of regularity, but these are commonly on a much larger scale than truly crystalline structures.

5.2.2. POLYMER CRYSTAL STRUCTURES

The conformations of representative polymer chains have been discussed and illustrated in Chapter 3 (Section 3.9.6) and will not be treated further here, as our main concern is with the morphology of polymer crystals.

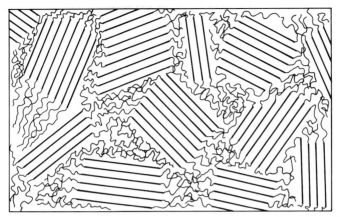

FIG. 5.1. The fringed micelle model of semicrystalline polymers (from Flory, 1953). Reprinted from Paul J. Flory: "Principles of Polymer Chemistry." Copyright © 1953 by Cornell University. Used by permission of the publisher, Cornell University Press.

Polymers exhibit a variety of different habits. These range from relatively simple single crystals, grown from solution, to complex spherulitic structures grown from the melt. We shall discuss both.

When the crystalline nature of polymers was first realized, it was recognized that the crystals were extremely small and, judged from density and heat of fusion, very imperfect as well. In addition, crystalline polymers retained polymeric physical behavior, being flexible and elastic though perhaps somewhat less so than in the rubbery amorphous state. To account for this behavior, the concept of the "fringed micelle" was proposed (Gerngross *et al.*, 1930). This is illustrated in Fig. 5.1. Here, the crystalline regions are of much smaller extent than the lengths of the polymer chains, any one of which may traverse several "micelles" or crystallites. The micelles were assumed to be surrounded by an amorphous network, which they tended to stiffen without eliminating pliablility and elasticity. However, further observations of crystalline polymers by electron and optical microscopy showed that this view could not be maintained. The actual structures were quite different. They were both more complex and more interesting.

5.2.3. Degree of Crystallinity

Even though the chains may be entirely regular in structure, polymers never crystallize completely. The degree of disorder far surpasses that corresponding to the occasional vacancies and dislocations found in crystals of small molecules. The meaning of the amorphous phase is not so straightforward conceptually as that implied in the fringed micelle model. Nevertheless, the notion of degree of crystallinity persists and is most commonly arrived at quantitatively by comparing actually observed quantities to those calculated for the usually unattainable ideal state of complete crystallinity. The properties usually employed are the *density* and *heat of fusion*. These quantities do not distinguish the variety of crystalline textures that can be present, and so their use has limitations but is nevertheless of considerable value in making comparisons in a well-defined context. Defining the degree of crystallinity as χ, it is found that most polymers vary in χ from less than 0.5 to 0.95. Even for single crystals of polyethylene, χ is usually no more than 0.9. For high pressure, branched polyethylene χ is about 0.5. Thus, semicrystalline polymers are always characterized by considerable disorder.

5.3. Crystallization from Solution

5.3.1. CHAIN FOLDING

In order to crystallize a polymer from solution, one selects a solvent in which the polymer dissolves at high temperature but in which it is insoluble, or nearly so, at room temperature. The solution is cooled to a desired lower temperature and crystallization is allowed to proceed. Since the crystal habit varies with temperature, it is best not to employ the simpler procedure, appropriate for small molecules, of simply allowing the solution to cool to room temperature, since this could give rise to a range of crystal types.

In the case of linear polyethylene, typical conditions would involve an approximately 0.01% solution of the polymer at 70–80°. The product of such crystallization takes the form of thin, apparently flat platelets or lamellae a few micrometers in long dimensions and approximately 10 nm in thickness, as shown in Fig. 5.2. The lamellar thickness increases with the temperature of crystallization. It is found that the chain or *c* axis of the molecules is oriented across the thickness of the platelet rather than in its plane. Since the molecules may be of the order of 1000 nm or more in length, the surprising and important conclusion to be drawn is that the chains must be folded many

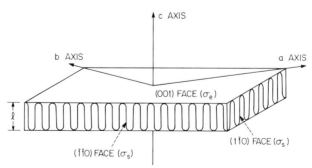

Fig. 5.2. Schematic representation of crystalline lamella showing folded chains, crystal faces, and axes (H. D. Keith and F. J. Padden, Jr., by permission).

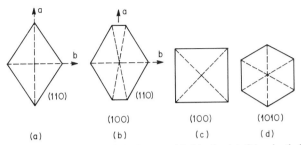

FIG. 5.3. Schematic illustrations of single crystal habits for (a), (b) polyethylene; (c) poly-1-pentene; and (d) polyoxymethylene; representing sectorization of crystals (from Magill, 1977).

times. This was first realized by Storks (1938), who observed by electron diffraction that in thin films of *gutta percha* or *trans*-1,4-polyisoprene made by evaporation of chloroform solutions "the macromolecules are folded back and forth upon themselves" since they were oriented normal to the surface of the film but were too long to be accomodated within its thickness. This discovery by Storks was not appreciated at the time and the full breadth and significance of the hypothesis remained to be rediscovered and developed nearly two decades later by Keller (1957).

Solution-grown polymer single crystals may vary in shape, exhibiting sectors that are associated with their growth. This sectorization is illustrated in Fig. 5.3, in which are also designated the crystallographic planes on which growth occurs. Although represented in Fig. 5.2 as planar for simplicity, lamellar crystals often take complex forms of hollow pyramids (Fig. 5.4). Viewed along the *c* axis such crystals would appear lozenge-shaped.

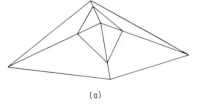

(a)

FIG. 5.4. Schematic representations of non-planar crystals. The lines drawn within the sectors represent the intersection of the surface of the pyramid with a plane normal to the pyramid axis (from Bassett *et al.*, 1963).

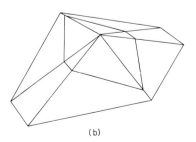

(b)

What is seen under a microscope as a planar lamella may actually represent a collapsed or partially collapsed pyramid (Fig. 5.5). In such hollow pyramids, the stems of the molecules remain parallel to the pyramid axis and are therefore inclined to the sloping faces of the pyramid. This implies that the folding in each sector must be staggered.

FIG. 5.5. A single crystal of polyethylene. It corresponds in form to Fig. 5.3b and is actually a collapsed hollow pyramid, as indicated by the pleats in the center (from H. D. Keith and F. J. Padden, Jr., by permission).

5.3.2. NATURE OF THE FOLD SURFACE

The nature of the surface at which the polymer chains emerge, fold, and reenter has been the subject of considerable controversy. On the one hand, the appearance of the seemingly planar and regular surfaces of lamellar crystals, with well-defined edges, corners, and sectors, suggests that the chains fold back with minimal loops and probably with adjacent reentry, as this would seem to minimize surface irregularity. Such a structure is pictured in Fig. 5.6a. On the other hand, Flory (1962, see also Flory and Yoon, 1978) has proposed that the fold surface has the form of a "switchboard," with reentry occurring more or less at random, and with loops much larger than the minimal size. There is at present no conclusive evidence for either picture. In favor of the latter is the fact that the density

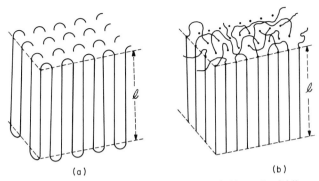

FIG. 5.6. Models of chain morphology in a single crystal: (a) regular folding with adjacent reentry; (b) irregular "random switchboard" fold surface.

of polyethylene single crystals (on which most of the work has been done) is well below the x-ray value, corresponding to amorphous surface layers of about 1–2 nm thickness. In favor of the former, aside from appearance, is the infrared spectroscopic study of Krimm (Bank and Krimm, 1969a,b; Krimm and Ching, 1975). This concerns the interaction of neighboring chains in the polyethylene unit cell (Chapter 3, Fig. 3.30). Krimm et al. employed mixtures of polyethylene and perdeuteropolyethylene. Splittings of the rocking bands at 720 and 731 cm^{-1} and of the bending bands at 1463 and 1473 cm^{-1} occur in crystalline polyethylene (Chapter 3, Fig. 3.8); and there are corresponding splittings in perdeuteropolyethylene. On mixing the polymers, these splittings will persist if adjacent reentry predominates, whereas random reentry would lead to perdeutero chains with $-CH_2CH_2-$ neighbors and vice versa and would be expected to weaken the splitting. Krimm and co-workers interpreted their results to indicate that adjacent reentry does indeed predominate, but their conclusions were weakened by the possibility of segregation of the recrystallizing chains aside from questions of the mode of reentry. Polyethylene and perdeuteropolyethylene have appreciably different melting points, and segregation on crystallization is a possibility (Kloos et al., 1974). In addition there is evidence for it from neutron scattering results on mixed polyethylene and perdeuteropolyethylene (Schelten et al., 1974). The question of chain conformation at the fold surface, therefore, cannot be regarded as answered at present.

5.3.3. NATURE OF THE LAMELLAR GROWTH PROCESS

The exact manner in which polymer chains add to the growth faces of lamellar crystals is, as we have seen, subject to discussion and uncertainty. One of the best established relationships is that between the crystallization temperature or degree of undercooling and fold period. We have seen (p. 320) that as the temperature is increased, or the degree of undercooling decreased, the lamellar thickness increases. In Fig. 5.7a are shown data of Nakajima and Hamada (1972) for polyoxymethylene in a variety of solvents. When the degree of undercooling, $(T_m - T_c)$ where T_m is the melting temperature in contact with the solvent, is used rather than the temperature of crystallization T_c, it can be seen (Fig. 5.7b) that the many curves for polyoxymethylene form instead a single curve. Similar results are observed for poly(4-methyl-1-pentene) and polyethylene. Thus the fundamental variable is the degree of undercooling. The nature of the solvent also has some effect, thermodynamically "poorer" solvents generally promoting the growth of larger and seemingly more perfect crystals with fewer defects.

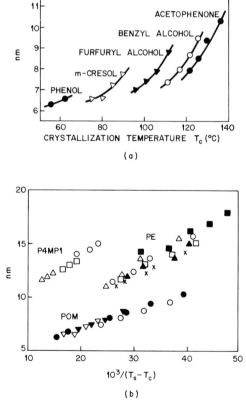

FIG. 5.7. (a) The dependence of lamellar thickness in polyethylene on crystallization temperature T_c in several solvents. (b) Plot of $10^3/(T_m - T_c)$ for poly(4-methyl-1-pentene) (P4MP1), polyoxymethylene (POM), and polyethylene (PE) (from Nakajima and Hamada, 1972).

Theories have been advanced to explain why chains fold, why the folding is uniform (or at least fairly uniform) and why the fold length varies with temperature in the manner observed. These theories fall into two classes: equilibrium and kinetic. A lattice-dynamic theory has been offered by Peterlin (Peterlin and Fischer, 1960; Peterlin, 1960; Peterlin *et al.*, 1962), which explains certain features of the process but implies a reversibility that is not normally observed. According to kinetic theories, the structure of lamellar crystals is one that maximizes the crystallization rate and is not the lowest free energy state. The critical step in determining the fold period is not the primary nucleation that gives birth to the crystal, but rather it is the secondary nucleation along the growing crystal face and the degree of undercooling prevailing there.

We outline briefly here the essential content common to most theories of single crystal formation (Lauritzen and Hoffman, 1960; Frank and Tosi, 1961; Price, 1959, 1961; Lauritzen and Passaglia, 1967; for a convenient discussion, see Keller, 1968). We assume an existing lamellar crystalline substrate of finite fold thickness but of infinite extent in the long dimension. The new chain to be deposited on the growing crystal face is also infinite in length. Let each deposited chain segment be represented by a square prism with end dimensions equal to a, as shown in Fig. 5.8. The laying down of this segment increases the crystal surface by an area equal to that of the side surfaces and end surface of the prism. The resulting increase in free energy is greater than the reduction of free energy resulting from the attachment of the new chain, and so the free energy keeps increasing as the new chain is deposited along a straight line. If the chain now bends back on itself, there will be a large increase in free energy associated with the bend but its continued deposition along side the previous segment will create no

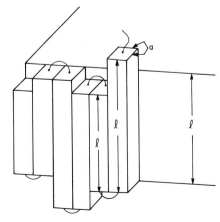

FIG. 5.8. Schematic diagram for the model of a chain-folded polymer molecule on the lateral face of a chain-folded lamella (Lauritzen and Passaglia, 1967).

new side surface, and therefore will be accompanied by a decrease of free energy. The net free energy at this point will be greater than that of the system at the start of the deposition but will be decreased further if now another fold occurs followed by further deposition. If l is the length of the deposited stem, the change in free energy on deposition of the first fold is

$$\Delta g = 2al\sigma_s + 2a^2\sigma_e - a^2l\,\Delta G \tag{5.1}$$

where σ_s is the surface free energy of the side surface and σ_e that of the end (σ_e being thus the fold surface free energy), while ΔG is the free energy of crystallization of unit volume of polymer. ΔG may be expressed in terms of the heat of fusion Δh and the melting temperature T_m° of the infinitely extended chain crystal

$$\Delta G = \Delta h(T_m^\circ - T_c)/T_m^\circ \tag{5.2}$$

The first two terms on the right-hand side of Eq. (5.1) give the increase in free energy arising from the creation of new surface, while the third is the decrease resulting from the formation of the crystal. If the folds are uniform and consequently no new side surface is formed after the initial deposition, the decrease in free energy for each depositing fold will be $[a^2l\Delta h(T_m^\circ - T_c)/T_m^\circ] - 2a^2\sigma_e$. When these terms are equal we have

$$l_{min} = 2\sigma_e T_m^\circ/\Delta h(T_m^\circ - T_c) \tag{5.3}$$

where l_{min} now represents the shortest stem that can maintain itself, i.e., the length at which the free energy of crystallization just balances that of the formation of the fold. However, when $l = l_{min}$ there is no driving force for further addition. Growth requires $l > l_{min}$, but this entails a larger energy barrier (whose height is linear in l) when adding the first stem. It can be shown (Lauritzen and Hoffman, 1960; see also Keith, 1973) that the maximum rate of addition is realized at length $l^* = l_{min} + (kT/a\sigma_s)$, at which reduced rate of addition of the first stem and increased rate of attachment of subsequent stems combine in an optimum way.

Each crystallizing molecule must overcome the activation energy barrier to deposit one stem of favorable length on the surface, after which it folds and adds further stems in both directions on the surface. We see that

$$l^* = \frac{2\sigma_e T_m^\circ}{\Delta h(T_m^\circ - T_c)} + \frac{kT}{a\sigma_s} \tag{5.4}$$

Typical values for the growth of polyethylene crystals are $\sigma_e = 60\,\text{erg cm}^{-2}$, $\sigma_s = 10\,\text{erg cm}^{-2}$, $T_m^\circ = 400°\text{K}$, $T_m^\circ - T_c = 20°$ and $\Delta h = 63\,\text{cal gm}^{-1}$, so that $l^* \simeq 10\,\text{nm}$. The first term in Eq. (5.4) is much larger than the second so that it is predicted, in agreement with experiment, that l^* is inversely proportional to $T_m^\circ - T_c$.

A striking property of chain-folded crystals is that on being annealed below their melting points they exhibit an increase of fold length l. The process may take place with crystals in contact with an inert liquid or in the dry state. It is not reversible: on being held at a lower temperature, the fold length will not again decrease. Data from Fischer and Schmidt (1962a,b) show an approximate linear increase with log time, the rate and l increasing with temperature (Fig. 5.9). In suspension, the increased fold length may involve successive dissolution and crystallization. In the dry state a different mechanism prevails, but its detailed nature is not clear.

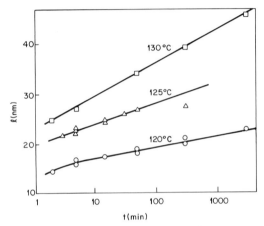

FIG. 5.9. Increase of lamellar thickness l as a function of temperature and time (Fischer and Schmidt, 1962).

5.3.4. MULTILAYERED CRYSTALS

So far we have considered only single-layered crystals, i.e., crystals that are one chain-folded-layer thick. Such crystals are typically formed at low degrees of undercooling. The growth of polymer crystals, however, is not limited to lateral propagation. The formation of multilayered crystals consisting of several superimposed chain-folded layers, all of equal thickness, is also frequently observed. These may occur at low undercoolings but are more typical of relatively high undercoolings. Their growth may originate from a screw dislocation (or several contiguous screw dislocations of the same sign) at the center, followed by growth to form spiral terraces. A multi-layered crystal of polyethylene is shown in Fig. 5.10.

FIG. 5.10. Electron micrograph of a metal-shadowed positive replica of multilayered crystalline growth in linear polyethylene (9000X). Spiral growth from a central screw dislocation can be observed (H. D. Keith and F. J. Padden, Jr., by permission).

5.4. Crystallization from the Melt

5.4.1. STRUCTURE OF SPHERULITES

The dominant form in which polymers crystallize from the melt is the *spherulite*. Spherulitic crystallization can occur in small molecule systems, particularly in the presence of impurities, but is relatively rare. Spherulites are aggregates of crystals with a radiating fibrillar structure. In polymers they are usually microscopic in size, being generally smaller at high undercoolings because more are nucleated per unit volume. They are commonly of the order of 100 μm or less in diameter. In Figs. 5.11 and 5.12, spherulites in isotactic polystyrene and polyethylene succinate are observed under the polarizing microscope between crossed polarizers. Being birefringent, they show up against a dark background of molten polymer in Fig. 5.11; in Fig. 5.12 crystallization is complete and the spherulites impinge on one another with more or less straight boundaries. In both figures, the spherulites are seen in cross section, having grown in two dimensions in a film of molten polymer between cover slips whose separation is small compared to the

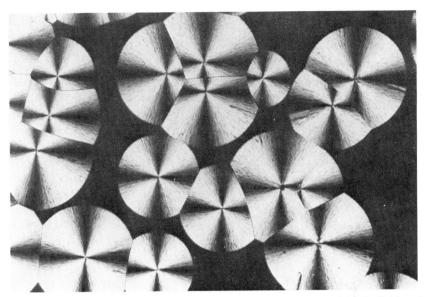

FIG. 5.11.   Spherulitic growth in a melt of isotactic polystyrene (115X) (H. D. Keith and F. J. Padden, Jr., by permission).

FIG. 5.12.   Spherulites in polyethylene succinate (115X) (F. J. Padden, Jr., by permission).

spherulite diameter. In bulk polymer their growth would of course take place in three dimensions.

The maltese cross pattern is highly characteristic, and reflects the birefringent nature of the polymer and the symmetric nature of the spherulite. The refractive index for light polarized with its electric vector along the radial direction of the spherulite is different from that for light polarized with its electric vector normal to the radial direction. As one turns the sample under the microscope, the cross pattern remains stationary, being determined by the positions of the polarizer and analyzer.

An important question is that of the orientation of the polymer chains in the spherulite. It has been answered both by noting the sign of the birefringence and by use of the microbeam x-ray diffraction technique. In the latter technique, a very narrow beam is used to irradiate a small segment along a radius of the spherulite. The resulting diffraction pattern shows that the molecular chains are generally normal to the radial direction, or nearly so. This unexpected finding was made before the chain orientation in single

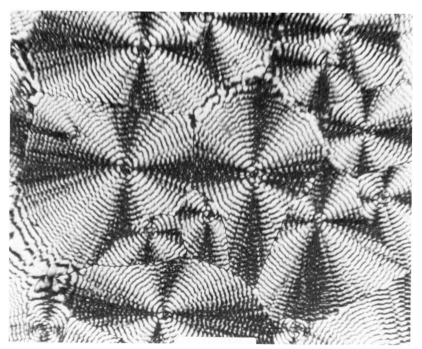

FIG. 5.13. Banded spherulitic structure in polyethylene (7000X) (F. J. Padden, Jr., by permission).

crystals (Section 5.3) had been established, and remained an unexplained anomaly for some years. We now realize that the radiating fibrils of the spherulite are actually elongated, chain-folded lamellae.

Spherulites often exhibit a more complex banded pattern, as shown in Fig. 5.13. The maltese cross pattern is still evident, but the radiating fibrillar appearance has been replaced by a series of very regular concentric bands. This banded appearance has been shown (Keith and Padden, 1959a,b; Price, 1959; Keller, 1959) to arise from the fact that the crystal orientation rotates about the radius. The regularity of the pattern shows that neighboring fibrils twist with the same period, with the same phase, and, within given sections, in the same helical sense.

Figure 5.14, which is a metal-shadowed electron micrograph of a replica of the surface of a banded spherulite of polyethylene, shows cooperatively twisting ribbonlike lamellae. The period of twist corresponds to the separation of bands seen in the optical microscope. The edges of the lamellae appear as steps a few tens of nanometers in thickness. Their structure at the molecular level is essentially the same as that of the lamellar single crystals discussed in Section 5.3. Spherulites that do not have a banded appearance exhibit the same radially oriented lamellar structure but without cooperative

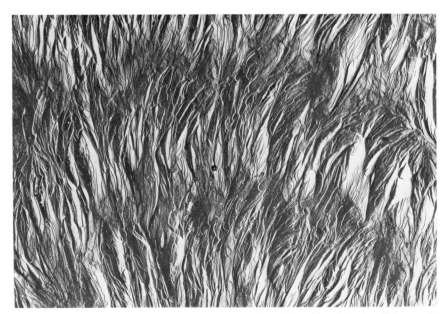

FIG. 5.14.   Electron micrograph of an etched positive replica of banded spherulitic growth in polyethylene (7000X) (H. D. Keith and F. J. Padden, Jr., by permission).

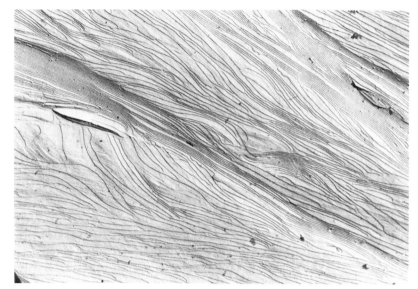

twist. Figure 5.15 shows a polyethylene sample exhibiting this mode of
growth. As with solution-grown crystals, the lamellar thickness of melt-
grown crystals increases as the degree of undercooling decreases.

5.4.2.   GROWTH OF SPHERULITES

Polymer spherulites do not begin their growth in a spherically symmetric
manner, but only attain this mode at later stages. Immediately after nucleation, the initial crystallite has the form as shown in Fig. 5.16a. Upon growing,
the multilayer crystal (see Section 5.3.4) splays out progressively as seen at
(b) and (c). These early forms are sometimes referred to as "hedrites" or
"axialites." They are not axially symmetric but rather have square (or poly-
gonal) forms when seen "flat on," i.e., at an angle of 90°; such forms are
illustrated in the bottom row. Sheaflike forms (d) represent a somewhat
later stage. At (e) an approximately spherical form is reached; this is still an
early stage, the final spherulite usually being much larger.

In order to fill space in the manner observed, spherulitic growth requires
(a) a fibrillar habit and (b) the occurrence of low-angle "branching" and
continued splaying of the crystallites as they grow. Theories have been
advanced to account for these features. A viable theory must also explain
the fact (c) that the spherulite radius increases linearly with time during the

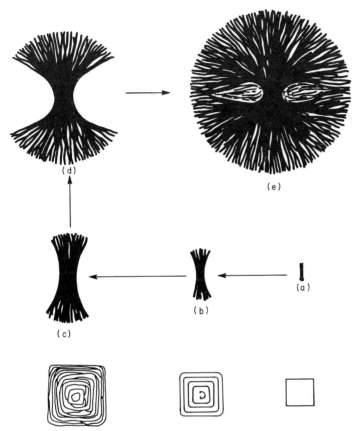

Fig. 5.16.   Stages in the growth of a spherulite. Growth begins at (a) with a small multi-layered crystal which splays out as it grows (b) and (c), reaching the sheaflike stage shown at (d), which then fills in to a spherulitic nucleus (e); the latter would normally grow further to considerably greater size. In the bottom row below (a), (b), and (c) are shown the appearance of these structures at a 90° viewing angle, i.e., normal to the crystal surface (adopted from H. D. Keith and F. J. Padden, Jr., by permission).

growth process. This indicates that growth is controlled not by the radial diffusion of latent heat outward or of crystallizable material inward (which should lead to a rate proportional to the square root of time), but rather by a nucleation process at the growing tips of the fibrils. The most comprehensive theory accounting for these observations is that of Keith and Padden (1963, 1964a,b). The principal question to be answered is why nucleation is so much faster at the fibril tips than on their lateral surfaces.

A condition for the formation of spherulites appears to be the presence of impurities in the melt. This has been clearly shown for small-molecule

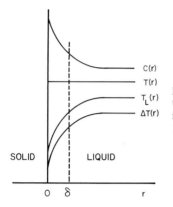

Fig. 5.17. Impurity concentration $C(r)$, ambient temperature $T(r)$, equilibrium liquidus temperature $T_L(r)$, and undercooling $\Delta T(r)$ near the surface of a crystal growing into a spherulite-forming melt (from Keith and Padden, 1963).

crystallization (Keith and Padden, 1963, and references therein). For polymer melts, chains too short or too irregular in structure to crystallize are assumed to act as "impurities." The growing polymer crystal rejects these impurities, the concentration of which accordingly builds up on the liquid side of the interface to a value higher than the average value for the melt as a whole. This impurity layer has a width of approximately $\delta = D/G$, where D is the diffusion constant and G is the growth rate. In Fig. 5.17 a growth front is represented in which the concentration of impurities $C(r)$ decreases as we move away radially from the front, reaching a steady value characteristic of the whole melt. Since the growth of polymeric crystals is slow, primarily because of high melt viscosity and small D, latent heat diffuses away without creating appreciable temperature gradients and the ambient temperature $T(r)$ may be taken as uniform through the impurity layer. The effective melting point $T_L(r)$ (or equilibrium liquidus temperature) and consequently the degree of undercooling $\Delta T(r)$ increase as shown as one moves away from the growth front. For example, if the melting point is $\sim 90°$ in the impurity layer and $100°$ outside the layer and the ambient temperature is $80°$, the undercooling $\Delta T(r)$ will be about $10°$ in the impurity layer and $20°$ outside it. The result is that any fortuitous irregularity or protuberance in the growth front will tend to be propagated because its tip will reach out into a region of greater undercooling where growth is more rapid. This is illustrated in Fig. 5.18, where impurities are represented as black dots. The proliferating irregularities thus assume a fibrillar character, the accumulating impurity between fibrils tending to discourage growth on their lateral surfaces. It can be shown that the fibrillar diameter should be of the order of magnitude of δ.

As with any crystallization from a viscous melt, spherulitic growth rates go through a maximum as a function of temperature, primarily because the rate of secondary nucleation increases as undercooling increases but is offset

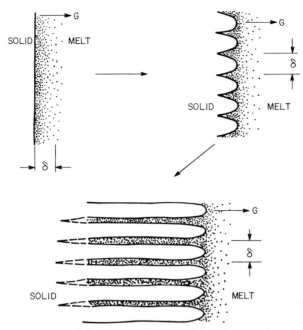

FIG. 5.18. Development of fibrous crystalline growth in polymer melt containing "impurities" (black dots) (from H. D. Keith and F. J. Padden, Jr., by permission).

at lower temperatures by slow transport processes in the melt. Taking advantage of this fact, it is possible to obtain polymer spherulites at the same value of G but differing values of D. As expected from the above discussion, the spherulites obtained at the higher temperature show a coarser structure. The addition of atactic polymer to a crystallizing melt causes a more open texture because the atactic material segregates between crystalline lamellae, reducing the overall crystallinity of the spherulite.

5.4.3. INTERCRYSTALLINE LINKS

The picture of a solid, semicrystalline polymer, which we have so far built up, is that of a mass of spherulites that are merely contiguous, apparently not bound together, and are themselves further composed of radiating lamellae which are not joined together except perhaps at branch points. This view is evidently too simple as it does not account for the mechanical properties of such polymers. Crystalline polymers are tough and can be extensively deformed under stress, whereas one might expect failure at interspherulitic boundaries and possibly within the spherulites as well. It has accordingly been suggested that neighboring lamellae are held

336   F. A. Bovey

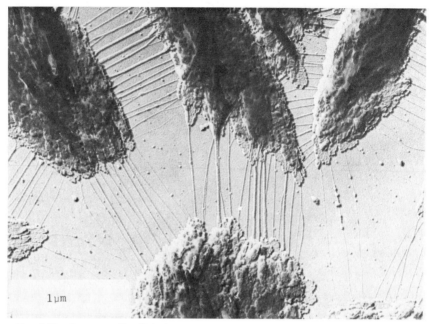

FIG. 5.19.   Intercrystalline links between the lateral edges of lamallae in spherulites of polyethylene grown from a solution in dotriacontane. The electron micrograph was taken after the
dotriacontane had been dissolved away (15,000X) (from H. D. Keith by permission of the
American Physical Society).

together by tie molecules that are embedded in both. Keith *et al.* (1966,
1971) provided convincing evidence of intercrystalline links by crystallizing
polyethylene in the presence of $n\text{-}C_{32}H_{66}$. The paraffin, having the basic
structure of polyethylene, was assumed not to interfere with its growth habit,
other than to produce a relatively open structure, and it could be dissolved
away with a suitable solvent. When this was done, a large number of links
joining the lateral faces of lamellae in spherulites were revealed, as shown in
Fig. 5.19. These bridges are clearly thicker than a single molecule, and it is
thought that groups of molecules are pulled together under stresses developed
in the crystallization process. Individual tie molecules are probably formed
as well but are too fine to be visible.

REFERENCES

Bank, M. I., and Krimm, S. (1969a). *J. Polym. Sci. Part A-2* **7**, 1785.
Bank, M. I., and Krimm, S. (1969b). *J. Appl. Phys.* **40**, 4248.
Bassett, D. C., Frank, F. C., and Keller, A. (1963). *Phil. Mag.* **8**, 1739.
Fischer, E. W., and Schmidt, G. F. (1962a). *Angew. Chem.* **74**, 551.

Fischer, E. W., and Schmidt, G. F. (1962b). *Angew. Chem. Int. Ed.* **1**, 488.
Flory, P. J. (1953). "Principles of Polymer Chemistry." Cornell Univ. Press, Ithaca, New York.
Flory, P. J. (1962). *J. Am. Chem. Soc.* **84**, 2857.
Flory, P. J., and Yoon, D. Y. (1978). *Nature (London)* **272**, 226.
Frank, F. C., and Tosi, M. (1961). *Proc. R. Soc. London Ser. A* **263**, 323.
Gerngross, O., Hermann, K., and Abitz, W. (1930). *Biochem. Z.* **228**, 409.
Keith, H. D. (1973). *Metall. Trans.* **4**, 2747.
Keith, H. D., and Padden, F. J. Jr. (1959a). *J. Polym. Sci.* **39**, 101.
Keith, H. D., and Padden, F. J. Jr. (1959b). *J. Polym. Sci.* **39**, 123.
Keith, H. D., and Padden, F. J. Jr. (1963). *J. Appl. Phys.* **34**, 2409.
Keith, H. D., and Padden, F. J. Jr. (1964a). *J. Appl. Phys.* **35**, 1270.
Keith, H. D., and Padden, F. J. Jr. (1964b). *J. Appl. Phys.* **35**, 1286.
Keith, H. D., Padden, F. J. Jr., and Vadimsky, R. G. (1966). *J. Polym. Sci. Part A-2* **4**, 267.
Keith, H. D., Padden, F. J. Jr., and Vadimsky, R. G. (1971). *J. Appl. Phys.* **42**, 4585.
Keller, A. (1957). *Phil. Mag.* **2**, 1171.
Keller, A. (1959). *J. Polym. Sci.* **39**, 151.
Keller, A. (1968). *Rep. Prog. Phys.* **31**, Part 2, 623.
Kloos, F., Go, S., and Mandelkern, L. (1974) *J. Polym. Sci. Polym. Phys. Ed.* **11**, 1091.
Krimm, S., and Ching, J. H. C. (1975). *Macromolecules* **5**, 209.
Lauritzen, J. I., and Hoffman, J. D. (1960), *J. Res. Nat. Bur. St.* **A64**, 73.
Lauritzen, J. I., and Passaglia, E. (1967). *J. Res. Nat. Bur. St.* **A71**, 261.
Magill, J. H. (1977). In "Treatise on Materials Science and Technology" (J. M. Schultz, ed.), Vol. 10. Academic Press, New York.
Nakajima, A., and Hamada, F. (1972). *J. Pure Appl. Chem.* **31**, 1.
Peterlin, A. (1960). *J. Appl. Phys.* **31**, 1934.
Peterlin, A., and Fischer, E. W. (1960). *Z. Phys.* **159**, 272.
Peterlin, A., Fischer, E. W., and Reinhold, C. (1962). *J. Chem. Phys.* **37**, 1403.
Price, F. P. (1959). *J. Polym. Sci.* **39**, 139.
Price, F. P. (1961). *J. Chem. Phys.* **35**, 1884.
Schelten, J., Wignall, G. D., and Ballard, D. G. H. (1974). *Polym. Preprints* **15**, 331.
Storks, K. H. (1938). *J. Am. Chem. Soc.* **60**, 1753.

GENERAL REFERENCES

Astbury, W. T. (1933). *"Fundamentals of Fiber Structure."* Oxford Univ. Press, London and New York.
Geil, P. H. (1963). "Polymer Single Crystals." Wiley, New York.
Hermans, P. T. (1949). "Physics and Chemistry of Cellulose," pp. 1–40. Elsevier, Amsterdam.
Hoffman, J. D., Davis, G. T., and Lauritzen, J. I. Jr. (1976). In "Treatise on Solid State Chemistry" (N. B. Hannay, ed.), pp. 497–614. Plenum Press, New York.
Khoury, F., and Passaglia, E. (1976). In "Treatise on Solid State Chemistry" (N. B. Hannay, ed.), pp. 335–496. Plenum Press, New York.
Mandelkern, L. (1964). "Crystallization of Polymers." McGraw-Hill, New York.
Sharples, A. (1966). "Introduction to Polymer Crystallization." St. Martin's Press, New York.
Wunderlich, B. (1973). "Macromolecular Physics," Vol. 1, Crystal Structure, Morphology, Defects. Academic Press, New York.

Chapter 6

PHYSICAL BEHAVIOR OF

MACROMOLECULES

S. Matsuoka and T. K. Kwei

6.1. Introduction

Macromolecules may crystallize or vitrify upon cooling, or become rubbery or even liquidlike upon heating. Stereoregularity is an obvious requisite for polymer crystallization. Both polyethylene and isotactic polypropylene are highly crystalline polymers. The noncrystalline portions of polymers form glasses upon further cooling. The glass-forming or glass-transition temperature is typically 1/2 to 2/3 of the melting temperature in degrees Kelvin, according to the empirical rule of Boyer (1954). Atactic polymers, such as polystyrene or poly(methyl methacrylate), will not crystallize because

339

they are too irregular to fit together. These polymers are found either in the glassy state or in the rubbery (molten) state. In summary, a polymer can be found either in an entirely rubbery state, a part crystalline and part rubbery state, a part crystalline and part glassy state, or a totally amorphous glassy state. It is important to recognize that mechanical and electrical properties are strongly influenced by the state of the polymer. Another important factor to consider in studying mechanical behavior of polymers is the presence or absence of chemical cross-links between molecular chains. In the absence of cross-links, macromolecules will creep slowly but indefinitely, while with cross-links, macromolecules stop creeping at a certain degree of deformation.

Macromolecules respond to an external force in two ways. Part of the response is an instantaneous one called *elastic*, and all of the mechanical energy put into the body is recovered upon the release of the external force. The other part of the response is a delayed one that involves dissipation of mechanical energy into heat as in the flow of a viscous fluid. The combined behavior is therefore termed *viscoelastic*. Viscoelastic behavior of macromolecules in the rubbery state is a combination of the rubberlike elasticity and the liquidlike viscosity arising from intersegmental friction. The rubberlike elasticity arises from the tendency of the macromolecules to retain the random conformations against the external force that tends to orient them. The intersegmental viscous contribution is much more sensitive to a temperature change than the elastic contribution, and therefore the overall temperature dependence of the viscoelastic behavior of macromolecules in the rubbery state is practically identical to the manner in which the viscosity of a liquid depends on temperature. A quantitative description of the way in which the viscoelastic behavior depends on temperature will be presented in detail in this chapter.

The viscoelastic behavior of glassy and crystalline polymers is much more complex than that of a rubbery polymer. First of all, the mechanical and thermodynamical properties continue to change long after the polymer has solidified, because macromolecules in the solid state are still readjusting themselves towards the more stable denser state. In crystalline polymers, the long persisting process known as "secondary" or "residual" crystallization will continue to increase density and alter mechanical properties. The process is accelerated at elevated temperature (but of course below the melting temperature), and the thermal treatment is called the annealing process. In short, the mechanical behavior of crystalline or semicrystalline polymers depends on the thermal history. Although glassy polymers are considered completely disordered, they eventually pack after a long time, and their mechanical properties are affected by annealing. Thus, the behavior of glassy polymers is also history-dependent.

Another important characteristic of both crystalline and glassy polymers is the dependence of viscoelastic behavior on the magnitude of strain. The characteristic relaxation time, which is a measure of how rapidly or how slowly a material responds to an external force, is decreased by several orders of magnitude under a strain of only a few percent. Any behavior in which a characteristic constant such as the relaxation time depends on an experimental variable such as the strain magnitude is termed *nonlinear* behavior, since the differential equation for the stress–strain and the rate of strain must include coefficients that are not constant. One of the objectives of this chapter is to describe how such nonlinear viscoelastic behavior that depends on thermal history may be interpreted in terms of macromolecular structure.

Polymeric solids can fail mechanically in a number of ways. When subjected to an increasing tensile force, polystyrene typically breaks in a brittle manner, while polycarbonate or polyethylene undergoes plastic deformation, increasing the length of the test specimens to several times the original dimensions. During deformation poly(methyl methacrylate) develops many tiny flaws called *crazes*. Polyethylene, which can dissipate a large amount of strain energy if strained in air, will sometimes crack brittlely in a dilute solution of a detergent. This phenomenon is termed *environmental stress cracking*. Stress concentration, induced by incorporating a sharp notch or a sharp corner in a plastic part, often becomes a nucleation point for the formation of a crack, which propagates in time to a size large enough to cause a catastrophic failure. Plastics can fail under repeated cycles of loading, even if the magnitude of the stress is not large enough to cause a failure the first time. This phenomenon is known as *fatigue*, and is associated with slow crack or craze growth in the body as it is repeatedly stressed.

All of this complex time-dependent mechanical behavior of macromolecules is related to the way in which molecular motions depend on temperature, strain, thermal history, solvent, and many other variables that affect the dynamics and thermodynamics of macrocolecules. Studies of molecular motions thus constitute a basis for the analysis of macromolecular behavior. Dielectric studies can unravel the type of molecular motions that are likely to be prevalent at different temperatures and pressures. Nuclear magnetic resonance studies can also elucidate the nature of molecular dynamics and are particularly useful for polymers that are essentially nonpolar and therefore not suitable for dielectric studies. Many types of viscoelastic tests are also helpful in understanding what must go on in plastics at the molecular level. It is the aim of this chapter to explore what is known in regard to the relationship between mechanical and electrical behavior and the molecular structure of macromolecules.

6.2.   Viscoelastic Properties of Macromolecules in the Rubbery State

6.2.1.   INTRODUCTION

Viscoelastic properties of macromolecules in the rubbery state may be considered as contributions from rubberlike elasticity and from fluidlike viscosity. The nature of the elastic behavior of a rubber is quite different from that of a solid material. The thermodynamic and statistical mechanical aspects will be described in the next section of this chapter. The viscosity contribution to the viscoelasticity diminishes with rising temperature in a manner consistent with that observed for many organic liquids. The flow properties of polymeric melts are greatly influenced by the nature of inter-actions between the long macromolecular chains and are unique among all substances that can be made to flow under external stresses.

6.2.2.   RUBBERLIKE ELASTICITY

When a body of vulcanized rubber, whose dimensions are as shown in Fig. 6.1, is pulled with a force of F dynes, the body will elongate in the direc-tion of the pull and contract in the lateral direction. Here, we have chosen a piece of vulcanized rubber in order to discuss only the elastic properties of macromolecules. If the body is deformed slowly enough, it is always at equilibrium as the stress and strain are increased, and all of the mechanical energy imposed on the body will be recovered with the release of the stress. If the dimensions of the piece of rubber are accurately measured, it will be found that the lateral contraction will exactly make up for the longitudinal

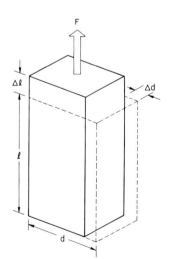

FIG. 6.1.   Deformation of a body in tension under force F; extension by Δl in the direction of force is partially compensated by the contraction Δd in the two lateral directions.

extension in such a way that the total volume will remain constant at all times, i.e.,

$$V = ld^2 = (l + \Delta l)(d - \Delta d)^2 \tag{6.1}$$

where V is the volume, l is the length, and d is width of the sample. The ratio of the lateral contraction to the longitudinal extension is defined as μ, Poisson's ratio, and in the case where there is no net change in the volume, Poisson's ratio is 0.5. The fact that the volume and the density of a piece of rubber remain nearly constant under tension is an important characteristic of macromolecules in the rubbery state. The observation that the body will change its *shape* far more readily than its *volume* under stress has led to the terminology "incompressible" to describe the mechanical behavior of rubber. This terminology is misleading, because in reality it does not mean that the specific volume of rubber remains unaffected under hydrostatic pressure. In fact its compressibility is comparable to that of many liquids, and far greater than those of metals and glasses. The terminology incompressible for describing a rubber is also objectionable because metals and glasses, which are *not* "incompressible," *increase* in volume in tension. Thus rubbery behavior should best be termed *undilatable* rather than incompressible.

If the piece of rubber is wrapped with heat-insulating material to minimize heat transfer to and from the ambient atmosphere, the temperature of the body will be found to rise as it is pulled in tension. After the sample body is held in tension for some time to let the temperature equilibrate with the atmosphere and is then released to recover to the original dimensions, the temperature will be found to decrease. Thus, the tensile deformation is exothermic. This is a consequence of heat generated when macromolecules are pulled and oriented. Even though the rubbery molecules at this stage are still far from attaining the crystalline order (which they will if pulled to a sufficiently large elongation), the heat content of macromolecules decreases as the degree of order is increased. The heat generated in this particular case is not the frictional heat associated with dissipation of mechanical energy, since the body will take in heat as the stress is relieved.

Thermodynamics of macromolecules with externally imposed order has been extensively studied by many scientists. The heat of orientation ΔH must accompany the corresponding change in entropy ΔS because the net change in free energy ΔG must be equal to the mechanical work done in order to satisfy the condition of equilibrium:

$$\Delta G = \Delta H - T\,\Delta S \tag{6.2}$$

Thus the entropy, which will decrease with orientation, plays an important factor in the elastic behavior of rubbery molecules. It can be shown through a statistical mechanical analysis that macromolecules can take the greatest number of conformations when they are under no stress. Under the stress, the average shapes of molecules become more skewed from the unstressed random conformations, and the number of possible conformations will decrease. The entropy of macromolecular chain S is described by the formula

$$S = k \ln W \tag{6.3}$$

where k is the Boltzmann constant and W is the number of possible conformations, so that we obtain for the entropy change ΔS with deformation:

$$\Delta S = k \ln(W_1/W_0) \tag{6.4}$$

where the subscripts 0 and 1 refer to the initial condition of no strain and the condition of the oriented state under stress, respectively.

Rubberlike elasticity is thus the consequence of molecular arrangements. This is quite different from the elasticity of ordinary solids such as metals, glasses, crystals, and wood, where the elasticity, or the resistance to deformation under external force, arises from the distortion of the intermolecular potential fields. When the crystal lattice of diamond is distorted under stress, the interatomic distance is changed and the potential energy is raised. When the stress is relieved, the atoms in the lattice will return to the original positions of the potential minima. The driving force for the return to the original position is the internal energy of an atom with respect to its neighbors that are held together by interatomic bonds. Conformational entropy discussed previously has nothing to do with this case. For macromolecules under strain, the intermolecular potential energy remains nearly constant with or without strain. The elastic driving force is therefore entirely from conformational entropy, that is, from the tendency for macromolecules to randomize in order to attain the maximum entropy and the minimum free energy. For this reason, rubber is often called the entropy spring. The entropy spring becomes stiffer at higher temperature, since the tendency to randomize becomes stronger with more vigorous segmental Brownian motions like those of molecules in a liquid. This is in contrast with the behavior of most solid bodies in which the potential energy is weakened and the stiffness is diminished at higher temperatures as a consequence of thermal expansion, which sets atoms further apart. All the above discussion on elasticity can be simply stated in terms of a thermodynamic equation of state for stress σ

$$\sigma = (dA/d\varepsilon)_T = (dU/d\varepsilon)_T - T(dS/d\varepsilon)_T \tag{6.5}$$

where A is Helmholtz free energy, U is the internal energy such as those associated with the interatomic potential energy in crystals, and ε is the strain. For diamond, $T(dS/d\varepsilon)_T$ is zero, and for rubber, $(dU/d\varepsilon)_T$ is zero.

6.2.3. FLOW VISCOSITY OF POLYMER MELTS

In the last section, the equilibrium elastic behavior of vulcanized (mildly cross-linked) macromolecules was described. In this section, the steady flow behavior of uncross-linked macromolecules in the amorphous rubbery or molten state (above the glass transition temperature or the melting point) will be described.

Most uncross-linked polymers will, upon heating, eventually become fluidlike and will flow readily. This property is essential for extruding or molding of plastics. The viscosity of the melt is a measure of the resistance to flow under external force, just as the elasticity discussed in the last section is the resistance to deformation under external force. Unlike the elastic body, a fluid body has no definite shape and its mechanical properties are described not in terms of deformation but in terms of the rate of deformation. The stress that is commonly described in fluid dynamics is shear stress. When a force of F dynes is applied to a plane with area A cm^2 in the direction *parallel* to that plane, the shear stress, F/A dynes per square centimeter, will cause the layers of fluid to move. The rate of strain in shear at a moving layer is defined by the velocity of that layer v divided by the distance from the stationary layer, y, as shown in Fig. 6.2. The ratio of the shear stress to the rate of strain is the viscosity η, i.e.,

$$F/A = \eta(v/y) \tag{6.6}$$

The viscosity can be measured in the fluid being sheared between two rotating plates or being extruded through a capillary tube. The latter technique is particularly simple and convenient, and a quick qualitative comparison of melt viscosity among various thermoplastic compounds can be made by this technique. A standard test for the melt flow index (ASTM, 1978) is to measure the weight of plastic extruded through a capillary die during a given time interval while a weight is placed on top of the plunger pressing the reservoir of the heated plastic melt. Since a fluid with low viscosity will flow rapidly through the capillary, a high melt flow index figure would mean

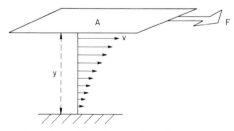

FIG. 6.2.   Shear flow of liquid under force F acting on the area A.

a low viscosity. In industry, a plastic material is specified in terms of both minimum and maximum melt flow index figures; the minimum to assure easy processibility and the maximum for high durability.

The viscosity of a polymer melt depends on the molecular weight. This is because the flow of a long macromolecule requires coordination among the segments of the molecular chain. Thus the viscosity of polymers should increase *in proportion to* the chain length, i.e., the molecular weight. Unfortunately, the influence of molecular weight on the melt viscosity in polymers is not quite so simple because of the complex interactions between the molecules, commonly termed chain entanglement. When the chains are sufficiently long, a pull on one of them will be transmitted to others as with a bundle of threads that are entangled with each other. The effect of molecular weight on viscosity will be greatly enhanced when there are such intermolecular interactions, and the viscosity will not be proportional to the molecular weight. Empirically it has been determined that the viscosity depends on the 3.4th power of the molecular weight, i.e.,

$$\eta = cM^{3.4} \tag{6.7}$$

where c is a constant and M is the molecular weight. The above equation holds only above a certain critical molecular weight M_b that depends on the nature of the monomeric units making up the polymer chains. This is illustrated in Fig. 6.3. The critical molecular weight for entanglement M_b is an important parameter, not only for the determination of rheological properties in the molten state, but with regard to viscoelastic properties and brittle–ductile failure criteria in the solid state, as will be discussed in the latter part of this chapter. Table 6.1 lists the critical entanglement molecular weight for various polymers.

In Fig. 6.3, the effect of the increasing rate of flow on viscosity is also included. As the macromolecular chains are forced to flow at a faster rate, the chains are induced to slip past each other more easily. As a consequence a lesser effect is felt from the entanglement, and the viscosity vs. molecular weight relationship approaches that without entanglement, i.e., the relationship observed below the critical molecular weight M_b.

The effect of temperature on viscosity is another important aspect of melt rheology. In the temperature range more than 100°C above the glass transition temperature, the melt viscosity varies according to the Arrhenius formula with constant activation energy:

$$\eta = \eta_0 \exp \frac{H^*}{R}\left(\frac{1}{T} - \frac{1}{T_0}\right) \tag{6.8}$$

where the subscript 0 refers to some reference state at temperature T_0, and H^* is the flow activation enthalpy, the values of which range for various

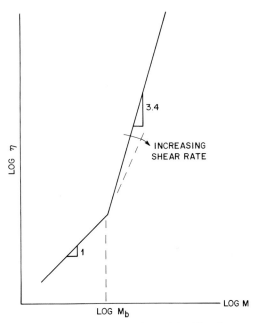

FIG. 6.3. Zero shear viscosity versus molecular weight. The change of the slope on the doubly logarithmic plot is observed at the molecular weight above which entanglement takes place.

TABLE 6.1
CRITICAL MOLECULAR WEIGHT M_b OF POLYMERS

Polymer	Critical molecular weight, M_b	Molecular weight of flow unit, M_0	$\dfrac{M_b}{M_0}$	Method
Polystyrene	37,000	52	380	Fractions
Poly(methyl methacrylate)	10,000	50	100	Stress relaxation
Polyisobutylene	17,000	28	310	Fractions
Polydimethylsiloxane	29,000	37	500	Viscosity
Poly(vinyl acetate)	22,500	43	260	Viscosity
Polyethylene	4000	14	140	Viscosity
Natural rubber	5000	17	240	Compliance
Poly(ethylene oxide)	6000	15	140	Compliance
Polycarbonate	13,000	21	620	Viscosity

polymers from several to a few tens of kilocalories per mole of the flow units. As the temperature is lowered, however, the apparent flow activation energy no longer remains constant but begins to rise. In this temperature regime, therefore, the thermally activated flow process cannot be considered

348 S. Matsuoka and T. K. Kwei

as a suitable model. The concept of free volume has been introduced to explain this phenomenon.

The origin of this concept is a simple one. Vacant sites, or holes, mixed in with liquid molecules will play a significant role in reducing the flow viscosity. An analogy is made with a case of people riding in an elevator. In an elevator that is completely full nobody would be able to move, but if one person walks out people can now move past each other as liquid molecules would in shear. Thus the vacant sites, whose number is only a fraction of the total number of molecules or segments, can enhance the segmental mobility and reduce viscosity considerably. A relationship between the probability for the segmental jump versus the concentration of holes has been theoretically derived by Bueche (1959), for example, to predict the empirically obtained equation of Doolittle (1952):

$$\ln(\eta/\eta_0) = B(1/f - 1/f_0) \qquad (6.9)$$

where B is a constant of nearly unity, and f is the fractional free volume V_f/V, which is shown in Fig. 6.4. The free volume in the liquid state, V_f,

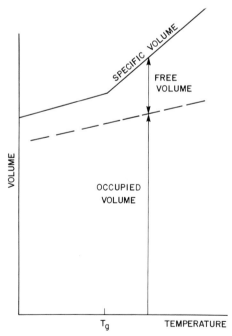

FIG. 6.4. Schematic diagram for free volume and occupied volume versus temperature following the Doolittle formula. T_g is the glass transition temperature.

increases with temperature more rapidly than the "occupied" volume including the thermal expansion due to anharmonicity of vibrations. The temperature dependence of viscosity is thus described through the increase in free volume with a nearly constant thermal expansion coefficient of its own, α_f, i.e.,

$$f = f_0 + \alpha_f(T - T_0) \tag{6.10}$$

It is found that this free volume model is a powerful tool in explaining phenomenological aspects of the viscoelasticity of macromolecules above the glass transition temperature, not only with respect to temperature dependence but also with respect to the addition of solvents, which tends to increase free volume, or to pressure, which tends to decrease free volume.

Below the glass transition temperature, the free volume is not considered to be zero, but the rate of molecular rearrangements has become so slow that the vacant sites are no longer able to contribute toward the ease of flow. Interestingly, the amount of fractional free volume that corresponds to this transition is one fortieth ($=0.025$) for many polymers. This is the so-called universal iso-free volume state for the glass transition. The role of the frozen-in free volume in mechanical behavior of glassy polymers will be discussed toward the end of this chapter.

6.2.4. VISCOELASTICITY OF MACROMOLECULES IN THE RUBBERY STATE: STRESS RELAXATION

In Sections 6.2.2 and 6.2.3, the elastic and viscous properties of macromolecules have been discussed separately. The combination of these two aspects will generate a general picture of the viscoelastic behavior of flexible macromolecular chains. For elastic behavior, the stress $\sigma = F/A$ is related to the strain $\varepsilon = \Delta l/l$, by the equation

$$\sigma = E\varepsilon \tag{6.11}$$

where E is called the modulus of elasticity, or the Young's modulus in case of tensile deformation. Now for viscous behavior, the stress σ is related to the rate of strain $\dot{\varepsilon} = d\varepsilon/dt$ by the formula

$$\sigma = \eta\dot{\varepsilon} \tag{6.12}$$

where η in the case of tension is the elongational viscosity. One of the ways for combining these two equations would be to consider that the elastic and the viscous stresses are additive, i.e.,

$$\sigma = \sigma_{\text{elastic}} + \sigma_{\text{viscous}} = E\varepsilon + \eta(d\varepsilon/dt) \tag{6.13}$$

Another way to combine the two equations would be to add the rates of strain, i.e.,

$$d\varepsilon/dt = (d\varepsilon/dt)_{\text{elastic}} + (d\varepsilon/dt)_{\text{viscous}} = (1/E)(d\sigma/dt) + (\sigma/\eta) \qquad (6.14)$$

The first model is called the Voigt model, and the second is called the Maxwell model. These two models are shown in terms of springs for elasticity and dashpots for viscosity in Fig. 6.5. Experiments for viscoelastic measurements are usually designed to follow either transient or steady-state response to the imposed stress (or strain). Two common examples of the transient experiments are stress relaxation and creep experiments. In a stress relaxation experiment, a sample is strained suddenly at time $t = 0$, and the stress is measured as a function of time t. Take as an example, the Maxwell model described by Eq. (6.14), and shown in Fig. 6.5b. Since the strain is fixed at $\varepsilon_0 = $ constant at $t = 0$ and thereafter, $d\varepsilon/dt$ is always zero, and Eq. (6.14) becomes a first-order homogeneous differential equation

$$(1/E)(d\sigma/dt) + (\sigma/\eta) = 0 \qquad (6.15)$$

and its solution is given by the equation

$$\sigma = \sigma_0 \exp(-Et/\eta) \qquad (6.16)$$

where σ_0 is the initial value of stress.

The ratio η/E is a constant with the dimension of time, and is defined as the relaxation time τ or

$$\tau = \eta/E \qquad (6.17)$$

The stress decays with the rate constant of the reciprocal of the relaxation time. If the viscosity η is great, the relaxation time is long and the stress decays slowly. The relaxation time may be defined as the time it takes for the stress to decay to $1/e$ of the initial value.

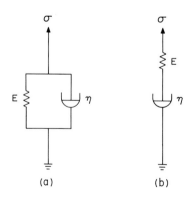

Fig. 6.5. Two simplest linear viscoelastic models where the spring (elastic element) and the dashpot (viscous element) are in parallel as in (a) Voigt model, and in series as in (b) Maxwell model.

The creep experiment is carried out by imposing the stress at time $t = 0$, and measuring the strain as a function of time while the stress is held constant. The Voigt model described by Eq. (6.13) and in Fig. 6.5a can be used as a convenient model for creep behavior. Since the stress is constant, $\sigma = \sigma_0$ and

$$\sigma_0 = E\varepsilon + \eta(d\varepsilon/dt) \qquad (6.18)$$

This differential equation can be solved, and the solution is

$$E\varepsilon/\sigma_0 = 1 - \exp(-Et/\eta) \qquad (6.19)$$

The quantity η/E is called the retardation time for creep experiment, but its physical meaning is essentially the same as the relaxation time in the stress relaxation experiment.

Before discussing steady-state viscoelastic properties, the stress relaxation experiments will be discussed in more detail in terms of actual experimental data. As a rule, stress relaxation tests are run at many temperatures. Typical data on an amorphous polymer above the glass transition temperature are shown in Fig. 6.6, where the stress divided by the strain is plotted against logarithmic time for polyisobutylene.

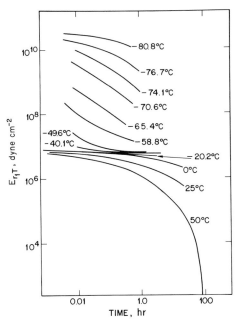

Fig. 6.6.   Log relaxation modulus versus log time for polyisobutylene obtained by Tobolsky and Catsiff (1956).

In Section 6.2.3, the effect of temperature on the viscosity of liquids and polymer melts was formulated in terms of an exponential function of the reciprocal of free volume fraction, $\exp(1/f)$. The variation in viscosity with temperature turns out to be far greater than the change in elasticity, and thus the relaxation time $\tau = \eta/E$ depends on the temperature essentially in the same manner as the viscosity. An important principle can be derived from all of the points discussed above. It is called the *time–temperature superposition* principle or the principle of reduced variables. Let us assume that two relaxation curves have been obtained at T_1 and T_2, where $T_1 < T_2$. The relaxation time τ_2 at T_2 is shorter than τ_1 at T_1 following the equation

$$\tau_2 = \tau_1 \exp(1/f_2 - 1/f_1) \tag{6.20}$$

and the stress relaxation data follow the equations

$$\sigma_1(t) = \sigma_0 \exp(-t/\tau_1) \tag{6.21}$$

and

$$\sigma_2(t) = \sigma_0 \exp(-t/\tau_2) = \sigma_0 \exp\left[-(\tau_1 t/\tau_2)/\tau_1\right] \tag{6.22}$$

But Eq. (6.22) is precisely equal to σ_1 in Eq. (6.21) if the time t is substituted by $\tau_1 t/\tau_2$, i.e.,

$$\sigma_2(t) = \sigma_1(\tau_1 t/\tau_2) \tag{6.23}$$

Substitution of t by $\tau_1 t/\tau_2$ on the logarithmic time axis means a shift of the time-dependent function (σ in this case) by $\log(\tau_1/\tau_2)$ towards shorter times, as shown in Fig. 6.7. This amount of shift for the relaxation stress from one temperature to another is precisely

$$\ln a_T = \ln(\tau_2/\tau_1) = (1/f_2) - (1/f_1) \tag{6.24}$$

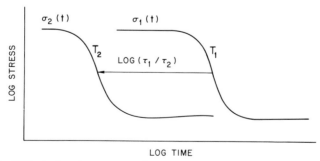

FIG. 6.7. Shift of relaxation curves as the temperature is increased from T_1 to T_2 is illustrated schematically.

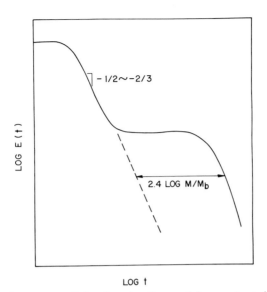

LOG t

FIG. 6.8. The "master curve" for the relaxation modulus constructed from data such as those of Fig. 6.6 taken at various temperatures by shifting according to the principle of time–temperature superposition.

from Eq. (6.20), a_T is defined as the shift factor following the convention adapted by Ferry (1970).

Thus from a series of experimental curves such as shown in Fig. 6.6, two kinds of basic information can be obtained by shifting these curves to super-impose on one another. First, a "master" relaxation curve to encompass an enormous range of time well beyond experimental possibility is obtained, as shown in Fig. 6.8. Second, the temperature dependence of the fractional free volume f can be obtained from the shift factors, from which the free volume and the occupied volume as functions of temperature can be calculated, as shown in Fig. 6.4. In general, this calculated fractional free volume is nearly a linear function of temperature, i.e.,

$$f = f_0 + \alpha_f(T - T_0)$$

which is identical to Eq. (6.10). When an estimate of f at the glass transition temperature T_g, is made, it is found to be nearly equal to 0.025 for many polymers, as has been mentioned in the previous section of this chapter. The values for f_g and α_f are tabulated for several polymers in Table 6.2.

As may be readily recognized from the shape of the relaxation curve in Fig. 6.8, the relaxation curves for real polymers cannot be fitted with an equation for a simple relaxation time such as those described above. The

TABLE 6.2
VALUES FOR FRACTIONAL FREE VOLUME (f_g) AND
THERMAL EXPANSION COEFFICIENT (α_f)

Polymers	$T_g \, °K$	f_g	$\alpha_f (10^{-4}) \, °K^{-1}$
Polyisobutylene	205	0.026	2.5
Poly(vinyl acetate)	305	0.028	5.9
Polystyrene	373	0.033	6.9
Poly(dimethylsiloxane)	150	0.071	10.3
Poly(methyl methacrylate)	388	0.013	1.7

relaxation process in polymers is a sum of many relaxation processes each of which is characterized by a different relaxation time, i.e.,

$$\sigma(t) = \sum_i \sigma_{0i} \exp(-t/\tau_i) \tag{6.25}$$

A polymer molecule can in fact undergo many kinds of relaxation processes, ranging from a motion involving only a few segments to those involving hundreds or thousands of segments. As the viscosity was shown to depend on the molecular weight, so does the relaxation time. The characteristic relaxation time for the motion of a few segments is short compared to that of a large number of segments. Thus there is a wide distribution of relaxation times for polymers, and the stress relaxation data encompasses many decades of time, as shown in Fig. 6.8.

The phenomenon of molecular entanglement, which has been mentioned with regard to the melt viscosity, extends the distribution of relaxation times toward longer times. Referring to Fig. 6.8 again, after the modulus initially relaxes to about 10^6–10^7 dynes cm^{-2}, there is a long period of time where the stress remains constant. The value of the modulus at this point is typical of the rubbery state, and the length of time t_e that it remains constant has been found to depend on the molecular weight according to the formula

$$t_e = c(M/M_b)^{2.4} \tag{6.26}$$

Here again, the critical molecular weight M_b for the formation of the entanglement network has been found to be an important parameter. When $M < M_b$, t_e is zero and the stress does not halt but continues to relax as shown by the dotted line in Fig. 6.8. The time t_e corresponds to the time required for "disentanglement," and the 2.4th power of M in Eq. (6.26) is consistent with the 3.4th power for viscosity described in Eq. (6.9), since for a molecular weight of less than M_b the viscosity and the relaxation time are proportional to (the first power of) the molecular weight.

The initial part of the stress relaxation process prior to the "disentangle-ment" process is independent of molecular weight (unless the molecular weight becomes so small that the concentration of the chain ends become significant, and T_g becomes lower). In this region in time, relatively small numbers of segments are involved in relaxation, and the modulus relaxes from the high value typical of the glassy state, or 10^{11} dynes cm^{-2}, toward the typical value of the rubbery state. This time range is, therefore, the time range where a transition is occurring from glassy to rubbery behavior in the polymer. Near T_g this transition would occur after waiting for a long time. If the experiment is run at $50°C$ above T_g, this relaxation would occur during the first fraction of a microsecond (even if the stress could be raised instantaneously), and the sample would appear rubbery throughout the experiment. The principle of time–temperature superposition thus sug-gests time as a variable in defining the glass transition, an important concept in understanding the physics of the glassy state.

6.2.5. Viscoelasticity of Macromolecules in the Rubbery State: Frequency Response

In Section 6.2.4, the stress relaxation process was discussed in detail. It can be shown mathematically that from the stress relaxation data of a viscoelastic body any other time-dependent stress–strain relationships may be calculated. For example, the stress–strain relationship at a constant rate of strain or the dynamic mechanical response as a function of frequency may be obtained from stress relaxation data as shown in Fig. 6.8. Let the strain $\varepsilon(t)$ be any function of time, as shown in Fig. 6.9. We shall now attempt to calculate the stress as a function of time as a result of the strain history of $\varepsilon(t)$ from time zero to time t. To do this, the strain $\varepsilon(t)$ is considered as an accumulation of small step functions that have taken place at various points in time so that

$$\varepsilon(t) = \sum_{x=0}^{x=t} |\delta\varepsilon| \cdot u(X) \tag{6.27}$$

where $u(X)$ is the unit step function such that

$$u(t) = 0, \quad t < X; \qquad u(t) = 1, \quad t \geq X \tag{6.28}$$

This procedure of dividing the strain into many minute increments makes it possible to calculate the corresponding stress as the cumulative sum of many stress relaxation experiments that started at various times X, where $0 < X < t$, i.e.,

$$\delta\sigma = E(t - X)\delta\varepsilon \tag{6.29}$$

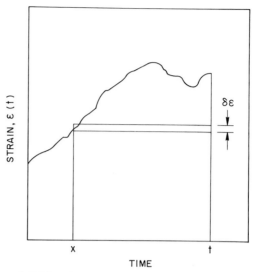

FIG. 6.9. Scheme of dividing the strain history up to time t into increments to apply the convolution integral to obtain the stress at time t.

or

$$d\sigma = E(t - X)(d\varepsilon/dX)\,dX \qquad (6.30)$$

where $E(t - X)$ is the stress relaxation modulus that has been under strain of $\delta\varepsilon$ for the time duration starting at X and lasting to time t. The stress $\sigma(t)$ is obtained by integrating Eq. (6.30),

$$\sigma(t) = \int_0^t E(t - X)(d\varepsilon/dX)\,dX \qquad (6.31)$$

This equation is called the convolution integral and is a useful formula in dealing with time-dependent behavior in general.

As an illustration, stress–strain curves will be calculated by using this integral for the Maxwell model of Eq. (6.14) whose relaxation modules can be derived from Eq. (6.16):

$$E(t) = E_0 \exp(-t/\tau) \qquad (6.32)$$

Substitution of Eq. (6.32) into the convolution integral, Eq. (6.31), will result in

$$\sigma(t) = E_0 \dot{\varepsilon}\tau(1 - \exp(-t/\tau)) \qquad (6.33)$$

where $\dot{\varepsilon} = d\varepsilon/dt$, is the rate of strain. Since $t = \varepsilon/\dot{\varepsilon}$ when the constant rate of strain is applied, this can immediately be converted into the stress–strain relationship,

$$\sigma(\varepsilon) = E_0 \dot{\varepsilon}\tau(1 - \exp(-\varepsilon/\dot{\varepsilon}\tau)) \qquad (6.34)$$

The corresponding stress–strain curves are plotted in Fig. 6.10. By increasing the rate of strain $\dot{\varepsilon}$, both the initial slope (apparent modulus) and the final maximum stress (for viscous flow) increase. The same result is obtained also by increasing the relaxation time. In an actual polymer, the relaxation time is increased by lowering the temperature as discussed previously. A decrease in temperature or an increase in the strain rate will have similar effects on the mechanical behavior of polymers. The polymer shifts toward a stiffer and harder appearance, approaching the glassy behavior.

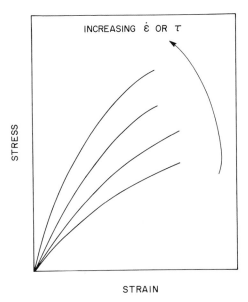

FIG. 6.10. Stress–strain behavior of a viscoelastic body with different rates of strain and/or different relaxation times.

For another illustration, let us consider, with the same Maxwell model, a case where the strain is varied as a periodic function of time, or

$$\varepsilon = \varepsilon_0 \sin \omega t \qquad (6.35)$$

where ε_0 is the strain amplitude and ω is the circular frequency, 2π times frequency in hertz. Substituting Eq. (6.35) again into the convolution integral of Eq. (6.31), the steady-state stress $\sigma(t)$ is calculated by extending the initial time to $-\infty$,

$$\sigma(t) = \frac{E_0\omega^2\tau^2}{1 + \omega^2\tau^2}\,\varepsilon_0 \sin \omega t + \frac{E_0\omega\tau}{1 + \omega^2\tau^2}\,\varepsilon_0 \cos \omega t \qquad (6.36)$$

which is the periodically alternating dynamic stress corresponding to the periodic sine function of Eq. (6.35) for the strain. The stress function alternates with the same frequency as the strain, but consists of two components, $\sin \omega t$, which is in phase with strain and $\cos \omega t$, which is out of phase from strain by $\pi/2$ or $90°$. The coefficient for the first term,

$$E'(\omega) = E_0\omega^2\tau^2/(1 + \omega^2\tau^2) \qquad (6.37)$$

is the so-called in-phase modulus or the storage modulus and is a monotonically increasing function of frequency. Comparing with the case of the stress–strain curve illustrated previously, increasing the frequency will have a similar effect on modulus as increasing the strain rate, and the behavior may be described as tending toward stiffer, more nearly glasslike behavior.

The coefficient for the second term for the stress as described in Eq. (6.36),

$$E''(\omega) = E_0\omega\tau/(1 + \omega^2\tau^2) \qquad (6.38)$$

is the so-called loss modulus, since it is a measure of the amount of strain energy dissipated into heat by a unit volume of the sample per cycle. The loss modulus will undergo a maximum when the frequency matches the reciprocal of the relaxation time, i.e.,

$$\omega\tau = 1$$

as can be calculated by taking the derivative of $E''(\omega)$ with respect to ω and setting it to zero. The time-dependent change from rubbery to glassy behavior is now more clearly defined through the introduction of the frequency at which the loss maximum is observed. Again, this loss maximum should also be observed as a function of temperature when the frequency is held constant, since the change would occur in relaxation time τ as the temperature is changed. One other viscoelastic function which is frequently employed is the loss tangent, or tan δ. It is the ratio of the loss modulus to the storage modulus,

$$\tan \delta = E''(\omega)/E'(\omega) \qquad (6.39)$$

and is simply equal to $1/\omega\tau$ for the Maxwell model, which will not undergo a maximum at $\omega\tau = 1$. This is due to the mathematical limitation of the Maxwell model, and tan δ for real polymers *will* undergo a maximum as $E''(\omega)$ undergoes a maximum. In general, since there exists a broad distribution of relaxation times in real polymers, the maximum in the loss modulus, for example, shows a broad peak, but the essential features of dynamic properties as derived with the Maxwell model for the single relaxation time are true for real polymers as well.

Thus it is evident that the mechanical relaxation process is closely related to the nature of molecular motions. The relaxation time is a measure of the rate constant characteristic of a relaxation process involving a certain mode of motion of a certain number of segments. The method of detecting molecular relaxation is not limited to mechanical measurement. If a probe can be made that can detect signals reflecting segmental motions of macromolecules, the molecular response to an externally applied field can be measured to obtain similar results. The dielectric technique for measurement of molecular dynamics, particularly near the glass transition, is sometimes more advantageous than mechanical techniques because of the uniformity of the stress field, the lack of complexities arising from nonlinear behavior as in the viscoelasticity of the solids, and the availability of a wide frequency range. Dielectric data represent the motions of dipoles that are usually part of polymer chains, in response to an alternating electrical field. Since the motions of these dipoles reflect the dynamics of polymer molecules themselves, these dipoles serve as a probe for the study of molecular dynamics. Suppose a polymer containing dipoles is suddenly subjected to a constant electric field at $t = 0$. Dipoles that are initially randomly oriented must partially orient themselves along the newly imposed field. They cannot do so instantly, however, because they are part of the viscoelastic polymeric material. If the polymer is in a capacitor, the charge, which is a measure of the extent of dipole orientation, will increase while the displacement current, which is large initially, decreases as shown in Fig. 6.11. By again invoking a single relaxation time model, it can be shown that the charge Q will increase under the suddenly imposed voltage as

$$Q(t) = Q_\infty(1 - \exp(-t/\tau)) + I_\infty t \qquad (6.40)$$

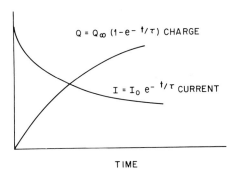

$$Q = Q_\infty (1 - e^{-t/\tau}) \text{ CHARGE}$$

$$I = I_0 e^{-t/\tau} \text{ CURRENT}$$

TIME

FIG. 6.11. Charge Q and current I versus time after the electrical potential is imposed at time $t = 0$ on a dielectric material.

where I_∞ is the steady-state current. This is analogous to the case of creep strain under constant stress for the Voigt model. In fact, there is an equivalence between mechanical and dielectric relaxation, where the stress corresponds to the voltage, the strain to the charge, and the rate of strain to the current. The dielectric constant is obtained from the charge divided by the voltage, and thus corresponds to the mechanical compliance, which is the strain divided by the stress and is dimensionally the reciprocal of the modulus. By assuming a sinusoidally varying voltage, the steady-state charge can be calculated using a convolution integral similar to Eq. (6.31). The charge will have two components, one in phase with the voltage and the other out of phase by $\pi/2$. The coefficient for the first is the in phase or capacitive part of the dielectric constant ε'. For a process with a single relaxation time τ,

$$\varepsilon'(\omega) = C/(1 + \omega^2\tau^2) \qquad (6.41)$$

where C is a constant that depends on concentration and strength of dipoles. The coefficient for the out of phase component is the dissipation factor ε'', which is resistive, corresponding to the displacement current. For the single relaxation process,

$$\varepsilon''(\omega) = C\omega\tau/(1 + \omega^2\tau^2) \qquad (6.42)$$

The functional form of $\varepsilon''(\omega)$ is exactly that of the loss modulus $E''(\omega)$, and will go through a maximum at $\omega\tau = 1$. Thus the dielectric loss maximum also depicts the glass to rubber transition. McCall's extensive data (McCall, 1969) show that mechanical and dielectric loss maxima at various temperatures for a given polymer overlap each other.

One may also deduce molecular motion from studies of nuclear magnetic resonance (Slichter, 1958; McCall et al., 1962; McCall, 1971). Whereas the dielectric method must depend upon the motion of permanent electric dipoles, the NMR method is especially suitable for studying the locations and motions of protons, and so it is highly useful for polymers.

Application of nuclear magnetic resonance in polymer science has been discussed in detail in Chapter 3, and it is briefly reiterated here. For the proton, the two permissible orientations in a magnetic field differ in energy by an amount that is proportional to the strength of the net magnetic field, i.e.,

$$h\nu_0 = 2\mu(H_0 \pm H_{loc}) \qquad (6.43)$$

where h is the Planck constant, ν_0 is the frequency of the radiation, μ is the magnetic moment, H_0 is the magnetic field which gives a resonance, and H_{loc} is the local magnetic field experienced at a given proton from the other protons around it. For a typical laboratory magnetic field of 10,000 G, ν_0

is 42.577 MHz. The resonance range is broad for a solid, corresponding to a spread of some 5 to 20 G because of the distribution of local magnetic fields. When molecules or segments of molecules move sufficiently rapidly and extensively, a given proton travels through an environment whose variations appear to be smeared out. The result is that all the protons experience about the same average field, the deviations of H_{loc} appear smaller, and the resonance envelope becomes narrower. The frequencies of the motions that cause noticeable narrowing of the resonance lie in the range of 10^4–10^5 Hz (Gutowsky and Pake, 1950). A common kind of study is to measure the width of the NMR response as a function of the temperature, and to identify transitions by noting the temperature regions in which the resonance narrows markedly. In Fig. 6.12 the line width, expressed as its second moment, is shown to narrow markedly for polystyrene near its glass transition temperature.

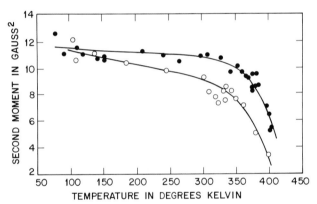

FIG. 6.12. Illustration of the line narrowing in polystyrene as the temperature is increased above the glass transition temperature of $\sim 370°C$ ●, isotactic; ○, atactic (after Slichter, 1958).

The local field H_{loc}, given in Eq. (6.43) has a strength of about 5 G at a distance of 1 nm, as compared to the field strength of an average laboratory magnet of 10,000 G. The local fields therefore may contribute a modest but significant amount to the total field experienced by a given nucleus. Thus the spread in the resonant magnetic field or, equivalently, in the resonant frequency, arises from interactions between nuclear spins. One may write this frequency spread in terms of a characteristic time T_2:

$$1/T_2 = \pi \, \delta v = 2\pi \mu \, \delta H / \eta \qquad (6.44)$$

where δv and δH are the spread in frequency and the resonant magnetic
field resulting from the spin–spin interaction, and T_2 is called the spin–spin
relaxation time.

At equilibrium in the presence of an external magnetic field, the relative
populations of the states are described by the Boltzmann relation

$$N_-/N_+ = \exp(\Delta E/kT)$$

where N_+ and N_- are the numbers of protons per unit volume in the higher
and the lower energy states and ΔE is the difference in the energy levels.
Suppose the sample is irradiated with a radiofrequency field at the frequency
v_0. Protons in the lower state will absorb energy and will consequently
transfer to the upper state. Simultaneously, protons in the upper state will
emit energy, thereby transferring to the lower state. The populations of the
two states will not, however, become equal because energy is transferred
from the excited nuclei to their surroundings, or the "lattice." This energy
transfer process is described by a characteristic time constant, T_1, which
is called the *spin–lattice relaxation time*. From the definition, one notes
that a long T_1 indicates an insufficient exchange of energy from the nuclear
spins to the lattice. Molecular motions that occur near the frequency of
nuclear magnetic resonance are the ones that couple most closely with the
spin system and produce the spin–lattice exchange of energy. According to
the theory for simple liquids (Bloembergen *et al.*, 1948), the minimum in
T_1 occurs when the correlation frequency is of the order of the nuclear
resonance frequency. Also T_1 and T_2 are predicted to coincide when the
motion is rapid enough. The T_1 minimum is shown in Fig. 6.13 for natural

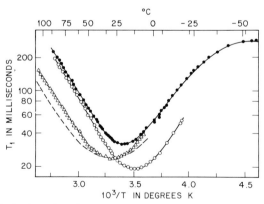

FIG. 6.13. Spin–lattice relaxation time of rubber exhibits a minimum near T_g ●, unvulcanized (28 MHz); ○. unvulcanized (20 MHz); △, vulcanized (20 MHz); ---, calculated
(after Slichter, 1961).

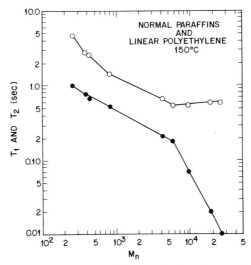

FIG. 6.14. Spin–lattice relaxation time T_1 (○) and spin–spin relaxation time T_2 (●) for polyethylene melts of various molecular weight (after McCall *et al.*, 1962).

rubber. The molecular weight dependence of T_1 and T_2 has been determined for polyethylene melt by McCall *et al.* (1962), as shown in Fig. 6.14. There is an abrupt change in the slope of the curves of both T_1 and T_2 versus logarithmic molecular weight at the number average molecular weight of about 6000, which is particularly interesting, as the critical molecular weight for entanglement of polyethylene is not too far from this value.

6.3. Glassy and Crystalline Polymers

6.3.1. EFFECTS OF THERMAL HISTORY ON THE GLASSY STATE

In Section 6.2, it has been emphasized that the glass transition depends on temperature and time or frequency. The usually quoted values of T_g pertain to the quasistatic condition corresponding to the frequency range near 1 Hz. The time-dependent nature of the glass transition may also be illustrated with an isothermal volume decrease with time, when the polymer, originally in the rubbery state, is suddenly brought to a temperature below T_g. Kovacs (1963) measured the rate of volume contraction for several glass forming polymers. An example with poly(vinyl acetate) is shown in Fig. 6.15. The straight line portions of the curves in the semilogarithmic plot indicate a tendency for the rate of contraction to become slower as contraction proceeds with time. Since the amount of fractional free volume f should

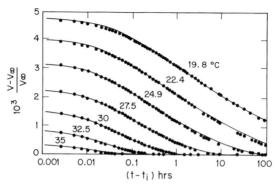

FIG. 6.15.   Isothermal volumetric contraction of poly(vinyl acetate) (after Kovacs, 1963).

decrease as the volume contracts, the relaxation time should increase as should the viscosity. If the rate of volume contraction is controlled by a transport coefficient such as viscosity η, an equation for the rate of decrease of the fractional free volume should be as follows:

$$-(1/f)(df/dt) = (C/\eta) = (C/\eta_0)\exp - |(1/f) - (1/f_0)| \qquad (6.45)$$

where the subscript 0 refers to a reference state at time t_0. The solution of this differential equation is an exponential integral, which has been shown to fit the curves obtained by Kovacs, shown in Fig. 6.16. The temperature range of Kovacs' experiments extends to more than 30°C below the usual value of T_g, indicating that the glass will continue to contract toward the equilibrium volume of the extrapolated liquidus line, though the rate will continue to slow down exponentially with time. Equation (6.45) also implies that the volume decrease will no longer be possible when the fractional free volume reaches zero, though in actuality the contraction rate will become so slow that it can be considered unchanging when f is less than 0.025.

According to the empirical arguments presented in the previous section, the occupied volume should correspond to about 1/40 less than the volume at T_g. If we assume for $\alpha_f (= df/dT)$ an average value of 5×10^{-4} per °C, the temperature at which all of the free volume vanishes *at equilibrium* would be about 50°C below T_g, as indicated by T_2 in Fig. 6.16. Gibbs and DiMarzio (1958) have treated the problem of an equilibrium polymeric glass theoretically, using a technique based on statistics of polymer chain conformation, to show that the conformational entropy would reach zero at a temperature of 20°C to 50°C below T_g.

The concept of diminishing free volume in a polymeric glass can also be convincingly demonstrated by observation of the change in relaxation time with thermal history. In this case creep experiments are conducted on several

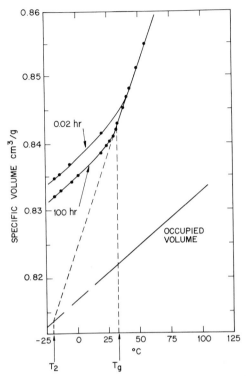

FIG. 6.16. Isochronal volume versus temperature curves constructed from the curves of
Fig. 6.15.

samples of rigid poly(vinyl chloride), all at 50°C but annealed for 1, 17, 170,
and 2000 hr prior to the start of the experiment. As shown in Fig. 6.17,
each curve is shifted along the logarithmic time axis by approximately a
decade, indicating that each decade of waiting period before the experiment
results in an increase of the relaxation time by a decade. This is exactly
what Eq. (6.45) predicts. The estimated amount of fractional free volume at
the beginning of each experiment is also shown in the same figure. Moreover, these curves exhibit a relaxation spectrum having the shape of $-\frac{1}{2}$
power in time, a result predicted (Rouse, 1953) for rubbery polymers. Thus at
temperatures of 20°C or 30°C below T_g, the relaxation processes in polymeric glasses still follow the same type of dependence on fractional free
volume operative for the rubbery state above T_g. The important departure
of mechanical behavior of polymers in the glassy state from that in the
rubbery state seems to arise then not from any dramatic shift in the deformation mechanisms or from the state of aggregation of molecular segments,

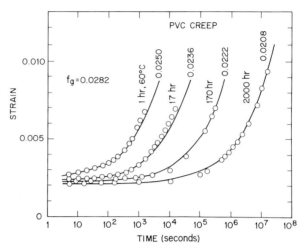

FIG. 6.17.   Creep curves of unplasticized poly(vinyl chloride) all at the same temperature under the same stress but with various thermal history so that the fractional free volume at the start (indicated for each curve) are different (after Matsuoka *et al.*, 1978).

but from a change in the apparent temperature coefficient of the free volume. Below T_g the isochronal curves in Fig. 6.16 show that the fractional free volume remains approximately constant regardless of temperature, but diminishes slowly with time.

The fractional free volume f is an empirical parameter evaluated from relaxation experiments. It is not derived directly from volumetric measurements, but its physical relationship to the classical concept of holes has been inferred. In fact, various models for free volume other than this empirical one (Bondi, 1964; Macedo and Litovitz, 1965; Simha and Boyer, 1962; Turnbull and Cohen, 1961) will give different values for the fractional free volume. One of the weaknesses of other free volume models is the over-simplified but implicit assumptions as to how the occupied portion of the total volume should behave. In most models this "occupied volume," that is, the total volume minus the free volume, is assumed to exhibit the same thermal expansion coefficient and the same compressibility as those of the glassy state. This notion is implicit in the following relation between T_g and hydrostatic pressure (Goldstein, 1963):

$$dT_g/dP = (\beta_1 - \beta_g)/(\alpha_1 - \alpha_g) \qquad (6.46)$$

where β is the compressibility and α the thermal expansion coefficient, and the subscripts l and g refer to the states above and below T_g, respectively. Experimentally, Eq. (6.46) more often fails than not. If the fractional free volume, and not the free volume based on the assumed behavior of the occupied portion, is treated as the determining parameter, the rate of T_g

increase with pressure should in fact depend on the ratio of the compressibility and the thermal expansion coefficient of the *fractional* free volume, or

$$dT_g/dP = \beta_f/\alpha_f \tag{6.47}$$

where $\beta_f = -(1/f)(df/dp)$ and $\alpha_f = (1/f)(df/dT)$. The fractional free volume at elevated pressures can be evaluated, again not from the volumetric measurements, but from relaxation experiments at high pressures above T_g at that pressure. Experimental data evaluated in this way show that the compressibility of the fractional free volume is considerably smaller than the compressibility of the rubbery state minus the compressibility of the glassy state, i.e., not $\beta_f = \beta_1 - \beta_g$, but

$$\beta_f = \beta_1 - \beta_0, \quad \text{and} \quad \beta_f < \beta_1 - \beta_g \tag{6.48}$$

Thus the occupied volume, derived from the relaxation experiment rather than estimated from the glassy behavior, exhibits a much higher compressibility than the glass. The thermal expansion coefficient of the occupied volume, on the other hand, is about the same as that of the glassy state. While Eq. (6.46) fails, another equation derived in exactly the same way but based on the enthalpic changes, i.e.,

$$dT_g/dP = TV(\alpha_1 - \alpha_g)/\Delta C_p \tag{6.49}$$

has been found successful, as can be seen from Table 6.3. This equation does not contain the questionable quantity $(\beta_1 - \beta_g)$ (O'Reilly, 1962). Therefore an empirically derived parameter f, the fractional free volume, rather than the free volume V_f alone, can better characterize the experimental results.

One might reasonably expect that the occupied volume should depend on the configurational characteristics of macromolecules. For example, isotactic polymers would have a smaller occupied volume than syndiotactic

TABLE 6.3

DEPENDENCE OF T_g ON p, SHOWING
AGREEMENT WITH EQ. (6.49)

Polymer	$\dfrac{dT_g}{dp}\left(\dfrac{°C}{atm}\right)$	$\dfrac{TV\,\Delta\alpha}{\Delta C_p}\left(\dfrac{°C}{atm}\right)$
Poly(vinyl acetate)	0.022	0.025
Polyisobutylene	0.024	0.024
Poly(vinyl chloride)	0.016	0.030
Glycerol	0.004	0.004
n-Propanol	0.007	0.005
Selenium	0.013	0.011
Salicin	0.005	0.005
B_2O_3	0.020	0.027

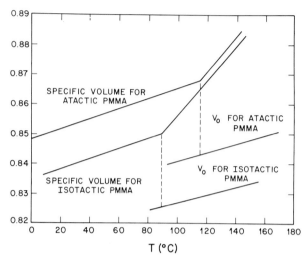

FIG. 6.18.   Illustration of how the difference in occupied volume of isotactic and atactic poly(methyl methacrylate) would change the T_g (after Ishida *et al.*, 1967).

or even atactic species. The specific volumes of these polymers in the rubbery states, on the other hand, are about the same. The net results are that the syndiotactic molecules show a higher T_g than the isotactic (Ishida *et al.*, 1967), as schematically shown in Fig. 6.18. The same argument may be extended to T_g's of stiff versus flexible chains. Stiffer chains tend to have a greater occupied volume than flexible chains, but the difference in the total volume in the rubbery state does not compensate for it, and the stiffer chains in general tend to show higher T_g's.

At the glass transition point, not only the thermal expansion coefficient but other temperature coefficients of thermodynamic quantities are observed to undergo a discrete change. Petrie (1972) has followed specific heat as a function of temperature for glass-to-liquid transitions involving glasses with different thermal histories. For a given rate of heating in the differential scanning calorimeter, a C_p versus T curve may show a typical discrete stepwise change or an apparent endotherm depending on the rate of heating and the thermal history. The total amount of the endotherm is greater for the more thoroughly annealed glass. Petrie reasoned that this peak is the consequence of a rate effect wherein the well-annealed glass with less free volume and more restricted mobility apparently takes a longer time to attain the rubbery state, as illustrated in Fig. 6.19. The integral of the C_p versus T curve is a measure of the increase in enthalpy that occurred during the transition from the glassy to the rubbery state. The greater the difference, the smaller the initial enthalpy of the glass. The enthalpy of the glass thus

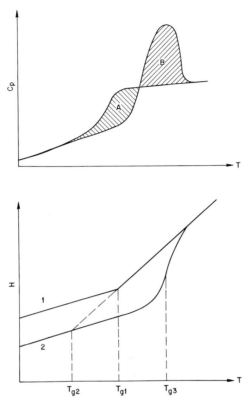

FIG. 6.19. C_p and excess enthalpy versus temperature near T_g. The area (B–A) corresponds to the difference in enthalpy between glass 1 and glass 2. Glass 2 with longer relaxation time would relax at higher temperature T_{g_3} than T_{g_1}, but its glass transition temperature should be extrapolated to the lower temperature T_{g_2} (after Matsuoka and Bair, 1977).

measured has been shown by Sharonov and Volkenshtein (1967) to decrease with time according to the equation:

$$-dH/dt = \Delta H \exp(-\Delta H/KT) \qquad (6.50)$$

According to this equation, the rate of decrease of enthalpy with time is decreased as the enthalpy decreases. This result is strikingly similar to the formulation used to describe the decrease of the fractional free volume (Eq. (6.45)). If we regard that portion of the enthalpy corresponding to f as excess enthalpy, rather than that enthalpy in excess of the liquidus enthalpy, the analogy between the two equations becomes even stronger.

In order to estimate the amount of enthalpy change associated with the amount of free volume change, two glassy states, glass 1 and glass 2, which have been subjected to different thermal histories, will be considered. Glass 2 will be designated as having been more extensively annealed than glass 1, so that glass 2 would contain the lesser amount of free volume and enthalpy. The difference in Gibbs free energy G between these two states is approximately given by the equation

$$G_1 - G_2 = H_1 - H_2 - T(S_1 - S_2) \qquad (6.51)$$

where H and S are the enthalpy and entropy, and T is the absolute temperature.

The contribution from the free volume toward the excess entropy is formulated in terms of the entropy of mixing between the occupied and the vacant sites.

When the amount of free volume is small as compared to the total volume, the entropy of mixing, S_v, can be approximated by the formula:

$$S_v \approx -Rf \ln f \qquad (6.52)$$

for 1 mole of the sites.

If we identify the occupied sites with polymer segments, there should be an increase in conformational entropy with the increased probability of mixing of the segments with vacant sites. If one is able to estimate the relative contributions from the free volume and the conformational probability, then the total excess entropy can be calculated from the volume contribution described by Eq. (6.52) by using such a ratio. For crystallizable polymers, relative contributions from volume and conformations toward the entropy of fusion may be estimated from the rise in the melting point (Matsuoka, 1962; Karasz, 1967), or from theoretical calculations involving chain flexibility (Tonelli, 1974). The ratio of the total entropy increase to the part of entropy which corresponds to the volume increase varies from the low value of 2.5 for polytetrafluorethylene to the high value of 5 for poly(ethylene terephthalate), and centers around 3 for many other polymers. The difference in the total excess entropy between glass 1 and glass 2 can be estimated by multiplying the volumetric term described by Eq. (6.52) by this ratio, ϕ. Thus,

$$S_1 - S_2 = -\phi R(f_1 \ln f_1 - f_2 \ln f_2)/M \qquad (6.53)$$

where the entropy, S is in entropy units per gram, and M is the molecular weight of the segmental unit. For vinyl-type addition polymers, the average molecular weight of one segmental unit is taken to be one half that of the monomeric unit. The enthalpy difference between glass 1 and glass 2 is related to the heat of formation of the additional vacant sites, and this difference should be much greater than the difference in the Gibbs free energy.

At or near T_g, it is approximately given by the equation

$$H_1 - H_2 \approx T_g(S_1 - S_2) = -T_g \phi R(f_1 \ln f_1 - f_2 \ln f_2)/M \qquad (6.54)$$

6.3.2. VISCOELASTIC BEHAVIOR OF GLASSY POLYMERS

The viscoelastic behavior of a glassy polymer is more complex than that of a rubbery polymer. In the glassy state, the relaxation time depends not only on the temperature, but also on the strain magnitude and the thermal history, such as the extent of annealing, as discussed in Section 6.3.1. The yield stress also depends on the rate of strain and the thermal history. While the relaxation modulus can be approximated with a power function of time t^{-m}, the exponent m depends on the temperature. Such complex *nonlinear* behavior found in polymeric glasses, however, is surprisingly uniform from polymer to polymer at comparable temperatures with respect to their respective glass transition temperatures. We show that the concept of free volume, which has been found useful in characterizing viscoelastic behavior of polymers above T_g, can be also useful in explaining these phenomenological aspects below T_g.

The analysis of nonlinear viscoelasticity of polymers in the glassy state presented here is based on a modified extension of well-known theories on relaxation phenomena in the rubbery state. In the rubbery state, the mechanical relaxation time depends on temperature but not significantly on the strain magnitude, so the viscoelastic functions are linear, that is the spring and the dashpot in the Maxwell model remain constant regardless of the strain magnitude. The distribution of relaxation times has been described in molecular terms with models involving the rubberlike entropic elasticity combined with the segmental flow viscosity. According to these theories, the broad distribution of relaxation times arises from the large variation in possible modes of dynamic motions with a long chain molecule. The model described by Rouse (1953), for example, predicts a slope of $-\frac{1}{2}$ for the plot of the logarithmic relaxation modulus against the logarithmic time.

Phenomenological aspects of the temperature dependence of relaxation processes in the rubbery state have been discussed in Section 6.2. The shift of relaxation time with temperature in the rubbery state has been described by considering the thermal expansion of the fractional free volume. When the temperature of the polymer in the rubbery state is lowered, the relaxation time becomes greater, and finally the polymer undergoes the glass transition, which can be noted, for example, by the dynamic loss modulus maximum. The time-dependent nature of the glass transition has been emphasized. The measured value for the glass transition temperature depends on the frequency if determined dynamically, or on the rate of the temperature change if determined volumetrically or thermoanalytically.

Kovacs' experiment (p. 363) constituted a departure from this free volume theory in a minor way, since it dealt, in Eq. (6.45), with the free volume in a nonequilibrium state, assuming the effect on the relaxation time to be the same as that in the equilibrium state. The solution of Eq. (6.45) is the exponential integral

$$\int_f^{f_0} \left(\exp\left[\frac{1}{f} - \frac{1}{f_0} \right] \middle/ f \right) df = \frac{t}{\tau_0}$$

where the subscript 0 refers to the initial condition. It is a characteristic of this equation that $\log \tau$ versus $\log t$ is a straight line with a slope of unity, as shown in Fig. 6.20. Thus it specifies the very simple way in which the relaxation time τ shifts according to the time spent for "annealing," namely, the relaxation time increases exactly in proportion to the elapsed time whatever the annealing temperature. The test for this predicted behavior was made by a series of creep measurements on poly(vinyl chloride) samples with different thermal histories, as was shown in Fig. 6.17.

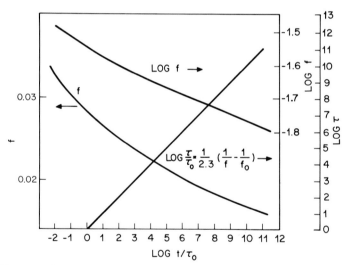

FIG. 6.20. Fractional free volume f and the characteristic relaxation time τ versus log time of annealing. The log f curve has the same slope, ~ -0.02, as the log relaxation modulus curves for many polymers. The relaxation time increases proportionately with annealling time (after Matsuoka et al., 1978).

At temperatures near T_g a large amount of shift in relaxation time can be observed during a relatively short period of thermal treatment. At lower temperatures the effects of annealing will become manifest much more slowly, yet the relaxation time increases in proportion to the lapsed time in the same manner as at the higher temperature, as predicted. The results of these studies on several polymers with relaxation, creep, and dielectric experiments have been reported in detail (Matsuoka, 1978; Johnson, 1978; Struik, 1977).

As the temperature is lowered further from the range where annealing experiments are often conducted, several gradual but significant changes in mechanical behavior will occur. The logarithmic plot of the relaxation curves becomes flatter until the slope reaches a typical value of −0.02, as shown in Fig. 6.21. The polymer dilates under tensile deformation as Poisson's ratio decreases from 0.45 to 0.375. At the same time, viscoelastic behavior of the polymeric glass becomes more and more strain dependent as the tendency to dilate under tensile strain increases, as can be noted from Fig. 6.22.

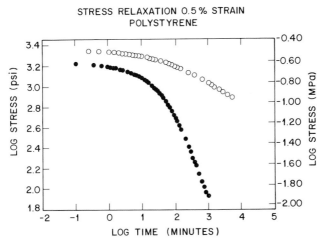

FIG. 6.21.   Stress relaxation in polystyrene at two different temperatures: ○, 56°C, 15 hr anneal at 57°C; ●, 90°C, 15 hr anneal at 90°C (after Matsuoka *et al.*, 1978).

A model has been proposed (Matsuoka *et al.*, 1978) in which the strain-induced dilation would result in a greater free volume fraction:

$$f = f_0 + \varepsilon(1 - 2\mu) \tag{6.55}$$

where ε is the tensile strain, and μ is Poisson's ratio. This f is then substituted into the free volume formula of Eq. (6.20). The predicted magnitude of the shift with strain agrees well with experimental data. It has also been shown that the excess enthalpy, which decreases with increasing annealing time, would increase after the polymer has been mechanically deformed beyond yield (Matsuoka *et al.*, 1973).

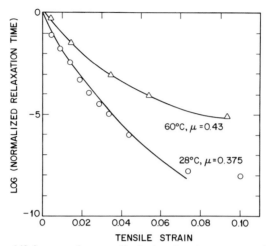

FIG. 6.22.   The shift factor or the normalized relaxation time versus strain in ABS. μ is the Poisson ratio. The lines are then theoretically calculated from the WLF equation using the strain-induced dilatation to account for the increase in fractional free volume (after Matsuoka *et al.*, 1973).

Stress relaxation curves for injection molded polycarbonate are shown in Fig. 6.23. A pronounced effect of strain on the relaxation modulus is evident from the curves, which shift with the strain. How far in the shorter time range the modulus will continue to increase is a question that has a direct bearing on the impact properties. The sound velocity in polycarbonate continues to increase with the frequency in the megahertz range. The tensile modulus of 589,000 psi (log E is 5.77!) is calculated from the data of Patterson (1976) at the equivalent frequency of 5.4 GHz. It seems, therefore, that the relaxation modulus is likely to continue to increase for a considerable number of decades toward the shorter time.

According to this free volume model, the logarithmic modulus curves are shifted with strain in two ways: toward the left along the logarithmic time axis, due to the shortening of the relaxation time, as described above, and downward due to the decrease in the elastic constant, because of the lowered Helmholtz elastic energy term $(dA/d\varepsilon)_T$. Both of these shifts have previously been evaluated from the relaxation data on polystyrene and acrylonitrile–butadiene–styrene (ABS) plastic composite system. The validity of these shift factors may be tested by comparing the relaxation data with the stress–strain curve for polycarbonate. For a linear viscoelastic body, the superposition of stress–strain data is always valid, and the convolution integral has been derived in Section 6.2.5

$$\sigma(t) = \int_0^t E(t - X)(d\varepsilon/dX)\,dX$$

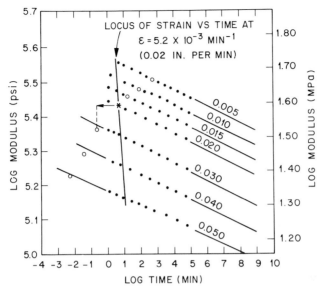

FIG. 6.23. Relaxation modulus versus time for polycarbonate. The points are the experimental data which cover up to 1 year. The numbers on the curves denote the strain (after Matsuoka *et al.*, 1978).

where $\sigma(t)$ is the stress at t, and X is the dummy variable with the dimension of time. For the constant rate of strain $\dot{\varepsilon}$

$$\sigma(\varepsilon) = \dot{\varepsilon} \int_0^{\varepsilon/\dot{\varepsilon}} E(t - x) \, dX$$

In the case of nonlinear viscoelasticity, such as for the glassy polymers, the same procedure is not always valid. However, the convolution integral can still be used to evaluate the stress at a constant rate of strain if the strain and time dependences of the relaxation modulus are explicitly known. The shortening of the relaxation time with the increase in the strain magnitude at a constant rate of strain would be equivalent to decreasing the rate of strain in the linearly viscoelastic body, and we obtain for the nonlinear case,

$$\sigma(\varepsilon) = \int_0^{\varepsilon/\dot{\varepsilon}} E((\varepsilon/\dot{\varepsilon}) - X, \varepsilon)\dot{\varepsilon} \, dX \qquad (6.56)$$

For the relaxation process for a polymeric glass such as polycarbonate, as shown in Fig. 6.23, the plot of logarithmic relaxation modulus vs. logarithmic time is very nearly a straight line, i.e.,

$$E(t, \varepsilon) = E(t_0, \varepsilon)(t/t_0)^{-m} \qquad (6.57)$$

where $-m$ is the slope of such a curve, and

$$E(t_0/a_\varepsilon, \varepsilon) = E(t_0/a_{\varepsilon_0}, \varepsilon_0)\psi^{\varepsilon_0 - \varepsilon} \qquad (6.58)$$

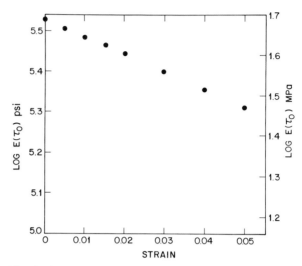

FIG. 6.24. The elastic constant dependence on strain for polycarbonate (after Matsuoka *et al.*, 1978).

where ψ is the contribution of the strain to the elastic constant, and a_ε is the contribution of the strain to the relaxation time, i.e.,

$$a_\varepsilon = a_{\varepsilon_0} \exp\left[\frac{1}{f(\varepsilon_0) + (\varepsilon - \varepsilon_0)(1 - 2\mu)} - \frac{1}{f(\varepsilon_0)} \right] \qquad (6.59)$$

The empirically observed dependence of the elastic constant multiplier $\psi^{-\varepsilon}$ on the strain is plotted in Fig. 6.24. The elastic force constant is proportional to the derivative of Helmholtz energy with respect to strain, as discussed in Section 6.2.2

$$(\partial A/\partial \varepsilon)_T = (\partial U/\partial \varepsilon)_T - T(\partial S/\partial \varepsilon)_T \qquad (6.5)$$

and the decrease in this elastic constant with strain is likely to be associated with the greater excess entropy. Since the excess enthalpy should increase with the excess entropy as explained in Section 6.3.1, the elastic constant $\psi^{-\varepsilon}$ should decrease in proportion to the increase in the excess enthalpy, and the excess enthalpy should increase with strain. This exothermic response to the strain is demonstrated by measuring an actual temperature as shown in Fig. 6.25. The result indicates that the enthalpy in fact does increase with tensile deformation, and the amount of the increase of enthalpy per unit strain has been estimated from the rate of the temperature drop. The total amount of the increase up to the yield was found to be in fair agreement with the increase in excess enthalpy of an annealed sample after having been deformed beyond the yield point. A similar temperature drop was also

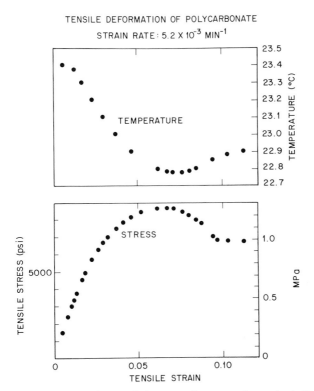

FIG. 6.25. The temperature drop and the stress rise versus the strain, indication of the enthalpy increase with the strain. This stress–strain curve was compared with the values calculated from the relaxation data as shown in Fig. 6.23 (after Matsuoka and Bair, 1977).

observed while a cylindrical sample was sheared by twisting with one end free to move axially. This latter result strongly suggests that a similar shift in the relaxation time could occur with shear deformation where dilation is in fact occurring.

Substitution of Eqs. (6.57)–(6.59) into Eq. (6.56) gives the equation of state for a polymeric glass:

$$\sigma(\varepsilon) \cong E(t_0/a_{\varepsilon_0}, \varepsilon_0)\psi^{\varepsilon_0 - \varepsilon}(a_\varepsilon/a_{\varepsilon_0})^m \varepsilon \tag{6.60}$$

for $m \ll 1$, and $t_0 = \varepsilon_0/\dot{\varepsilon}$. The stress calculated from the relaxation data using Eq. (6.60) is compared with the actual stress–strain data. Starting with a reference strain, e.g., $\varepsilon_0 = 0.02$, the shift factors $a_\varepsilon/a_{\varepsilon_0}$ and $\psi^{\varepsilon - \varepsilon_0}$ are calculated, and the calculated relaxation modulus at each strain, $E(\varepsilon/\dot{\varepsilon}, \varepsilon)$ must agree with the value of the stress divided by the strain from the stress–strain

data. It can be shown that the stress–strain curve in Fig. 6.25 can be reproduced exactly from relaxation data similar to those shown in Fig. 6.24 but
with the thermal history equivalent to 70°C for 10 h.

According to this free volume model, the relaxation spectrum can be
shifted by more than five decades while the strain is increased from 0 to
5%. The irreversible plastic deformation should occur easily when the
relaxation time is short enough to be comparable to the reciprocal of the
strain rate. Thus, when the product of the rate of strain and the average
relaxation time reaches the order of unity, i.e.,

$$\dot{\varepsilon}\tau = \dot{\varepsilon}\tau_0 \exp[1/f_0 + \varepsilon(1 - 2\mu) - 1/f_0] \approx 1 \tag{6.61}$$

the yield phenomenon should occur. Thus an increase in *either* the rate of
strain by 1 decade *or* the annealing period by 1 decade will result in the
same increase in the yield stress, as shown in Fig. 6.26. Moreover, the *slope*

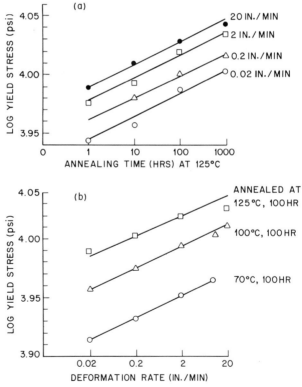

FIG. 6.26.   The yield stress (a) versus annealing time and (b) versus strain rate for polycarbonate showing the same slope as the relaxation modulus and the free volume versus time.

of the plot of the logarithmic yield stress vs. the logarithmic strain rate (or vs. logarithmic annealing time) is approximately equal to m in Eq. (6.58), that is, the magnitude of the slope of the logarithmic relaxation modulus versus logarithmic time curve in Fig. 6.23. The magnitude of the slope m for the relaxation curve increases with temperature, as shown in Fig. 6.27. Thus viscoelastic data for a polymeric glass, obtained at different temperatures, cannot be superimposed by shifting along the time axis as can be done in the rubbery state.

The analysis presented above pertains only to the stress–strain up to the yield point. Beyond the yield point the plastic deformation continues with little further change in volume. The amount of plastic deformation that can take place before rupture depends on the molecular weight and, as Gent (1970) has pointed out, failure after this stage might take place by cavitation under multiaxial tension.

Beyond the point of deformation, where the relaxation time has become sufficiently short and the apparent modulus has become sufficiently small *with respect to the rate of deformation*, volumetric relaxation by means of lateral contraction would occur. As soon as the free volume returns to the original level for the glass and the drawn portion is able to revert back to the glassy state, the adjacent region will begin to undergo the same deformation cycle, and thus the necking zone propagates into the unnecked zone.

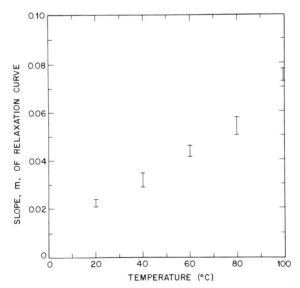

FIG. 6.27. The slope of log relaxation modulus versus log time varies with temperature, approaching the asymptotic value near 0.02 at low temperature.

The propagation of the necked region can continue without rupture only as the molecular weight is sufficiently high. In the quasistatic deformation process, the minimum molecular weight required to prevent such a rupture has been shown to coincide with the critical molecular weight for formation of an entanglement network, as shown by the solid lines in Fig. 6.28. Gent (1970) considers the same rupture process in terms of microscopic cavitation, such as has been observed in the catastrophic failure of rubbery materials under triaxial tension. The critical molecular weight related to this process is again shown to coincide with the critical molecular weight for entanglement.

The contribution of chain entanglement to viscosity diminishes with the rate of deformation. The critical molecular weight for ductile behavior must increase if the strain rate is increased. Under very high rates of strain, such as under impact test conditions, the critical molecular weight required for a ductile, high energy break is substantially higher, as shown by a broken line in Fig. 6.28. The notch radius, which increases the stress concentration and the microscopic rate of strain ahead of the propagating crack, shows the further increase in the critical molecular weight for ductile deformation. The process of fracture will be discussed in more detail in Section 6.3.5.

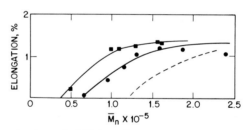

FIG. 6.28.   Elongation beyond yield for polystyrenes: (■) sample with narrow molecular weight distribution; (●) sample with broader distribution; (———) tests at a higher rate of strain (Gent, 1970).

6.3.3. CRYSTALLINE POLYMERS

When the strain is kept below a few percent, the macroscopic mechanical behavior of a crystalline polymer is very much like that of the glassy polymers discussed in Section 6.3.2. As shown in Fig. 6.29, the relaxation spectrum is flat, and depends on the strain magnitude, as is the case with glassy polymers. It dilates under tensile stress at about the same rate as glassy polymers, so that the typical value of the Poisson ratio for crystalline polymers is also 0.37. As will be mentioned in Section 6.3.4, there is a transition known as the α_c transition related to the relaxation process of segments in the crystalline regions, and it is the last dynamic transition encountered as the temperature is raised. Thus, phenomenologically the α_c transition of a crystalline polymer

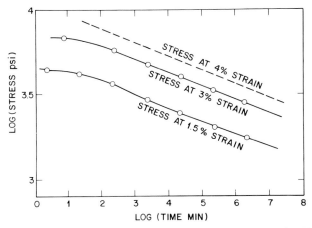

Fɪɢ. 6.29. Stress relaxation of polyoxymethylene (Matsuoka *et al.*, 1978).

is similar to the α_a or glass transition of an amorphous polymer. When the shift of the relaxation spectrum with strain is substituted in Eq. (6.56) for the stress–strain curve, yield stresses are again obtained that are in agreement with experimental data. Thus the behavior leading to the yield phenomena in the crystalline structure can be considered as a strain-induced α_c transition. Crystalline polymers can be annealed at elevated temperatures to become more rigid; the enthalpy and the specific volume both decrease. The annealed crystalline polymers exhibit longer relaxation times and higher yield stresses. The enthalpy returns to the quenched level when it is deformed beyond the yield point.

The analogies thus drawn between the macroscopic behavior of crystalline and glassy polymers are surprising, particularly when one considers the existence of the amazing variety of morphological microstructures in crystalline polymers, as discussed in Chapter 5. One is accustomed to treat an amorphous glassy polymer as a homogeneous body and a crystalline polymer as a complex composite of microstructures whose individual mechanical properties vary widely. Evidently the macroscopic behavior observed is the result of averaging over all of these structural features and the average behavior turns out to be quite similar to that of a homogeneous body, when deformation is small.

Differences between crystalline polymers and glassy polymers due to the microstructure become evident in mechanical behavior during large deformation. For example, different parts of a spherulite begin to show different deformation patterns under uniaxial stress, as shown in Fig. 6.30 (Keith and Padden, 1959), and under biaxial stress (Hopkins and Baker, 1959; Haas

FIG. 6.30.   Micrograph of polypropylene during large uniaxial deformation (after Keith and Padden, 1959).

1965), as shown in Fig. 6.31. Since the most important engineering properties are those related to failure or fracture behavior, the microstructure of crystalline polymers is a very important physical aspect of macromolecules.

In the range of deformation where yield or crazing begins to occur, a crystalline polymer begins to orient in a complex manner. Keller (1957) observed that during deformation of polyethylene, crystals are oriented along the b-axis first, then along the a axis and finally along the c axis. Stein and co-workers (1964) have demonstrated a series of steps of orientation, starting from microscopically homogeneous (affine) deformation, in which crystalline and amorphous regions may orient at different strain levels. Kwei and co-workers (1967) have demonstrated the anistropic mechanical properties of crystalline structures by measuring the dynamic mechanical properties of the two dimensionally grown "transcrystalline" morphology. Crystallizable polymers with high molecular weight draw to form a very strong fiber structure. Their ability to undergo a large elongation before breaking is enhanced by the absence of long chain branches, the presence of crystal defects and amorphous regions to add plasticity, and freedom from impurities or gels.

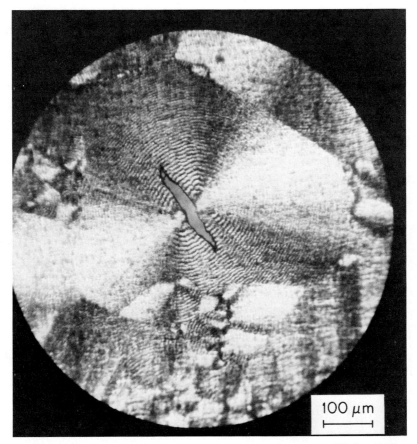

FIG. 6.31.   Polyethylene spherulite under biaxial stress (after Haas and Keller, 1965).

6.3.4.   MULTIPLE TRANSITION PHENOMENA

It is a well-known fact that polymers undergo other transitions besides the glass transition. Perhaps the most important of the transitions that take place in the solid state, insofar as the molecular dynamics are concerned, is the so-called β transition in amorphous glass and the γ transition in a semicrystalline solid. These transitions are best studied through measurements of mechanical or dielectric loss maxima as a function of either temperature or frequency, and by the T_1 minima or line narrowing in nuclear magnetic resonance studies.

The β transition in amorphous glasses has several characteristics that are distinct from the glass transition. First, the temperature coefficient of the β transition is typically about 10 kcal, whereas the glass transition follows

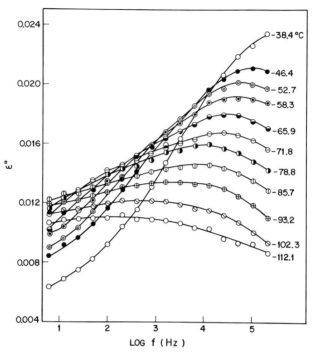

F_IG. 6.32.   Dielectric loss dispersion in polycarbonate in the β transition region whose principal feature of the increasing intensity with the temperature is clearly observable (after Matsuoka and Ishida, 1966).

the free volume formula, resulting in an *apparent* activation energy in excess of 100 kcal mole^{-1}. Secondly, the intensity of the β loss peak increases with temperature, while that of the glass transition decreases, as shown by the dielectric data in Figs. 6.32 and 6.33. The areas under the two dielectric loss peaks are complementary to each other so that the sum of the two transitions will constitute the quantity $\varepsilon_s - \varepsilon_\infty$, where ε_s is the static dielectric constant of the rubbery state, and ε_∞ in this case is the optical dielectric constant plus all the contributions from infrared absorptions. The difference between the static dielectric constant ε_s of the rubbery state and ε_∞ of the glassy state follows the Onsager relationship (Frohlich, 1958):

$$\varepsilon_s - \varepsilon_\infty = \frac{3\varepsilon_s}{2\varepsilon_s + \varepsilon_\infty}\left(\frac{\varepsilon_\infty + 2}{3}\right)^2 \frac{4\pi N}{3kT}\mu_v^2 \qquad (6.62)$$

where μ_v is the dipole moment and N is the concentration of dipoles. The total of the two loss peaks is in fact found experimentally to decrease with T as an inverse function of T, as shown in Fig. 6.34.

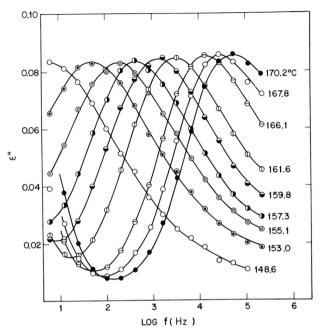

FIG. 6.33. Dielectric dispersion in polycarbonate near the main (α) glass transition region. The loss intensity is much greater than that of the β dispersion so that the intensity is nearly independent of the temperature (after Matsuoka and Ishida, 1966).

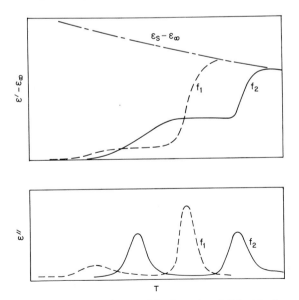

FIG. 6.34. The temperature dependmnce of the intensity of dielectric dispersion involving multiple transitions is illustrated schematically, where the frequency f_2 is greater than f_1 (after Matsuoka et al., 1977).

385

There are a number of ways for describing a solidlike environment in which a polar polymer segment may undergo a transition. The condition for an absorption or loss maximum is that the motional frequency of the segment or the dipole reorientation be comparable to the frequency of the measurement. One of the convenient ways of interpreting dielectric data is provided by the so-called site model. A site is a specific dipole orientation relative to its environment. There may exist a number of such preferred directions for a dipole with respect to its immediate neighbors and each site orientation may be characterized by its energy relative to the site of lowest energy. In general there is also an energy barrier between the discrete orientations. The origin of these energy barriers may be the electrical field created by the surrounding dipoles, or intra- and intermolecular potential energies, against which the dipole-supporting framework must move. If the energy differences between sites is ignored the calculation of the temperature dependence of the loss intensity for this model follows the classical work of Fröhlich (1958) on the order–disorder transition in polar crystals. A treatment more specifically related to polymer structure has been developed by Hoffman and co-workers (1966). Assuming that a two-site model contains many of the qualitative but essential features of the more complicated structures that would lead to a multiple site model, the probability W of finding a dipole in the higher energy well of the two sites may be obtained:

$$W/(1 - W) = \exp(-\Delta E/kT) \qquad (6.63a)$$

where ΔE is the energy difference between the two sites. At low temperatures when $\Delta E/kT \gg 1$, it can be shown that

$$\varepsilon_s - 1 = \frac{3\varepsilon_s}{2\varepsilon_s + 1} \, 4\pi N \frac{4\mu^2}{kT} \, 2\exp\left(-\frac{\Delta E}{kT}\right) \qquad (6.63b)$$

which predicts that the β loss intensity should increase with reciprocal temperature as observed. Since the apparent activation energy for the β process is far less than that of the main glass transition process, the difference in the frequencies for the two loss maxima becomes smaller with increase in temperature. The multiple transition behavior, however, is not limited only to polymers. Johari (1973) has shown that in chlorobenzene–cis-decalin the α and β curves merge and continue along the β line, as shown in Fig. 6.35. Thus the β process also takes place in the equilibrium liquid state above T_g. Since the curve of the log frequency vs. $1/T$ for the α process has such a steep slope, the temperature at which the α and β processes merge is considered to be primarily dictated by the glass transition temperature. Such curves are shown in Fig. 6.36 for polycarbonate and poly(ethylene terephthalate).

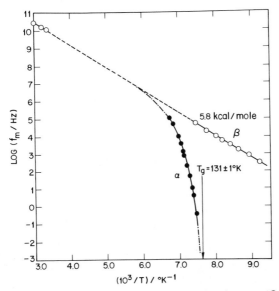

FIG. 6.35. An illustration of a close relationship that exists between (●) the α or main glass transition and (○) the β or low-temperature transition in chlorobenzene–*cis*-decalin (after Johari, 1973).

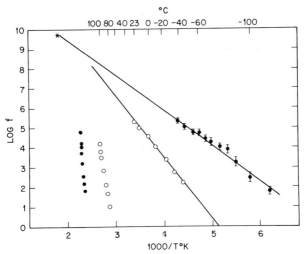

FIG. 6.36. A plot similar to Fig. 6.35 (●) for polycarbonate and (○) for poly(ethylene terephthalate) (after Matsuoka and Ishida, 1966). The asterisk value is deduced from Brillouin light scattering by Patterson (1976).

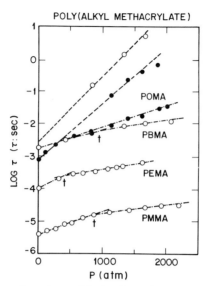

FIG. 6.37. The pressure dependence of the (– – –) α and (– ·–) β processes in polyalkyl meth-acrylate with side chains of various lengths; PMMA = methyl, PEMA = ethyl, PBMA = n-butyl, POMA = n-octyl (after Sasabe and Saito, 1968).

Factors that determine the magnitude of T_g were discussed earlier in this chapter; chain flexibility and the size of the occupied volume in com-parison to the total volume in the rubbery state were singled out as especially important. Clearly, different factors must be sought for the β process. For example, T_β for polycarbonate is always lower than T_β for poly(ethylene terephthalate) but the reverse is true for the T_g values.

The effect of pressure on the β process is much less than on the glass transition process, as shown in Fig. 6.37. The intensity and characteristic frequency of the β peak are affected very little by the thermal history (annealing) of the glass, and so the excess enthalpy evidently has no appre-ciable effect on the β process. On the other hand, polycarbonate densified under a pressure of 2000 atm has shown a marked decrease in the β peak, as shown in Fig. 6.38. This clearly implies that the greater α peak is caused by efficient packing, not achievable at atmospheric pressure, and a decreased number of segments are able to participate in the β relaxation process in the dense material. This special effect of high pressure is probably related to the high compressibility of the occupied volume discussed in Section 6.3.1, and indicates that the β process is affected by packing of the occupied volume and not by the free volume, which affects mechanical behavior.

We have thus far discussed the multiple transitions in amorphous glassy polymers, where the α or glass transition and the β or low-temperature

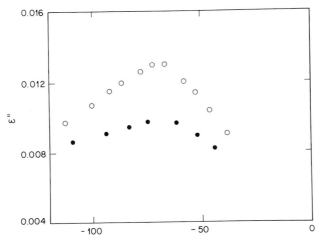

FIG. 6.38. The β peak of polycarbonate as ($\bigcirc$) molded and ($\bullet$) pressure-densified (Matsuoka, unpublished data).

relaxation are closely related to each other. In semicrystalline polymers, there are typically three main transitions. The highest temperature transition, called α according to the convention of naming the transitions from the highest temperature, is a transition associated with the crystalline structure. The next in the temperature scale is the β transition, which is associated with the glass transition of the amorphous regions. The β transition for a semicrystalline polymer corresponds to the α or the glass transition in an amorphous polymer. Takayanagi (1965) has proposed calling the β transition in crystalline polymers the α_a process for this reason, denoting it to mean the highest temperature transition for the amorphous region. According to this convention, the crystalline transition is to be called α_c to mean the highest temperature transition in the crystalline region.

The γ transition in semicrystalline polymers is phenomenologically quite similar to the β transition in the amorphous glass discussed before. Its apparent activation energy is about 10 kcal and the loss intensity increases markedly with temperature, as shown in Fig. 6.39. In this case the three transitions, α, β, and γ complement each other as compared to the two transitions for glassy polymers.

The γ process is also found in highly crystalline polymers such as linear polyethylene in which the β or α_a process is not resolved, as is shown in Fig. 6.40. A β-type relaxation process, involving some restricted motions of the glassy solid, must occur in the crystalline solids as well.

Takayanagi's (1965) convention calls for β_a and β_c for the two types of γ transitions. For a given polymer, the two processes occur at very closely

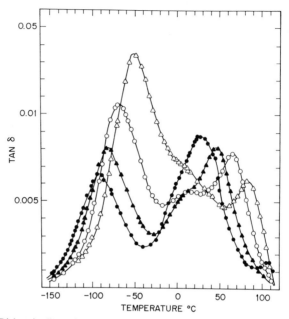

FIG. 6.39.   Dielectric dispersion of chlorinated linear polyethylene (3.19 mole% Cl). The β or α_a peak is best resolved at 10 kHz, but is totally absent for a sample with less than 2 mole % chlorination. ●, 100 Hz; ▲, 1 kHz; ○, 10 kHz; △, 100 kHz (after Matsuoka *et al.*, 1972).

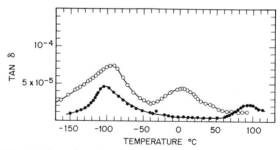

FIG. 6.40.   Dielectric dispersion of (○) branched and (●) linear polyethylenes at 10 kHz. Loss arise from very low concentrations of dipoles, primarily carbonyl groups (after Matsuoka *et al.*, 1972).

spaced intervals in frequency and temperature and are often difficult to separate. The β_c process usually takes place at a slightly higher temperature than the β_a process. In linear polyethylene, crystallized under a pressure of 1500 atmospheres, the γ (or β_c since there is no β_a) loss peak is practically absent as shown in Fig. 6.41. This is reminiscent of the densified poly-

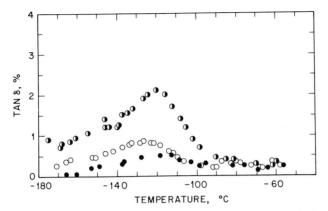

FIG. 6.41. The dependence of dynamic mechanical and loss peak on morphology for linear polyethylene. ◑, melt crystallized; ○, solution crystallized; ●, high pressure crystallized (Matsuoka, unpublished data).

carbonate mentioned before, though one is glassy and the other crystalline. The near perfect morphology of the extended chain crystals of the polyethylene sample may be such that all of the relaxation process may take place at the α_c transition.

The α_c transition in a crystalline polymer is similar in some aspects to the glass transition of amorphous polymers. It is the low frequency, high-temperature transition of a partially relaxed (through the γ process) structure. Dielectric data show that there are no additional loss peaks above this transition. In particular, there are no loss peaks at the melting point where the crystalline structure undergoes the most drastic change. So far as the relaxation processes are concerned, the final transition has taken place at the temperature of the α transition, though this temperature is still substantially below the melting point, particularly at the low frequencies of about 100 Hz.

Polymer crystals above T_α can be drawn without whitening, and indeed they feel more like a rubbery polymer. Yet an x-ray scan shows the strong diffraction pattern of a polymer crystal, and the density is also typical of the crystalline structure. High-pressure oriented, high-modulus polyethylene produced by Southern and Porter (1970) is processed above the α transition.

Hoffman et al. (1966) proposed a model for the composite α process that is supported by a wide variety of dielectric, mechanical, and morphological data on n-paraffins, their dipolar derivatives, and polymers with well-defined morphological parameters. The model consists of two overlapping mechanisms involving motions of chain folds and reorientation (with translation) of chains in the crystalline interior. The activation energy for the correlation frequency of the α_c relaxation is predicted to be 25 kcal for polyethylene.

The results of dielectric studies on chlorinated (to different concentrations of up to 4 mole %) (Matsuoka *et al.*, 1972) or slightly oxidized (Ishida and Yamafuji, 1965) linear polyethylene samples agree with this value.

The mechanical relaxation behavior of a composite structure such as a semicrystalline structure will invariably present challenges to quantitative analyses (Takayanagi *et al.*, 1963; Halpin, 1971; Ward, 1974). For example, the intensity of the mechanical loss modulus for α_c relaxation is influenced not only by the number of segments involved but also the frequency and temperature dependence of the elastic modulus. Although qualitative assignments of the loss peaks to various transitions are possible, discussions involving the absolute intensities of these loss peaks should be regarded as less reliable than those from the dielectric data in general.

In addition to the three main transitions discussed, there have been several more loss peaks found by mechanical tests (Hiltner and Baer, 1972). Origins of some have been attributed to the trace of a solvent and others to orientation (Dahl and Müller, 1961) and other molecular constraints (Ishida *et al.*, 1963). The rotation of the methyl group has been detected by NMR (McCall and Slichter, 1957). In all of the cases involving multiple transitions, each process has its own temperature dependence of the intensity as well as of the characteristic frequency, but the sum of the intensities of all these transitions must add up to the change in the in-phase modulus or the dielectric constant from the low temperature, high frequency, unrelaxed state to the high temperature, low frequency, relaxed state.

6.3.5.   FAILURE AND FRACTURE IN SOLID POLYMERS

An accurate prediction of failure is one of the ultimate goals for application of the basic understanding of macromolecular behavior under stress. Almost all existing theories related to solid polymers are based on the mode of failure in which a crack or craze propagates when either the strain energy or the stress concentration at the advancing tip exceeds the work required to create the open surfaces (Fig. 6.42). The work required to create the crack surface includes not only the surface energy (the energy of the exposed surface minus the energy of the closed surface) but also the energy that is dissipated during the crack opening process. In polymeric materials with sufficiently high molecular weight, the amount of this plastic energy that is dissipated into heat is often several times greater than that of the elastic, recoverable energy. In general the strength of polymers is derived from the toughness, or the ability to exhibit plastic flow behavior. For example, stress–strain curve B in Fig. 6.43 characterizes a sample of greater toughness than one with the higher modulus of curve A, or even the higher yield stress curve A'. Both A and A' illustrate failure without the desirable plastic

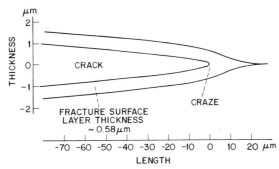

FIG. 6.42. Schematic diagram of advancing front of craze and crack (after Kambour, 1966). In unbroken craze, stress is approximately zero for this configuration.

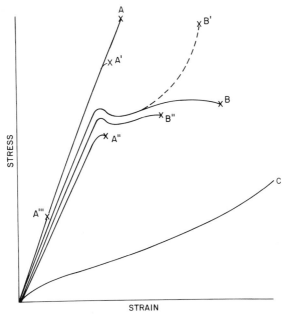

FIG. 6.43. Stress–strain curves of hard brittle solids A, A′, A″, A‴, ductile solids B, B′, B″, and soft rubbery material C. Some crystallizable polymers strain-harden as in B′, a characteristic suitable for forming a strong fiber.

deformation, often called post-yield elongation. By comparison, a rubbery material, curve C, can be stretched to several times its unstressed length, but the modulus is several orders of magnitude lower than that of the polymeric solids and the energy required is again small.

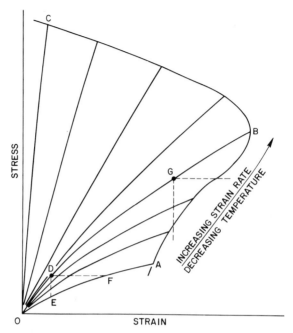

FIG. 6.44. Stress–strain diagram of amorphous polymer exhibiting typically rubbery behavior at slow strain rate and/or high temperature and glassy behavior at high strain rate and/or low temperature (after Smith and Stedry, 1960).

The stress–strain curve beyond the yield point does not describe a true stress–strain relationship, since the elongation is a measure of how much of the sample has undergone the necking process. The ultimate elongation, or the total elongation before the sample breaks, is a statistical average of how much drawing can be sustained by the necking region. It was theorized in Section 6.3.3 that during the necking process the macromolecules undergo a strain-induced transition and are at least momentarily in the rubbery state. The ultimate equilibrium strength of a rubber is expected to be proportional to the number of load holding chains, or the number of effective network chains (Figs. 6.44 and 6.45) (Smith and Stedry, 1960). The strength of uncross-linked macromolecules at a given rate of deformation depends on the entanglement network. The higher the rate of deformation, the greater is the number of molecules able to participate in bearing the load, and the higher the stress the network will sustain. In terms of linear viscoelasticity this effect is described by Eq. (6.31) for the convolution integral in Section 6.2.5 and can be evaluated from relaxation data. Smith (1968) has indeed found that curves of tensile strength vs. the rate of strain could be

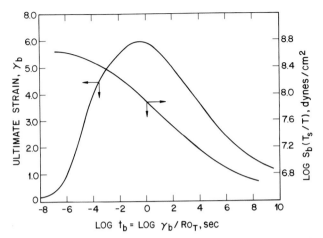

Fig. 6.45.   Dependence of the strength and ultimate strain of amorphous polymer on time and temperature (expressed in terms of reduced time) (after Smith and Stedry, 1960).

shifted according to the free volume formula. Thus the molecular weight dependence of elongation can be qualitatively explained on the basis of the strength of the entangled network as it undergoes plastic deformation.

Stress–strain data, particularly beyond yield, thus gives a measure of the work required to create a new surface in the material. For a crack to grow, this energy must be less than the energy concentrated ahead of the crack.

A theory of crack propagation based on this theory was proposed by Griffith (1924) and extensively applied to polymeric solids by Berry (1961). It can be summarized in an algebraic form; for a crack to propagate, the following conditions must prevail:

$$(\partial U / \partial C) + (\partial \Gamma / \partial C) < 1 \qquad (6.64)$$

where U is the recoverable strain energy, Γ is the energy required to create the unit area of crack surface, and C is the crack surface. Since the strain energy is relieved as the crack propagates, $\partial U / \partial C$ is less than zero, and when its magnitude exceeds that of $\partial \Gamma / \partial C$, the crack will continue to grow. As previously mentioned, the energy to be dissipated by plastic deformation is a large part of the surface energy Γ in tough polymers. The strain energy U is only the recoverable portion of the work done to the body, and depends on relaxation time. Schapery's theory (1975) is one of the few theories on fracture which take into account the time-dependent nature of the strained area in the material. The critical crack size C_{crit}, for failure is given by

$$C_{crit} = 8\Gamma E_0 / \pi \sigma^2 \qquad (6.65)$$

where E_0 is the initial modulus, and σ is the applied stress. The time at which the crack size C reaches C_{crit} is the failure time, and it is independently calculated from relaxation data. Good agreement has been reported between theory and data for urethane rubber.

The effect of oxidation, which is discussed in Chapter 7, is chiefly reflected in the time to failure under different stress levels; oxidation decreases Γ by reducing the molecular weight and hence by elimination of plastic deformation beyond yield.

When the molecular weight is less than the critical magnitude, plastics should become weak and brittle, but this may or may not be true depending upon the stress level. When there is an environment adverse to the polymer, for example, a dilute solution of detergent coated on polyethylene, the relaxation time of the polymer at the crack tip is drastically reduced and the time-to-fail is greatly shortened, often so much so that diffusion of the cracking agent becomes the rate controlling process. The slope of stress versus time is much greater in such a case (Marshall and Williams, 1973); see Fig. 6.46. Annealing shortens the failure time also, while multiaxial stress limits the mode of failure to cavitation; the plastic energy dissipation process is severely curtailed, again reducing the surface energy Γ and hence decreasing the time to failure.

Clearly, the mode of failure, and therefore also the strength of a material, depend on all of the stress components. Sternstein (1972) has demonstrated that in poly(methyl acrylate) crazes preferentially develop in regions where

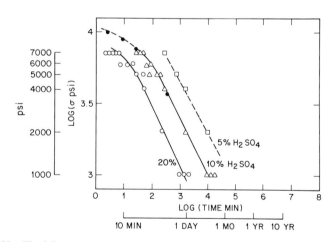

FIG. 6.46. The failure stress versus time for polyoxymethylene in environment of sulfuric acid solutions with various concentration (Marshall and Williams, 1973).

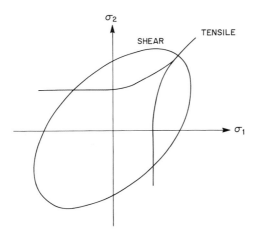

Fig. 6.47.   The failure criterion (after Sternstein, 1972).

the stress is more nearly biaxial. His analysis leads to the maximum shear yield as a failure criterion for polymeric solids, $\sum(\sigma_i - \sigma_j)$, where σ's are the principal stresses, and the tensile yield by another curve, as shown in Fig. 6.47. Depending on the magnitudes of the stress components, the mode of failure is determined as tensile or shear.

The brittle to ductile transition for failure may be viewed from a somewhat different standpoint in the light of the discussions presented in this section. A high rate of strain will raise the yield stress, and fracture occurs instead of ductile plastic deformation. A notch of small radius would have the same effect because the rate of strain in the critical region is high. Annealing, on the other hand, will have an effect on increasing the relaxation time, and will raise the yield stress so that brittle fracture can occur. Increasing the molecular weight will increase the fracture energy without raising the yield stress, thus favoring yielding rather than brittle failure. The effects of the notch radius, molecular weight and annealing on the impact strength of polycarbonate are shown in Fig. 6.48.

The mechanism of rubber reinforcement is closely related to all aspects of mechanical behavior discussed thus far. The incorporation of rubber spheres in a rigid glassy matrix, through a chemically bonded interface, enables a styrene–acrylonitrile copolymer, called ABS, to avoid the propagation of sharp cracks. The presence of the sphere induces the otherwise rigid and brittle matrix of acrylonitrile–styrene copolymer to undergo ductile deformation, increasing Γ in Eq. (6.64) or the total work required to create the crack. When a plate of ABS is scored with a razor blade, however, the plate

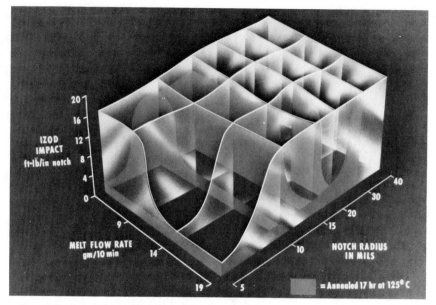

Fig. 6.48. Impact strength versus molecular weight (melt flow index), rate of strain (notch radius), and thermal history (annealing time for polycarbonate) (after Ryan, 1978).

fails easily without going through a typical craze formation or whitening. In this case the radius of the razor cut is smaller than the size of the rubber spheres, and a crack can propagate through the sample without blunting.

6.4. Transport

6.4.1. Introduction

In the preceding sections, the phenomenological aspects of mechanical, dielectric, and nuclear magnetic relaxation processes in polymer solids are brought together under a unified view of molecular and chain dynamics. The same molecular considerations are applicable to the transport phenomena of small molecules in polymer films. The subject has received increasing attention in recent years because it has important bearings on protective coatings as well as biomedical applications, e.g., controlled-release drugs. On one hand, the segmental mobility of the macromolecular chain governs the transport of diffusing species; but on the other hand, astute choice of diffusant as "molecular probe" produces information pertaining to the architecture and mobility of chain segments.

The basic equation for the description of diffusion in isotropic media is Fick's first law, which relates flux J, the rate of transfer of mass Q per unit cross-sectional area A, to the concentration gradient $\partial c/\partial x$ and the diffusion constant D:

$$J = Q/AT = -D(\partial c/\partial x) \tag{6.66}$$

By considering the mass balance of a volume element in which diffusion takes place, the rate of change of concentration $\partial c/\partial t$ is obtained as

$$\partial c/\partial t = D(\partial^2 c/\partial x^2) \tag{6.67}$$

This is known as Fick's second law. Note that the equations governing mass transport and heat transfer are completely analogous and solutions of Eq. (6.67) for different boundary conditions are readily available.

6.4.2. EXPERIMENTAL METHODS

Two experimental methods are usually employed in the determination of the diffusion constant, the permeation method and the sorption (or desorption) method. In the permeation method, let us consider diffusion through a plane sheet of thickness l whose surfaces $x = 0$, $x = l$ are maintained at constant concentrations c_1 and c_2, respectively. At steady state, the concentration at every point in the sheet remains invariant with time but changes linearly with l from c_1 to c_2. The flux J is simply

$$J = D(c_1 - c_2)/l \tag{6.68}$$

and D can be computed readily from a knowledge of J, l, and surface concentrations c_1 and c_2.

Frequently, however, the surface concentrations are not known, for example, in the diffusion of gases and salt solutions. The flux is then more conveniently expressed in terms of external concentrations of the diffusant. In gas diffusion, the external pressures p_1 and p_2 on the two sides of the membrane are often used,

$$J = P(p_1 - p_2)/l \tag{6.69}$$

and the constant P is called permeability constant. If Henry's law is applicable,

$$c = SP \tag{6.70}$$

where S is the solubility constant, then it follows that P is the product of D and S

$$P = DS \tag{6.71}$$

The unit of D is square centimeter per second. The unit of P in gas diffusion is cubic centimeters (gases STP) per centimeter (thickness) per square centimeter (cross-section) per second per centimeter Hg (pressure difference) although other units are also used.

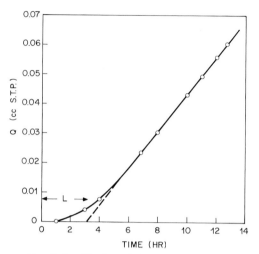

FIG. 6.49.   The flux of diffusant versus time.

An alternate approach in the determination of D by the permeation method is to study the time for a steady state to be reached. If the sheet is initially free of diffusant and the diffusant is introduced to one side of the sheet but continuously removed from the other side ($c_2 = 0$), the amount Q passing through the sheet at time t is given by

$$\frac{Q_t}{lc_1} = \frac{Dt}{l^2} - \frac{1}{6} - \frac{2}{\pi^2} \sum^{\infty} \frac{(-1)^n}{n^2} \exp\left(-\frac{Dn^2\pi^2 t}{l2}\right) \tag{6.72}$$

The steady state is reached at large t and the exponential terms become negligible. The portion of the Q_t versus t plot (Fig. 6.49) representing the steady state is linear, and obeys Eq. (6.73)

$$Q_t = \frac{Dc_1}{l}\left(t - \frac{l^2}{6D}\right) \qquad \text{at large } t \tag{6.73}$$

The intercept of the extrapolated straight line on the time axis is the time lag L, given by

$$L = l^2/6D \tag{6.74}$$

The diffusion constant is computed from L and P is computed from the steady value of Q. The solubility constant S is then calculated by the use of Eq. (6.71). A typical gas permeability apparatus is shown in Fig. 6.50.

The sorption method is most suitable for experimental arrangements in which the total uptake of the diffusant can be monitored by weighing, radiotracer, and other appropriate techniques. In the study of vapor sorption,

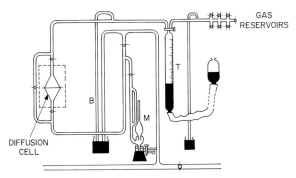

FIG. 6.50.   Gas permeability apparatus.

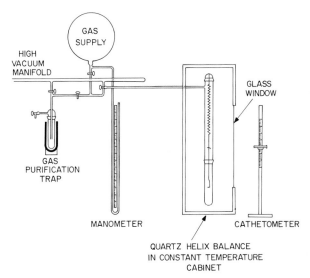

FIG. 6.51.   Vapor sorption apparatus.

weighing has proved to be a simple and accurate method with the use of quartz spring or an electrobalance (Fig. 6.51). The weight gain M_t for a sheet of thickness l surrounded by the vapor is

$$\frac{M_t}{M_\infty} = 4\left(\frac{Dt}{l^2}\right)^{1/2}\left[\frac{1}{\pi^2} + 2\sum_{n=0}^{\infty} (-1)^n i \operatorname{erf} c \frac{nl}{2(Dt)^{1/2}}\right] \qquad (6.75)$$

where M_∞ is the equilibrium sorption. The initial slope of the plot of M_t/M_∞ versus $(t/l^2)^{1/2}$ is linear, according to Eq. (6.75). Typical experimental data are shown in Fig. 6.52. The diffusion constant D is calculated from the slope by

$$D = (\pi/16)(\text{slope})^2 \qquad (6.76)$$

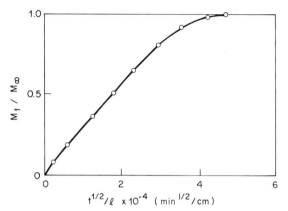

FIG. 6.52. Time-dependent sorption data.

6.4.3. DIFFUSION OF GASES

The diffusion constants of simple gases such as He, H_2, Ne, O_2, N_2, CO, and CO_2 in many common polymers above their glass transition temperatures fall in the range of $10^{-6}–10^{18}$ cm^2 sec^{-1}. The effect of temperature on D obeys the Arrhenius relationship,

$$D = D_0 \exp(-E_D/RT) \qquad (6.77)$$

where E_D is the activation energy and D_0 the preexponential factor. Deviations from Eq. (6.77), however, have been noted over wide temperature ranges.

The measured value of E_D is usually between 8 to 15 kcal/mole for simple gases. The magnitude is obviously large when compared to the activation energy of flow in gaseous medium. The energy is acquired by the diffusant and the surrounding medium of polymer segments for the creation of a "hole" in the polymer structure to allow the diffusant molecule to move from one equilibrium position to another. The cooperative motion of more than one polymer segment is required to create a "hole" of the proper size. Thus, the transport process in polymers is governed by the same molecular dynamics that underlies other relaxation processes.

When all the gas diffusion data are assembled, it is found that the activation energy E_D can be factored into two parameters, one characteristic of the diffusant and the other of the polymer:

$$E_D = g(\text{gas})f(\text{polymer})$$

The parameter g increases with increasing size of the gas molecule; the correlation lies between d and d^2, where d is the diameter of the gas molecule. Secondly, ln D_0 increases hand in hand with E_0/RT. Consequently, the

TABLE 6.4
FORCE AND GIBBS FREE ENERGY VALUES AT 30° C[a]

Polymer	F value[b]
Poly(vinylidene chloride)	0.094
Poly(ethylene terephthalate)	0.050
Poly(caprolactam)	0.10
Poly(vinyl butyral)	2.5
Cellulose acetate	2.8
Polyethylene (low density)	19
Polybutadiene	64.5
Natural rubber	80.8

Gas	G value
Nitrogen	1.0
Oxygen	3.8
Hydrogen sulfide	21.9
Carbon dioxide	24.2

[a] Rogers (1962).
[b] In cubic centimeters at STP per centimeter thickness per square centimeter cross-sectional area per second at a pressure differential of 1 cm Hg.

diffusion constant D can also be factorized. For any given pair of gases, the ratio of D is approximately constant, regardless of polymer. Similarly, for any given pair of polymers, the ratio of D values is also approximately constant, regardless of gas.

The solubility of gases in polymers correlates well with molecular forces expressed in terms of a Lennard-Jones force constant, boiling point or critical temperature. A factorization scheme can also be devised for solubility.

It is natural, from the factorization of both D and S, that the permeability constant P can also be factorized:

$$P = G(\text{gas})F(\text{polymer})$$

Typical values of F and G are shown in Table 6.4.

6.4.4. DIFFUSION OF VAPORS AND LIQUIDS

Organic molecules comparable to the size of the polymer segment require for their diffusion the cooperative motion of many segments. They also tend to swell the polymer and alter the free volume of the system. These special features are best illustrated in the diffusion of organic vapors or liquids above the T_g of poly(vinyl acetate), as shown in Fig. 6.53.

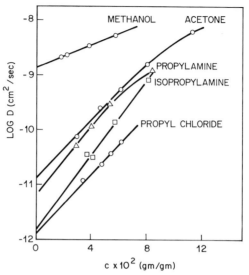

FIG. 6.53. Diffusion constant of various diffusants against concentration in poly(vinyl acetate) (Kokes and Long, 1953).

The diffusion constants are, as a rule, functions of both the temperature and the concentration of the organic diffusant c in the system. In many systems, the concentration dependence is of the exponential form. The interpretation of the temperature and concentration dependence has met with success with the use of free volume theory. The basic assumptions of the free volume approach of Fujita *et al.* (1960) are as follows. The mobility m_d of the diffusant relative to the polymer is

$$m_d = A_d \exp(-B_d/f) \tag{6.78}$$

where A_d is a constant, f is the average fractional free volume, and B_d has the same meaning as in Eq. (6.9). The mobility m_d is connected to the thermodynamic diffusion constant D by Eq. (6.79):

$$D = RTm_d \tag{6.79}$$

When the volume fraction of the diffusant v_d in the system is small, the free volume of the mixture may be written in the form

$$f(T, v_d) = f(T, 0) + \beta(T)v_d \tag{6.80}$$

where β, a function of T, represents the effectiveness of the diffusant to increase the free volume of the polymer. The value of D at zero diffusant concentration, $D(0)$, and the ratio $D/D(0)$ are given by

$$\ln(D(0)/RT) = \ln(A_d - B_d)/f(T,0) \tag{6.81}$$

$$\frac{1}{\ln[D/D(0)]} = \frac{f(T,0)}{B_d} + \frac{[f(T,0)]^2}{[B_d \beta(T) v_d]} \tag{6.82}$$

In both equations the term $f(T,0)$ appears which can be independently evaluated from viscosity measurements. Equation (6.81) predicts a linear relation between $\ln(D(0)/RT)$ and $1/f(T,0)$. Equation (6.82) predicts a linear relation between $1/\ln(D/D(0))$ and $1/v_d$ with a slope of $[f(T,0)]^2[B_d l(T)]$. Both predictions have been demonstrated in several cases. The concentration dependence of D in polymer–organic diffusant systems can be attributed, therefore, to the change in the average free volume of the polymer due to the presence of the diffusant. Like other relaxation processes, diffusion is controlled by the mechanism of segment motion.

Diffusion of organic vapors and liquids at temperatures below the glass transition of the polymer often deviates from Fick's law. A sharp boundary separates the outer swollen shell from the inner glassy core. If the swelling power of the solvent is modest or the polymer is highly cross-linked, the boundary advances at a constant velocity and the weight gain increases linearly with time. This is a limiting case called case II diffusion. Cracks and crazes in polymers often accompany the diffusion of solvent in glassy polymers.

6.4.5. DIFFUSION OF WATER AND SALTS

Finally, brief mention is made of the diffusion of water. In some hydrophilic polymers such as nylon 66, the peremeability constant tends to increase sharply at high vapor pressures (Fig. 6.54). The large increase arises only in part from the expected increase of D with c. The second contributing factor is the rapid increase in equilibrium sorption at high vapor pressures.

For several less hydrophilic polymers, a rather unusual behavior has been observed. The diffusion constant decreases markedly with concentration in polydimethylsiloxane and polyurethane; a phenomenon explained by postulating clustering of water molecules in the polymer that reduces the mobility of the diffusant.

The permeability of an inorganic salt is very low because the solubility of the electrolyte in the organic polymer is usually insignificant. When a pressure ΔP, larger than the osmotic pressure π, is applied to a salt solution against a polymer membrane, the effluent contains only a minute amount of the electrolyte. This is the basis of reverse osmosis.

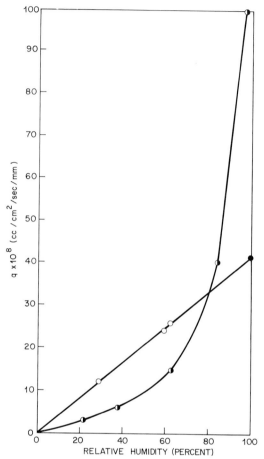

FIG. 6.54. Water permeability against humidity. ●, liquid–water values; ◑, Nylon 66; ○, Mylar A.

REFERENCES

ASTM (1978). Standard Method of Measuring Flow Rates of Thermoplastics by Extrusion Plastometer, D1238-70, Annual Book of ASTM Standards, Part 27, Plastics. American Society for Testing Materials, Philadelphia.

Berry, J. P. (1961a). *J. Polym. Sci.* **50**, 107.

Berry, J. P. (1961b). *J. Polym. Sci.* **50**, 313.

Biot, M. A., (1954). *J. Appl. Phys.* **25**, 1385–1391.

Bloembergen, N., Purcell, E. M., and Pound, R. V. (1948). *Phys. Rev.* **73**, 679.

Blyler, L. L. Jr., and Daane, J. H. (1976). Private communication.

Bondi, A. (1964). *J. Polym. Sci., Part A-2* 3159.
Boyer, R. F. (1954). *J. Appl. Phys.* **25**, 825.
Bueche, F. (1956). *J. Chem. Phys.* **24**, 269.
Bueche, F. (1959). *J. Chem. Phys.* **30**, 748.
Cohen, R., and Turnbull, D. (1959). *J. Chem. Phys.* **31**, 1164.
Dahl, W., and Muller, F. H. (1961). *Z. Elektrochem.* **65**, 652.
Doolittle, A. K. (1952). *J. Appl. Phys.* **23**, 236.
Drott, E. E., and Mendelson, R. A. (1970). *J. Polym. Sci. Part A-2* **8**, 1361.
Ferry, J. D. (1970). "Viscoelastic Properties of Polymers," 2nd ed., p. 314. Wiley, New York.
Ferry, J. D., and Stratton, R. A. (1960). *Kolloid-Z.* **171**, 107.
Fox, T. G., and Flory, P. J. (1954). *J. Polym. Sci.* **14**, 314.
Fröhlich, H. (1958). "Theory of Dielectrics." Oxford Univ. Press, London and New York.
Fujita, H., Kishimoto, A., and Matsumoto, K. (1960). *Trans. Faraday Soc.* **56**, 424.
Gent, A. N. (1970). *J. Mater. Sci.* **5**, 925.
Gibbs, J. H., and DiMarzio, E. A. (1958). *J. Chem. Phys.* **28**, 373.
Goldstein, M. (1963). *J. Chem. Phys.* **39**, 3369.
Griffith, A. A. (1924). *Proc. Int. Congr. Appl. Mech. Delft* 55.
Gutowsky, H. S., and Pake, G. E. (1950). *J. Chem. Phys.* **18**, 168.
Haas, T. W., and Keller, A. (1965). Private communication.
Halpin, J. C. (1971). *Quad. Ing. Chim. Ital.* **7**(12), 173.
Hildebrand, J. H. (1959). *J. Chem. Phys.* **31**, 1423.
Hiltner, A., and Baer, E. (1972). *J. Macromol. Sci. Phys.* **6**(3), 545.
Hoffman, J. D., Williams, G., and Passaglia, E. (1966). *J. Polym. Sci. Part C* **14**, 173.
Hopkins, I. L., and Baker, W. O. (1959). *Kunststoffe* **49**, 26.
Ishida, Y., and Yamafuji, K. (1965). *Kolloid Z.* **202**, 26.
Ishida, Y., Shimada, K., Takayanagi, M., and Yamafuji, K. (1963). *Rep. Prog. Polym. Phys. Jpn.* **6**, 237.
Ishida, Y., Togami, S., and Yamafuji, K. (1967). *Kolloid-Z Z. Polym.* **221**, 1, 16.
Johari, G. P. (1973). *J. Chem. Phys.* **58**, 1766.
Johnson, G. E. (1978). *Preprints, Organic Coatings and Plastics Division* (*Amer. Chem. Soc.*) **38**, 350.
Kambour, R. P. (1966). *J. Polym. Sci. Part A-2* **4**, 349.
Karasz, F. E., and Jones, L. D. (1967). *J. Phys. Chem.* **71**, 2234.
Kauzmann, W. J., and Eyring, H. (1940). *J. Am. Chem. Soc.* **62**, 3113.
Keith, H. D., and Padden, F. J. (1959). *J. Polym. Sci.* **41**, 525.
Keller, A. J. (1957). *Phil. Mag. Ser. 8* **2**, 1171.
Kokes, R. J., and Long, F. A., *J. Am. Chem. Soc.* **75**, 6142 (1953).
Kovacs, A. J. (1963). *Fortschr. Hochpolym.-Forsch* **3**, 394.
Kwei, T. K., Schonhorn, H., and Frisch, H. L. (1967). *Appl. Phys.* **38**, 2512.
Macedo, P. B., and Litovitz, T. A. (1965). *J. Chem. Phys.* **42**, 245.
Marshall, G. P., and Williams, J. G. (1973). *J. Mater. Sci.* **8**(1), 138–40.
Marshall, G. P., and Williams, J. G. (1974). *Pure Appl. Chem.* **39** (1-2), 275–285.
Matsuoka, S. (1962). *J. Polymer Sci.* **57**, 569.
Matsuoka, S., and Bair, H. E. (1977). *J. Appl. Phys.* **48**, 4058.
Matsuoka, S., and Ishida, Y. (1966). *J. Polym. Sci. Part C* **14**, 247.
Matsuoka, S., Roe, R. J., and Cole, H. F. (1972). "Dielectric Properties of Polymers" (F. E. Karasz, ed.), p.255. Plenum Press, New York.
Matsuoka, S., Aloisio, C. J., and Bair, H. E. (1973). *J. Appl. Phys.* **44**, 10, 4265.
Matsuoka, S., Bair, H. E., Bearder, S. S., Kern, H. E., and Ryan, J. T. (1978). *Polymer Eng. Sci.* **18**, 1073.

McCall. D. W. (1969). "Molecular Dynamics and Structures of Solids" (R. S. Carter and J. J. Rush, eds.), National Bureau Standards. Spec. Publ. 301, Washington, D.C.

McCall, D. W. (1971). *Accounts Chem. Res.* **4**, 223.

McCall, D. W., and Slichter, W. P. (1957). *J. Polym. Sci.* **26**, 171.

McCall, D. W., Douglass, D. C., and Anderson, E. W. (1962). *J. Polym. Sci.* **59**, 301.

Mendelson, R. A. (1964). *SPE Trans.* **5**, 34.

O'Reilly, J. M. (1962). *J. Polym. Sci.* **57**, 429.

Patterson, G. D. (1976). *J. Polym. Sci.* **14**, 143.

Petrie, S. E. B. (1972). *J. Polym. Sci. Part A-2* **10**, 1255.

Porter, R. S., Knox, J. P., and Johnson. J. F. (1968). *Trans. Soc. Rheol.* **12**:3, 409.

Rogers, C. E., Meyer, J. A., Stannett, V., and Szwarc, M. (1962). TAPPI Monogr. Ser. No. 23.

Rouse, P. (1953). *J. Chem. Phys.* **21**, 1272.

Ryan, J. T. (1978). *Polymer Eng. Sci.* **18**, 264.

Sabia, R. (1964). *J. Appl. Polym. Sci.* **8**, 1951.

Sasabe, H., and Saito, S. (1968). *J. Polym. Sci. Part A-2*, **6**, 1401.

Sasaguri, K., and Hoshino, S. (1964). *J. Appl. Phys.* **35**, 47.

Schapery, R. A. (1975). *Int. J. Fracture* **11**, 141–159, 369, 549.

Sharanov, Yu. A., and Volkenshtein, M. V. (1962). *In* "The Structure of Glasses" (E. A. Porai-Koshits, ed.), Vol. 6, p. 62. Consultants Bureau, New York.

Simha, R., and Boyer, R. F. (1962). *J. Chem. Phys.* **37**, 1003.

Slichter, W. P. (1961). *Rubber Chem. Technol.* **34**, 1574.

Smith, T. L. (1958). *J. Polym. Sci.* **32**, 99.

Smith, T. L., and Dickie, R. A. (1969). *J. Polym. Sci. Part A-2* **7**.

Smith, T. L., and Stedry, P. (1960). *J. Appl. Phys.* **31**, 1892.

Southern, J. H., and Porter, R. S. (1970). *J. Macromol. Sci.-Phys.* **B4**, 541.

Stein, R. S. (1958). *J. Polym. Sci.* **31**, 327, 335.

Stein, R. S. (1959). *J. Polym. Sci.* **34**, 709.

Sternstein, S. S. (1972). *J. Appl. Phys.* **43**, 4370.

Struik, L. C. E. (1977). TNO Central Laboratorium Communication No. 565.

Takayanagi, M. (1965). *Proc. Int. Congr.*, 4th.

Takayanagi, M., Harima, H., and Iwata, Y. (1963). *Prog. Polym. Phys. Jpn.* **6**, 113.

Tobolsky, A. V., and Catsiff, E. (1956). *J. Polym. Sci.* **19**, 111.

Tonelli, A. E. (1974). *Anal. Calorimetry* **3**, 89.

Treloar, L. R. G. (1958). "The Physics of Rubbery Elasticity," p.26. Oxford Univ. Press, London and New York.

Turnbull, D., and Cohen, M. H. (1961). *J. Chem. Phys.* **34**, 120.

Wall, W. T. (1942). *J. Chem. Phys.* **10**, 132, 485.

Ward, J. M. (1974). *Polymer* **15**(6), 379.

Westover, R. F., and Vroom, W. I. (1969). *SPE J.* **25**, No. 8, 58.

Williams, J., Shohamy, E., Reich, S., and Eisenberg, A. (1975). *Phys. Rev. Lett.* **35**, 14, 951.

Williams, M. L. (1957). *J. Appl. Mech.* **24**, 109.

Chapter 7

REACTIONS OF MACROMOLECULES

L. D. LOAN AND F. H. WINSLOW

7.1. Introduction

Much of our understanding of polymer chemistry is based upon the very basic assumption that the reactivity of a given group is unaffected by the size of the molecule of which it forms a part. Perhaps the most obvious example of the use of this assumption is in the analysis of the kinetics of addition polymerization where all polymer radicals are assumed to have

ISBN 0-12-119755-7

equal rates of reaction with monomer molecules. Thus, in principle, the reactivity of chemical groups in polymer molecules is the same as in small molecules, and it might be thought that there is no need for special study. However, many apparent anomalies have been observed due to complications related to the polymeric environment. These include neighboring group effects, limited diffusion of reactants, and morphology. Such considerations lead to a set of conditions for equality of reactivity in low molecular weight and polymeric systems. They may be listed (Alfrey, 1964) as follows:

(1) Reactions must take place in homogeneous solution.
(2) Only one polymer group may react at any step.
(3) An appropriate low molecular weight comparison must be used.

As an indication of some of the complexities observed in polymer systems, a few examples will be given. First, in the hydrolysis of the copolymer of acrylic acid and p-nitrophenyl methacrylate, it was found that the formation of p-nitrophenol took place much faster than in corresponding low molecular weight compounds. This was found to result from neighboring group interactions as follows:

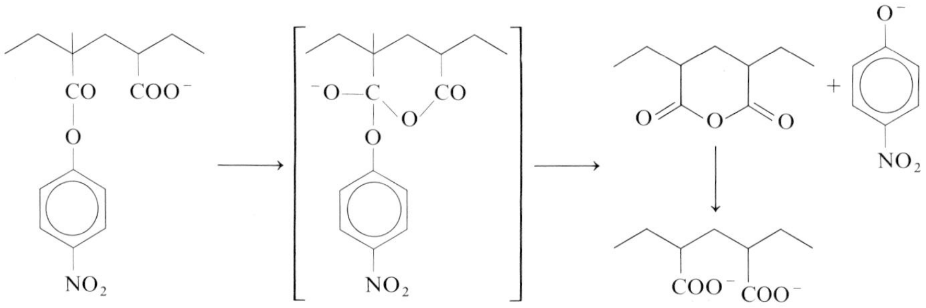

$$
\begin{array}{ccc}
\text{CO} \quad \text{COO}^- & \text{[} \quad ^-\text{O}-\text{C} \quad \text{CO} \quad \text{]} & \text{O=} \quad \text{O} \quad \text{=O} \quad + \quad \text{C}_6\text{H}_4\text{O}^-\text{NO}_2 \\
\text{O} & \text{O} & \\
\text{C}_6\text{H}_4\text{NO}_2 & \text{C}_6\text{H}_4\text{NO}_2 & \text{COO}^- \quad \text{COO}^-
\end{array}
$$

That such a mechanism is important may also be demonstrated in low molecular weight materials such as succinic acid monoesters (Morawetz, 1964).

A similar reaction is, of course, an important commercial reaction in polymeric systems. The base-catalyzed alcoholysis of poly(vinyl acetate) is used to produce poly(vinyl alcohol), which interestingly enough is a polymer based on a nonexistent monomer. A second reaction is also used commercially to convert poly(vinyl alcohol) to poly(vinyl acetal) by the acid-catalyzed addition of an aldehyde, most commonly formaldehyde or butyraldehyde.

The reaction here involves two neighboring hydroxyl groups. It is somewhat surprising that this reaction goes almost to completion with very few

OH   OH   OH   OH   O   O
                                              \   /
HCHO   CH₂

unreacted hydroxyl groups remaining. It might have been expected that, where neighboring groups react in pairs, a significant number of isolated hydroxyl groups would remain unable to react in the manner illustrated. However, the reaction is reversible and by interchanges all sites are able to react.

A similar but irreversible reaction, the dehalogenation of poly(vinyl chloride) by metallic zinc, was studied by Marvel *et al.* (1939). Here, as expected, some chlorine remains in the product after exhaustive reaction.

Cl   Cl   Cl   Cl

The expected number of isolated groups has been calculated statistically (Flory, 1939) using the assumption of equal reactivity. It was thus shown that a fraction $1/e^2$ (or about 13%) of the original chlorine will never be eliminated. The experimental data of Marvel agree very well with this prediction. Other reactions where neighboring groups are of importance include the dehydrochlorination of poly(vinyl chloride), which will be considered elsewhere in this chapter.

Another factor limiting the reactions of polymers is the morphology of the substrate. In partially crystalline materials there is often no diffusion of the reactant into the crystalline regions while the amorphous regions are readily accessible. An example of this is the oxidation of polyethylene by nitric acid. As the reaction proceeds, the degree of crystallinity is substantially increased as a result of both the removal of the amorphous material by oxidation and also the continued crystallization of chains thereby freed from entanglement.

7.2. Mechanochemical Reactions

It is possible to apply substantial stresses to a molecule of giant size. This often happens during the shearing mixing of rubbers and grinding of glassy materials. When the stress becomes excessive the molecular chain is broken (Kauzmann and Erying, 1940), producing a pair of free radicals that may take part in subsequent reactions. In the presence of oxygen the first reaction probably will result in the formation of peroxy radicals. Such radicals have been detected by electron spin resonance measurements.

Perhaps the most important commercial application of this type of reaction is the mastication of natural rubber (Bristow and Watson, 1963). As obtained from the tree, polyisoprene has a molecular weight of several million, which is far too high for easy processing. The molecular weight is usually reduced to perhaps 500,000 by a process known as mastication. This involves heavy shearing on a two-roll mill or a mixer, often in the presence of *peptizers*. The chemical reactions in this process have been studied in some detail and have been shown to be mechanochemical in nature. The breakdown as a function of mastication temperature is shown in Fig. 7.1. Initially the number of bonds broken decreases with a rise in temperature. This is a natural result of the decrease in viscosity that reduces the amount of energy transferred to the polymer molecules. The increase in bond breakage at still higher temperatures is a result of the onset of oxidative breakdown initiated by the free radicals produced by shearing. The overall efficiency of the breakdown is often increased by the addition of peptizers. These are of two varieties, radical traps to stabilize the free radicals formed and thus prevent recombination, and pro-oxidants which accelerate the higher temperature reactions.

Mechanochemistry is also a useful approach to block copolymer formation. If, for instance, natural rubber is swollen with methyl methacrylate and subsequently extruded, the free radicals formed mechanically initiate polymerization of the monomer. The resulting block copolymer of isoprene

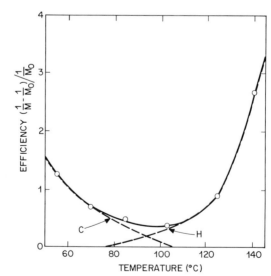

FIG. 7.1.   Breakdown as a function of mastication temperature. M_0 is molecular weight before treatment; M, molecular weight after treatment (Bristow and Watson, 1963).

and methyl methacrylate has some interesting properties and was for a time used as a substitute for crepe rubber in items such as shoe soles.

7.3. Cross-Linking Reactions

One of the most important reactions in which polymers take part is cross-linking. The previous chapters have dealt largely with polymers made up of discrete molecules. Under the appropriate conditions of shear and temperature, these molecules are free to move and slide over one another. Some mention was also made, however, of a fundamentally different kind of polymer in which these single molecules are joined together by cross-links. The presence of covalent cross-links, of course, precludes movement of one molecule past another, and thus these materials will not flow unless, of course, temperature or shear conditions are severe enough to cause fracture of covalent bonds. The presence of these cross-links causes the materials to show a rubberlike behavior—an ability to recover their original shape after a modest deformation; and at sufficiently high temperatures they show the kind of retractile properties we commonly associate with a rubber band.

On the other hand, it is of interest that the internal cross-linking of rubber latex particles into discrete gelled domains decreases the viscosity of the rubber (Baker, 1949; Blow, 1971).

Since a completely cross-linked polymer is essentially one giant molecule, it is insoluble in all solvents. This makes its characterization difficult; it is no longer possible to measure properties like molecular weight by solution methods. However, one of the most important properties, the number of cross-links in the network, can be measured by an analysis of the stress–strain curve of the network. The analysis makes use of the statistical theory of rubber elasticity. The basic ideas behind this theory have been presented in Chapter 6, and it is thus not necessary to repeat them in detail here. However, the outcome may be stated in the form of the following basic equation relating stress and strain in any network to the number of cross-links present (Treloar, 1975; Flory, 1953; Mullins and Thomas, 1963):

$$fv_r^{1/3}/A_0 = 2N_CRT(\lambda - \lambda^{-2})$$

where f is the applied force, v_r the volume fraction of rubber (unity when the sample is not swollen), A_0 the initial cross-sectional area, N_C the number of cross-linked units, R the gas constant, T the temperature, and λ the extension ratio. The multiplier N_CRT is often denoted by C_1. In this form the equation accurately predicts the behavior of rubber in compression. In elongation, however, experimental results are found to deviate somewhat from the theoretical curve, and in this case a semiempirical modification has been made. This involves the introduction of a second constant, C_2 (Flory, 1953; Mullins and Thomas, 1963, Mark, 1975). The new equation is

$$fv_r^{1/3}/2A_0 = (C_1' + C_2/\lambda)(\lambda - \lambda^{-2})$$

The structural significance of C_2 is not clearly understood. While it is, strictly speaking, no longer theoretically correct to identify C_1' with the number of cross-links in the network because of the similarity of the two equations, it is still often used in this way.

When polymer networks are brought into contact with a liquid that would have dissolved the uncross-linked precursor molecules, some of the solvent is absorbed and the network becomes swollen. After a sufficiently long time, an equilibrium degree of swelling is observed, and at this time the free energy decrease associated with the mixing of solvent molecules and polymer is balanced by the increase associated with the expansion of the network. A theoretical treatment has been formulated that relates the equilibrium degree of swelling with cross-link density:

$$-[\ln(1 - v_r) + v_r + \chi v_r^2] = (2C_1 v_1/RT)(v_r^{1/3} - v_r/2)$$

where v_1 is the molar volume of the solvent.

Again, we shall not go into the details of the derivation (Flory, 1953). The solvent-polymer interaction coefficient χ (see Chapter 4) may be measured from osmometric data on uncross-linked molecules, but it is obviously more directly determined from a series of swelling experiments on networks of known C_1 and v_r.

This simplified account has neglected many irregularities. In an exact treatment corrections must be made for features such as chain ends, elastically inactive chains, and chain entanglements. For a detailed treatment of these finer points of the theory, the reader is referred to more specialized texts (Flory, 1953; Treloar, 1974, 1975).

The largest practical use of cross-linked networks is in the field of vulcanized rubber, where the term vulcanization is a common synonym for cross-linking. In the preparation of the rubber network, it is obviously necessary to introduce a carefully controlled number of cross-links. With too few, permanent deformation of the network can occur; with too many, the strength of the vulcanizate first falls but rises again. Highly cross-linked rubber is a very hard, rigid material called ebonite or hard rubber. The aim in vulcanization is to introduce an optimum number of cross-links, usually about one cross-link for every 100 monomer units. The detailed chemical means of introducing these cross-links will be discussed later. In many vulcanizations the number of cross-links first increases as a function

of time and then decreases as a result of cross-link breakdown in a phenomenon generally known as *reversion*. An ideal rubber compound is one
that can be stored safely, processed easily, and cured readily without reversion. In other words the cured properties of all commercial compounds
should be relatively insensitive to cure time.

7.3.1.   CROSS-LINKING BY PEROXIDES

Cross-linking by peroxides, because of its apparent chemical simplicity,
has been widely studied. A simple chemical scheme for cross-linking by
di-*tert*-butyl peroxide is shown below (PH = polymer, P· = polymer radical).

$$(CH_3)_3COOC(CH_3)_3 \rightarrow 2(CH_3)_3CO\cdot$$

$$(CH_3)_3CO\cdot \rightarrow CH_3COCH_3 + CH_3\cdot$$

$$(CH_3)_3CO\cdot \text{ (or } CH_3\cdot) + PH \rightarrow (CH_3)_3COH \text{ (or } CH_4) + P\cdot$$

$$P\cdot + P\cdot \rightarrow P - P$$

This particular cross-linking agent was used in the first detailed examination
of the validity of the stress–strain equations discussed in the last section
(Bateman *et al.*, 1963). In the study natural rubber was used as the base
polymer and the number of cross-links introduced by the peroxide was
measured by both physical and chemical means. In order to make this
chemical determination accurate, all possible side reactions were studied.
These include such things as breakdown of the *tert*-butoxy radical, scission of
the polymer chain, and recombination of initiating free radicals. The results
of the overall investigation showed that the number of cross-links measured
by the two methods agreed very well provided that adequate corrections
were made for physical imperfections in the network, such as chain ends,
elastically inactive chains, and entanglements.

Following this early investigation, the peroxide cross-linking of many
different polymers has been studied (Loan, 1967; Wood, 1976). Remarkable
differences have been found between different materials. Although the simple
scheme outlined above is perfectly adequate for natural rubber and leads to
the formation of one cross-link for each initial peroxide molecule, other
rubbers can show far more efficient cross-linking. Indeed in *cis*-polybutadiene
each peroxide molecule can give rise to more than ten cross-links provided
that all traces of antioxidant are removed. With another common rubber,
ethylene–propylene copolymer, less than one cross-link is formed for each
peroxide molecule decomposed, and this is explained as resulting from a
mixture of scission and cross-linking produced by the alkoxy radicals.

Although the chemistry of peroxide cross-linking is now well understood, peroxides are only used to a very limited extent in commercial rubber technology. The usual commercial peroxide, dicumyl peroxide, gives rise to cumyloxy radicals, and these behave very similarly to the *tert*-butoxy radicals referred to earlier. It is only used where cured rubbers with a very low permanent set are required or where the presence of sulfur in the vulcanizate would prove deleterious by causing staining or corrosion.

7.3.2. CROSS-LINKING BY SULFUR

Vulcanization by sulfur was the original method discovered by Goodyear for cross-linking polymer molecules. His simple process consisting of heating rubber and sulfur together required many hours to give substantial cross-linking. It was soon discovered that inorganic activators, notably zinc oxide, were able to approximately halve the time necessary for cross-linking. Later, organic bases were also introduced, and a new era in rubber vulcanization started in 1921 with the discovery of the accelerating effects of mercaptobenzothiazole, which when used together with zinc oxide, stearic acid, and sulfur, gives what might be termed a modern rubber compound with a cure time of about 40 min at 140°C. The number of accelerators now available is large—the benzothiazoles, thiurams, and thiocarbamates being the most common. In spite of its long commercial use, sulfur vulcanization

$$\underset{\substack{\text{mercaptobenzo-}\\\text{thiazole}}}{\text{mercaptobenzothiazole structure (CSH)}} \qquad \underset{\substack{\text{tetramethyl-}\\\text{thiuram disulfide}}}{(CH_3)_2NCSSCN(CH_3)_2} \qquad \underset{\substack{\text{zinc dimethyl}\\\text{dithiocarbamate}}}{Zn[SCN(CH_3)_2]_2}$$

is not completely understood from a chemical point of view. Recent work has made tremendous advances in our understanding (Bateman *et al.*, 1963; Morrell, 1971), but much remains to be done. As might be expected from the rather complicated vulcanization recipes mentioned, the mechanism is complex. Cross-links of many different structures are formed; and the discovery of chemicals capable of breaking down cross-links of specific structure has enabled us to show the structure of a typical, unaccelerated cure. As may be seen, sulfur is wastefully used in such a network; indeed, an overall average of perhaps 40 to 50 sulfur atoms per cross-link is found. Conventional accelerated cures show a better ratio of approximately ten sulfur atoms per cross-link, and recently developed, high-efficiency cure systems give networks with only four or five. The chemical reactions that occur in the accelerated cross-linking of unsaturated rubbers involve first a reaction of sulfur and accelerator with the activator to form a sulfurating agent. This then reacts with the polymer after which sulfur–sulfur bonds switch to form cross-links that eventually shorten to give the final vulcanizate.

7.3.3. CROSS-LINKING BY HIGH-ENERGY IRRADIATION

Most polymeric materials are affected in some way by high-energy irradiation (Bovey, 1958; Charlesby, 1960; Chapiro, 1962; Dole, 1973). The predominant reactions induced are cross-linking and scission, but elimination reactions, such as the evolution of hydrogen chloride by poly(vinyl chloride), are also observed. The factors that determine which type of reaction will predominate are not entirely understood, but polymers with doubly substituted backbone carbon atoms frequently degrade by chain scission while most others cross-link. Generally, polymers behave in the same way in the presence of both peroxides and high-energy radiation, which is not surprising as the reactions induced in both cases are mainly free radical in nature. Table 7.1 gives a short list of polymers together with the main reaction mode of each.

TABLE 7.1

EFFECT OF ELECTRON RADIATION ON VARIOUS POLYMERS

Cross-linking	Scission
Polyethylene	Polyisobutylene
Polystyrene	Poly(α-methylstyrene)
Chlorinated polyethylene	Poly(methacrylonitrile)
Polybutadiene	
Poly(dimethylsiloxane)	

In many of these cases, although the net effect is one of cross-linking or scission, both reactions occur simultaneously. Where this occurs, some uncross-linked material is always present and may be dissolved and extracted by a suitable solvent. The quantity of this so-called *sol* fraction may be calculated statistically by assuming that cross-linking and scission reactions occur randomly throughout the network. In this way the following equation has been derived (Charlesby, 1960):

$$s + s^{1/2} = p_0/q_0 + 1/q_0 n_1 r$$

where s represents the sol fraction, p_0 and q_0 the fraction of monomer units cross-linked and broken, respectively, n_1 the degree of polymerization (DP), and r the radiation dose.

The chemical processes involved in cross-linking and scission are similar to those in peroxide-induced reactions

$$-CH_2-CHR-CH_2-CHR- \longrightarrow -CH_2-\overset{\cdot}{C}R-CH_2-CHR-$$

$$-CH_2-CR-CH_2-CHR$$
$$\quad\quad\quad\,|$$
$$-CH_2-\overset{\cdot}{C}R-CH_2-CHR$$

$$-CH_2-CR=CH_2 + \cdot CHR-$$

In some systems the cross-linking reactions occur too slowly and are either outweighed by scission or require high-radiation doses for adequate reaction. Certain chemical modifiers have been widely used to alleviate this problem and are now used commercially as cross-linking enhancers. All of the materials used are multifunctional, having several reactive double bonds. Incorporation of a typical enhancer, trimethylol propane trimethacrylate, into poly(vinyl chloride) leads to a manyfold increase in the efficiency of cross-linking. These same materials also find use in the peroxide cross-linking field when hard, tightly bound networks are required. The mechanism of this enhancement has been investigated in the case of poly(vinyl chloride) and tetraethylene glycol dimethacrylate. A graft copolymerization of the methacrylate onto the poly(vinyl chloride) is apparently involved (Salmon and Loan, 1972) but is somewhat complicated by phase separation of the component blocks (Bair et al., 1972).

Polymer reactions initiated by ultraviolet or higher energy radiation also play a major role in integrated circuit technology. Using a masking technique or a controlled electron beam, an intricate pattern is formed in a polymer coating on a substrate such as silicon. Exposed areas of the film, called a *resist*, either cross-link (negative resist) or degrade (positive resist) during exposure to the radiation. The image is developed by solvents that remove uncross-linked or degraded polymer and the pattern is then cut into the substrate with an acid or other etchants as shown in Fig. 7.2.

A common negative photoresist is partially cyclized polyisoprene containing a diazide cross-linking agent. The azide decomposes in ultraviolet light to form free radicals that subsequently cross-link the polymer. Typical classes of positive electron beam resists are poly(methyl methacrylate) and aliphatic polysulfones (Bowden, 1977).

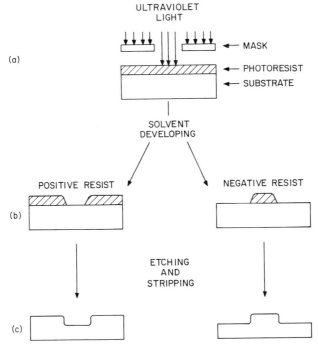

FIG. 7.2. Schematic diagram showing the function of a polymer resist.

7.4. Oxidative Degradation

Before proceeding to a detailed discussion of this subject, it is perhaps necessary to elaborate to some extent on what the polymer chemist means by degradation. It may be best defined as any process that leads to a deterioration of any physical property of a polymer. From a chemical point of view, therefore, it is a very nonspecific term since any one or more of a wide variety of chemical reactions may be involved.

A few examples will amply illustrate this point. Polyethylene, when exposed to high temperatures in the presence of oxygen, becomes brittle. This physical change results from the combination of oxidative scission and crystallization of chain segments in disordered regions of the semicrystalline polymer. Polybutadiene under similar conditions degrades by an oxidative reaction leading to a weak highly cross-linked polymer network. Poly(vinyl chloride) may during processing or high temperature aging become colored, and thus in appearance may be said to have suffered degradation although its mechanical properties may be unimpaired. This discoloration results from a dehydrochlorination reaction that leads to sequences of conjugated double bonds in the polymer backbone and these are powerful chromophores.

The higher molecular weight fraction of a polymer is usually less stable since reactive groups such as structural irregularities occur more often in larger chains. Although it is, in principle, possible to stabilize polymers by removal of such irregularities and adventitious low molecular weight impurities, this is not usually feasible as degradative reactions are often chain reactions of great kinetic length. Thus very small concentrations of initiators can lead to high rates of degradation. However, since chain reactions are involved, small concentrations of chain terminating agents can greatly reduce the rate of degradation and, as a result, the development of stabilizers (Hawkins, 1972) has been a key factor in the rapid growth and diversification of the polymer industry. This importance of the chain character of degradation and the use of stabilizers is perhaps best exemplified by the autoxidation of polyolefins, which is treated in detail in the following section.

7.4.1. MECHANISM OF AUTOXIDATION

Thermal oxidation of polyolefins proceeds by an autocatalytic three-step mechanism involving the radical chain reactions outlined in the following simplified reaction sequence:

Initiation

$$\text{Polymer} \rightarrow RO_2\cdot \tag{7.1}$$

Propagation

$$RO_2\cdot + R'H \xrightarrow[\text{slow}]{k_p} ROOH + R'\cdot \tag{7.2}$$

$$R'\cdot + O_2 \xrightarrow[\text{fast}]{} R'O_2\cdot \tag{7.3}$$

Termination

$$2RO_2\cdot \xrightarrow{k_t} \text{unreactive products} \tag{7.4}$$

Degradation begins when a polymer system absorbs enough energy to form a peroxy radical $RO_2\cdot$, which is sufficiently reactive to abstract hydrogen from a polymer chain to yield a hydroperoxide group $ROOH$ and a polymeric radical $R'\cdot$ (reaction (7.2)). The latter then reacts rapidly with oxygen to produce another peroxy radical and the cycle repeats itself many times before termination occurs. The hydroperoxide products also decompose (reaction (7.5)) forming more free radicals which accelerate the oxidation rate as shown in Fig. 7.3.

$$ROOH \xrightarrow{\text{heat}} RO\cdot + \cdot OH \tag{7.5}$$

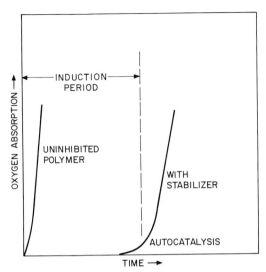

FIG. 7.3.   Oxygen absorption during thermal oxidation.

Eventually, the rate declines owing to the depletion of reactive sites. Since reaction (7.2) is the rate controlling (slowest) step, the overall rate is

$$-d[O_2]/dt = k_p[RO_2\cdot][RH] \qquad (7.6)$$

In the so-called steady-state reaction, where the rates of formation and destruction of radicals are equal, the rates of reactions (7.1) and (7.4) may be equated to give

$$[RO_2\cdot] = (R_i/2k_t)^{1/2} \qquad (7.7)$$

where R_i is the rate of production of radicals by reaction (7.1). The overall rate is then

$$-d[O_2]/dt = k_p(R_i/2k_t)^{1/2}[RH] \qquad (7.8)$$

The reactions discussed so far are the initial changes that determine the rate of oxygen absorption and the rate of oxidation. Various physical changes also accompany oxidation. The surface tackiness of some oxidized rubbers is usually the result of a greatly decreased molecular weight. Chain scission originating at hydroperoxide groups is the main cause of this decrease.

$$-CH_2CHCH_2- \rightarrow -CH_2CHCH_2- \rightarrow -CH_2CHO + \cdot CH_2- \qquad (7.9)$$
$$\quad\quad\ | \quad\quad\quad\quad\quad\quad |$$
$$\quad\quad OOH \quad\quad\quad\quad\quad O\cdot$$

In semicrystalline polymers this drop in molecular weight leads to embrittlement. Although the primary products of hydroperoxide homolysis are alcohols, aldehydes, and ketones, these finally oxidize to carboxylic acids, which play a major part in metal-catalyzed oxidation. Indeed, a poorly stabilized polyolefin insulation on copper wire will often show a greenish coloration at the polymer metal interface as a result of copper carboxylates being formed. These reactions enhance the diffusion of copper into the polymer and thus facilitate copper-catalyzed oxidation discussed in the following section.

7.4.2. MECHANISM OF ANTIOXIDANT ACTION

To reduce the oxidation rate of polyolefins, it is clearly necessary to interfere with this series of reactions in some very efficient way. One approach has been to shorten the reaction chain by introducing a chemical with a greater affinity than a polyolefin for the peroxy radical. Such materials are called antioxidants (AH). They react with peroxy radicals to form a relatively inactive radical A · by reaction (7.10):

$$RO_2\cdot + AH \xrightarrow[\text{fast}]{\text{very}} ROOH + A\cdot \tag{7.10}$$

Typical antioxidants are hindered phenols and aromatic amines such as

Phenols have become more popular because amines discolor and some appear to be carcinogenic. Usually phenols can stop more than one free radical chain,

$$+ R''OOH \tag{7.11}$$

When the resonance-stabilized aryloxy radical is protected by bulky electron-releasing *tert*-butyl groups in the 2 and 6 positions it can combine easily with another peroxy radical but cannot combine readily with molecular oxygen or with another aryloxy radical. Also, it cannot abstract hydrogen from the polymer substrate and initiate a new radical chain at a significant rate.

Although the autoxidation rate is reduced, it is not stopped entirely by the protective action of phenols. Hydroperoxides still accumulate, and eventually their breakdown to free radicals accelerates decomposition of the polymer. Consequently, the best stabilizer systems include a combination of antioxidants with agents that suppress homolytic breakdown of hydroperoxides.

The normal mode of hydroperoxide decomposition is homolytic scission to free radicals. The reaction can be catalyzed by certain transition metal ions, especially those of copper, cobalt, and manganese. To reduce the rate of free radical formation, two classes of supplementary protectants are used. The first catalyzes the decomposition of hydroperoxides to nonradical products, while the second sequesters metal ions. Examples of peroxide decomposers are sulfur-containing compounds and organic phosphites:

$$(C_{12}H_{25}OCCH_2CH_2)_2S \qquad (\langle\bigcirc\rangle\!-\!O\text{-}\!)_3P$$
$$\underset{O}{\overset{\|}{}}$$

Common types of metal chelators are oxamide derivatives (Hansen *et al.*, 1964) and related compounds such as

$$\langle\bigcirc\rangle\!-\!CH\!=\!NNH\!-\!\underset{O}{\overset{\|}{C}}\!-\!\underset{O}{\overset{\|}{C}}\!-\!NHN\!=\!CH\!-\!\langle\bigcirc\rangle$$

Combinations of stabilizers can be either antagonistic or synergistic. For instance, mixtures of amines and phenols are nearly always less effective than the same concentrations of the separate components. Also, some phenols promote the copper-catalyzed breakdown of hydroperoxides, and carbon black reduces the effectiveness of most, but not all, antioxidants. Noteworthy exceptions are the thiobisphenols and other sulfur-containing substances that form synergistic combinations with acidic blacks as indicated in Fig. 7.4.

While we have considered the chemical properties necessary in the various types of stabilizers used, other properties are also necessary in commercial stabilizers. Ideally they should be stable, nontoxic, colorless, inexpensive, and highly compatible with the polymer. To be effective, they must also have enough solubility and mobility to readily reach reaction sites without being lost from the polymer. Effectiveness varies in accordance with the length of the induction period shown schematically in Fig. 7.3.

7.4.3. CHEMICAL STRUCTURE AND REACTIVITY

Oxidation rates of polymers at temperatures below 150°C depend on the reactivities of the peroxy radicals formed and on the dissociation energies of

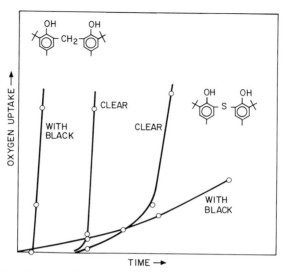

FIG. 7.4. Effect of carbon black on oxygen uptake showing the synergism with a thio-bisphenol and antagonism with the corresponding methylene-bridged phenol.

available carbon–hydrogen bonds in the polymer substrate. Polymers with carbon–hydrogen bonds of the following types may be expected to react in the order:

$$\text{allylic} > \text{tertiary} > \text{secondary} > \text{primary}$$

Outstanding exceptions to the rule are polystyrene and polyisobutylene in which benzylic and secondary hydrogens, respectively, are shielded by relatively inert phenyl and methyl groups.

Since the dissociation energy of the oxygen–hydrogen bond in hydro-peroxides is about 90 kcal/mole, peroxy radicals react with substances with similar or lower hydrogen bond strengths such as those listed in Table 7.2. As might be expected, polymers most resistant to oxidation either have no hydrogen at all or have it only in the form of unreactive methyl and phenyl groups.

Heteroatoms affect the strength of neighboring carbon–hydrogen bonds mainly by modifying the polar properties of transition states. Since the peroxy radical is electrophilic, the oxidation of ethers, aldehydes, amines, and sulfides occurs through abstraction of hydrogen atoms on carbons adjacent to the unshared electron pair on the heteroatom. Conversely, electron-withdrawing groups tend to stabilize neighboring hydrogens even though the radicals resulting from hydrogen removal are stabilized by electron delocalization. However, electronegative groups can favor oxidation by promoting carbanion formation in the presence of bases.

TABLE 7.2
BOND DISSOCIATION ENERGIES

	Bond energy (kcal/mole)
C—C Bond	
CH_3—CH_3	88
$(CH_3)_3C$—$C(CH_3)_3$	68
R—H Bond	
⬡—H	104
$CH_3CH_2CH_2$—H	98
$(CH_3)_2CH$—H	95
$(CH_3)_3C$—H	91
ROO—H	90
CH_3C—H ‖ O	88
CH_2=$CHCH_2$—H	85
$\cdot CH_2CH_2$—H	39

Disordered regions in polymers above the glass temperature are accessible to oxygen. Surface regions, of course, are most readily available, and in turn, are most vulnerable to oxidative degradation. The crystalline regions of a polymer are normally immune to oxidation because of their impermeability to molecular oxygen, but in the case of poly(4-methyl-1-pentene), the crystal structure is sufficiently open to allow oxygen diffusion and hence oxidation. The relative susceptibility of various polymers to thermal oxidation is shown in Table 7.3.

7.4.4. FACTORS IN WEATHERING

The deterioration of synthetic polymers during outdoor exposure is largely an oxidation process initiated by ultraviolet light of wavelengths between 290 and 400 nm. The rate of this oxidation, in temperate zones, varies as much as tenfold with the seasons and depends on film thickness, temperature, light intensity, chemical structure, and the concentration and efficiency of sensitizing defects or impurities.

The data in Table 7.3 show wide differences in weather resistance exhibited by polymeric materials. Poly(vinyl chloride) and polymers with aromatic groups discolor by forming conjugated systems of double bonds, while chain scission reactions cause chalking and embrittlement of many plastics and softening in some rubbers. Polytetrafluoroethylene and related fluoro- polymers require virtually no protection. However, the commonly used,

TABLE 7.3

RELATIVE RESISTANCE TO DETERIORATION[a]

	Pyrolysis	Autoxidation	Photo-oxidation	Ozone attack	Combustion (oxygen index)	Hydrolysis	Moisture absorption (%)[b]
Polyethylene	H	L	L_B	H	0.18	H	<0.01
Polypropylene	M	L	L_B	H	0.18	H	<0.01
Polystyrene	M	M	L_D	H	0.18	H	0.03–0.10
Polyisoprene	H	L	L_S	L	0.18	H	L
Polyisobutylene	M	M	M_S	M	0.18	H	L
Polychloroprene	M	L	M	M	0.26	H	M
Polyoxymethylene	L	L	L_B	M	0.16	L	0.25
Poly(phenylene oxide)	H	L	L_D	M	0.28	M	0.07
Poly(methyl methacrylate)	M	H	H	H	0.18	H	0.1–0.4
Poly(ethylene terephthalate)	M	H	M_D	H	0.25	M	0.02
Polycarbonate[c]	M	H	M_D	H	0.25	M	0.15–0.18
Poly(hexamethylenediamine-co-adipic acid)	M	L	L_B	H	0.24	M	1.5
Polyimide[d]	H	H	L_D	H	0.51	M	0.3
Poly(vinyl chloride)	L	L	L_D	H	~0.40	H	0.04
Poly(vinylidene chloride)	L	M	M_D	H	0.60	H	0.10
Polytetrafluoroethylene	H	H	H	H	<0.95	H	<0.01
Poly(dimethyl siloxane)	H	H	M	H	—	M	0.12
ABS resin	M	L	L_D	H	0.19	H	0.20–0.45
Cellulose triacetate	L	L	L_B	H	0.19	L	1.7–6.5

[a] Resistance ratings are high (H), medium (M) and low (L); subscripts indicate mode of weathering: embrittlement (B), discoloration (D) and softening (S).

[b] Gross (1975).

[c] Poly(4,4'-isopropylidenediphenylene carbonate)

[d] Poly[N,N'-(p,p'-oxydiphenylene)pyromellitimide]

low-cost, polymers including polystyrene, the polyolefins, and poly(vinyl chloride), are among the least stable outdoors.

Photoreactions in polyolefin films seem to be confined, for the most part, to a surface layer less than a micrometer thick. The rate in polyethylene increases several fold over the temperature range from $10°$ to $50°C$ but is negligible below room temperature and at wavelengths above 340 nm. Photooxidations of polyethylene produce more gel, less ketones, and more ester and carboxylic acid groups than are formed during thermal oxidation. The rate of photooxidation is usually monitored by measuring change in color, brittleness, tensile strength, infrared spectrum, or by oxygen consumption.

7.4.5. MECHANISMS OF PHOTOOXIDATION

Since paraffinic hydrocarbons do not absorb sunlight, the photoreactions of polyolefins have been ascribed to sensitizing carbonyl and hydroperoxide groups, unsaturation, catalyst fragments, aromatic hydrocarbons, and other contaminants (Rånby) and Rabek, 1975). It was further assumed that these chromophores were located at or near film surfaces where rates were at a maximum.

Burgess (1953) proposed that Norrish chain scission reactions were responsible for development of the strong carbonyl and vinyl bands in the infrared spectrum of photooxidized polyethylene. There are two types of Norrish reactions:

Type I

$$R'CCH_2CH_2CH_2R \xrightarrow{h\nu} R'C\cdot + \cdot CH_2CH_2CH_2R \qquad (7.12)$$
$$\parallel \qquad\qquad\qquad\qquad \parallel$$
$$O \qquad\qquad\qquad\qquad\quad O$$

Type II

$$R'CCH_2CH_2CH_2R \xrightarrow{h\nu} R'CCH_3 + CH_2{=}CHR \qquad (7.13)$$
$$\parallel \qquad\qquad\qquad\qquad \parallel$$
$$O \qquad\qquad\qquad\quad\;\; O$$

Studies of the photolysis of ethylene–carbon monoxide copolymers (Hartley and Guillet, 1968) shows that process II accounted for 90% of the chain breaks at room temperature. About the same time Trozzolo and Winslow (1968) suggested an initiation mechanism involving singlet oxygen as an important intermediate in polyethylene weathering.

$$^3(RCR) + O_2 \longrightarrow RCR + {}^1O_2 \qquad (7.14)$$
$$\quad\parallel \qquad\qquad\qquad \parallel$$
$$\quad O \qquad\qquad\qquad O$$

$$^1O_2 + RCH_2CH{=}CH_2 \xrightarrow{h\nu} RCH{=}CHCH_2OOH \qquad (7.15)$$

$$RCH{=}CHCH_2OOH \longrightarrow \text{radical reactions} \qquad (7.16)$$

According to their proposal, ultraviolet light is absorbed by carbonyl or other chromophoric groups and the energy of the excited groups is transferred to oxygen molecules. The resulting excited oxygen species react with vinyl or similar unsaturated groups forming hydroperoxides that then initiate free radical reactions by mechanisms already discussed in previous sections.

The mechanism assumes that the juxtaposition of acetyl and vinyl groups from a Norrish II reaction would favor the singlet oxygen reaction sequence. At least, the proposal is in keeping with known singlet oxygen reactions, with the photooxidation products from the polymer, and with the known effectiveness of singlet oxygen quenchers as photostabilizers.

Polyolefin photooxidations are autocatalytic, possibly because the oxidation products act as sensitizers. The sensitizers or light-absorbing chromophores may be formed on film surfaces by ozonolysis, or may be deposits of aromatic contaminants from the atmosphere. So far, the initiation processes have remained obscure.

Mechanisms for chromophore formation in polystyrene (Grassie and Weir, 1965) and poly(vinyl chloride) (Geddes, 1967) are also speculative, but more is known about photolyses of some aromatic polyesters (Cohen et al., 1971) and polycarbonates (Humphrey et al., 1973) that develop protective surface films by rearranging to photostable groups.

7.4.6. PHOTOSTABILIZATION

In order to understand photostabilization, we should first consider the basic mechanisms involved in photooxidation. The initial step is the electronic excitation of a polymer molecule, either by direct light absorption or by energy transfer from a light absorbing species. This excited molecule A* may then lose its excess energy by radiation emission without undergoing chemical change, or it may decompose.

Excitation

$$A \rightarrow A^* \tag{7.17}$$

Emission

$$A^* \rightarrow A + \text{light or heat} \tag{7.18}$$

Degradation

$$A^* \rightarrow \text{decomposition products} \tag{7.19}$$

This process may clearly be suppressed or inhibited by interfering with the reaction sequence. A light screen prevents light from penetrating into the polymer and thus greatly reduces reaction (7.17), while a deactivator reacts with an excited molecule to remove its excess energy and dissipate it harmlessly.

Deactivation

$$A^* + D \rightarrow A + D^* \rightarrow A + D + \text{light or heat} \qquad (7.20)$$

The best light screens are carbon blacks with an average particle size of about 20 nm. Finely divided iron oxide is also effective but can promote degradation of poly(vinyl chloride) and some other polymers at elevated temperatures. Where the color of carbon black is objectionable, dispersions of the rutile form of titanium dioxide are often used in conjunction with deactivators for stabilizing polymers in outdoor applications.

Photodeactivation takes place when process 11 is much faster than reaction 10 and when D readily loses its excitation energy. Two common types of colorless deactivators are

o-hydroxybenzophenones   *o*-hydroxyphenylbenzotriazines

Here, R represents an electron-releasing alkyl or alkoxyl group, which improves both compatibility with, and energy dissipation of the polymer. Deactivators are usually referred to as ultraviolet absorbers but they actually function mainly as energy transfer agents. Various nickel chelates of oximes or bisphenols are especially efficient deactivators and singlet oxygen quenchers. These protectants are not efficient radical chain stoppers, nor are radical chain stoppers very effective photostabilizers. Examples (Winslow, 1977) of synergistic combinations of radical chain stoppers with light screens or ultraviolet absorbers have, however, been reported indicating the participation of several different reactions.

7.4.7. OZONE CRACKING OF RUBBER

The surface reactions of ozone have no significant effect on the appearance or mechanical properties of most plastics and are often used for making film surfaces suitable for printing applications. On the other hand, ozone causes cracking in stretched vulcanized rubbers with unsaturation in the main polymer chain. Cracks form at strain levels above 5% and grow in a direction perpendicular to the strain direction. Crack growth rates vary with temperature, sample modulus, and network density. The inherent resistance of a few rubbers to ozone is shown in Table 7.3.

Evidently cracks result from the scission of double bonds in the main polymer chains. Experiments on low molecular weight olefins have shown that a labile adduct is first formed by reaction with ozone. The adduct then rearranges to an ozonide. The full reaction sequence is

$$RCH{=}CHR' \xrightarrow{O_3} \underset{\text{molozonide}}{RCH{-}CHR'} \tag{7.21}$$

$$\underset{}{R{-}CH{-}CHR'} \xrightarrow{a} RCHO + \underset{\underset{\text{zwitterion}}{}}{O^-} \xrightarrow{b} \underset{\text{ozonide}}{RCH \quad CHR'} \tag{7.22}$$

If the polymer chain is under tension during reaction (7.22a), the broken chain ends separate before rearrangement to the ozonide can occur.

7.4.8. ANTIOZONANTS

Rubbers may be protected from ozone attack by two types of additives: microcrystalline waxes and N,N-dialkyl-p-phenylenediamines. The waxes diffuse (bloom) to the surface and form a thin layer that is impervious to ozone. They are very effective in the absence of flexing. Under the dynamic conditions encountered in tires, a combination of wax and a substituted p-phenylenediamine is generally more effective. The exact mechanism by which the phenylenediamine derivatives protect rubber against ozone is not known. They have been variously described as ozone scavengers, as ozone shields, as cross-linking agents to counteract scission reactions, and as prooxidants that generate a soft protective surface film.

As mentioned ozone cracking only occurs in rubbers containing backbone unsaturation. A logical method of preventing it is therefore to use rubbers with no such unsaturation, but since the reactivity associated with this unsaturation is needed for sulfur vulcanization, a new family of rubbers, ethylene–propylene terpolymers, has been developed with occasional pendant groups carrying the unsaturation. One of the early versions was a terpolymer of ethylene, propylene, and hexadiene:

$$\underset{}{{+}CH_2{-}CH_2{+}_{x'}} \qquad {+}CH_2{-}CH{+}_y, \qquad \text{and} \qquad {+}CH_2{-}CH{+}_z$$

7.5. Hydrolytic Degradation and Biodegradation

Many polymers in everyday use are subjected to an atmosphere containing water, and some must withstand long-term immersion in this ubiquitous liquid. Hydrocarbon polymers are inert to moisture and contain only very small amounts of it in solution at equilibrium. Even here, however, hydrophilic impurities or unusual processing conditions may lead to higher concentrations of water and may affect certain electrical properties in very demanding applications.

Polymers such as nylon and cellulose, which contain polar groups, accept larger amounts of water and at room temperature this water provides some plasticization, reducing stiffness, hardness, and yield strength. At higher temperatures and relative humidities, hydrolysis can cause significant degradation. Such conditions often exist during processing and care must be taken to ensure that sensitive polymers such as polycarbonates and polyesters are adequately dried before processing. Table 7.3 shows the moisture susceptibility of several polymers.

In warm climates at relative humidities above 70%, microorganisms attack naturally occuring polymers and some synthetic materials. Numerous types of bacteria and fungi generate enzymes that hydrolyze peptide and glycosidic linkages and form water-soluble products. Resistance of these materials to biodegradation can be increased by cross-linking in the case of natural rubber or by the acetylation of cellulosic polymers. Biocides such as phenols or organic compounds containing copper, mercury, or tin may also be added. Polyolefins, poly(vinyl chloride), polycarbonates, acrylic plastics, and many other polymers are essentially unaffected by long-term burial at various depths in acidic or alkaline soils.

7.6. Thermal Degradation

Although thermal oxidation is the most common cause of polymer degradation, pure thermal degradation is important also, especially in the case of poly(vinyl chloride). As in oxidative degradation, a variety of chemical reactions occur during thermal degradation. They can be classified into four main types. In the first, the main polymer chain is reduced in length by the sequential elimination of monomer units. For obvious reasons this is often called unzipping or depolymerization and is exemplified by poly(methyl methacrylate), which was incidentally the first polymer in which the mechanism of thermal degradation was subjected to detailed study. Another common reaction is random chain scission. This reaction is shown by many polymers, including polyethylene, poly(methyl acrylate), and polystyrene.

It is sometimes combined with some unzipping at the reactive broken chain ends. The third type of degradation involves reactions of substituents while the polymer backbone remains more or less intact. Again, sequential reactions are often observed, sometimes with the elimination of a small molecule as in the loss of hydrogen chloride from poly(vinyl chloride). More complex condensation reactions also occur. We shall now deal in turn with each of these types by using one notable example of each. For a much more detailed account of the mechanism of polymer degradation than is intended here, the reader is referred to a recent book by Bamford and Tipper (1975).

7.6.1. DEPOLYMERIZATION

This type of degradation results when the main polymer chain is ruptured above its ceiling temperature. If the reactivity of the resultant chain end is low and the polymer molecule contains no easily abstractable atoms, transfer reactions are unlikely and unzipping occurs.

$$-CH_2-C(CH_3)-CH_2-\dot{C}(CH_3) \rightarrow -CH_2-\dot{C}(CH_3) + CH_2=C(CH_3) \quad (7.23)$$
$$\overset{|}{COOCH_3} \quad \overset{|}{COOCH_3} \quad \overset{|}{COOCH_3} \quad \overset{|}{COOCH_3}$$

In a detailed study of the degradation of poly(methyl methacrylate), Grassie and Melville (Grassie, 1956) measured rates of monomer formation and molecular weight decrease, showing that the predominant mode of degradation was unzipping or depolymerization. In the ideal case depolymerization leads to monomer production with no change in molecular weight of the residue, while in random main chain scission little monomer would be expected in the initial stages of degradation. It is thus possible to draw up a simple set of reactions to explain the process:

Initiation

$$R + P \rightarrow P \cdot \quad \text{Rate } R_i \quad (7.24)$$

Propagation

$$P \cdot \xrightarrow{k_p} P \cdot + M \quad (7.25)$$

Termination

$$P \cdot + P \cdot \xrightarrow{k_t} P \quad (7.26)$$

where R indicates an initiating species, probably a free radical; P, a polymer; M, a monomer; and P·, a polymer radical. Using a steady-state assumption, the initial rate of monomer production may be expressed as:

$$dM/dt = k_p(R_i/2k_t)^{1/2}$$

This is obviously a greatly simplified view. We have neglected both zip length and molecular chain length. Similarly, we have neglected chain scission reactions that would provide polymer radicals. Chain end initiation is also possible as are many transfer reactions.

A detailed analysis of poly(methyl methacrylate) degradation has shown that at temperatures less than 270°C the zip length is long and initiation is predominantly at chain ends. At higher temperature random initiation becomes more important.

A number of other methacrylates decompose in this fashion as might be expected from their structure. However, some degrade by olefin elimination, notably those having a beta hydrogen in the ester group and thus rightly belong in Section 7.6.3.

$$\begin{array}{c} \text{CH}_3 \\ | \\ -\text{CH}_2-\text{C}- \\ \underset{\text{C}}{|} \text{H}-\text{CH}_2 \\ \underset{\text{O}}{\diagup}\underset{\text{O}-\text{C}-\text{CH}_3}{\diagdown} \\ | \\ \text{CH}_3 \end{array} \longrightarrow \begin{array}{c} \text{CH}_3 \\ | \\ -\text{CH}_2-\text{C}- \\ | \\ \text{COOH} \end{array} + (\text{CH}_3)_2\text{C}{=}\text{CH}_2 \qquad (7.27)$$

Another important class of polymers that unzip are the polyaldehydes. Here the mechanism is probably not free radical in nature. The most widely studied polymer has been polyformaldehyde (Loan and Winslow, 1972) and here initiation can easily occur at a terminal hydroxyl group.

$$-\text{CH}_2\text{OCH}_2\text{OCH}_2\text{OH} \rightarrow -\text{CH}_2\text{OCH}_2\text{OH} + \text{HCHO} \qquad (7.28)$$

The commercial polymers are stabilized either by careful esterification or etherification of the hydroxyls or by incorporation of blocking groups, such as ethylene oxide units in the chain.

Polystyrene also unzips to some extent during degradation, although here only about 40% of the original polymer is converted to monomer. Poly(α-methylstyrene) is converted almost completely to monomer probably because it has inaccessible secondary hydrogens, an absence of tertiary hydrogen atoms, a less reactive radical chain end, and a lower heat of polymerization.

7.6.2. RANDOM SCISSION

The considerations are quite different in the case of polyethylene. Here

$$-\text{CH}_2-\text{CH}_2-\text{CH}_2-\text{CH}_2\cdot$$

the polymer radical has a structure that is both highly reactive and is surrounded by an abundance of secondary hydrogen. Transfer reactions are therefore very important as shown by the major decomposition products that are best explained by an intramolecular "backbiting" mechanism (Tsuchiya and Sumi, 1968) previously described in Chapters 2 and 3.

$$
\begin{array}{ccc}
 & \text{CH}_2\!-\!\text{CH}_2 & \\
& \diagup \qquad \diagdown & \\
-\text{CH}_2\text{CH}_2\overset{|}{\text{CH}} & \qquad \text{CH}_2 & \longrightarrow \\
& \diagdown \quad \diagup & \\
\text{H} & \cdot\text{CH}_2 &
\end{array}
\qquad
\begin{array}{c}
\text{CH}_2\!-\!\text{CH}_2 \\
\diagup \qquad \diagdown \\
-\text{CH}_2\text{CH}_2\overset{|}{\text{CH}} \qquad \text{CH}_2 \\
\diagup \\
\text{CH}_3
\end{array}
\qquad (7.29)
$$

$$-\text{CH}_2\text{CH}_2\text{CH}\!=\!\text{CH}_2 + \text{CH}_3\text{CH}_2\text{CH}_2\cdot \qquad\qquad -\text{CH}_2\cdot + \text{CH}_2\!=\!\text{CHCH}_2\text{CH}_2\text{CH}_2\text{CH}_3$$

7.6.3. UNZIPPING OF SUBSTITUENT GROUPS

This is, perhaps, the most important thermal degradation mechanism since it is the primary breakdown process for poly(vinyl chloride) (PVC). Like all thermoplastics, PVC needs to be processed at about $200°\text{C}$ and under these conditions it loses hydrogen chloride quite rapidly and is transformed into a deeply colored polyene polymer.

The overall chemical reaction may be written as

$$-\text{CH}_2\text{CHClCH}_2\text{CHCl}- \overset{\Delta}{\longrightarrow} -\text{CH}\!=\!\text{CH}\!-\!\text{CH}\!=\!\text{CH}- + 2\text{HCl} \qquad (7.30)$$

Although this appears quite simple and straightforward, the reaction is in reality anything but simple. After some twenty years of study, the mechanism is still in doubt though certain aspects of it are now universally accepted. The loss of hydrogen chloride is catalyzed by free hydrogen chloride and, in industrial situations, this catalytic effect is reinforced by the presence of metal chlorides such as ferric chloride formed by the reaction of hydrogen chloride with processing equipment. Many of these chlorides are also strong catalysts of dehydrochlorination. This catalysis is often suppressed by the incorporation of a few percent of an acid absorber.

One of the early mechanisms was proposed by Winkler (Geddes, 1967). It is a free radical reaction involving a chlorine atom as a chain carrier

$$-\overset{\cdot}{\text{C}}\text{H}\!-\!\text{CHCl}\!-\!\text{CH}_2\!-\!\text{CHCl} \longrightarrow -\text{CH}\!=\!\text{CH}\!-\!\text{CH}_2\!-\!\text{CHCl}- + \text{Cl}\cdot \quad (7.31)$$

$$\downarrow$$

$$-\text{CH}\!=\!\text{CH}\!-\!\overset{\cdot}{\text{C}}\text{H}\!-\!\text{CHCl}- + \text{HCl}$$

Although often cited, it is not widely accepted and it seems certain that at least some degradation also results from ionic mechanisms that would account for the catalytic effect of hydrogen chloride and Lewis acids.

One very interesting feature of the degradation is that PVC appears to be far less stable than would be expected from a study of low molecular weight model compounds. This has often been attributed to the presence of unstable groups in the polymer which initiate the dehydrochlorination. Such groups could obviously arise from irregularities in the polymerization reaction and many attempts have been made to identify possible labile groups. This search is handicapped by the very low concentration of unstable groups; however, recent work using high resolution NMR has shown very encouraging results in that chloromethyl groups have been detected (Bovey *et al.*, 1975) and strong indications of the presence of tertiary and other labile chlorine atoms have been obtained.

In addition to the simple approach to stabilization, reagents that are believed to react with the labile initiating groups have also been used as stabilizers. The most important of these are perhaps barium and cadmium carboxylates and organotin compounds. These materials are all believed to work by exchanging carboxylate or mercaptide groups with labile chlorine and the detailed mechanism has been proposed by Frye and co-workers (Geddes, 1967; Loan and Winslow, 1972; Ayrey *et al.*, 1974). Recent work has supported their general ideas and shows that only very small amounts of such labile centers are present. The thermal stability of poly(vinyl chloride) is compared to that of other polymers in Table 7.3.

7.6.4. STRUCTURE AND THERMAL STABILITY

The pyrolytic properties of polymers depend ultimately on bond energies, some of which are listed in Table 7.2. In turn, these dissociation energies depend on resonance effects, on steric strain induced by bulky neighboring groups, and occasionally on the rigidity of their own or adjacent valence structures. When activation energies are favorable, relatively strong bonds, such as C–OH in cellulose and C–Cl in poly(vinyl chloride), can break to form products with higher dissociation energies, i.e., H_2O and HCl. The relative importance of these factors on polymer stabilities is indicated by the examples in Figs. 7.5 and 7.6. Steric strain from crowded methyl groups makes polyisobutylene less stable to heat than polyethylene. On the other hand, polytetrafluoroethylene is unusually stable and, unlike polyethylene, breaks down almost entirely to monomer because the C–F bonds are strong enough to prevent much chain transfer, even at $500°C$. A rigid cross-linked network shifts the pyrolytic decomposition of polytrivinylbenzene to a temperature range well above that of linear polystyrene and leads to some

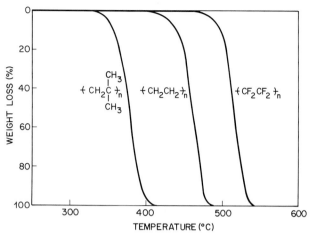

FIG. 7.5. The influence of substituent methyl groups and fluorine on the stability of polyethylene. The samples were heated at a rate of $100°C/hr$.

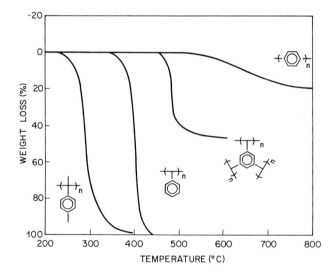

FIG. 7.6. Thermal stability of cross-linked and linear aromatic-type polymers. The samples were heated at a rate of $100°C/hr$.

carbon formation as well (Winslow and Matreyek, 1956). In comparison, diamond, the most compact nonaromatic valence network, rearranges rapidly to a more inert graphitic structure only above $1400°C$. However, the stability of diamond is unique among nonaromatic substances. Most heat-resistant polymers, other than silicones and fluorocarbons, have wholly aromatic chains like the poly(p-phenylene) in Fig. 7.6

7.6.5. HEAT-RESISTANT POLYMERS

Few polymers have both the chemical stability and the high melting or glass temperatures necessary for withstanding long-term exposure to air at temperatures above 200°C. Those that can meet these rigorous requirements usually consist of linear chains of recurring aromatic or heterocyclic units linked together by ether, ester, amide or sulfone groups. Rigid benzenoid rings stiffen chains and raise transition temperatures, whereas ether and other flexible links improve mechanical properties and processibility. The relation of these structural features to the thermal properties of some typical polymers is shown in Table 7.4. Note that melting points increase with the concentration of phenylene units and especially with the frequency of the more symmetrical p-phenylene groups. For instance, flexible thioether links make poly(phenylene sulfide) readily moldable above 300°C while the complete absence of flexible links is responsible for poly(p-phenylene) being intractable and infusible at all temperatures.

TABLE 7.4
POLYMER STRUCTURE AND TRANSITION TEMPERATURES

	Repeat unit	T_g (°C)	T_m (°C)	Reference
1	$[-NH(CH_2)_6NHC(CH_2)_4C-]_n$ (two C=O)	80	265	(Black, 1970)
2	$[-NH—\bigcirc—NHC—\bigcirc—C-]_n$ (two C=O)	273	~380	(Black, 1970)
3	$[-NH—\bigcirc—NHC—\bigcirc—C-]_n$ (two C=O)		~500	(Black, 1970)
4	$[-\bigcirc-S-]_n$	85	285	(Schooler, 1976)
5	$[-\bigcirc-]_n$		>500 dec.	(Stille and Gilliams, 1971)

From the outset, processing problems have been a major concern in research on high temperature polymers. Polypyromellitimides, the early standard for commercial high temperature polymers, are most conveniently made by a two-step procedure involving a soluble intermediate suitable for forming films.

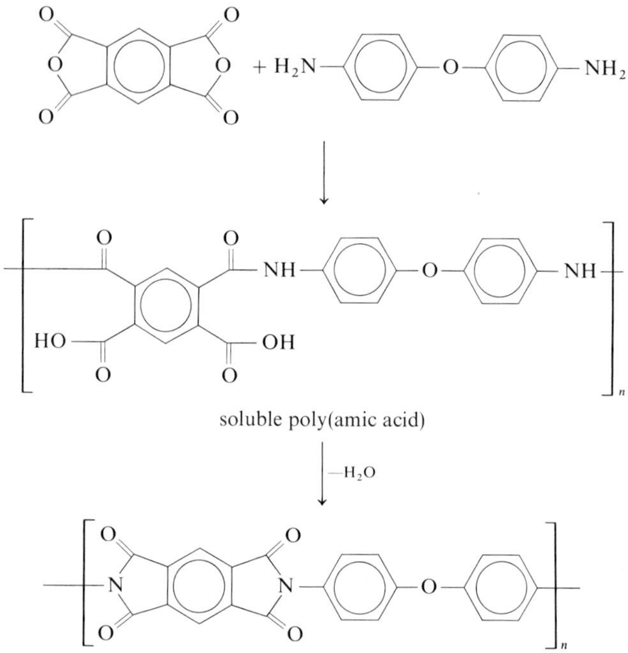

soluble poly(amic acid)

insoluble polyimide

Bis(4-aminophenyl) ether and pyromellitic dianhydride react at room temperature in dimethylacetamide to form a soluble precursor that loses water on heating to produce the aromatic polyimide (Sroog, 1976). Dehydration is never complete. Films of the polymer retain flexibility in air for more than one year at $275°C$ but gradually become brittle above $400°C$. Thermal stabilities of this polyimide and similar polymers are compared in Fig. 7.7. A related polymer, devoid of hydrogen, is far more resistant to oxidation. The

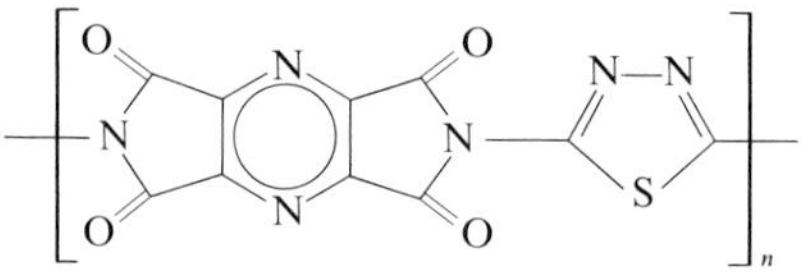

polymer is remarkably stable in air at temperatures up to $600°C$ (Hirsch, 1969). These wholly aromatic polyimides are highly flame resistant, converting to chars with only moderate weight loss.

Polyaramides have perhaps the most unusual and interesting properties. The rodlike chains of poly(1,4-benzamide) form liquid crystal solutions above some minimum molecular weight and critical concentration. The polymer can be prepared from p-aminobenzoyl chloride in a dialkylamide

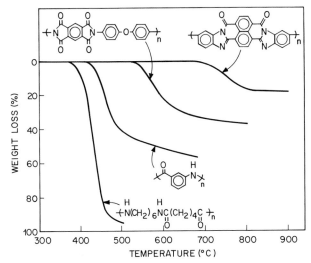

FIG. 7.7. Thermal stabilities of various polyimides.

solvent at temperatures below 30°C. It has been reported (Kwolek *et al.*, 1976) that fibers with a tensile strength of 20 gm/denier (37×10^4 psi) have been made by dry spinning the anisotropic phase in a tetramethylurea–LiCl solution of the polymer. The fibers are actually stronger than steel on a weight basis.

7.6.6. LADDER AND PYROLYTIC POLYMERS

Double-stranded polymers should be more stable than their corresponding single-strand structures since complete rupture of a ladder molecule requires at least two bond breaks per ring (Tessler, 1966), But observed gains in thermal stability have been rather moderate, possibly because a bond break sharply reduces the dissociation energies of adjacent bonds (see Table 7.2) and thereby lessens the likelihood of cage recombination of the radical pair. The postulate obviously does not apply to acenes, such as naphthalene and its higher homologs whose stability varies inversely with annelation or the number of fused rings.

A *cis*-syndiotactic poly(phenylsilsesquioxane) (p. 440, top) has been synthesized (Brown, 1963) and seems to have a wormlike conformation in benzene or ethylene dichloride solution. Silicones are somewhat susceptible to hydrolysis and the single-strand derivatives break down at high temperatures to volatile cyclic products. The rodlike ladder polymer (BBL), derived from 1,2,4,5-tetraaminobenzene and 1,4,5,8-naphthalene tetracarboxylic acid, dissolves with difficulty in strongly protonating solvents and rapidly forms disordered aggregates from solution (Arnold and Van Dusen, 1969).

A few spiropolymers have been described (Bailey and Volpe, 1967) but spiro structures are far more stable and occur more frequently in inorganic than in organic substances.

Although the carbonization of polymers has been studied extensively, reaction mechanisms are not well understood yet. Above 200°C poly-(vinylidene chloride) loses hydrogen chloride to yield a polyene, which then gradually converts to a rigid aromatic network, presumably by Diels–Alder reactions that lead to elimination of more hydrogen chloride. Carbon fibers are made from poly(acrylonitrile) by a two-step process consisting of low temperature preoxidation followed by pyrolysis. It has been postulated that short ladder sequences of heterocyclic rings are formed during the oxidative step. However, since the polymer contains 10–20% oxygen after pretreatment, the expression customarily used to describe the process represents an over-simplification.

Watt (1972) produced carbon fibers with exceptionally high tensile strength ($> 10^5$ psi) by oxidizing the polymer under stress. Orientation in the original fiber seemed to be retained in its carbon replica. The light-weight, heat-resistant material meets the requirements for use in rocket nozzles and heat shields.

7.7   Flammability

In an age of increasing usage of polymers, it is of extreme importance that we understand the way in which they burn. Such knowledge should provide guidance towards more efficient flame-retardant techniques.

The processes involved in a burning polymer must include thermal degradation of the polymer to volatile, flammable fragments and the heat for this decomposition is obviously provided by the flame. We can outline the burning process as follows:

$$
\text{polymer} \xrightarrow{\qquad} \text{flammable products} \xrightarrow{\ O_2\ } \text{flame} \xrightarrow{\qquad} CO_2, H_2O, \text{other products}
$$

with the feedback labelled "heat" returning to the polymer step.

Although this is only a general picture of the overall burning process, it is useful in pinpointing steps where flame retardants might conceivably interfere. Two major possibilities exist, the first in the condensed phase and the second in the gaseous flame.

Before going into more detail on burning and flame retardancy, it is perhaps convenient to introduce a popular test used to measure oxygen index. The procedure measures the lowest oxygen concentration necessary for continued burning. A polymer sample of standard size is mounted in a vertical glass tube up which passes a slow stream of gas. The gas is an oxygen–nitrogen mixture in which oxygen concentration may be varied. The limiting *oxygen index* is the minimum oxygen concentration at which steady combustion will take place. This index may thus be used as a rough measure of the relative flammability of different polymers, some of which are listed in Table 7.3. The oxygen index may also be used to measure the effect of various additives and may thus be used in mechanistic studies.

One of the easier distinctions may be made between condensed phase and gas phase inhibition. If oxygen index measurements are made using the alternative gas phase compositions, oxygen–nitrogen and nitrous oxide–nitrogen, and similar results are seen for a given variation, condensed phase inhibition is probably involved. If differences are seen, flame inhibition may be occurring. Experiments of this type indicate that bromine and antimony oxide–halogen systems function by a flame inhibition process whereas fillers such as aluminium oxide trihydrate act in the condensed phase. Some fillers merely physically cool the polymer, whereas the others eliminate some nonflammable molecule in an endothermic reaction and cool the polymer in this way. The high oxygen index of poly(vinyl chloride) is probably partly due to the endothermic loss of hydrogen chloride but other reactions also may be involved.

The field of polymer flammability is still very active and our knowledge of it quite limited. A recent review (Warren, 1972) surveys the field thoroughly and is recommended for further study.

REFERENCES

Alfrey, T. Jr. (1964). "Chemical Reactions of Polymers" (E. M. Fettes, ed.), pp 1–9. Wiley (Interscience), New York.

Arnold, F. E., and Van Deusen, R. L. (1969). *Macromolecules* 2, 497.

Ayrey, G., Head, B. C., and Poller, R. C. (1974). *J. Polym. Sci. Macromol. Rev.* 9, 1.

Bailey, W. J., and Volpe, A. A. (1967). *Polym. Prepr.* 8, 292–299.

Bair, H. E., Matsuo, M., Salmon, W. A., and Kwei, T. K. (1972). *Macromolecules* 5, 114.

Baker, W. O. (1949). *Ind. Eng. Chem.* 41, 511.

Bamford, C. H., and Tipper, C. F. H. (eds.) (1975). "Degradation of Polymers." Elsevier, Amsterdam.

Bateman, L., Moore, C. G., Porter, M., and Saville, B. (1963). "The Chemistry and Physics of Rubberlike Substances" (L. A. Bateman, ed.), Chapter 15. McLaren, London.

Black, W. B. (1970). *Trans. N.Y. Acad. Sci.* 32, 765.

Blow, C. M. (1971). "Rubber Technology and Manufacture," Institution of the Rubber Industry, London.

Bovey, F. A. (1958). "The Effects of Ionizing Radiation on Natural and Synthetic High Polymers." Wiley (Interscience), New York.

Bovey, F. A., Abbås, K. B., Schilling, F. C., and Starnes, W. H. Jr. (1975). *Macromolecules* 8, 437.

Bowden, M. J. (1977). Critical Reviews in Solid State Science (in press).

Bristow, G. M., and Watson, W. F. (1963). *In* "The Chemistry and Physics of Rubberlike Substances" (L. Bateman, ed.), Chapter 14. McLaren, London.

Brown, J. F. Jr. (1963). *J. Polym. Sci. Part C* 1, 83.

Burgess, A. R. (1953). *Nat. Bur. Std. U.S. Circ.* 525, 149.

Chapiro, A. (1962). "Radiation Chemistry of Polymeric Systems." Wiley (Interscience), New York.

Charlesby, A. (1960). "Atomic Radiation and Polymers." Pergamon, Oxford.

Cohen, S. M., Young, R. H., and Markhart, A. H. (1971). *J. Polym. Sci. Part A-1* 9, 3263.

Dole, M. (ed.) (1973). "The Radiation Chemistry of Macromolecules." Academic Press, New York.

Flory, P. J. (1939). *J. Am. Chem. Soc.* 61, 1518.

Flory, P. J. (1953). "Principles of Polymer Chemistry." Cornell Univ. Press, Ithaca, New York.

Frye, A. H., and Horst, R. W. (1959). *J. Polym. Sci.* 40, 419.

Geddes, W. C. (1967). *Rubber Chem. Technol.* 40, 177.

Grassie, N. (1956). "Chemistry of High Polymer Degradation Processes." Butterworths, London.

Grassie, N., and Weir, N. A. (1965). *J. Appl. Polym. Sci.* 9, 999.

Gross, S. (ed.) (1975). "Modern Plastics Encyclopedia," Vol. 52. McGraw-Hill, New York.

Hansen, R. H., Russell, C. A., DeBenedictis, T., Martin, W. M., and Pascale, J. V. (1964). *J. Polym. Sci. Part A* 2, 587–609.

Hartley, G. H., and Guillet, J. E. (1968). *Macromolecules* 1, 165–170.

Hawkins, W. L. (ed.) (1972). "Polymer Stabilization." Wiley (Interscience), New York.

Hirsch, S. S. (1969). *J. Polym. Sci. Part A-1* **7**, 15–22.

Humphrey, J. S. Jr., Shultz, A. R., and Jaquiss, D. B. G. (1973). *Macromolecules* **6**, 305–314.

Kauzmann, W., and Eyring, H. (1940). *J. Am. Chem. Soc.* **62**, 3113.

Kwolek, S. L., Morgan, P. W., Schaefgen, J. R., and Gulrich, L. W. (1976). *Polym. Prepr.* **17**, 53–58.

Loan, L. D. (1967). *Rubber Chem. Technol.* **40**, 149.

Loan, L. D., and Winslow, F. H. (1972). *In* "Polymer Stabilization" (W. L. Hawkins, ed.), Chapter 3. Wiley (Interscience), New York.

Mark, J. E. (1975). *Rubber Chem. Technol.* **48**, 495.

Marvel, C. S., Sample, J. H., and Roy, M. F. (1939). *J. Am. Chem. Soc.* **61**, 3241.

Morawetz, H. (1964). *In* "Chemical Reactions of Polymers" (T. Alfrey, Jr., ed.), Chapter 1. Wiley (Interscience), New York.

Morrell, S. H. (1971). "Rubber Technology and Manufacture" (C. M. Blow, ed.). Butterworths, London.

Mullins, L., and Thomas, A. G. (1973). *In* "The Chemistry and Physics of Rubberlike Substances" (L. Bateman, ed.), Chapter 7. McLaren, London.

Rånby, B., and Rabek, J. F. (1975). "Photodegradation, Photo-oxidation and Photostabilization of Polymers." Wiley (Interscience), New York.

Salmon, W. A., and Loan, L. D. (1972). *J. Appl. Polym. Sci.* **16**, 671.

Schooler, G. N. (1976). "Modern Plastics Encyclopedia, 1976–1977," p. 81. McGraw-Hill, New York.

Sroog, C. E. (1976). *J. Polym. Sci. Macromol. Rev.* **11**, 161–208.

Stille, J. K., and Gilliams, Y. (1971). *Macromolecules* **4**, 515–517.

Tessler, M. M. (1966). *J. Polym. Sci. Part A-1* **4**, 2521.

Treloar, L. R. G. (1974). *Rubber Chem. Technol.* **47**, 625.

Treloar, L. R. G. (1975). "Physics of Rubber Elasticity." Oxford Univ. Press (Clarendon), London and New York.

Trozzolo, A. M., and Winslow, F. H. (1968). *Macromolecules* **1**, 98–100.

Tsuchiya, Y., and Sumi, K. *J. Polym. Sci. Part B* **6**, 359.

Warren, P. C. (1972). *In* "Polymer Stablization" (W. L. Hawkins, ed.), Chapter 7. Wiley (Interscience), New York.

Watt, W. (1972). *Carbon* **10**, 121.

Winslow, F. H., and Matreyek, W. (1956). *J. Polym. Sci.* **23**, 315–324.

Winslow, F. H. (1977). *Pure Appl. Chem.* **49**, 495.

Wood, L. A. (1976). *J. Res. Nat. Bur. Std.* **80A**, 451.

Chapter 8

BIOLOGICAL MACROMOLECULES

F. A. Bovey

8.1. Introduction

In this chapter we shall discuss the structure and properties of the major classes of polymers occurring in nature. Knowledge of the structural details and functions of these macromolecules has grown at such a stupendous pace in the last two decades that our treatment is necessarily very incomplete. A selected list of general references is given at the end of the chapter for readers who wish a more complete discussion of the topics discussed here. We shall include in our discussion not only the naturally occurring materials, but also synthetic macromolecules and model compounds, which are of interest in themselves and have been of great importance in understanding the behavior of their more complex natural analogs.

445

8.2. Polypeptides

Polypeptide chains may be regarded as formed by the condensation of α-amino carboxylic acids with splitting out of water (8.1). This process can

$$
\underset{\text{NH}_2\overset{|}{\text{C}}\text{HCO}_2\text{H}}{\overset{R_1}{|}} + \underset{\text{NH}_2\overset{|}{\text{C}}\text{HCO}_2\text{H}}{\overset{R_2}{|}} \longrightarrow \underset{\text{NH}_2\overset{|}{\text{C}}\text{HCONH}\overset{|}{\text{C}}\text{HCO}_2\text{H}}{\overset{R_1 \qquad R_2}{|\qquad\quad|}} + \text{H}_2\text{O} \tag{8.1}
$$

continue in the same manner as the polycondensations discussed in Chapter 2, Section 2.7, yielding long chains. Although polypeptides[†] can be prepared in this way (and are thought by some to have been so formed on the primordial earth), the method is generally not a practical one owing to competing side reactions of decarboxylation, deamination, and cyclization. The most common laboratory synthesis is carried out by the N-carboxy anhydride process, first discovered by Leuchs (1906) and later developed by Katchalski and his co-workers (1964). N-carboxy anhydrides (frequently abbreviated NCA) are highly reactive substances. Their polymerization is initiated by bases such as amines, sodium methoxide, or sodium hydroxide in organic solvents and is accompanied by release of CO_2; in Eq. (8.2), the basic catalyst is represented by $B:^{\ominus}$. Primary and secondary amides yield polymers with rather low degrees of polymerization, 10–200; tertiary amines and sodium methoxide are preferred for the preparation of material with higher degrees

$$
\begin{array}{c}
\text{R}-\text{CH}-\text{C} \overset{O}{\underset{O}{\diagdown}} \\
\mid \qquad\qquad\quad + \text{B:}^{\ominus} \\
\text{NH}-\text{C} \overset{}{\underset{O}{\diagdown}}
\end{array}
\longrightarrow
\left[
\begin{array}{c}
\overset{B}{\overset{|}{}} \\
\text{R}-\text{CH}-\text{C}-\text{O:}^{\ominus} \\
\mid \qquad\qquad\diagdown O \\
\text{NH}-\text{C} \overset{}{\underset{O}{\diagup}}
\end{array}
\right]
\longrightarrow
\begin{array}{c}
\overset{B}{\overset{|}{}} \\
\text{R}-\text{CH}-\text{C}=\text{O} + \text{CO}_2 \\
\overset{|}{\text{}} \\
^{\ominus}\text{:NH}
\end{array}
\tag{8.2}
$$

$$
\begin{array}{c}
\overset{B}{\overset{|}{}} \\
\text{R}-\text{CH}-\text{C}=\text{O} \\
\overset{|}{\text{}} \\
^{\ominus}\text{:NH}
\end{array}
+ n
\begin{array}{c}
\text{R}-\text{CH}-\text{C} \overset{O}{\underset{O}{\diagdown}} \\
\mid \qquad\qquad \\
\text{NH}-\text{C} \overset{}{\underset{O}{\diagdown}}
\end{array}
\longrightarrow
\tag{8.3}
$$

$$
\underset{\text{B}-\text{C}-\text{CHNH}}{\overset{O \quad R}{\overset{\parallel \quad |}{}}} \left[\underset{\text{C}-\text{CH}-\text{NH}}{\overset{O \quad R}{\overset{\parallel \quad |}{}}} \right]_n \underset{\text{C}-\text{CHNH:}^{\ominus}}{\overset{O \quad R}{\overset{\parallel \quad |}{}}} + n\text{CO}_2
$$

of polymerization, 200–10,000, corresponding to molecular weights of 20,000–1,000,000. Support for the mechanism shown is provided by the observation that the initiator is consumed and is found at the ends of the chains. The growing chains do not react with each other and, in the absence of termination reactions, would constitute a "living polymer" system with a

[†] Some authors refer to the synthetic materials as *polyamino acids*, reserving the term polypeptides for naturally occurring substances. We shall not make this distinction here.

correspondingly narrow molecular weight distribution determined by the ratio of monomer to initiator (M/I) (Chapter 2, Section 2.2.4). However, the molecular weight distributions are in fact often very broad; the average degree of polymerization may exceed M/I; and the molecular weight does not necessarily increase in proportion to the reaction time. The kinetics are likewise not simple, showing an autocatalytic behavior. It is therefore evident that the reaction is not a simple one, but the complexities are such that we shall not review them here; the reader is referred to the review by Szwarc (1965). Regardless of these complications, the method is one of primary utility for preparation of homopolypeptides and random copolypeptides of high molecular weight.

For the step-by-step synthesis of specific peptides and polypeptides, the use of blocking or protective groups for the amino function and of activating groups for the carboxyls is of great importance. Common amino blocking groups are the carbobenzoxy group (abbreviated Z), which may be removed by catalytic hydrogenation, and the *tert*-butyloxycarbonyl (*t*-Boc) group.

$$\underset{NH_2-\overset{\overset{\displaystyle R}{|}}{C}H-CO_2H}{} \xrightarrow{C_6H_5CH_2OCOCl} \underset{C_6H_5CH_2OCONH-\overset{\overset{\displaystyle R}{|}}{C}H-CO_2H}{}$$

$$\underset{NH_2-\overset{\overset{\displaystyle R}{|}}{C}H-CO_2H}{} \xrightarrow[\text{(azide)}]{t\text{-BuOCON}_3} \underset{t\text{-BuOCONH}-\overset{\overset{\displaystyle R}{|}}{C}H-CO_2H}{}$$

The carboxyl group may be blocked by esterification, methyl or benzyl esters being common, or for synthetic purposes by a more reactive ester group such as phenyl. The amino-blocked, activated ester may then be allowed to react with another amino acid or peptide having a free amino group at the amino terminal:

$$t\text{-BuOCONH}-\overset{\overset{\displaystyle R_1}{|}}{C}HCO_2Ph \;+\; NH_2-\overset{\overset{\displaystyle R_2}{|}}{C}H-CO\cdots \longrightarrow$$

$$t\text{-BuOCONH}-\overset{\overset{\displaystyle R_1}{|}}{C}H-CONH-\overset{\overset{\displaystyle R_2}{|}}{C}H-CO\cdots$$

The *t*-Boc group can then be removed by acid hydrolysis. In the absence of the amino-protecting group, the first amino acid derivative would react with itself. Such blocking groups, if allowed to remain, enhance the solubility of the peptides in organic solvents. The "deblocked" charged groups tend to enhance solubility in aqueous solvents.

A convenient class of reagents are the *carbodiimides*, which can form the peptide bond directly by removal of the elements of water:

$$\cdots NH-\overset{\overset{\displaystyle R_1}{|}}{C}H-CO_2H + NH_2-\overset{\overset{\displaystyle R_2}{|}}{C}H-CO_2CH_3 \xrightarrow[\text{(dicyclohexylcarbodiimide)}]{C_6H_{11}N=C=NC_6H_{11}}$$

$$\cdots NH-\overset{\overset{\displaystyle R_1}{|}}{C}H-CONH-\overset{\overset{\displaystyle R_2}{|}}{C}H-CO_2CH_3 + C_6H_{11}NHCONHC_6H_{11}$$

During all such synthetic reactions, functional groups in the side chains, such as amino or carboxyl, must also be blocked and the blocking groups later removed.

One of the features of these reactions that makes the synthesis of long polypeptides arduous is the necessity of isolating and purifying the product of each step. This can be obviated and the process greatly speeded up by *solid-state synthesis* (Merrifield, 1969). In this method, the polypeptide is assembled stepwise while one end of it is attached to an insoluble solid support. The polypeptide is thus itself insolubilized and may be purified at each step by relatively rapid processes of filtration and washing. The solid phase is a cross-linked styrene–divinylbenzene gel in the form of beads that are chloromethylated. The chloromethyl group is then reacted with a *t*-Boc-amino acid as the initial step. The last step terminates the synthesis at the dipeptide stage, but it is possible to carry it forward for many scores of steps, producing long biologically active peptides and even proteins (ribonuclease).

8.2.1. CHAIN STRUCTURE

The side chain R in polypeptides may be varied at will, but we shall confine our attention principally to the amino acids that occur in proteins. These chain units, together with the name of the amino acid and the standard three-letter abbreviation, are listed in Table 8.1. Although the D chirality is

TABLE 8.1
THE COMMON AMINO ACID CHAIN UNITS

Name	Symbol	Structure
Alanine	Ala	CH_3 | $—NHCH_\alpha CO—$
Arginine	Arg	$HN{=}C{\diagup}^{NH_2}_{\diagdown NH}$ | $CH_{2\delta}$ $CH_{2\gamma}$ | $CH_{2\beta}$ | $—NHCH_\alpha CO—$
Asparagine	Asn	$CONH_2$ | $CH_{2\beta}$ | $—NHCH_\alpha CO—$
Aspartic acid	Asp	CO_2H | $CH_{2\beta}$ | $—NHCH_\alpha CO—$
Cysteine	Cys	SH | $CH_{2\beta}$ | $—NHCH_\alpha CO—$
Cystine	Cys	$S{\bigg]}_2$ | $CH_{2\beta}$ | $—NHCH_\alpha CO—$
Glutamic acid	Glu	CO_2H | $CH_{2\gamma}$ | $CH_{2\beta}$ | $—NHCH_\alpha CO—$
Glutamine	Gln	$CONH_2$ | $CH_{2\gamma}$ | $CH_{2\beta}$ | $—NHCH_\alpha CO—$
Glycine	Gly	$—NHCH_{2\alpha}CO—$

TABLE 8.1 (*Continued*)

Name	Symbol	Structure

Hydroxyproline — Hyp

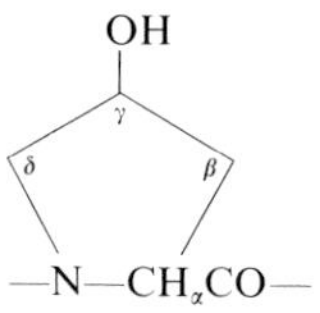

$$OH$$

(ring structure with positions γ, δ, β)

$$-N-CH_2CO-$$

Histidine — His

(imidazole ring, positions 4, 5, 3, 1, 2, N, NH)

$$CH_{2\beta}$$
$$-NHCH_aCO-$$

Isoleucine — Ile

$$CH_{3\gamma'} \qquad CH_{2\gamma}CH_{3\delta}$$
$$CH_\beta$$
$$-NHCH_aCO-$$

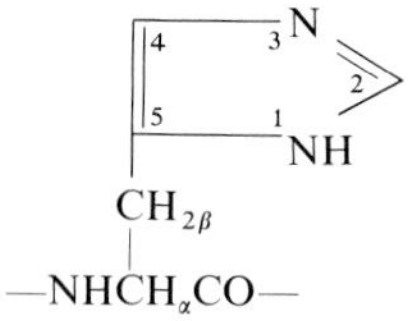

Leucine — Leu

$$CH_{3\delta'} \qquad CH_{3\delta}$$
$$CH_\gamma$$
$$CH_{2\beta}$$
$$-NHCH_aCO-$$

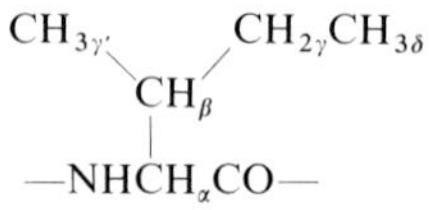

Lysine — Lys

$$CH_{2\gamma}CH_{2\delta}CH_\varepsilon NH_2$$
$$CH_{2\beta}$$
$$-NHCH_aCO-$$

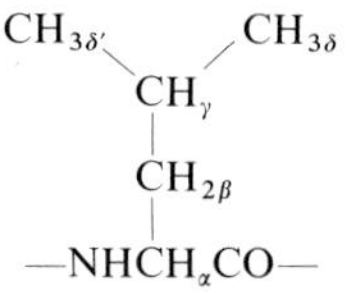

Methionine — Met

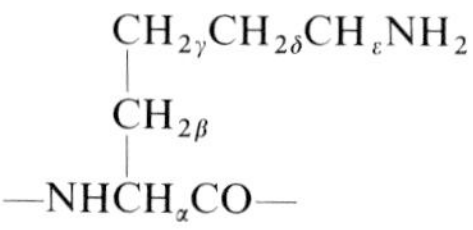

$$CH_3$$
$$S$$
$$CH_{2\gamma}$$
$$CH_{2\beta}$$
$$-NHCH_aCO-$$

Ornithine — Orn

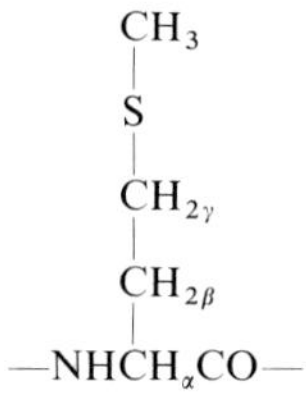

$$CH_{2\gamma}CH_{2\delta}NH_2$$
$$CH_{2\beta}$$
$$-NHCH_{2a}CO-$$

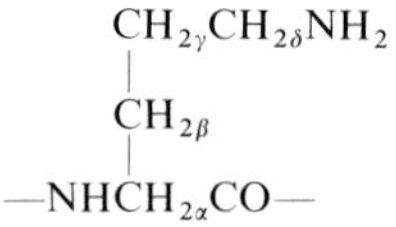

Phenylalanine — Phe

(benzene ring, positions ζ, ε, δ, γ)

$$CH_{2\beta}$$
$$-NHCH_aCO-$$

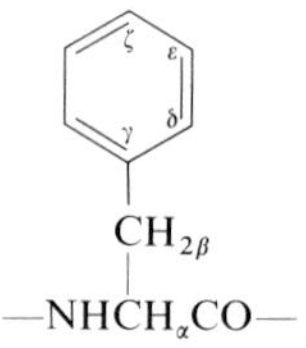

TABLE 8.1 (*Continued*)

Name	Symbol	Structure
Proline	Pro	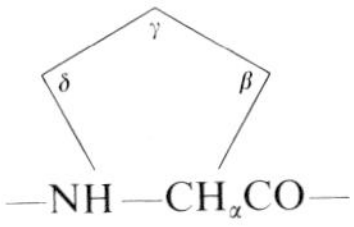 $-NH-CH_\alpha CO-$
Serine	Ser	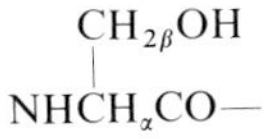 $CH_{2\beta}OH$ $\|$ $NHCH_\alpha CO-$
Threonine	Thr	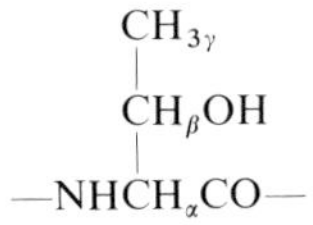 $CH_{3\gamma}$ $\|$ $CH_\beta OH$ $\|$ $-NHCH_\alpha CO-$
Tryptophan	Trp	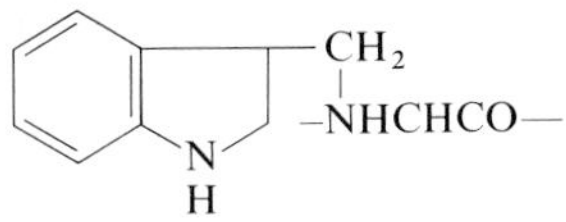 CH_2 $-NHCHCO-$
Tyrosine	Tyr	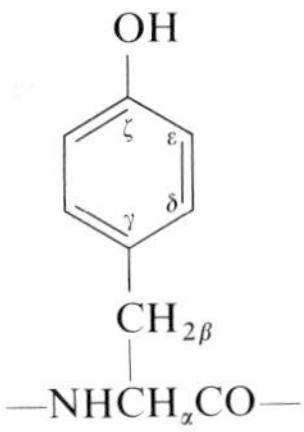 OH $CH_{2\beta}$ $\|$ $-NHCH_\alpha CO-$
Valine	Val	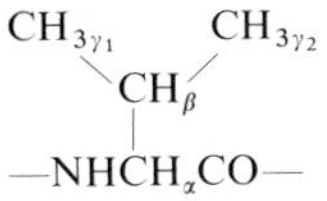 $CH_{3\gamma_1}$ $CH_{3\gamma_2}$ CH_β $\|$ $-NHCH_\alpha CO-$

not unknown among naturally occurring amino acids, the L enantiomer overwhelmingly predominates:

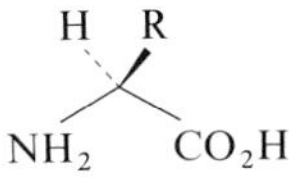

(Threonine has two asymmetric centers, the α and the β carbons; the configuration at the latter is L. The isomer with the D configuration at C_β is called allothreonine.)

If the polyamino acid chain has been prepared by synthetic means, the primary structure is obviously known, with the exception of possible errors in a long chain preparation by the solid-state method. For chains of natural origin, the amino acid sequence must be determined. We shall consider methods for doing this in Section 8.3.1.

Polypeptide chains have a directional sense, and this is by convention taken as running from left to right with the amino end or "N-terminus" represented at the left and the carboxyl or "C-terminus" at the right. Protein and polypeptide residues are numbered in this direction. It perhaps goes without saying that the direction of the sequence has a marked effect on the properties of the polypeptide. Thus, for example, Tyr-Gly-Pro-Phe is a quite different substance from Phe-Pro-Gly-Tyr. For peptide chains that play a biological role such differences are particularly critical.

8.2.2. CONFORMATION OF POLYPEPTIDES

The conformation of a polypeptide chain is determined if the three rotational angles ϕ, ψ, and ω are specified for each amino acid residue:

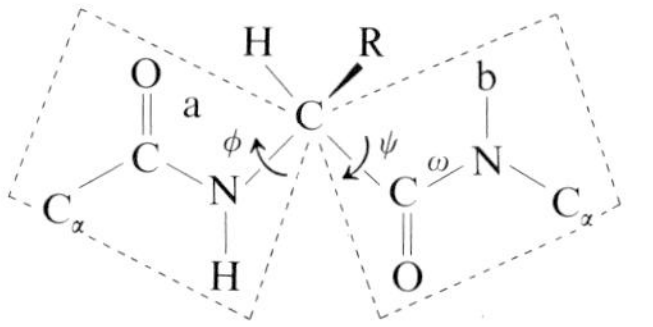

The amino acid residue is pictured here as part of a chain in a planar zigzag conformation. The ϕ and ψ angles are defined as increasing in the sense indicated by the arrows. The peptide groups are planar, and so changes in ϕ and ψ can be viewed as describing rotations of the planar units a and b about the α-carbon as a common swivel point. The angle ω describes the state of the peptide bond, which may be either cis or trans:

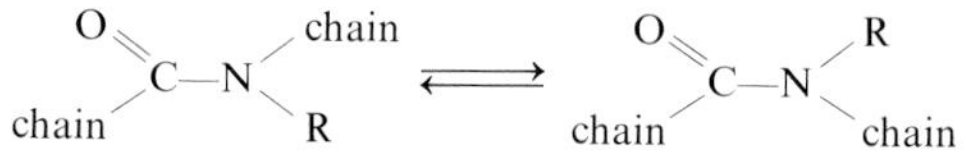

For polyamino acid chains in which R = H, the trans conformation is normally strongly preferred, as in the planar zigzag represented above. If R is an alkyl group, the energy difference between these conformations is much smaller and can depend on the solvent. Such residues are termed *imino* acids. Proline is the only common imino acid, but *sarcosine*, N-methyl glycine, occurs in certain natural peptides.

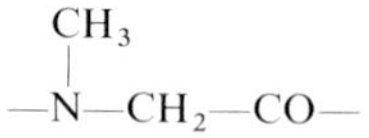

$$\mathrm{CH_3}$$
$$|$$
$$-\mathrm{N-CH_2-CO-}$$

The convention for angles of a polypeptide (Kendrew *et al.*, 1970 a–c) defines the angles of the planar zigzag as $-180°$ for ϕ and ψ and $180°$ for ω;

in a cis peptide bond, $\omega = 0°$. In older literature (prior to 1970), one finds other conventions, most commonly that in which all the angles of the planar zigzag are $0°$.

Since the planar trans conformation of the peptide bond is strongly preferred, a poly-α-amino acid chain actually has only two degrees of freedom, ϕ and ψ. In such chains, there is usually no significant steric interaction between the side chains of neighboring residues, although neighboring peptide NH protons and carbonyl oxygen atoms may (on paper) approach well within the sum of the nonbonded contact radii. One may therefore represent each type of amino acid residue by its characteristic contour map in which the steric interaction energy is plotted as a function of ϕ as abscissa and ψ as ordinate, as first pointed out by Ramachandran et al. (1963). An approximate general map is shown in Fig. 8.1. The white regions, only about 7.5% of the conformational space, are fully allowed assuming that the atoms act as hard spheres. The larger, hatched regions, about 15% of the space, are permitted if the atoms are assumed to be somewhat compressible. Since each residue of a helical conformation has the same values of ϕ and ψ, a helix may be represented by a single point on the map. (The helices and other structures indicated in Fig. 8.1 will be discussed later.)

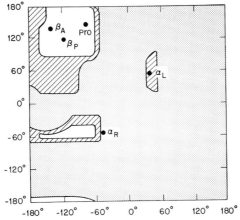

FIG. 8.1. A generalized L-amino acid energy contour diagram.

More sophisticated maps are prepared by computer summation of all pairwise atomic interactions, commonly using a "6–12" potential for the van der Waals and dispersion forces. To these are added torsional potentials; these are quite low except for the amide bond itself, for which the barrier to rotation is ~ 20 kcal. This barrier is not involved in the calculation, however,

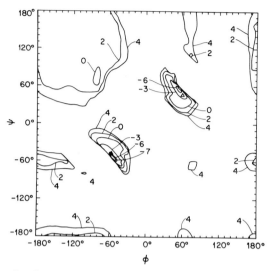

FIG. 8.2. Ramachandran map for poly-L-alanine; absolute energies are shown in kilocalories per mole. ▲, right-handed α-helix; △, left-handed α-helix (Brant, 1968).

since this bond is normally assumed to be trans. Electric dipoles or point monopole interactions must also be included, and hydrogen bonds, commonly C=O · · · NH–, if present. In Fig. 8.2 is shown such a refined map for poly-L-alanine, as calculated by Brant (1968). It will be noted that it resembles Fig. 8.1 in a general way; other amino acid maps are similar but differ in detail.

The best known poly-α-amino acid conformation, which occurs in nature as well as in synthetic polypeptides, is the *α-helix*, first deduced by Pauling *et al.* (1951) from model building and x-ray data. This structure is shown in Fig. 8.3. It has 18-fold symmetry, i.e., the main chain repeats exactly every 18 amino acid residues. There are 3.6 residues per turn and a residue translation of 0.149 nm, i.e., as we pass from one point in one residue to the

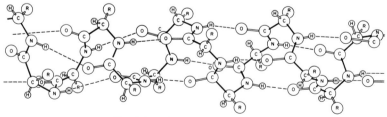

FIG. 8.3. The right-handed α-helix.

corresponding point in the next, we move 0.149 nm along the helix axis. There are hydrogen bonds from the carbonyl group of one residue to the NH of the fifth residue "ahead," i.e., in the N- to C-terminal direction. For L-amino acids, the right-handed α-helix represented in Fig. 8.3 is commonly preferred, being of somewhat lower energy than the left-handed form. In Fig. 8.2, it occupies a deep narrow cleft at $\phi \simeq -58°$, $\psi \simeq -47°$. The left-handed helix ($\phi \simeq +58°$, $\psi \simeq +47°$) corresponds to a cleft nearly as deep. The rather delicate balance of forces that dictates this chirality can be upset in certain cases, as for example in poly-β-benzyl-L-asparate

$$CO_2CH_2C_6H_5$$
$$CH_2$$
$$+NHCHCO+$$

which has a left-handed helix (Karlson *et al.*, 1960).

The values of ϕ and ψ vary somewhat for right-handed α-helices, depending upon the side chains and the environment, but are usually within $\pm 5°$ of those for poly-L-alanine (see also Table 8.2). Quite distorted α-helices occur in some proteins, however, notably in myoglobin and hemoglobin.

Other conformations of significance are *β-pleated sheet* structures, which are analogous to the hydrogen-bonded sheets of polyamides (Chapter 3, Section 3.9.6.). These may be (a) parallel ($\phi = -120°$, $\psi = 115°$) or (b) antiparallel ($\phi = -140°$, $\psi = 135°$), as shown in Fig. 8.4; their "pleated" character will be evident in the discussion of silk fibroin in Section 8.3.2.6.

The antiparallel β-conformation occurs in several cyclic structures and is the basic structure of silk (Section 8.3.2.6).

Of poly-α-imino acids, the most carefully studied is poly-L-proline. The prolyl unit can exist in both cis and trans conformations:

cis   trans

Poly-L-proline is known to exist in two distinct forms in both the solid state and in solution (Harrington and von Hippel, 1961; Carver and Blout, 1967). In water and organic acids it exists as a levorotatory (at 589 nm; see Section

456

F. A. Bovey

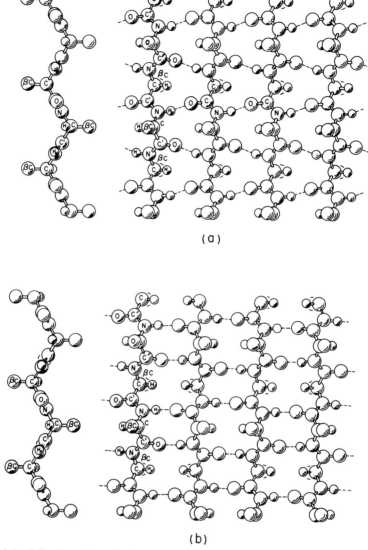

(a)

(b)

Fig. 8.4. β Conformations of polypeptide chains (a) parallel-chain pleated sheet; (b) anti-parallel-chain pleated sheet (from Pauling and Corey, 1951).

8.2.3.2) form designated as form II. As normally prepared by the polymerization of the N-carboxy anhydride and precipitation from pyridine, it exists as a dextrorotatory form called form I, stable in aliphatic alcohols but unstable in water and organic acids. Upon dissolving in these latter solvents, this form

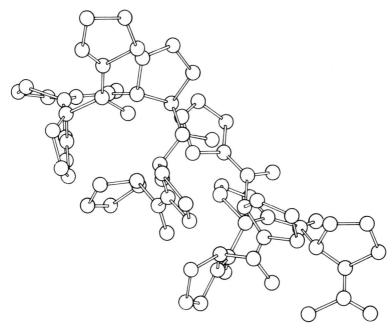

FIG. 8.5.   Poly-L-proline form I helix.

mutarotates over a period of several hours to give form II. This process may be followed by ORD and CD observations (see Section 8.2.3.2) and by NMR (Section 8.2.3.5). X-ray diffraction measurements on the crystalline polymers show that form I is a right-handed helix with a residue translation of 0.185 nm and with peptide bonds in the cis conformation (Traub and Shmueli, 1963). This structure is shown in Fig. 8.5. It has approximately tenfold symmetry and is almost as tightly wound as the α-helix. Form II is a left-handed helix with threefold symmetry and the much greater residue translation of 0.312 nm, nearly double that of an α-helix and 87% of that of a fully extended planar zigzag. The peptide bonds are in the trans conformation. The x-ray structure (Cowan and McGavin, 1955; Sasisekharan, 1959) is shown in Fig. 8.6.

Other poly-α-imino acids of interest are poly-L-γ-hydroxyproline, poly-sarcosine (poly-N-methylglycine) (a), and poly-N-methyl-L-alanine. The conformational behavior of poly-L-γ-hydroxyproline (and its acetyl derivative)

$$\left[\begin{array}{c} CH_3 \quad\quad O \\ | \quad\quad\quad || \\ N-CH_2-C \end{array} \right]_n$$

(a)

is much like that of poly-L-proline. Polysarcosine does not attain a regular

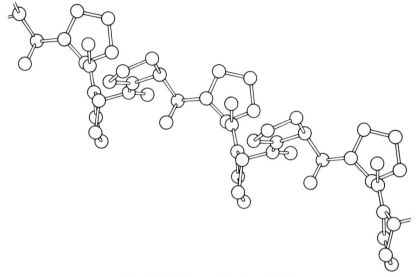

FIG. 8.6.   Poly-L-proline form II helix.

helical conformation but appears to consist of a nearly random sequence of cis and trans peptide linkages. Poly-N-methyl-L-alanine is an approximately threefold right-handed helix (Goodman and Fried, 1967) in most solvents.

In Table 8.2 are summarized the main chain angles for the polypeptide conformations we have discussed.

8.2.3.   OPTICAL AND SPECTROSCOPIC PROPERTIES OF POLYPEPTIDES

Many of the structural conclusions discussed in the previous section have been drawn from studies of the optical and spectroscopic characteristics of polypeptides in solution or as films. *Ultraviolet* and *infrared* absorption spectroscopy are widely used in the study of polypeptides. Even more important, since these polymers contain chiral centers, are (ORD) and (CD). In this section, we describe these methods and their usefulness. We shall also deal briefly with *fluorescence* methods and with the type of information that NMR can yield.

8.2.3.1.   *Ultraviolet Absorption*

Absorption in the near UV is observed for polypeptides containing the aromatic amino acids, phenylalanine, tyrosine, and tryptophan. Phenylalanine

TABLE 8.2
SOME POLY-L-AMINO ACID MAIN CHAIN ROTATIONAL ANGLES[a]

	ϕ	ψ	ω	ϕ', the H–H dihedral angle for NH—C_αH
Planar zigzag	-180	-180	180	120
α-Helix, right-handed	-58	-47	180	118
α-Helix, left-handed[b]	58	47	180	~ 0
β, Antiparallel	-140	135	180	160
β, Parallel	-120	115	180	180
Poly-L-pro I	-83	158	0	—
Poly-L-pro II	-78	149	180	—

[a] These angles (in degrees) are expressed in terms of the 1970 convention (Kendrew et al., 1970a–c). They are not absolutely exact (except for the planar zigzag) but are meant rather to be illustrative.

[b] The angles for a left-handed α-helix of D residues are the same, except that $\phi' = 118°$.

shows the weakest absorption ($\varepsilon \simeq 100$–150) with a maximum at about 250 nm; the band often shows vibrational fine structure. Tyrosine absorbs more strongly by nearly an order of magnitude, the maximum being at about 280 nm. Tryptophan shows a four- to fivefold greater extinction with a maximum also at about 280 nm, and may exhibit fluorescence (Section 8.2.3.3). The spectra of all of these chromophores exhibit wavelength and intensity shifts with changes in their environments. The most marked is that exhibited by tyrosine about pH 9, which shows a shift to longer wavelength (red shift) and enhanced absorption upon ionization of the hydroxyl group. Smaller but still significant shifts with solvent, conformation, and presence of nearby charges have been useful in the study of the structure and unfolding of proteins and will be discussed in Section 8.3.3.1.

The peptide groups themselves absorb quite strongly in the far UV, exhibiting maxima at about 200 nm. This band is sensitive to conformation in both position and strength, but particularly in the latter. When the peptide groups become arranged in such a manner that their transition dipoles are stacked in an ordered array, a decreased absorption, or *hypochromic* effect, is expected (Tinoco et al., 1962). Such an effect is shown very clearly in Fig. 8.7 for poly-L-glutamic acid in aqueous solution at pH 4.9. Under these conditions it exists as an α-helix; when the pH is raised to 8.0, the coulombic repulsion of the charged carboxyl groups disrupts the helix, and the extinction coefficient increases to the value characteristic of isolated or disordered peptide chromophores.

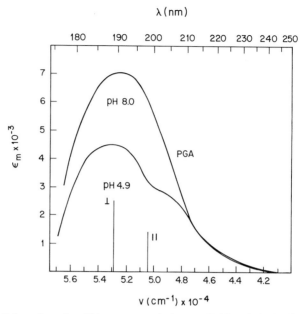

Fɪɢ. 8.7. Poly-ʟ-glutamic acid in aqueous solution at pH 4.9 and pH 8.0 (from Tinoco, I., Halpern, A., and Simpson, W. I., "Polyamino Acids, Polypeptides, and Proteins" (M. A. Stahmann, ed.), University of Wisconsin Press, 1962).

8.2.3.2. *Optical Rotatory Dispersion and Circular Dichroism*

In a monochromatic, *linearly polarized* wave train, the electric and magnetic vectors at any point oscillate sinusoidally in planes perpendicular to each other with a frequency $\bar{v} = c/\lambda$, where c is the velocity and λ the wavelength of the light. The electric vector E is our primary concern. It is convenient and realistic to regard E as the resultant of two equal vectors rotating in opposite directions with frequency $\bar{v}$. In Fig. 8.8, the wave train is directed toward the observer; the clockwise vector is designated E_R and the counterclockwise vector E_L. At any instant, E_R and E_L always make equal angles with the axis X representing the plane of polarization. E_R taken alone represents the electric field of a *right-circularly polarized* wave train, while E_L represents that of a *left-circularly* polarized wave train. By use of a so-called *quarter-wave plate* it is possible to resolve linearly polarized light into its circular components. (The application of circularly polarized light will be discussed a little later.)

The vectors E_R and E_L will make equal angles with X only if the right- and left-circularly polarized components travel with equal velocities, i.e., if the

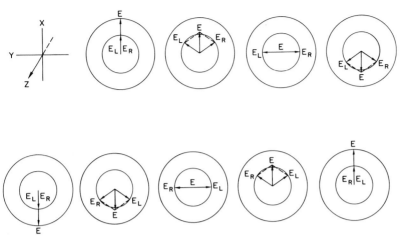

FIG. 8.8.   Plane polarized light expressed as the resultant of counterrotating vectors E_L and E_R.

refractive index of the medium is the same for both. It is a fundamental characteristic of optically active media that this is in general not the case, i.e., they exhibit *circular birefringence*. If for example, E_R travels faster than E_L, i.e., $n_R < n_L$, it can be readily shown (Fig. 8.9) that although the components remain in phase, the resultant E will be tipped clockwise through

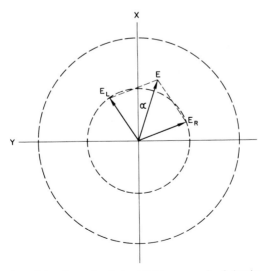

FIG. 8.9.   Rotation of the plane of polarized light as a result of circular birefringence.

the angle α. Conversely if $n_L < n_R$, E will be rotated counterclockwise. Quantitatively, it is found that

$$[\alpha]_\lambda = (1.8 \times 10^{10}/\lambda)(n_L - n_R) \tag{8.4}$$

where α is expressed in degrees per decimeter of path length and λ in nanometers (nm); the latter must be specified because of the dependence—often strong—of optical rotation on wavelength (p. 464). For solutions, the *molar rotation* is given by

$$[M]_\lambda = [\alpha]_\lambda M/100c \tag{8.5}$$

where M is the molecular weight and c the concentration of the optically active solute in grams per cubic centimeter of solution. In reporting results on polymers, it is customary to express the results in terms of the molar rotation per monomer unit; if these are not all alike, this quantity is termed the *mean residue rotation*. The units of $[M]_\lambda$ are *degrees square centimeters per decimole*.

Because of the sensitivity of modern instruments, which automatically scan and record the optical rotation as a function of λ, it is customary to employ path lengths of the order of 1 mm or less rather than the decimeter lengths of visual instruments, and to measure rotations in millidegrees. We then have

$$[M]_\lambda = [\alpha']_\lambda/[M]l \tag{8.6}$$

where $[\alpha']_\lambda$ is the rotation in millidegrees, $[M]$ is the molar concentration, and l the path length in millimeters.

In addition to exhibiting different refractive indices for right- and left-circularly polarized light, an optically active medium will also exhibit different *absorptions* for these components of a linearly polarized beam. For the difference in extinction coefficients we have

$$\Delta\varepsilon = \varepsilon_L - \varepsilon_R = \log(I_R/I_L)/[M]l \tag{8.7}$$

where I_R and I_L are the intensities of the right- and left-circularly polarized components emerging from a 1-cm thickness of the solution. The significance of the differing extinction coefficients may be seen in Fig. 8.10. The vectors E_L and E_R are now unequal in length. Let us suppose that the left-circular component is the more strongly absorbed, E_L being then shorter than E_R. Let us further suppose than n_R and n_L are equal. (We shall see shortly that this does not necessarily contradict our assumption of an optically active medium.) The tip of the vector E, the resultant of E_L and E_R, now no longer oscillates linearly but instead traces an ellipse, the semimajor axis of which is $E_L + E_R$; the semiminor axis is $E_R - E_L$. Since $E_R - E_L$ is usually very small, the ellipse is a very elongated one. The emerging light is no longer

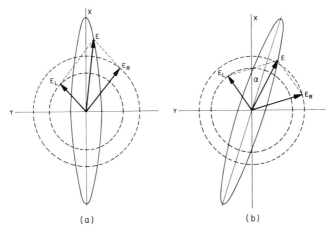

(a)   (b)

FIG. 8.10.   Positive circular dichroism at (a) the Cotton band center (no rotation of the resultant E); (b) at longer wavelength than the band center (rotation of E to the right).

linearly polarized but *elliptically polarized*, albeit very weakly. The *ellipticity* is defined as the angle whose tangent is the ratio of the minor to the major axis of the ellipse. For a 1-cm path, the *molar ellipticity* is given by

$$[\theta]_\lambda = 3300(\varepsilon_{\text{L}} - \varepsilon_{\text{R}}) \qquad (8.8)$$

The molecular origin of optical activity has been well understood in a general way for many years. To rotate the plane of polarized light and to exhibit circular dichroism, a molecule must develop, in the course of the electron oscillations induced by the impinging wave train, both an electric moment (leading to absorption) and a *magnetic moment* that has a finite component in the direction of the electric moment. Mechanistically, one can picture the electron as proceeding in a twisting path as it oscillates, thus acting like a current in a solenoid and generating a magnetic moment. It can be shown that such a twisting path is possible only in a molecule that cannot be superimposed on its mirror image.

The strength of the rotation and circular dichroism of a particular electronic transition will be proportional to the product of the electric and magnetic moments associated with the transition. It is possible for a transition to have a weak electric moment and a strong magnetic moment, i.e., to be weak in absorption but strong in rotation and circular dichroism. Such a transition is referred to as "magnetic dipole allowed"; a transition strong in absorption but weak in rotation and circular dichroism is termed "electric dipole allowed."

Both circular birefringence and circular dichrosim depend strongly on wavelength, as already indicated. Classically, one may regard the response of the electrons in a molecule to an impinging field as analogous to the forced oscillations of a mechanical system. As the resonant frequency of such a system is traversed, it will exhibit in-phase *dispersive* and out-of-phase or *absorptive* responses. In Fig. 8.11 is shown the dependence of optical rotation (dispersive) and circular dichrosim (absorptive) upon wavelength in the region of an absorption band. An absorption band characterized by optical activity is termed a *Cotton band.* Optical rotatory dispersion (or ORD) bands have zero intensity at the band center but extend with measurable intensity far out in both directions from the band center. Thus, rotations have in the past been commonly measured at the wavelength of the D line of sodium (589 nm), which is very far removed from most of the important chromophores in biological polymers. The circular dichroism (CD) spectrum, on the other hand, is a miniature reproduction of the absorption spectrum and can be observed only in its immediate neighborhood. The Cotton band shown in Fig. 8.11 is *positive*: the rotation is positive on the long wavelength side and $\varepsilon_L - \varepsilon_R$ is positive at all points. For a *negative* band, both curves would be inverted.

In earlier work, before ORD and CD spectrometers operating in the ultraviolet were available, the wavelength dependence of α in the visible (and near UV) was laboriously analyzed and fitted to either one-term or two-term relationships known as *Drude equations*:

$$[M]_\lambda = a_c \lambda_c^2 / (\lambda^2 - \lambda_c^2) \tag{8.9}$$

$$[M]_\lambda = a_0 \lambda_0^2 / (\lambda^2 - \lambda_0^2) + b_0 \lambda_0^4 / (\lambda^2 - \lambda_0^2)^2 \tag{8.10}$$

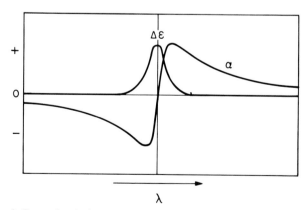

FIG. 8.11. A Cotton band observed (a) by optical rotatory dispersion and (b) by circular dichroism.

Equation (8.9) can be employed if there is only one predominant Cotton band, particularly if it is in the far-UV (220–185 nm) or "vacuum"-UV ($\lambda < 185$ nm) range, but if two or more strong bands are present, and particularly if they are of opposite sign, the two-term form must be used and the results expressed in terms of the adjustable coefficients λ_0, a_0, and b_0, and the position and intensity of unobserved bands deduced from their sign and magnitude. Although long outmoded experimentally, occasional mention of these parameters may still be encountered in the literature, particularly in connection with the α-helix–random coil transition. We shall not employ them further in this discussion.

The CD and ORD spectra of a molecule are related to each other by the Kronig–Kramers theorem (Moscowitz, 1960) and contain the same information. In CD spectra, overlapping bands are more readily disentangled and they are therefore generally easier to interpret. In modern practice, ORD spectra are relatively rarely reported and common use is made of recording spectrometers that generate only CD spectra. In our subsequent discussion, we shall deal entirely with CD.

In Fig. 8.12 are shown the absorption and CD spectra characteristic of a polypeptide composed of L-amino acid residues in the α-helical conformation. These spectra are of poly(γ-methyl-L-glutamate),

$$CO_2CH_3$$
$$|$$
$$(CH_2)_2$$
$$|$$
$$\text{—}[NHCHCO]_n\text{—}$$

where n is about 400, in hexafluoroisopropanol, a solvent that supports the α-helix (Bovey 1969, pp. 110–111; Holzwarth and Doty, 1965). Other helix-supporting solvents are chloroform, trifluoroethanol, and in general solvents that are not too strongly hydrogen bonding. Strong acids such as trichloroacetic or trifluoroacetic, which can compete with the internal hydrogen bonds of the helix, will disrupt them, giving the random coil conformation by shifting the hydrogen bonding equilibrium to the right:

$$\underset{\text{acid}}{HA \cdots HA} + \underset{\text{helix}}{-C{=}O \cdots HN} \rightleftarrows -C{=}O \cdots HA + HA \cdots HN \qquad (8.11)$$

The assignment of the bands is based in part on the theoretical studies of Moffitt (1956a,b), who predicted that ORD and CD bands of opposite sign should be generated by regular arrays of interacting chromophores because of resonance coupling of their electric dipole transition moments, a phenomenon termed "exciton splitting." It was further predicted that the band at longer wavelength should be polarized parallel to the helical axis

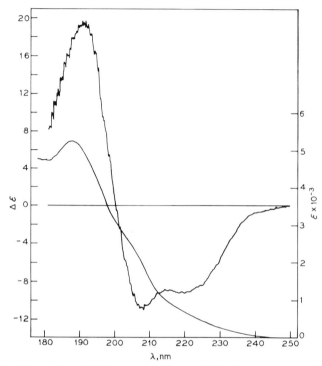

FIG. 8.12. Absorption and CD spectra of poly-γ-methyl-L-glutamate, $2 \times 10^{-2} M$ in hexafluoroisopropanol (from Bovey, 1969, p. 111).

and that the band at shorter wavelength should be polarized perpendicularly to the helical axis. These bands are termed $\pi-\pi^*_{\parallel}$ and $\pi-\pi^*_{\perp}$, as shown in Fig. 8.7, and are both electric- and magnetic-dipole allowed. These predictions have been amply borne out experimentally (for reviews, see Urnes and Doty, 1961; Yang, 1964; Walton and Blackwell, 1973, Chapter 6). The exciton-split bands appear at 190 and 205 nm. There is in addition a negative band at 225 nm, not predicted by Moffitt, corresponding to an $n-\pi^*$ transition (Schellman and Oriel, 1962; Tinoco et al., 1963). This is strongly optically active but very weak in absorption, appearing only as a foot of the UV curve in Fig. 8.12.

The principal contribution to the optical activity of a polypeptide chain in the α-helical conformation is the helix itself, rather than the asymmetric centers of the residual amino acids. This is clearly demonstrated by the CD spectrum of the left-handed helix of poly-β-benzyl-L-asparate (Karlson et al., 1960; Bovey, 1969, p. 125), in which the bands are of opposite sign although the amino acids are of the same chirality as in Fig. 8.12. The strong conformational dependence is further shown in Fig. 8.13. Here (a) shows the CD spectra of poly-α-L-glutamic acid at high pH and of poly-L-lysine at low pH, both representing random coils; (b) shows these polypeptides in the α-helical form at approximately neutral pH, and is essentially similar to the CD spectrum in Fig. 8.12; and (c) shows the spectrum of poly-L-lysine in the β-conformation, readily induced in this polypeptide by heating at high pH. These characteristic spectra are often of great diagnostic power, although there are many complications. For example, additional bands appear in the ORD and CD spectra of aromatic polypeptides (Goodman *et al.*, 1968); these we cannot discuss in detail here.

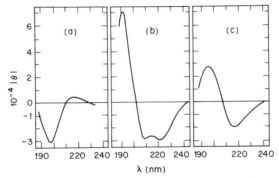

FIG. 8.13. The dependence of the CD spectra of polypeptides on conformation. (a) Random coil; (b) α-helix; (c) β form (from Morawetz, 1975).

Because the optical properties of forms I and II of poly-L-proline are so different, the transformation can be readily followed by observing them as a function of time. Indeed, this intriguing phenomenon was first discovered by noting that when form I is dissolved in water, the optical rotation at 589 nm changes over a period of hours from strongly positive to even more strongly negative (Kurtz *et al.*, 1956). The ORD and CD spectra are considerably more informative. In Fig. 8.14 are shown the UV and CD spectra in tri-fluoroethanol as a function of time. The initial CD spectrum (a) consists of two principal bands with very strong molar ellipticities: a positive band at 215 nm and a weaker negative one at 199 nm. (There is also a very weak negative band at ∼230 nm.) As isomerization proceeds, the positive band becomes weaker and the initial negative band is replaced by a somewhat stronger negative band (c) at 206 nm. Even upon complete conversion to form II, a weak positive band remains at ∼226 nm in both trifluoroethanol and water (Carver *et al.*, 1966). The temperature coefficient of the reaction

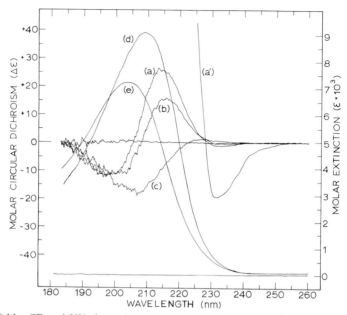

FIG. 8.14. CD and UV absorption spectra of poly-L-proline; $10^{-2} M$ in trifluoroethanol for CD and $10^{-3} M$ for absorption. CD curves (a), (b), and (c) were run at ~15 min, 58 min, and 19 hr after initial dissolution of poly-L-proline in form I. The UV curves (d) and (e) were run at 15 min and 19 hr, respectively. Curve (a') is similar to curve (a) but run with ten times greater path length in order to show the weak negative band (from Bovey, 1969; Bovey and Hood, 1967).

indicates an activation energy of about 20 kcal, corresponding to the cis–trans barrier. The absorption spectrum (d and e) decreases in intensity and exhibits a distinct blue shift.

Theoretical interpretation indicates the dominant contribution is the exciton splitting of the $\pi-\pi^*$ peptide transition (Pysh, 1967). As the transformation proceeds, intensity moves from the red-shifted exciton band, polarized parallel to the helix axis, to the perpendicularly polarized blue band.

8.2.3.3. Fluorescence

The fluorescence spectra of polypeptides have been studied extensively both for their intrinsic interest and as models for proteins. The three amino acids that absorb in the near UV are also those that fluoresce: tryptophan, with an emission maximum at 348 nm, tyrosine at 303 nm, and phenylalanine at 282 nm—all at substantially longer wavelength than their absorption maxima and corresponding to transitions from the first singlet excited state to the various vibrational levels of the ground electronic state. The quantum yields are highly dependent on the nature of the environment—particularly whether aqueous or organic—and are sensitive to the presence of quenching agents, such as amino and carboxylate groups (for a review, see Longworth, 1970).

Of particular significance is the occurrence of nonradiative dipole–dipole transfer of energy between a chromophore acting as donor and another acting as acceptor. For this to occur, the emission spectrum of the donor must overlap the absorption spectrum of the acceptor. The efficiency E of transfer is given by (Förster, 1960):

$$E^{-1} = 1 + (r^6/8.8 \times 10^{-25} Jn^{-4} K^2 Q) \tag{8.12}$$

where J expresses the extent of overlap between the emission spectrum of the donor and the absorption spectrum of the acceptor, K^2 is a function of the geometric relationship of the chromophores ($K^2 = \frac{2}{3}$ for random orientation), r is their separation, and Q is the fluorescence quantum yield of the donor when no acceptor is present. It is evident that E furnishes a highly sensitive measuring stick for intergroup distance, being dependent on the inverse sixth power of r. The correctness of this dependence was tested using proline oligomers in the extended helical form II conformation (see Section 8.2.2) with appropriate donor and acceptor groups at each end (Stryer and Haugland, 1967; Gabor, 1968).

8.2.3.4. Vibrational Spectroscopy

Infrared spectroscopy has proved an effective tool for the study of polypeptides, particularly in the solid state, mainly because (a) it permits ready detection of hydrogen bonds, and (b) it provides direct information concerning the orientation of absorbing groups by observations of dichroism (Chapter 3, Section 3.2.3) in appropriately oriented specimens. The most important bands are those of the peptide group: NH stretch at $\sim 3400 \, \text{cm}^{-1}$, moving to ~ 3300 when hydrogen bonded; C=O stretch at $\sim 1650 \, \text{cm}^{-1}$, called the *amide I* band; and NH deformation (wagging in the plane of the peptide group) at $1520–1550 \, \text{cm}^{-1}$, called the *amide II* band. The latter two decrease by about $25–30 \, \text{cm}^{-1}$ upon hydrogen bond formation, but are conformation-dependent and so less readily interpreted. The NH stretch band is customarily employed for hydrogen bond measurement. Table 8.3 shows the frequencies and dichroism of these bands in the α-helix and in β structures. The designations $\parallel$ and $\perp$ refer to the orientation of the transition moment (approximately coincident with the path of nuclear motion) with respect to the α-helical axis or the main-chain direction in the β structure. These observations are entirely in accord with the accepted structures and in

TABLE 8.3

FREQUENCIES AND DICHROISM OF PEPTIDE INFRARED BANDS
IN α AND β STRUCTURES[a]

Band	Frequency (cm^{-1})		Dichroism	
	α	β	α	β
N—H stretch	3300	3300	∥	⊥
C=O stretch (amide I)	1660	1640	∥	⊥
N—H deformation (amide II)	1545	1525	⊥	∥

[a] From Schellman and Schellman, 1964.

fact formed one of the strongest early pieces of evidence for them (Mizushima, 1954; Bamford *et al.*, 1956).

Figure 8.15 shows (at two intensities) the IR spectra of oriented films of α-helical poly-γ-benzyl-L-glutamate in the region of the amide I and amide II bands (Tsuboi, 1962). Analysis of the dichroic intensities shows that the angle between the amide I transition moment and the helix axis is ∼39°, whereas for the amide II transition moment, it is ∼74°. It is noteworthy that the amide I band is actually split, the parallel and perpendicular components being at slightly different frequencies.

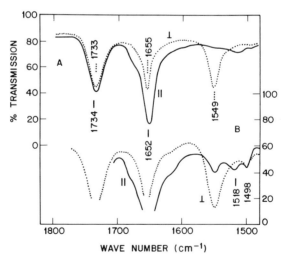

FIG. 8.15.   Infrared spectra of oriented films of α-helical poly-γ-benzyl-L-glutamate in the region of the amide I and amide II bands. Spectrum B is from a film thicker than used for A; (——) electric vector parallel to fiber axis; (· · ·) electric vector perpendicular to fiber axis (from Tsuboi, 1962).

In β-pleated sheet structures, a splitting of the peptide transitions can occur into parallel and perpendicular components with slightly different energies, an effect analagous to but much smaller than the splitting of the peptide π–π* electronic transition observed in the spectra of the α-helix.

It was for many years believed that chiral optical behavior was not to be expected in the infrared. However, since the electronic distribution in a molecule follows the variations of nuclear positions, electronic moments contribute to vibrational transitions and chiral properties of infrared bands should be detectable although weak. Such properties have in fact been observed for certain optically active compounds (Holzwarth et al., 1974) and have been theoretically predicted for polypeptides (Miyazawa, 1960; Miyazawa and Blout, 1961; Deutsche, 1970). For the latter, no experimental observations have been reported but the potential richness and usefulness of their infrared CD spectra have stimulated efforts in this direction that will no doubt give results in time, particularly with the availability of Fourier transform spectrometers of high sensitivity.

Raman spectroscopy (Section 3.2) is quite widely employed in the study of polypeptides (Koenig, 1972). Because these chains have no centers of symmetry, the mutal exclusion principle does not prevail, and the infrared and Raman spectra generally show the same lines, although often with markedly different intensities. A primary advantage over infrared is the low interference of water, the entire region from 2000 to 200 cm^{-1} being available. It is thus much better adapted to the study of aqueous solutions. Raman also shows greater sensitivity to the vibrational modes of homonuclear bonds such as C–C, C=C, and S–S, in which the dipole moment shows little change. There are in addition a variety of low frequency modes that are highly sensitive to conformation. A number of polypeptides have been observed in α-helical, β-pleated sheet, and random coil form. (Neither infrared nor Raman spectroscopy is particularly diagnostic or helpful in the study of the last of these.) The α-helical amide II band (N–H distortion) tends to be weak and broad in the Raman spectrum.

Raman spectroscopy can be particularly useful in showing whether a polypeptide has the same conformation in the solid state—as established by x ray—and in aqueous solution. By this means it has been demonstrated (Rippon et al., 1970) that this is in fact the case for form II of poly-L-proline and poly-γ-hydroxy-L-proline.

8.2.3.5. Nuclear Magnetic Resonance

The principal value of NMR in polypeptide studies is the observation of conformational change. It is of very limited value in the determination of primary structure, i.e., amino acid sequence and stereochemical configuration, which in any event is usually known from the method of synthesis or if unknown, as in the case of natural products, can be determined in other and better ways (see Section 8.3.1). Proton NMR has a longer history of use but, as in the investigation of synthetic polymers (Chapter 3, Section 3.3), ^{13}C NMR has claimed an increasing share of attention. It cannot yet, however, lay a valid claim to being more powerful than proton NMR.

The α-helix-to-random-coil transition (Section 8.2.2) is accompanied by highly characteristic behavior that is similar for both ^{1}H and ^{13}C (Bovey, 1975). In helix-supporting solvents such as chloroform, the proton lines are very broad owing to aggregation, the width increasing with molecular weight. Figure 8.16 shows the 300-MHz spectra of a 5% solution of a poly-γ-benzyl-L-glutamate of $\overline{DP} \simeq 35$ in CDCl$_3$ with increasing proportions of trifluoroacetic acid (TFA). The peptide NH resonance appears at $\sim 8\delta$, the aromatic protons at $\sim 7.2\delta$ and the α-CH at about 4δ; side-chain benzyl and β,γ resonances appear near 5δ and 2δ, respectively. The lines narrow abruptly upon addition of 5% of TFA; the chain is still in the α-helix conformation, but the aggregates are disrupted. At about 12% TFA, a second peak is observed in the α-CH region, well downfield from the helix resonance and arising from the random coil form; a corresponding resonance appears in the NH region but upfield from the main helix peak. As the TFA content is increased, the random coil resonances increase at the expense of the helix resonances and the other peaks shift in position and become narrow. Thus, the transition can be quantitatively followed. From the usual kinetic interpretation of NMR spectra, the well-separated helix and coil resonances would appear to indicate that this equilibration is a rather slow process. However, there is abundant independent evidence showing that in fact it is extremely fast, with lifetimes of each form of less than a microsecond. The double-peaked spectrum is actually due to the fact that in the transition regime, the residues near the ends of the chain have a greater probability of being in a random coil form than those near the middle (Lifson and Roig, 1961; Zimm and Bragg, 1959). It can be shown (Ullman, 1970; Nagayama and Wada, 1973) that under these circumstances separate resonances are to be expected and that the effect will be more marked if the molecular weight is poly-disperse, as is normally true.

Scalar couplings between protons have been of marked importance in the determination of the solution conformations of small peptides, i.e., those containing up to 10 amino acid residues (Bovey 1972; Deber et al., 1976). The vicinal coupling of the NH and C$_\alpha$H protons shows the "Karplus-like" dependence on the dihedral angle ϕ' already described in Section 3.3, and

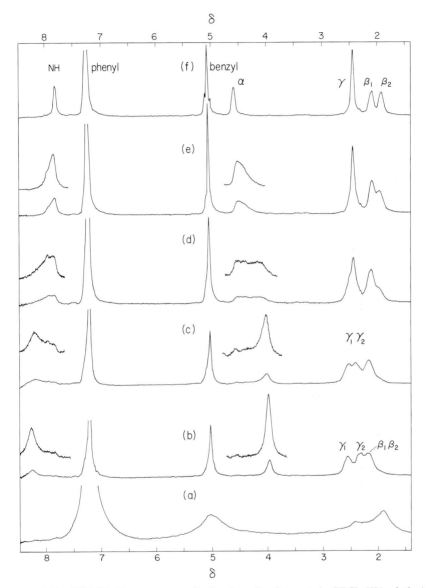

FIG. 8.16. 300 MHz Proton spectra of poly-γ-benzyl-L-glutamate in CDCl₃ (5% solution) at 32°. The volume per cent of trifluoroacetic acid added is (a) 0%; (b) 10%; (c) 12%; (d) 13%; (e) 16%; (f) 20% (from Bovey, 1975).

therefore provides information concerning the angle ϕ (Section 8.2.2) because of the relationship

$$\phi' = |60° - \phi| \qquad (8.13)$$

For small peptides this coupling is usually determined from the splitting of the NH resonances. In most long chain polypeptides this splitting is not resolved. The multiplets of the side-chain protons, however, can often be analyzed to yield information concerning their conformations.

8.3. Proteins

Proteins are found in nature in a vast variety of forms and are of crucial importance in the structure and function of living cells; they take their name from the Greek *proteios*—first or primary. They serve in *enzymatic catalysis*, in a variety of *structural functions* in cell walls, skin, and bone, in *transport and storage of oxygen*, as *hormones*, in *immune protection*, and in the control of *gene function*. Their principal structural elements are polypeptide chains, although they may be combined with fats as *lipoproteins* and with polysaccharides as *glycoproteins*. Molecular weights vary from ~ 4000 to several hundred thousand. Their molecules may consist of a single chain or of two or more chains joined by cystine disulfide bonds. *Globular proteins* consist of chains tightly intertwined to form a nearly spheroidal shape. In some more complex proteins these spheroidal units may themselves be joined together by noncovalent forces into larger structures of fairly precisely determined form. Enzymes often consist of single units, but many proteins, e.g. hemoglobin, are composed of several subunits. Immune proteins may be of quite elaborate architecture.

The molecular weights of proteins may be determined by the methods described in Chapter 4. By ultracentrifugation, mixtures of components of different molecular weight may be separated preparatively or analytically into their components. In a case where the protein consists of subunits, these may be split apart by the use of *denaturating* agents such as urea. Purification of the desired protein fraction can be done by *dialysis*, which removes small molecules or ions, or by *gel filtration*, which is the same in principle as gel permeation chromatography (Chapter 4, Section 4.5.5) but is performed with a carbohydrate polymer—a cross-linked dextran—as the stationary phase. Proteins may also be purified by *ion exchange chromatography*, in which sulfonated polystyrene beads bind the protein according to the sign and magnitude of its net charge, or by *electrophoresis*, in which separation is effected by migration in an electric field. By any of these means or by a combination of them, the material may be obtained as a single molecular entity. In this respect, the problem differs from those encountered with synthetic polymers, in which a single entity is seldom if ever obtained.

8.3.1. PRIMARY CHAIN STRUCTURE

In determining the composition of a protein, the first step is to subject it to hydrolysis in $6\,M$ HCl at 100–$110°$ for several hours to break it down into the free amino acids. The hydrolysate is then passed under standardized conditions through an ion exchange column. Buffers of increasing pH are used to elute the individual amino acids. The acidic amino acids (Table 8.1) appear first and the basic amino acids last, each appearing as a separate peak in a predictable position. The number of residues of each type is determined by the intensity of color produced in the eluate by reaction with

ninhydrin

ninhydrin, which reacts with all the amino acids to produce a blue color and with proline to give a yellow color. All these steps are now performed in automatic machines, the composition appearing as a recorded curve.

Another valuable technique, particularly for shorter peptides (which may for example be produced by the partial hydrolysis of proteins) is *paper chromatography* in which the stationary phase is similar to blotting paper. A drop of hydrolysate is placed on one corner of a rectangle of paper, which is then hung from a trough containing a solvent, often organic, and washed with this solvent. This spreads out the amino acids in sequence. The paper is then dried, turned by $90°$, and eluted with a second solvent, which separates the components of the initial linear chromatogram into a two-dimensional pattern. This is then developed by spraying with various reagents, including ninhydrin; each known amino acid appears in a known position.

A much more difficult task is the determination of the sequence of amino acids in a protein. The first sequence to be determined was that of the protein hormone *insulin*, carried out by Sanger in 1953 (Ryle *et al.*, 1955). This was a most significant accomplishment because it showed that a protein is in fact composed of identical molecules with a definite sequence, and because it laid the foundation for subsequent x-ray determination of the three-dimensional structure of proteins. Insulin consists of two amino acid sequences, an A and a B chain, joined by cystine disulfide bonds, with a third disulfide bond forming a six-residue loop, a total of 51 amino acids (p. 476).

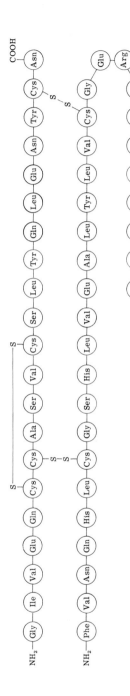

insulin sequence

The structure of insulin was determined by scission using a variety of specific enzymes and determining the N-terminal residues of the resulting peptides by labeling with reagents such as *fluorodinitrobenzene* (in which the fluorine is reactive) or *dansyl chloride*, which react with the amino groups to yield yellow and fluorescent derivatives, respectively. From the overlapping sequences, the overall structure could be logically worked out. A

fluorodinitrobenzene   dansyl chloride

more powerful method, because it can be repeated in cyclic fashion, is the *Edman degradation*. The uncharged amino group of the N-terminal residue is reacted with phenyl isothiocyanate and the resulting phenylthiocarbamoyl derivative reacted under mild acid conditions to yield a phenylthiohydantoin amino acid, which can be identified by paper chromatography:

Edman degradation

By this means, the sequences of polypeptides produced by specific enzymatic cleavage can be identified. Enzymes used for this purpose are *trypsin* and *chymotrysin.* Trypsin cleaves on the carboxyl side of the basic amino acids lysine and arginine:

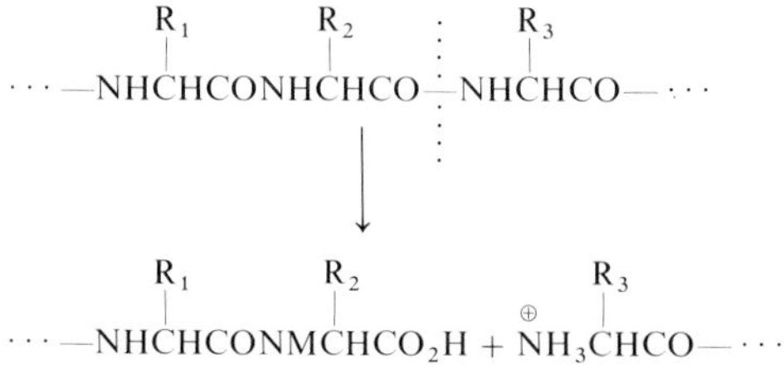

$$\cdots -NHCHCONHCHCO \vdots NHCHCO- \cdots$$

(with substituents R_1, R_2, R_3)

$$\downarrow$$

$$\cdots -NHCHCONMCHCO_2H + \overset{\oplus}{NH_3}CHCO- \cdots$$

(with substituents R_1, R_2, R_3)

Chymotrypsin cleaves on the carboxyl side of aromatic and certain other residues with bulky side chains. A useful nonenzymatic reagent is *cyanogen bromide*, CNBr, which cleaves on the carboxyl side of methionine residues to produce a lactone ring at the C-terminus:

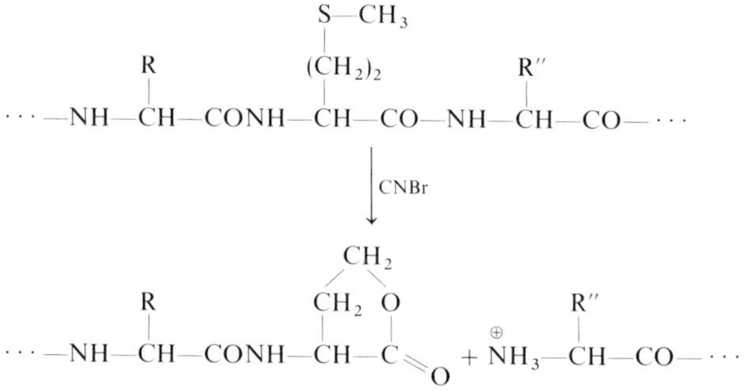

$$\cdots -NH-CH-CONH-CH-CO-NH-CH-CO- \cdots$$

with side chains: R ; $(CH_2)_2$ bearing $S-CH_3$; R''

$$\downarrow \text{CNBr}$$

$$\cdots -NH-CH-CONH-CH-C\underset{O}{\overset{}{\diagdown}}\diagup CH_2 + \overset{\oplus}{NH_3}-CH-CO- \cdots$$

with side chains: R ; CH_2 (lactone with O); R''

An automatic device called a *sequenator*, employing the Edman degradation together with chromatography, can sequence long peptides rapidly and may obviate the need to fragment the protein prior to analysis.

A complication is introduced by the disulfide units, since breakage of their peptide bonds would not release cystine units that hold two peptide chains together. These must therefore be cleaved by reduction with thiols or, more commonly, oxidized with performic acid:

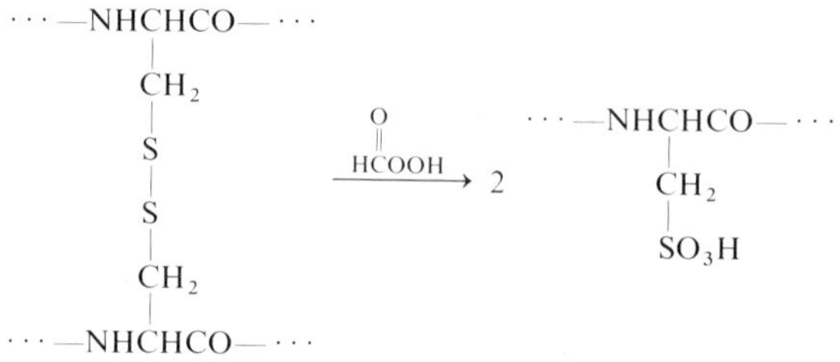

$$\cdots -NHCHCO- \cdots$$
$$| $$
$$CH_2$$
$$| $$
$$S$$
$$| $$
$$S$$
$$| $$
$$CH_2$$
$$| $$
$$\cdots -NHCHCO- \cdots$$

$$\xrightarrow[\text{HCOOH}]{O} \quad 2$$

$$\cdots -NHCHCO- \cdots$$
$$| $$
$$CH_2$$
$$| $$
$$SO_3H$$

Where more than one disulfide bond is present, linking two or more chains together, additional steps are necessary since cleavage of all the disulfides eliminates information as to how they were linked in the native molecule. These bonds may also be broken by reduction to thiols, i.e., cysteines, with reagents such as β-mercaptoethanol, but the latter upon oxidation to disulfides do not necessarily regenerate the cystine bonds in the correct pattern (Haggis *et al.*, 1964).

Proteins of a given function and type, and having the same name, do not always have exactly the same amino acid sequence. This may vary from species to species, having been altered by genetic mutations over the course of many millions of generations without materially affecting its function. The sequence may also vary by inheritable mutations within a given species. Of these, the most celebrated is the substitution of one amino acid—valine for glutamic acid—which leads to the pathological form of hemoglobin responsible for *sickle-cell anemia* (Section 8.3.2.1).

8.3.2. CONFORMATIONS OF PROTEINS

It is a fundamental postulate of protein science (although one slow to be understood) that the primary chain structure determines the three-dimensional structure, which is the basis of its function. The three-dimensional or *tertiary* structure is composed in part of α-helical portions and β-sheets—sometimes referred to as *secondary structure*—and in part of seemingly random structures that are nevertheless exactly the same in each molecule of a particular protein even though their conformational rationale may not be evident. Under the proper conditions, as first observed by Anfinsen for ribonuclease (for a review, see Anfinsen, 1973), the unfolded or *denatured* protein will spontaneously refold to reconstitute the native, active molecule.

The complete structures of about 30 proteins are now known from x-ray diffraction. We shall consider eight of these. Their structures are stabilized by a number of forces:

(i) *Cystine disulfide bonds*;

(ii) *Hydrogen bonds* involving peptide groups and the side chains of asparagine, glutamine, serine, and threonine (Table 8.1), which act as both donors and acceptors; hydrogen-donor side chains such as the guanidinium groups of arginine; groups such as the carboxyls of aspartic and glutamic acid and the hydroxyl of tyrosine, the hydrogen-bonding capability of which is lost at higher pH. Carboxyls can form *ion–dipole* bonds even when ionized, e.g. between CO_2^- and unionized tyrosine side chains;

(iii) *Ion–ion* (coulombic) and *dipole–dipole* forces;

(iv)   *Hydrophobic bonds*. Hydrophobic bonding is believed to be of great importance and requires a further word of explanation. The dissolving of a hydrocarbon in water is actually exothermic, e.g. $-2.6\,kcal/mole$ for methane, but nevertheless the solubility is very limited because the accompanying free energy is positive owing to an unfavorable entropy accompanying the process. The dissolved hydrocarbon molecule disrupts the hydrogen bonding of the water; in order to form the maximum number of hydrogen bonds, the water molecules rearrange themselves in a more ordered structure and must pay the price in decreased entropy. The net result is that oily molecules or groups tend to cluster together in water, since a large cavity requires less energy than a multiplicity of smaller ones. The hydrocarbon side chains of alanine, valine, leucine, isoleucine, and phenylalanine thus tend to associate not only because of the usual dispersion forces, which all molecules and groups experience, but because of hydrophobic forces.

The chain conformations of globular proteins are such that the hydrocarbon side chains cluster together in the center while the polar and ionizable groups are at or near the surface—much like an oil drop stabilized in water by surfactant molecules or like the micelles that the surfactant itself is capable of forming, with polar groups directed outward and fatty chains in association.

8.3.2.1.   Myoglobin and Hemoglobin

Heme proteins are present in all living organisms and are essential for oxygen transport and for controlling the path of electrons in oxidation–reduction processes. The heme group is present at the active site of all these proteins, but its function depends on the nature of the attached ligands, on the conformations of the surrounding polypeptide chains, and on the oxidation state of the iron atom at the center of the porphyrin ring. In myoglobin and hemoglobin the heme unit is a "protoporphyrin IX" group in which the iron is bound to four nitrogen atoms:

The iron is hexacoordinate and has two axial ligands. One is a histidyl residue of the polypeptide chain; the other varies with the function and state

```
      · · ·
          ╲─ CH ─╲ · · ·
             │
             CH₂
             │
             C ──── NH
             ║      │
          HC   ╲  ╱ CH
               N
             ·
             ·
    ╱──── Fe ────╱      (xy plane)
   ╱            ╱
             ·
             ·
```

of the protein and provides the active site. The heme iron is paramagnetic and varies in spin depending upon its state of oxidation and the nature of the sixth ligand.

Such a group, which constitutes the functional site of a protein, is termed a *prosthetic group*. Not being covalently bonded, the prosthetic group of hemoglobin and myoglobin can be reversibly removed and replaced. In the ferrous state, the heme units of myoglobin and hemoglobin are capable of binding molecular oxygen, becoming then diamagnetic.

The function of myoglobin is to store oxygen in muscle until it is required for metabolic oxidation. It is thus for obvious reasons especially plentiful in the flesh of diving mammals such as the whale and porpoise. The structure consists of a single chain of 153 amino acids (molecular weight of 17,800) and contains one heme unit. There are no disulfide cross-links. Sperm whale myoglobin was the first protein the structure of which was determined by x-ray diffraction (Kendrew *et al.*, 1960, 1961; Kendrew, 1963; Perutz, 1963). Its three-dimensional structure is shown in schematic form in Fig. 8.17. It is very compact, measuring approximately 2.5 × 3.5 × 4.5 nm, with no wasted volume within it. The amino acid sequence, not shown, varies somewhat from species to species, but is constant in the neighborhood of the heme unit. Its content of α-helix is exceptionally high, involving 121 of the residues. It is essentially a basket for the heme group, constructed from eight segments of helix lettered A through H as shown, connected by nonhelical regions. Within the protected nonpolar environment provided by the amino acid residues in its immediate neighborhood, the heme can be maintained in the ferrous state and exercise its oxygen-binding function; if the heme unit were freely exposed to water, the iron would be rapidly oxidized to the ferric state and this function would be lost. (The ferric state of the intact protein, called *ferrimyoglobin*, can be achieved with appropriate oxidizing agents; in this form the sixth ligand is a water molecule and oxygen is not bound.)

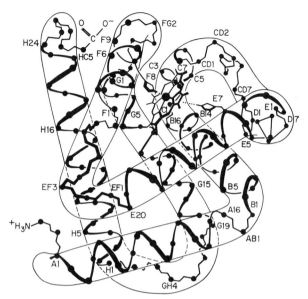

Fig. 8.17. The structure of myoglobin, showing only the α-carbon atoms. The α-helical regions are lettered from A through H, and the amino acid units numbered in each helix. The heme unit is shown at the upper center between histidines E7 and F8 (from Dickerson 1964).

Hemoglobin, which occurs in red blood cells, is composed of four polypeptide chains, a pair of α chains and a pair of β chains, which differ slightly from each other and from myoglobin, to which they have a strong family resemblance. Each of the four units contains a heme group in a hydrophobic pocket similar to that of myoglobin. The α chain has 141 amino acid residues and the β chain 146, giving a molecular weight of about 64,500. The four units are not covalently bonded together, but appear to be held largely by hydrogen and hydrophobic bonding. As in myoglobin, there are species variations in the chain sequences and in addition many variant and pathological human hemoglobins are known, one of which, sickle-cell hemoglobin, has already been mentioned (p. 479). The substitution of valine for glutamic acid in the β chains causes the hemoglobin to be abnormally insoluble under low oxygen pressure. The resulting crystallization collapses the erythrocyte into an inefficient, sicklelike form.

The packing of the α and β chains brings side chains of unlike units into close contact but permits very little contact between like units. There is one true twofold axis in the molecule and because of the close similarity of the α and β chains, there are two others that are nearly twofold axes; these are perpendicular to each other and to the true twofold axis. (For further structural details, see Dickerson and Geis, 1969 and Stryer, 1975.)

The function of hemoglobin is of course to bind oxygen in the lungs and release it in the tissues. Because of its multiple heme groups, the form of the oxygen-binding curve of hemoglobin differs from that of myoglobin in a characteristic and highly significant way. As shown in Fig. 8.18, myoglobin exhibits the expected simple hyperbolic relationship between oxygen pressure and fractional saturation, whereas hemoglobin shows a sigmoidal curve. This means that at moderate pressures, the affinity for oxygen increases with pressure. This characteristic enables hemoglobin to take up oxygen efficiently at the moderate pressure of the lungs but to give it up to myoglobin at the low oxygen pressure of the tissues. Upon loss of oxygen to yield deoxyhemoglobin, the subunits of the molecule slide past each other and rotate slightly so that the two hemes of the β chains move farther apart by about 0.65 nm—a change in *quaternary* structure. This conformational adjustment is closely associated with a cooperative interaction of the hemes that enables them to bind oxygen more strongly when the molecule is already partially oxygenated. Hemoglobin is thus able to communicate information within itself by conformational adjustments. Such a protein, of which this is the best understood example, is said to exhibit *allosteric* properties.

A further important characteristic of hemoglobin is that upon binding four oxygen molecules, it becomes a stronger acid and releases protons. Conversely, as the medium becomes more basic, the equilibrium shifts in

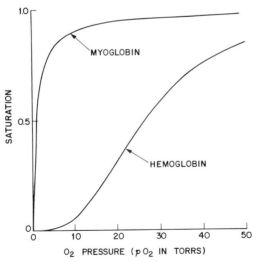

FIG. 8.18. Oxygen dissociation curves of myoglobin and hemoglobin. The fractional saturation of the binding sites is plotted versus the partial pressure of oxygen.

favor of the oxy form. This so-called *Bohr effect* enables deoxyhemoglobin to take up CO_2 in the veins and release it in the lungs.

8.3.2.2. Cytochrome c

Cytochrome *c* is essential to the function of all aerobic cells and is accordingly very widely distributed, being found in all plants and animals and in microorganisms of the type having a distinct nucleus (*eukaryotic*). Its role is to serve as an electron carrier, i.e., to be alternately reduced and oxidized, in the terminal stages of the process of oxidation of nutrients in the mitochondria of the cell. It has a single polypeptide chain of 104 residues and a molecular weight of 12,400. Its diameter is about 3.4 nm. It is, in short, an extraordinarily small and efficient electronic device for the sorting and channeling of electrons. Like myoglobin, it has a single heme unit, but, as might be expected from its very different function, it is bound to the polypeptide chain in a quite dissimilar manner. It is covalently bonded through two cysteine units by addition of their thiol groups to the heme vinyl double bonds. The amino acid sequence has been very thoroughly studied and is known for nearly 40 species. Although two-thirds of the amino acids are found to vary, 35 of them have remained entirely unchanged over many millions of years, and it must be supposed that these are vital to its correct functioning. Indeed, from a careful study of the variations, relationships between families of organisms can be determined and an evolutionary "family tree" worked out (see Dickerson and Geis, 1969, pp. 62–65).

Species variations occur in the sequence Lys-Cys-Ala-Gln-Cys-His-Thr shown (this sequence is found in horse, pig, bovine, sheep, and wheat cytochrome *c*), but the Cys-14, Cys-17, and His-18 residues (numbering from

the N-terminus) are never varied. The His-18 residue supplies one of the axial ligands. The sixth ligand is a methionine residue, Met-80:

$$\cdots\!-\!CH\!-\!\cdots$$

$$\begin{array}{c} | \\ CH_2 \\ | \\ C\!-\!NH \\ \| \quad \diagdown \\ HC \quad CH \\ \diagdown \! N \! \diagup \\ | \end{array}$$

Fe xy

$$\begin{array}{c} | \\ S \\ \diagup \quad \diagdown \\ CH_2 \quad CH_3 \\ | \\ CH_2 \\ | \\ \cdots\!-\!CH\!-\!\cdots \end{array}$$

(Met-80)

The three-dimensional structure of the molecule (Takano *et al.*, 1973) is shown in Fig. 8.19 and reveals an ovoid heme carrier with hydrophobic groups clustered about the heme group itself and with ionizable groups on the surface. The latter are grouped in a nonrandom manner—two positively charged areas with a negative area between them—which clearly must be

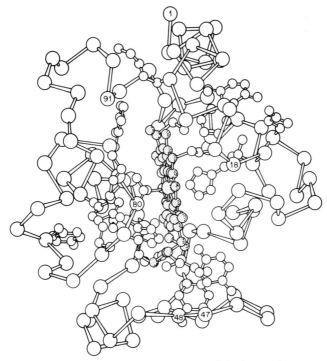

Fig. 8.19. The structure of cytochrome *c*, showing mainly the α-carbon atoms. The heme unit is nearly normal to the plane of the paper and is shown shaded (from Takano *et al.*, 1973; Stryer, 1975, p. 350).

intimately connected with the detailed positioning and function of the protein in the cell, enabling it to accept and hand on electrons to its neighbors in the electron-transport sequence.

8.3.2.3. Lysozyme

Lysozyme is the first enzyme to have its three- dimensional structure determined by x ray (see Dickerson and Geis, 1969, pp. 69–78). It is very widely distributed in bodily secretions, including tears and nasal mucus, and is particularly plentiful in hen egg whites. Hen egg white (HEW) lysozyme has a molecular weight of 14,600; there are 129 amino acids in a single chain with four disulfide cross-links. It is roughly ellipsoidal in form, measuring approximately $3.0 \times 3.0 \times 4.5$ nm. There is much less α-helix than in myoglobin and hemoglobin, but a considerable proportion of extended form and one region of antiparallel β structure. As expected, the interior is largely non-polar. It has no prosthetic group, being composed entirely of the common amino acid residues, but has a very deep cleft in one side. Its enzymatic function, which is to cleave the mucopolysaccharides of bacterial cell walls and thereby destroy them, is lodged in this cleft. The mucopolysaccharide substrate is a 1:1 alternating copolymer of N-acetylglucosamine (NAG) and N-acetylmuramic acid (NAM) joined by β-1,4 linkages:

NAG   NAM

Lysozyme acts at the oxygen linkage indicated by the dashed lines. *Chitin* (Section 8.5.3), which is poly-β-NAG, is cut at any such bond. NAG itself and the oligosaccharides $(NAG)_2$, $(NAG)_3$, and $(NAG)_4$ act as *competitive inhibitors*. Their effect is not irreversible, for they act in competitive equilibrium with the substrate. They have been used to study the mode of binding. Oligosaccharides of five or more units are cut beyond the fourth unit. $(NAG)_6$ occupies the entire binding site.

The basis of catalytic activity, which is maximal at pH 5, lies in the carboxyl groups of Glu-35 and Asp-52, which are situated at the appropriate distance of 0.3 nm from the oxygen indicated below at position 4 of the NAG unit on opposite sides of the ring. The donation of a proton to this oxygen by Glu-35 results in formation of a carbonium ion at position 4. This process is greatly assisted by the fact that binding forces this NAG 6-membered ring from a chair to a half-chair form, and by the presence of the negative charge at Asp-52, which remains ionized throughout the process. The cleavage is

completed when the carbonium ion reacts with OH^- from the H_2O solvent and the Glu-35 reacquires a proton. The cleaved products then diffuse away from the catalytic site. The distortion of the ring is crucial and is an example of a phenomenon that is probably general for most enzymes, i.e., *they act by stabilizing the substrate in the activated state appropriate to the reaction occurring.*

8.3.2.4. Ribonuclease

The function of ribonuclease is the hydrolytic scission of ribonucleic acid chains (Section 8.4.1). It is a single chain of 124 amino acid residues (with no tryptophans) and a molecular weight of 13,700, and thus is about the same size as lysozyme, which it further resembles in having four disulfide cross-links and an ellipsoidal form with a large groove in one side to accommodate the substrate (Kartha *et al.*, 1967; Wyckoff *et al.*, 1967; Dickerson and Geis, 1969, pp. 79–81). There are four histidine residues at positions 12, 48, 105, and 119. As a result of chain folding, His-12 and His-119, together with

Lys-7, are clustered together on one side of the groove, while Lys-41 is positioned on the other side. Chemical evidence and inhibitor-binding studies have shown that these residues are involved in the catalytic mechanism, but this has not been worked out in detail. The NMR spectroscopy of ribonuclease is described in Section 8.3.3.4.

8.3.2.5. Keratins

The keratins are a family of closely related proteins that occur in vertebrates in skin, hair, fur, nails, hooves, feathers, and a number of other protective structures. They contain most of the amino acids but with a content of glutamic and aspartic acids and of lysine and arginine such that nearly one-third of the side chains are charged. The solubility and aqueous solution properties thus depend quite strongly on pH.

The most extensive structural study has been given to wool fibers, beginning with the early x-ray pictures of Astbury (1938), which demonstrated two distinct chain conformations. In the unstretched or α form, wool fibers show the diffraction pattern of α-helices. These are extensively bonded together by cystine cross-links, which constitute about 10–12 mole % of the chains. In large part because of these disulfide cross-links, wool fibers are readily and reversibly extended by as much as twofold. At large extensions the intrahelical hydrogen bonds are broken and remade into interchain bonds as the structure goes over to an extended β form, reverting to the α form when stress is removed.

Although basically α-helical, the wool fiber has a very complex morphology with at least four levels of supermolecular organization. The α-helices are apparently themselves formed into triple superhelices of about 2.0 nm diameter, which constitute the *protofibrils*, the smallest unit observable in electron micrographs. These in turn are formed into *microfibrils* consisting of nine parallel protofibrils. Groups of microfibrils, embedded in an amorphous protein matrix, constitute the *macrofibril*. The latter are organized into *cells*, which are the largest building unit of the fiber itself. For a very clear description of the present state of knowledge, the reader is referred to Dickerson and Geis (Dickerson and Geis, 1969, pp. 37–40).

8.3.2.6. Silk Fibroins

The structure of silk fibers, in particular that of the species *Bombyx mori*, is entirely different from that of wool and appears to correlate with their great strength but much smaller extensibility. A conspicuous feature is the strong predominance of glycine (~ 45 mole %), alanine ($\sim 30\%$), and serine ($\sim 12\%$), corresponding to a six-unit repeat unit $(Gly-Ala-Gly-Ala-Gly-Ser)_n$. The basic structure is the extended antiparallel β sheet. To achieve the geometry required for efficient interchain hydrogen-bonding, the β sheets are crimped or pleated in such a way that, as shown in Fig. 8.20, the alanine and serine side chains all project on one side of the sheet. In this representation, we are looking perpendicular to the sheets. In front and back of each zigzag chain are other chains with zigzags "in phase" but running antiparallel to the one shown and hydrogen-bonded to it through $-C=O \cdots H-N-$ bonds perpendicular to the plane of the paper. This architecture provides strength with flexibility but, since the chains are already extended to their limit, little elasticity or "give." What extensivity there is, about 24%, arises from the $\sim 13\%$ of other amino acids with bulky side chains, principally Tyr, Val, Asp, and Asn; these cannot fit into the compact crystalline structure and result in a substantial amorphous content, about 40% in *B. mori* silk.

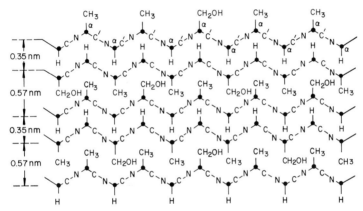

FIG. 8.20. The structure of *Bombyx mori* silk fibroin.

8.3.2.7. Collagen

Collagen is the primary structural protein in skin, cartilage, tendon, bone, teeth, and cell walls, and is very widespread, occurring in all multicelled organisms. The basic molecule, called *tropocollagen*, consists of three chains, each about 1000 amino acid residues in length, and wound about each other in a triple helical fashion. These are hydrogen-bonded to each other, giving a structure about 300 nm long and 1.5 nm in diameter with a molecular weight of 285,000. Two of the chains are identical in sequence, while the third is slightly different. One-third of the residues are glycine, with alanine, proline, and hydroxyproline each occurring to the extent of about 10–12 mole %. The sequence may thus be approximately represented as

$$(\text{Gly-X-Y})_{n \simeq 330}$$

Fig. 8.21.   The structure of collagen, showing three left-handed helices, each given a right-handed twist to form a threefold super-helix (from Dickerson and Geis, 1969).

where X is very likely to be proline and Y is very likely to be hydroxyproline. Each chain has a conformation very similar to the left-handed helix of poly-L-proline II (Fig. 8.6), but slightly less extended, and each helix is itself given a right-handed twist, forming a coiled coil (Fig. 8.21). (The structure is thus reminiscent of the keratin protofibril (Section 8.3.2.5).) The collagen fiber consists of tropocollagen molecules in a very regular staggered array, with each molecule out of register with its neighbor in the adjacent row by about one-quarter of its length, and with a large (40 nm) gap between the succeeding molecules, as shown in Fig. 8.22. This offset structure gives rise to the banded appearance of the collagen fiber under the electron microscope (Fig. 8.23). Because of the gaps, the tropocollagen molecules are not end-linked to each other, but nevertheless the interchain hydrogen bonds create a very strong structure. Collagen fibers can typically sustain tensile stresses up to 2×10^6 gm/cm^2.

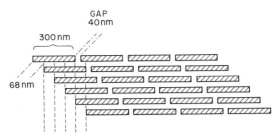

FIG. 8.22.   Schematic representation of the arrangement of tropocollagen molecules (bars) in collagen.

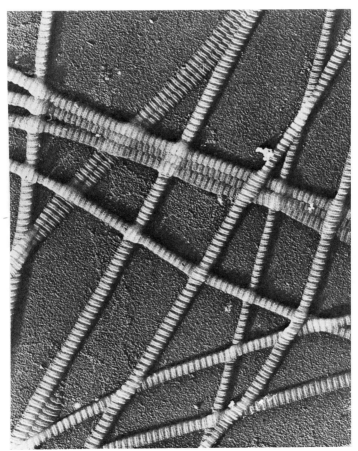

FIG. 8.23.   The banded structure of reconstituted collagen fibrils as seen by the electron microscope (courtesy of Dr. Jerome Gross).

It is a curious fact that the hydroxyproline units, which are vital to the interchain cross-linking, are not incorporated as such but arise from the enzymatic hydroxylation of the finished tropocollagen molecules. This enzyme apparently requires ascorbic acid (vitamin C) for activity. The distressing symptoms of scurvy—destruction of the gums and rupture of the walls of the capillary blood vessels—arise from a weakening of the collagen structure owing to insufficient interchain hydrogen-bonding.

In addition to hydrogen-bonded interactions, there are also covalent bonds between the component helices of the tropocollagen molecules and also between tropocollagen molecules. These are formed by a specific enzyme that oxidizes the terminal methylene groups of lysine side chains to aldehyde groups, which then form links by aldol condensation (Tanzer, 1973):

$$
\begin{array}{cc}
\vdots & \vdots \\
| & | \\
C{=}O & C{=}O \\
| & | \\
C_2H(CH_2)_3CHO \;+\; HCO(CH_2)_3C_2H \\
| & | \\
NH & NH \\
\vdots & \vdots
\end{array}
$$

$$
\downarrow
$$

$$
\begin{array}{cc}
\vdots & \vdots \\
| & | \\
C{=}O & C{=}O \\
| & | \\
C_2H(CH_2)_3{-}CH{=}C(CH_2)_3C_2H \\
| & | \\
NH & CHO \quad NH \\
\vdots & \vdots
\end{array}
$$

The cross-links are few in number in very young animals and their collagen is correspondingly more soluble and tractable; this circumstance has contributed much to recent advances in our knowledge of this protein. The number of cross-links increases with time and is believed by some investigators to be directly correlated with the symptoms of aging in humans.

8.3.3. SPECTROSCOPIC PROPERTIES OF PROTEINS

The spectroscopy of proteins has in some respects become less important than it was before detailed three-dimensional structures became known from x-ray diffraction. Much of the earlier work was done to support structural hypotheses, e.g. the environments of aromatic and ionizable groups and the extent of α-helical conformations, which were subsequently unequivocally demonstrated, or in some cases disproved, by x-ray. Optical spectroscopy, including absorption and CD (Sections 8.2.3.1 and 8.2.3.2), nevertheless retains some interest for proteins with strongly absorbing prosthetic groups, such as hemes, and for very large and complex proteins for which high resolution x-ray structures are not likely to be provided for a long time, if at all. Resonance Raman spectroscopy (Section 8.3.3.3) has been employed in the study of heme groups and other chromophores. High resolution NMR is of particular interest because it supplies a large number of parameters and can, for example, permit the titration of specific ionizable groups.

In the following discussion we shall emphasize those methods and results that extend and complement the x-ray information, rather than merely confirming it.

8.3.3.1. Electronic Absorption and CD–ORD Spectroscopy

It has long been known (Beavan et al., 1950, 1952) that the ultraviolet absorption maxima of proteins undergo shifts to shorter wavelengths ("blue shift") and slight increases in intensity when the native structure is disrupted, as by heat, extremes of pH, or high concentrations of urea. There are many possible contributions to this general effect, but by use of model systems it has been shown (Yanari and Bovey, 1960) that the dominant cause is the fact that a substantial fraction of the chromophores (aromatic side chains of phenylalanine, tyrosine, and tryptophan; see Section 8.2.3.1) are transferred from the interior of the protein to an aqueous environment. This is consistent with the "oil-drop" model for globular proteins (Section 8.3.2): the hydrophobic interior provides a medium of relatively high polarizability, whereas water has a low polarizability, as attested by its low refractive index. The stabilizing solvation energy in the excited state is decreased by the transition moment and by the greater dipole moment of this state; this effect will increase with the refractive index of the solvent. The electronic transition is thus of greater energy in a medium of low refractive index, resulting in the blue shift. Such observations were important in suggesting a correct general notion of the architecture of globular proteins well before their detailed three-dimensional structures were established.

From ORD and CD measurements, the α-helical content of proteins can be estimated, but it is evident from Fig. 8.13 that the presence of β structures will cause an appreciable perturbation, and it is very likely that the irregular but fixed-chain conformations that constitute a large proportion of most globular proteins also make a contribution that cannot properly be treated as if they were random coils. Aromatic chromophores likewise will contribute, particularly in the 200–240 nm region, and will do so to an unknown extent. In short, such structural estimates are necessarily highly approximate. Table 8.4 gives some examples, together with the α-helical content where this is known from x ray.

TABLE 8.4
ORD–CD ESTIMATES OF α-HELICAL CONTENT

Protein	Optical estimate	X-ray content
Tropomyosin	0.90	—
Hemoglobin	0.7–0.8	0.80
Myoglobin	0.7–0.8	0.80
Lysozyme	0.30	0.24
Ribonuclease	0.15	0.22
Chymotrypsinogen	0.10	—
Insulin	0.40	—
Bovine serum albumin	0.45	—
β-Lactoglobulin	0–0.10	—

8.3.3.2. Luminescence

The fluorescence and phosphorescence of proteins, like that of poly-peptides (Section 8.2.3.3), depend upon the presence of the aromatic amino acids phenylalanine, tyrosine, and tryptophan, in order of increasing wave-length of emission. Such measurements are significant because the quantum yield is highly sensitive to the conformational state of the protein. Since tryptophan tends to dominate the emission completely when present, pro-teins are divided into two classes (Teale, 1960) according to whether they have this amino acid or not. We have seen that ribonuclease and insulin do not have tryptophan, but this class is much smaller than the class of those that do. All proteins in the latter class have Phe and Tyr as well. Beyond this general-ization, there is no systematic way of treating the emission behavior of proteins; it is quite complex, making it necessary to treat each protein individually. For this reason, the subject is not discussed further here. The interested reader is referred to the excellent and comprehensive review of Longworth (1971).

8.3.3.3. Vibrational Spectroscopy

Infrared and Raman spectra of proteins have far too many lines to be fully resolved, but tend to group themselves into bands some of which can be assigned to specific amino acids. The peptide bands (see Section 8.2.3.4) of fibrous proteins yield some information of interest. For example, the amide A band in collagen is shifted from the normal value of $3300 \, \text{cm}^{-1}$ to approxi-mately $3330 \, \text{cm}^{-1}$ and is taken to indicate that in the NH $\cdots$ O=C hydrogen bonds of collagen, the N $\cdots$ O distance is greater than the usual 0.27 nm. The most significant application of vibrational spectroscopy to proteins, however, is the *resonance Raman* technique, by which particular vibrational bands may be enhanced and selected for study. This is done by exciting the Raman spectrum with a laser source in the electronic absorbing region of a

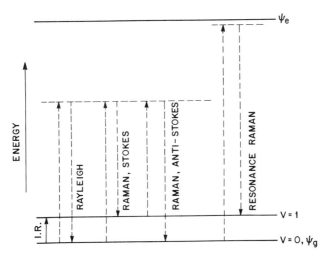

FIG. 8.24.   Energy level scheme indicating resonance Raman scattering.

protein chromophore, i.e., to an upper state close to an excited electronic state ψ_e, as shown in Fig. 8.24. (If the exciting radiation corresponds *exactly* to an electronic transition, the emission is normally termed *fluorescence*.)

The resonance Raman technique has been applied to hemoglobin and cytochrome *c* (Sections 8.3.2.1 and 8.3.2.2) in their various states of oxidation and ligand binding; these proteins absorb strongly in the visible and near UV region. Observation of the enhanced lines allows questions concerning the distribution of electrons in the heme ring and in the bound molecules to be answered. For example, the Raman bands of oxyhemoglobin appear consistent with a formulation not as $Fe^{2+}O_2$ but as $Fe^{3+}O_2^-$, the oxygen showing a stretching frequency characteristic of superoxide. For details, see the review of Spiro (Spiro, 1974).

8.3.3.4.   Nuclear Magnetic Resonance

The earliest reported proton spectra of proteins showed little detail because the low observing frequencies (40 or 60 MHz) did not permit sufficient separation of chemical shifts. At frequencies attainable with superconducting magnets (220–360 MHz), resolution is enormously enhanced and in many regions of the spectra resonances of individual protons can be readily observed.

The proton spectra of proteins all bear a strong family resemblance. It is the differences in detail, caused by substrate or inhibitor binding, chain folding and unfolding, and other structural alterations, that are important. Figure 8.25 shows the proton chemical shifts observed for all the common

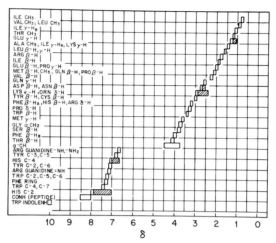

Fɪɢ. 8.25.   Proton chemical shifts in proteins and polypeptides in aqueous solution.

amino acids as reported for protein chains in a primarily aqueous environment, i.e., in the denatured, unfolded state (McDonald and Phillips, 1969). The deviations from these values observed in the native state are generally not large but sufficient so that the native and denatured spectra differ very noticeably. Particularly marked are the ring-current shifts observed for groups which in the folded state are positioned in the shielding regions of the ring currents of aromatic groups and heme rings. (Very large deviations, 10–15 ppm either upfield or downfield, owing to hyperfine interactions are observed for heme proteins in paramagnetic states.)

The least shielded protons (Fig. 8.25) are those of tryptophan indole NH (about 10δ), peptide NH ($8.0–8.5\delta$), and C-2 histidine and arginine NH protons ($7–8\delta$). Aromatic and C-4 histidine protons appear next, near $6.8–7.5\delta$, followed by a pronounced "window," which is observed at about $5–6.5\delta$ in all protein spectra. Between 6 and 4δ appear the broad and usually undifferentiated resonances of the α-CH protons; H_2O or HDO peaks commonly appear in about the same position. Then come a wide variety of side-chain methylene and methyl resonances. Most shielded, near 1δ, are the methyl groups of the aliphatic side chains of valine, leucine, and isoleucine.

Figure 8.26 shows the 360-MHz spectrum of native ribonuclease (0.01 M in D_2O with buffer, pH 7.5). The large, narrow resonance at $\sim 5\delta$ is that of residual HDO (Patel, 1976). At low field, in the region 7.8–8.2 are the C-2 protons of the four histidine residues discussed in Section 8.3.2.4. These are more clearly seen when the interfering peptide NH protons have been exchanged for deuterium. A particularly elegant feature of protein NMR spectroscopy is that it allows such ionizable residues to be individually

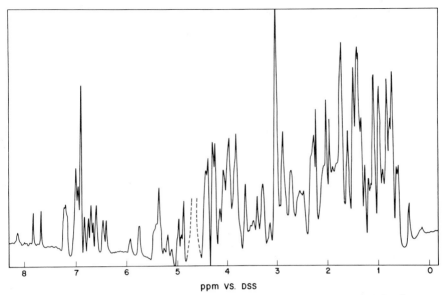

FIG. 8.26.   360 MHz Proton spectrum of ribonuclease in D_2O at pH 6.0 (D. J. Patel, unpublished observations).

titrated. When the pH is raised and the imidazolium group of histidine is

$$\tag{8.14}$$

deprotonated, the C-2 and C-4 proton peaks move upfield. By observing the effects of bound inhibitors, which should influence only the His-12 and His-119 residues at the site of enzymatic activity, and by other means that we cannot describe in detail here (Roberts and Jardetzky, 1970; Patel *et al.*, 1975), it has been possible to assign these resonance to specific histidines. The titration curves of the C-2 proton resonances (the C-4 protons tend to be masked by aromatic groups) are shown in Fig. 8.27. From these curves it is found that His-12 and His-119 have pK_a values of 6.1 and 6.5, respectively. Such information is clearly of significance in deducing the mechanism of action of this enzyme, which as yet has not been entirely clarified.

The proton NMR of proteins has been reviewed by Roberts and Jardetzky (1970). More recent reviews are those of Bovey (1972, Chapter XIV), James

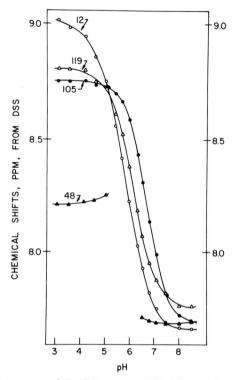

FIG. 8.27. Titration curves of the C-2 protons of histidine in ribonuclease. (The His-48 resonance broadens at the midpoint of transition and cannot be readily followed.)

(1975), Dwek (1973), and Wüthrich (1976). Carbon-13 NMR is also of interest and potential utility in the study of proteins. This field is reviewed in the monograph of Wüthrich and also by Komoroski et al. (1976, Vol. 2, Chapter 4).

8.4. Nucleic Acids

The nucleic acid chain is composed of pentose units connected by phosphate units in regular alternation. Purine and pyrimidine bases are attached to the sugar units in a sequence that provides the genetic code. The base–sugar–phosphate unit is termed a *nucleotide unit*:

Two classes of nucleic acids are recognized: those in which the sugar is D-ribose, called *ribonucleic acids* or RNA, and those in which the sugar is 2-deoxy-D-ribose, called *deoxyribonucleic acids* or DNA. DNA occurs in

D-ribose   2-deoxy-D-ribose

cell nuclei and is the genetic material that transmits genetic information from one generation to the next. (In some viruses, RNA plays this role.) The DNA replicates upon cell division and performs the essential function of serving as a template for the transcription of this information onto *messenger* RNA, or *m*RNA. The *m*RNA molecule then moves to the *ribosome*, a large complex nucleoprotein (that is, one having both polypeptide and polynucleotide chains) composed of two major subunits, one approximately twice the size of the other, containing *ribosomal* RNA, or *r*RNA. Of the two subunits, the larger contains two *m*RNA chains and about 35 protein chains; the other is composed of one *m*RNA and about 20 protein chains. The *m*RNA serves as an assembly line where amino acid residues are joined in proper sequence to form finished protein molecules. This process requires a third type of RNA, *transfer* RNA (*t*RNA, formerly designated as *soluble* or *s*RNA), which recognizes specific amino acids and adds them at the proper point as directed by the *m*RNA. (For a more complete discussion of these matters see "Biophysical Chemistry," a compilation of articles from the *Scientific American*; Bloomfield *et al.*, 1974; Stryer, 1975; see also Section 8.4.3).

   Messenger RNA composes only a small fraction of the RNA present in the cell cytoplasm, most of which is *t*RNA and *r*RNA. It is continually degraded by ribonucleases (Section 8.3.2.4) and regenerated. It is polydisperse in molecular weight, in the range 200,000–500,000 and higher. Ribosomal RNA is the predominant type; in bacteria it may constitute as much as 80% of the total. Its molecular weight varies from about 560,000 to 1,100,000 depending upon the ribosomal subunit in which it occurs. Transfer RNA molecules contain only 77–85 nucleotide units (molecular weights of 28,000 to 30,000). The detailed structure of *t*RNA molecules is considered in Section 8.4.1.

8.4.1. CHAIN STRUCTURE AND CONFORMATION

   The pyrimidine bases occurring in DNA are thymine and cytosine, called T and C, respectively. In RNA, uracil (U) almost entirely replaces thymine.

These initials, and those indicated below, are commonly employed as abbreviations not only for the bases themselves but also for the nucleotide units containing them in polynucleotide chains:

| pyrimidine | thymine (T) (5-methyl-2,4-dioxy-pyrimidine) | cytosine (C) (2-oxy-4-amino-pyrimidine) | uracil (U) (2,4-dioxypyrimidine) |

Other pyrimidine bases occurring in minor proportions in RNA are

dihydrouracil (U^h, D, or diHU) (5,6-dihydro-2,4-dioxypyrimidine)

5-methylcytosine (5 MeC or C^{5-Me}) (5-methyl-2-oxy-4-aminopyrimidine)

N^4-acetylcytosine (AcC)

The pyrimidine bases may exist in both lactim and lactam tautomers:

lactim   lactam

The lactam form, which is nonaromatic, strongly predominates except at very high pH and is the only form normally considered to be present. This is important in relation to interbase hydrogen bonding (p. 502) and also because pyrimidines as a consequence do not have ring currents and therefore do not produce the NMR shielding effects expected in the neighborhood of aromatic groups.

The purine bases occurring in DNA are adenine and guanine. Again, lactim forms can exist but are not considered to be present in significant proportion.

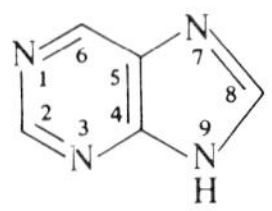

purine 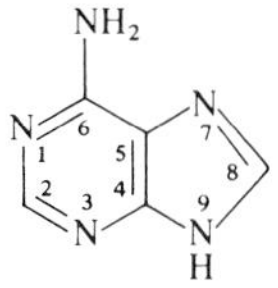   adenine (A)   guanine (G)
                    (6-aminopurine)   (2-amino-6-oxypurine)

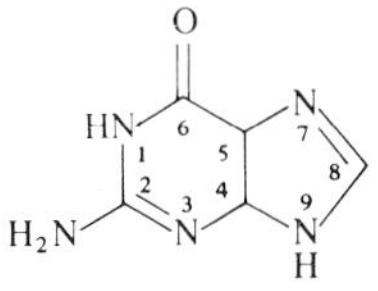

In RNA, a number of other purine bases occur in small but significant proportions. Some of these are

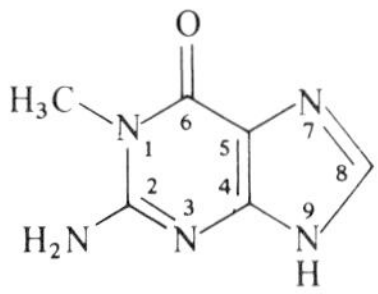

1-methylguanine N,N'-dimethylguanine
(G^m, G^{1Me}, MeG, 1MeG)   ($\underline{G}^m$, G^{2Me}, diMeG, 2MeG)

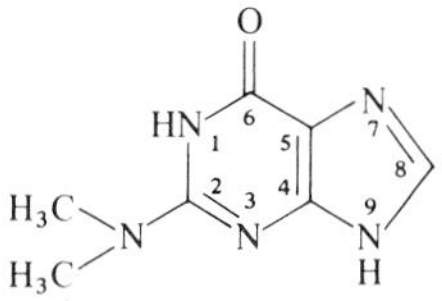

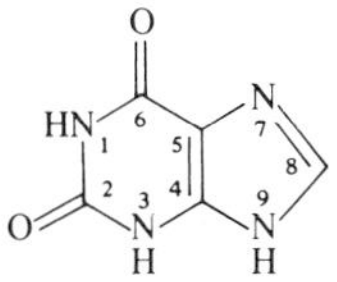

xanthine (X)   hypoxanthine (I)   1-methylhypoxanthine
(2,6-dioxypurine)   [6-oxypurine; the   (I^m, I^{Me}, 1MeI;
                         base of inosine]   the base of
                                                   1-methylinosine)

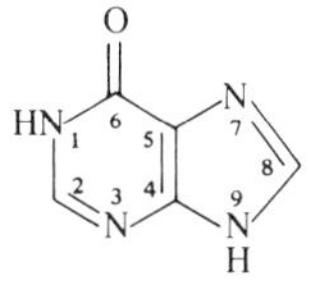

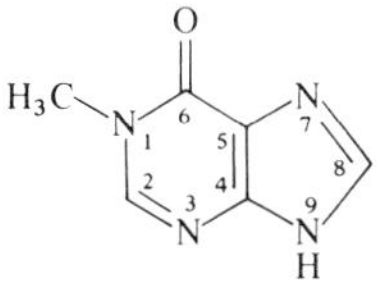

In DNA and RNA, the purine bases adenine and guanine are attached to the main chain by a bond between the 9-nitrogen of the purine and the 1-carbon (designated 1') of the pentose sugar. The pyrimidine bases cytosine and thymine are attached by their 1-nitrogen to the 1'-carbon of the pentose. The phosphate links between the pentose units connect the 5'-carbon of one to the 3'-carbon of its neighbor. The chain is conventionally written with the pentose 5'-OH unit at the left and the 3'-OH (and associated phosphate) at the right. The proportions and sequence of bases are of course not necessarily as represented in the segment of a DNA chain shown here:

Segment of DNA chain

In the DNA molecule two such chains are wound in a right-handed helical conformation about a common axis, as first deduced from x-ray fiber diagrams by Watson and Crick (Watson and Crick, 1953). As shown in Fig. 8.28, the two chains are linked by hydrogen bonds between the bases, which are on the inside of the helix. Their planes are perpendicular to the helix axis. The planes of the sugar rings (which are actually puckered) are approximately perpendicular to the base planes. The steric requirements are such that an adenine must be paired with a thymine at the same level in the other chain and a guanine with a cytosine:

All the bases are paired off in this manner along the entire helix, the chains being thus necessarily complementary in sequence. They also run in opposite directions, the sequence of main-chain atoms going one way in one chain and the opposite way in the other. The diameter of the helix is 2.0 nm. Adjacent bases are separated by 0.34 nm, each being rotated by 36° in relation to its neighbors. Thus 10 bases complete one turn of the helix which, apart from the differences among the bases, has tenfold symmetry. Upon heating, the twin helix is denatured; the hydrogen bonds are disrupted and the two chains form separate single strands. Upon cooling, the helix is partially reconstituted. For shorter sequences than those of the native molecules, the helix-to-single strand transition is entirely reversible (see Section 8.4.5).

The molecular weights of DNA molecules vary with the source, but are all extremely high. Thus, the polyoma virus (or SV40) has 4600 base pairs, corresponding to a molecular weight of approximately 3,000,000. If the

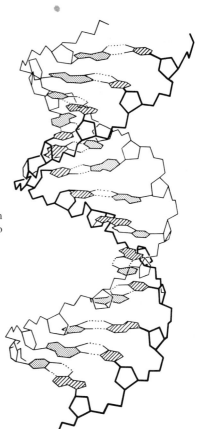

FIG. 8.28. The DNA double helix. The main chain of one is drawn more heavily so that the two chains can be more readily distinguished.

double helix were stretched out straight, it would be about 1.56×10^3 nm or 0.0016 mm in length. Other viral DNAs are from ten- to fiftyfold larger. Bacterial DNA may be of great length. That from *Escherichia coli* has 3,400,000 base pairs or a molecular weight 2.32×10^9; its contour length is 0.12 mm, a thousand times longer than the bacterium itself. A DNA from the fruit-fly *Drosophila melanogaster* has 6.2×10^7 base pairs, a molecular weight of 4.1×10^{12} and a contour length of 21 mm! Even the stresses of pouring or pipetting their solutions fragment such molecules readily, and so special precautions must be taken in order to measure their molecular weights correctly. Many DNA molecules are *circular* in form, in the sense that the chain is continuous. In their functioning state within the cell, all DNAs are necessarily very compactly folded and collapsed. They are thus not to be considered as entirely rigid rods and even in solution (Section 8.4.5) may show smaller end-to-end distances than expected for rods because of inherent flexibility and possibly also because of occasional defect regions where the bases do not mate correctly.

The nucleic acid chain fragments consisting of the base with the sugar attached are termed *nucleosides*. The DNA nucleosides may be visualized from the chain structure shown on p. 502. They are termed thymidine (T), deoxycytidine (dC), deoxyadenosine (dA), and deoxyguanosine (dG). Structures occurring in RNA are the *β-ribofuranosides*:

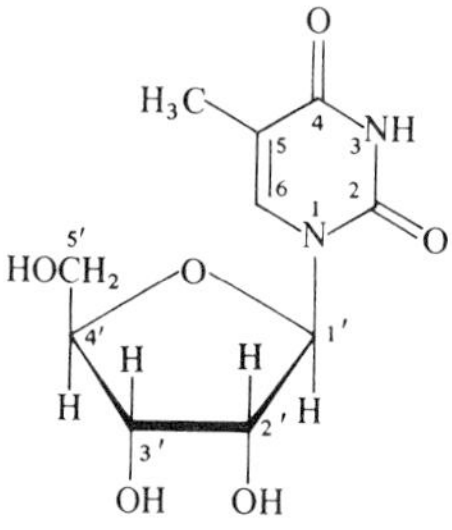

ribothymidine (rT) 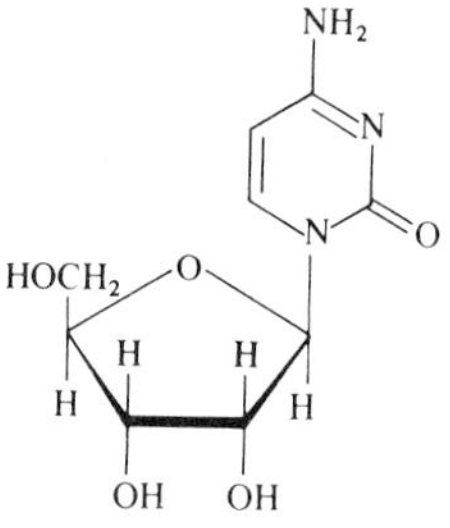   cytidine (C) 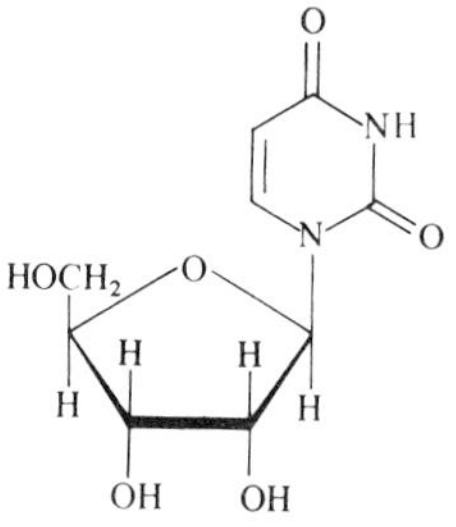   uridine (U)

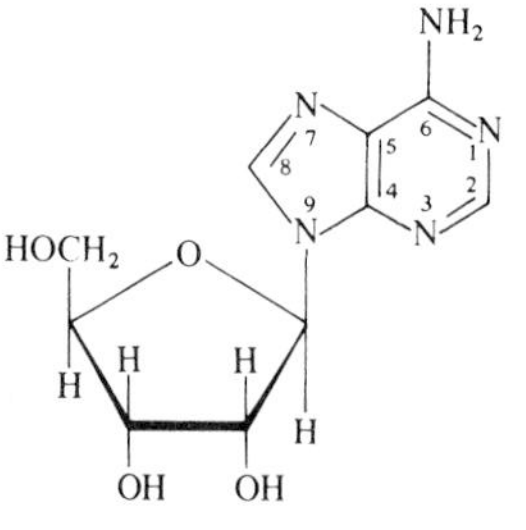

adenosine (A) 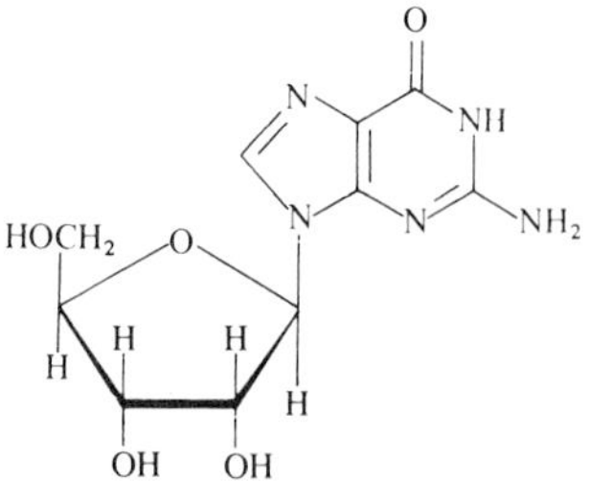   guanosine (G) 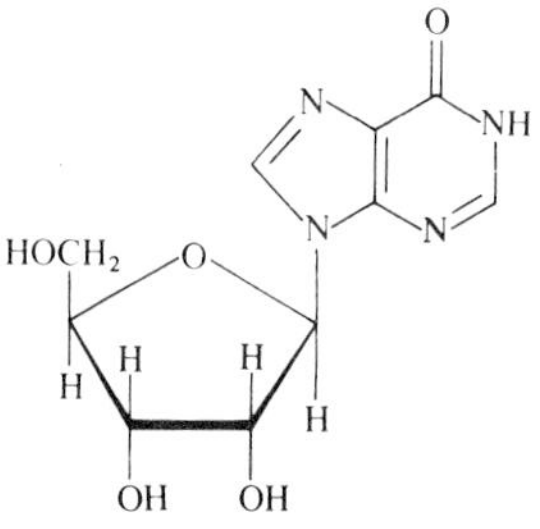   inosine (I)
                                                  (base is hypoxanthine)

In addition to the nucleosides of the several variant bases shown on p. 501, mention should be made of pseudouridine (ψ), which differs from uridine in having the uracil attached to the ribose at C-5. It appears to occur in all species of tRNA.

pseudouridine (ψ)

The *mononucleotides* are phosphoric acid monoesters of the nucleosides. The phosphate linkage in ribonucleotides may occur at the 2', 3', or 5' position; the last two correspond to the naturally occurring structural units in the RNA chain:

adenosine-3'-monophosphate
(3'-AMP)

adenosine-5'-monophosphate
(AMP), adenylic acid

The other 5'-monophosphates are

ribothymidine-5'-monophosphate
(rTMP) thymidylic acid

cytidine-5'-monophosphate
(CMP), cytidylic acid

uridine-5′-monophosphate
(UMP), uridylic acid

guanosine-5′-monophosphate
(GMP), guanylic acid

inosine-5′-monophosphate
(IMP), inosinic acid

The conformation of RNA is less certain than that of DNA. Although capable of forming double helices, as in certain viral RNAs, the single-stranded form is more common. Messenger RNA is single stranded. Native rRNA is combined with protein, as we have seen. Both probably have coiled structures with the bases "stacked" in a parallel array because of the lower free energy of this form, presumably a consequence of the π–π interaction of the rings, particularly between purines. Because of the predominance of the single-strand conformation, $1:1$ ratios of A to U and G to C (or of sterically similar minor bases) need not be maintained, and generally are not.

In some RNA molecules, both double-helical and single-stranded regions or loops may occur. Thus, tRNA molecules contain sequences that are at least partially complementary and others that are not, giving rise to the "clover-leaf" model, in which the recognized functional regions of the chain consistently appear in corresponding portions of the sequence. The primary structures of some 60 tRNA molecules are now known (sequencing methods are discussed in Section 8.4.2). Figure 8.29 shows the structures of two species, tRNA$_{\text{yeast}}^{\text{ala}}$ and tRNA$_{\text{yeast}}^{\text{phe}}$. Both sequences show the following common features: the 3′-OH terminal sequence ACC that binds the amino acid and is invariant in all tRNA molecules (the fourth unit is not always A); the TψC loop on the right; the anticodon loop below, consisting of three units, the

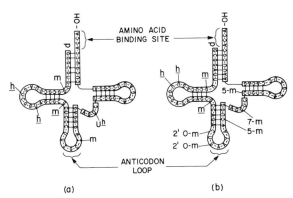

FIG. 8.29. Sequences and "clover-leaf" structures of (a) alanine *t*RNA of yeast and (b) phenylalanine *t*RNA of yeast.

next unit beyond it being always U; and the Uh or diHU loop on the right. The double helix or duplex regions of the molecule are indicated by lines representing base–base hydrogen bonds. The imperfection of these structures is exemplified by the UU pair in tRNA$_{yeast}^{ala}$, which cannot form hydrogen bonds. The three-dimensional structures of tRNA molecules, deduced from x-ray diffraction, reveal two segments of double helix, each corresponding to about one complete turn (Fig. 8.30). The molecule is roughly L-shaped

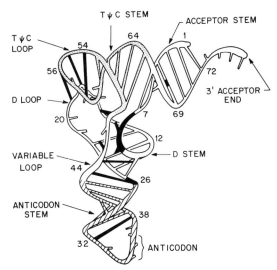

FIG. 8.30. Schematic representation of the three-dimensional structure of tRNA$_{yeast}^{phe}$ (from Rich, 1976). Copyright © 1976 by the American Association for the Advancement of Science.

with the amino acid binding site and anticodon loop far from each other, as suggested by the clover-leaf representation.

8.4.2. DETERMINATION OF SEQUENCE

The means for the determination of nucleic acid sequences are in general the same as those for sequencing proteins (Section 8.3.1), i.e., to break the long chains into manageable fragments by scission at known types of links, separation of the fragments, which are radioactively labeled, and the determination of their individual compositions and sequences. For polynucleotides, chemical reagents such as those described for proteins have only recently become available and primary use has been made of enzymes of established specificity. The first major progress was made with RNA molecules. Pancreatic ribonuclease (Section 8.3.2.4) catalyzes hydrolytic scission on the 3' side of U or C units to give fragments ending in Up or Cp. A variant ribonuclease from a mold, called ribonuclease T_1, scissions RNA on the 3' side only of G residues. The resulting oligonucleotides can be separated by chromatography and the overall sequence determined by fitting together the fragments. The task is made harder rather than easier by the fact that only four types of residues are involved, as the probability of multiple occurrence of specific sequences is much higher. The first complete sequencing of an RNA was that of alanine tRNA by Holley and his collaborators (see Holley, 1968).

Until very recently, the sequencing of DNA chains was a much more difficult task than that of RNA chains because of the lack of suitable enzymes for specific scission, and the only recourse (not always applicable) was to deduce the DNA structure from the transcribed RNA. The discovery of *restriction enzymes*, isolated from bacteria, has made the sequencing of DNA more facile than that of RNA (see Kolata, 1976). These enzymes recognize specific DNA sequences and cut the chain at both ends of the sequence. About 45 enzymes with different specificities are known. The fragments are then separated according to length by gel electrophoresis, which has been improved in resolution to the point where sequences of the order of 100 nucleotides in length can be separated if they differ by only one unit. After isolation these shorter fragments may be sequenced by use of chemical reagents that break the chain only at purines or only at pyrimidines, and which, under the right conditions, can also selectively break at A or G or at C or T. Another approach is to synthesize specific DNA fragments by replication from a known DNA (see Section 8.4.3), but to stop it at a specific point by omitting from the synthesizing system the nucleotide necessary to continue replication from that point. By omitting cytosine, for example, one obtains all sequences of the DNA concerned that end just before a cytosine. An example of the use of these methods is the sequencing of two lengths, each about 250 nucleotides long, of the single-stranded circular DNA of bacteriophage ϕX174, and the relating of these to the amino acid sequences of the proteins for which they code. The complete chain contains about 3000 nucleotides (Sanger, 1975).

8.4.3. Synthesis of Nucleic Acids

Synthetic RNA-like polynucleotides may be prepared by treating ribo-nucleoside diphosphates with the enzyme *polynucleotide phosphorylase* in the presence of magnesium ion:

adenosine-5'-diphosphate (ADP) polyadenylic acid (poly-A)

Polyuridylic acid (poly-U), polycytidylic acid (poly-C), and polyinosinic acid (poly-I) can be made in the same way from the corresponding nucleoside diphosphates. The conformation and solution behavior of these polymers will be discussed in Section 8.4.5. By using mixtures of different nucleoside diphosphates, random copolymers can be synthesized. These have proved of central importance in the working out of the *genetic code* (Section 8.4.4).

Corresponding DNA-like polymers can be prepared only if all four DNA bases—A, T, G, and C—are present as their nucleoside *triphosphates*, and if the enzyme *DNA polymerase* and single-stranded DNA are present, the latter serving as a primer. If, for example, the primer contains only A and T, the resulting polymer will contain only A and T. An analogous process occurs in the replication of DNA upon cell division.

Internucleotide bonds are formed by condensation of the 5'-phospho-monoester end group of a mono- or oligonucleotide with the 3'-hydroxyl end group of a second mono- or oligonucleotide, other functional groups being suitably protected as in polypeptide synthesis (Section 8.2). 5'-Hydroxyl groups are customarily protected using the mono-p-methoxytrityl group (p-CH$_3$O·C$_6$H$_4$(C$_6$H$_5$)$_2$C–) and 3'-hydroxyl groups by acetylation. The amino groups of adenine and cytosine rings are commonly blocked with N-benzoyl and N-anisoyl (p-$CH_3O \cdot C_6H_4CO$–) groups, respectively, while for guanine amino groups the N-isobutyryl group has been found preferable (Weber and Khorana, 1972). Because of less than complete yields at each step, the most practicable strategy for the synthesis of long chains is the chemical synthesis of oligonucleotide units of from 5 to 20 units length followed by their linking together using specific joining enzymes called *DNA ligases*. By these methods the gene of *Escherichia coli*, which codes for the production of tyrosine *t*RNA, was synthesized by Khorana and his colleagues. This *t*RNA is actually synthesized in the cell as a 126-unit chain, which is later chopped down by enzymes to 76 bases. They first synthesized the sequences for the 126-unit molecule and also the additional DNA sequences at the beginning and end of the main sequence known as the *promoter* and *terminator*, yielding finally the 206-unit double helix constituting the complete gene itself. This was found to be biologically active.

8.4.4. THE GENETIC CODE, PROTEIN SYNTHESIS, AND DNA REPLICATION

We have seen that the essential steps in protein synthesis are the following:

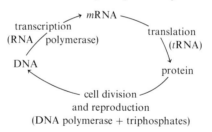

We now consider further some of the essential details of these processes. Of these, the most important is the genetic code by which a sequence of DNA bases becomes translated via *m*RNA into the protein molecules, chiefly enzymes. In this way the cell maintains its existence and its ability to divide and propagate. Although it was recognized over two decades ago that such a code must exist, its nature was unclear because the base sequences of DNA and *m*RNA molecules were unknown. It was evident that a two-base code of four kinds of bases could code for only 4^2 or 16 amino acids, so a three-base code, capable of 4^3 or 64 "words," seemed most likely even though capable of transmitting more information than required to encode 20 amino acids. It also appeared likely that the triads of bases constitute a nonoverlapping code, since otherwise genetic errors involving the replacement of only a single amino acid, as in sickle-cell hemoglobin (Section 8.3.2.1), would be hard to explain. However, the actual code remained obscure.

An important first step was the finding by Nirenberg (Nirenberg, 1973) that the addition of poly-U to a cell-free protein-synthesizing system caused the production of polyphenylalanine. The code for phenylalanine must then be UUU. By use of random ribonucleotide copolymer, synthesized as described in the previous section, the bases involved in coding for other amino acids could be deduced from the resulting polypeptides. The specific sequences of each triad were determined following the discovery that trinucleotides, such as pUpUpU, promoted the binding of specific *t*RNA molecules, in this instance phenylalanine *t*RNA, to the ribosome. All 64 of the possible trinucleotides could be synthesized, and from these the complete code was worked out. This is shown in Table 8.5. The code is highly redundant, only two amino acids, methionine and tryptophan, corresponding to a single base sequence, or *codon*, while arginine and leucine are coded for by six. Most of the redundancies reside in the third base of the codon, which is evidently less stringent in its requirements. The "stop" codons are signals for protein synthesis, which occurs in the direction from the N- to the C-terminus, to cease; these are read by specific enzymes called *release factors* rather than by a *t*RNA molecule. It is not at present clear why AUG and GUG are sometimes read as start signals and sometimes as nonterminal methionine and valine, respectively.

TABLE 8.5
THE GENETIC CODE

Amino acid	RNA base sequence	Amino acid	RNA base sequence
Ala	GCU GCC GCA GCG	Lys	AAA AAG
Arg	CGU CGC CGA CGU AGA AGG	Met	AUG
Asn	AAU AAC	Phe	UUU UUC
Asp	GAU GAC	Pro	CCU CCC CCA CCG
Cys	UGU UGC	Ser	UCU UCC UCA UCG
Gln	CAA CAG	Thr	ACU ACC ACA ACG
Glu	GAA GAG	Trp	UGG
Gly	GGU GGC GGA GGG	Tyr	UAU UAC
His	CAU CAC	Val	GUU GUC GUA GUG
Ile	AUU AUC AUA	Stop	UAA UAG UGA
Leu	UUA UUG CUU CUC CUA CUG	Start	AUG GUG

The amino acids do not recognize the *m*RNA codons as such but, as we have seen, are brought to the proper site by the *t*RNA molecules that have an anticodon region (Figs. 8.29 and 8.30) complementary to the binding site on the *m*RNA molecule. The *t*RNA molecules likewise cannot by themselves recognize the proper amino acid corresponding to their anticodon loops. The actual recognition of both the amino acid and its appropriate *t*RNA is dependent on specific enzymes for each amino acid, called *aminoacyl–tRNA synthetases*. This enzyme catalyzes the reaction of the amino acid with adenosine-5'-triphosphate (ATP) (analogous to adenosine-5'-monophosphate, shown on p. 505 but with a triphosphate group

$$\text{HO}-\overset{\overset{\displaystyle O}{\|}}{\underset{\underset{\displaystyle OH}{|}}{P}}-O-\overset{\overset{\displaystyle O}{\|}}{\underset{\underset{\displaystyle OH}{|}}{P}}-O-\overset{\overset{\displaystyle O}{\|}}{\underset{\underset{\displaystyle OH}{|}}{P}}-O-\cdots$$

at the 5' position) and the ribose of the terminal A on the *t*RNA amino acid binding site:

aminoacyl-*t*RNA

In this activated form, the amino acid is able to couple with its predecessor in the polypeptide chain being built up on the *m*RNA template.

After the synthesis of proteins and the development of the cell is complete, the cycle on p. 510 begins again (for cells that regularly divide such as those of bacteria) by replication of DNA. This is believed to occur in both directions on circular DNA (we recall that most natural DNA is circular) and to involve only short single strands, most of the DNA remaining double-helical. Both unwinding strands serve as templates for the synthesis of new DNA, which is formed in the 5' to 3' direction under the direction of DNA synthetase and, since the process is not always continuous, DNA ligases as well to join the fragments. One of the unwinding strands also serves for the synthesis of *m*RNA, as we have seen. The first cell division thus results in DNA helices, one strand of which arose from the original cell and one of which is newly generated; in the next generation, one-fourth of the newly formed DNA helices contain the original DNA, and so on (for details, see Stryer, 1975).

8.4.5. OPTICAL AND SPECTROSCOPIC PROPERTIES OF POLYNUCLEOTIDES

8.4.5.1. Electronic Spectra

One of the most widely employed methods for the study of conformational changes in polynucleotides is ultraviolet absorption. Figure 8.31 shows the spectra of the five principal bases as their ribonucleosides in neutral aqueous solution, the amino groups of A, G, and C being unprotonated (Voet, 1963; for a detailed discussion of band assignments see Bloomfield *et al.*, 1974). The purine nucleosides adenosine and guanosine show similar spectra; those of the pyrimidines cytidine, uridine, and thymidine likewise exhibit a family resemblance, their absorption bands being generally weaker than those of the purines. From these differences, it is possible by mathematical manipulation to gain an approximate notion of the composition of a DNA

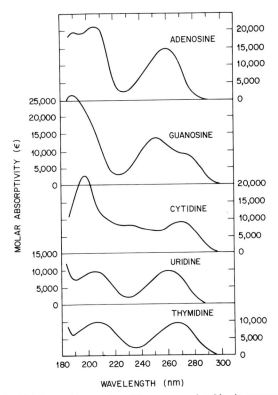

FIG. 8.31. The UV absorption spectra of five mononucleosides in aqueous solution (from Voet *et al.*, 1963; Bloomfield *et al.*, 1974, p. 24).

or RNA chain. However, the main power of UV absorption is in following the helix-to-single strand transition. When the bases are stacked in regular array in the duplex structure, there is a very marked hypochromic effect recalling that occurring in the α-helical state of polypeptides (Section 8.2.3), but even more marked. Thus, the molar extinction coefficient of an RNA double helix is approximately 6000, whereas that of the sum of the component nucleosides is nearly 12000. Single-strand RNA and DNA have absorptions corresponding closely to the latter, and therefore the measurement of absorbance at a band maximum as a function of temperature furnishes a convenient way of following the helix–coil transition or "melting" of the duplex structure. The term melting is appropriate, in that the process is a cooperative one. Initiation or nucleation involves "destacking" of bases and breaking of hydrogen bonds in the middle of the helix, a relatively improbable event; following this step, further unzipping is easier, and the transition has the typical sigmoid form shown in Fig. 8.32 for a bacterial DNA at nine different concentrations of KCl (Marmur and Doty, 1962). The melting point T_m is taken as the inflection point of the curve. It is evident that the presence of salt stabilizes the helix, a general phenomenon that is attributed to screening of the negative charges on the phosphate backbone. The fact that G–C base pair interactions involve three hydrogen bonds whereas A–U interactions involve only two (p. 502) would lead one to expect that double helices formed from poly-G and poly-C would exhibit higher melting temperatures than those formed from poly-A and poly-U. This is indeed the case. Analogous behavior is found for the self-complementary duplexes formed from the alternating copolymers poly(G–C) and poly(A–U) and for

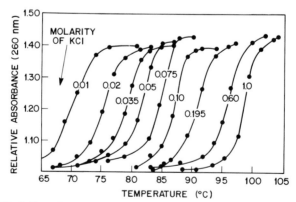

FIG. 8.32. The helix–coil transition of DNA from *E. coli* as a function of KCl concentration, measured by absorption at 260 nm (from Marmur and Doty, 1962; Bloomfield *et al.*, 1974, p. 295).

the corresponding DNA-like analogs poly(dG–dC) and poly(dA–dT); the latter differ in melting point by nearly 40°. The T_m values of natural DNA from different sources are found to vary from about 80 to 100° and to be approximately proportional to the content of G–C pairs.

The CD (and ORD) spectra of mononucleosides and mononucleotides arise from the asymmetry of the sugar residues, the base chromophores themselves being planar and having no asymmetric centers. The formation of a helical conformation makes a large contribution to the optical asymmetry, and as a consequence the CD spectra of DNA and RNA differ greatly both in band intensity and to a lesser extent in band position from those of the component mononucleosides. This is illustrated in Fig. 8.33 (Allen et al., 1972; Cantor et al., 1970). The UV spectra of double-helical DNA and RNA

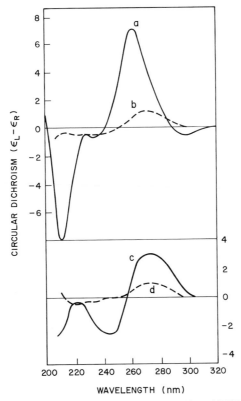

FIG. 8.33.   The circular dichroism of double-stranded DNA and RNA compared with their component mononucleosides (DNA data from Allen et al. (1972); RNA data from Samejima et al. (1968)). Nucleoside spectra calculated from the base composition and CD data of Cantor et al. (1970); also see Bloomfield et al. (1974, p. 134).

show little change with temperature in the region below the melting transition, whereas the CD spectra, being very sensitive to details of the helical conformation, show changes indicative of premelting structural alteration.

8.4.5.2.   Nuclear Magnetic Resonance

The study of native DNA by proton NMR is severely hampered by the phenomenon of dipolar broadening, which we have seen (Chapter 3, Section 3.3.3; Chapter 6, Section 6.2) to be general for polymers but which is particularly marked in nucleic acid spectra because of their high molecular weight and limited segmental mobility. At temperatures above the duplex–single-strand transition, a number of resonances can be distinguished and assigned (McDonald *et al.*, 1967). Much greater detail can be observed in the spectra of copolymers of more regular structure, partly because of the smaller number of different chemical shifts but chiefly because of the im-

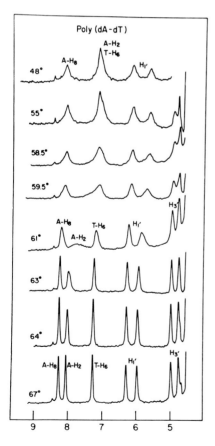

FIG. 8.34.   The temperature dependence of the 360 MHz proton NMR spectra of poly(dA–dT) in 0.1 *M* phosphate, D$_2$O, pH 6.3. (4.5 to 9 ppm vs. sodium 2,2-dimethyl-2-silapentane-5-sulfonate (DSS) as zero (Patel and Canuel, 1977)).

perfect nature of their helices. Thus, the duplex structure of the alternating copolymer poly(dA–dT) is formed by folding of the chain back on itself, requiring the existence of at least one loop. Figure 8.34 (Patel and Canuel, 1977) shows the 360-MHz proton spectra in D_2O of a poly(dA–dT) containing approximately 1000 base pairs as a function of temperature. Well before the helix–coil transition at $62°$, the H_2 and H_8 resonances of the adenine and the H_6 resonance of the thymine rings, as well as the nonexchangeable sugar ring protons, can be clearly resolved even though linewidths are of the order of 80–100 Hz. The lines sharpen at the transition and move to less shielded chemical shift positions. In contrast to optical data, NMR provides a detailed picture of the helix structure as it approaches and passes through the melting transition. Premelting phenomena, already alluded to in Section 8.4.5.1, can be clearly seen in the upward slopes of most of the chemical shift plots below $62°$.

8.5. Polysaccharides

The structures and conformations of polysaccharide chains have been determined almost entirely by classical chemical methods and x-ray diffraction. Optical rotation, particularly of the sugars corresponding to the monomeric units, has been a traditional method of observation with a long history, but circular dichroism and electronic absorption spectroscopy cannot normally be employed because the absorption maxima occur in the vacuum ultraviolet region, i.e., well below 180 nm. Vibrational spectroscopy has had some utility. NMR has been of considerable value in the study of sugar ring conformations, but its use for the study of polysaccharides has likewise been very limited, although carbon-13 spectroscopy appears promising (Komoroski et al., 1976, pp. 252–258).

The term polysaccharide includes a variety of chains, many (but not all) of which are composed of long sequences of glucose units. In *starch* and *cellulose* these are connected by α and β linkages, respectively:

α-1,4 linkage   β-1,4 linkage

The chain conformation of the 6-membered pyranose ring is dictated by steric requirements, most or all of the substituents being in the energetically preferred equatorial conformation.

8.5.1. CELLULOSE

Cellulose is the most abundant and widely distributed natural polymer and is the principal structural element of wood, constituting about half of the material of the cell wall. It has been estimated that the growth of plants corresponds to the synthesis of 10–100 billion tons of cellulose annually. Cotton is nearly pure cellulose and together with flax is the main source of cellulose for use as fiber. Wood cellulose occurs in association with *hemicelluloses*, built from β-1,4 xylose units, and *lignin*, which is a nonpolysaccharide of complex structure. Hydrolysis of cellulose yields *cellobiose* and *cellotetraose*, establishing the β-1,4-linked chain structure:

cellulose

X-ray fiber diagrams indicate that the length of the unit cell along the chain axis is about 1.03 nm, close to the length of the cellobiose unit. A number of forms of cellulose, differing somewhat in their x-ray dimensions, are recognized. The native form is called cellulose I. The principal variant is cellulose II, formed by dissolving or swelling native cellulose in strong aqueous sodium hydroxide. The process of *mercerization*, first discovered by Mercer in 1845, depends on the formation of cellulose II and results in an increase in the strength, luster, and dye receptivity of cotton fibers.

The degree of polymerization of native cellulose is probably 4000–5000. The process of isolation from associated materials by boiling in dilute alkali introduces some uncertainty in these figures, as it probably results in some chain scission. The structure of cellulose I proposed by Liang and Marchessault on the basis of infrared dichroism (and earlier x-ray fiber diagrams) (Liang and Marchessault, 1959) is shown in two views, perpendicular to the b and c axes, respectively, in Fig. 8.35.

Electron micrography of native cellulose from a variety of sources indicates the presence of *elementary fibrils* about 3.5 nm in diameter as the basic structural unit (Frey-Wyssling and Mühlethaler, 1963; Heyn, 1966). These commonly cluster into larger fibrils, up to 100 by 200 nm in transverse dimensions.

When cellulose is treated with aqueous sodium hydroxide and carbon disulfide, it yields cellulose xanthate, which forms a yellow viscous colloidal

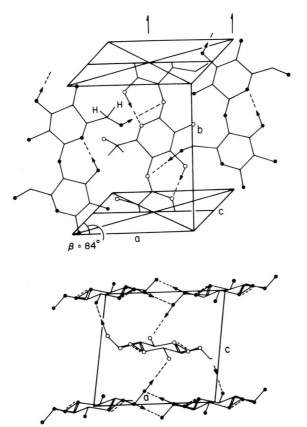

FIG. 8.35.   The structure of cellulose I (from Liang and Marshessault, 1959).

dispersion called *viscose*. When the viscose is forced through a spinnerette (a small orifice) into an acid bath, the cellulose is reformed as filaments of *viscose rayon*; the average degree of reaction is about one CS_2 for every two glucose units in the cellulose chain:

$$RONa + CS_2 \rightarrow RO\overset{\displaystyle S}{\overset{\displaystyle \|}{-C}}-ONa \qquad (8.15)$$

xanthate

$$RO\overset{\displaystyle S}{\overset{\displaystyle \|}{-C}}-ONa \xrightarrow[\text{spinning}]{\text{acid}} ROH \quad + CS_2 \qquad (8.16)$$

regenerated
cellulose

If extruded through a slit, a sheet of regenerated cellulose called *cellophane* is formed. This was once the principal transparent plastic film-forming material but both film and fiber have been largely displaced by more recent developments.

Cellulose *ethers*, formed by reaction of the hydroxyl groups with alkyl chlorides in the presence of alkali, are also of once great but now waning industrial significance as films and fibers, principally as the methyl, ethyl, and benzyl derivatives.

The first cellulose *ester* to be used commercially was the nitrate,

$$ROH + HONO_2 \;\overset{H_2SO_4}{\rightleftharpoons}\; RONO_2 + H_2O \tag{8.17}$$

discovered in 1838 and introduced commercially by Hyatt in 1870. The properties of cellulose nitrate depend on the degree of nitration. If all three hydroxyls are reacted, the result is *guncotton*, used in making smokeless powder. Unfortunately, the less reacted materials, though once enjoying widespread use as *pyroxylin* lacquers and in films, retain the characteristic high degree of flammability, and as a result are now entirely obsolete. Cellulose *acetate*, formed by reaction of cotton or wood fibers with a mixture of acetic acid, acetic anhydride, and sulfuric acid, retains the desirable

$$ROH + CH_3CO_2H \;\xrightarrow{\;\underset{\text{O}}{CH_3C}\text{--O--}\underset{\text{O}}{CCH_3},\, H_2SO_4\;}\; RO\underset{\text{O}}{C}CH_3 \tag{8.18}$$

film-forming properties of the nitrate together with low flammability. In the manufacturing process, it is customary to carry acetylation first to the triacetate stage, then age the solution to achieve some degradation of the chains, and then by addition of water to reverse the acetylation to approximately the diacetyl stage for optimum solubility characteristics and spinnability. Cellulose acetate fibers are still manufactured on a large scale despite the competition of newer materials. In practical use it is found that mixed acetate–butyrate or acetate–propionate esters have more desirable properties as films and molded plastics, showing less absorption of water together with greater impact resistance, compatibility with plasticizers, and solubility in appropriate solvents.

8.5.2. STARCH AND GLYCOGEN

Starch serves as a reserve source of carbohydrate in plants, while *glycogen* serves an analagous purpose in animals. Starch is composed of two components: *amylopectin*, which has mainly the α-1,4 structure already indicated, with varying amounts of α-1,6 branches, depending on the source; one such branch point is shown. The other component is *amylose*, which has nearly

amylopectin

exclusively the α-1,4 structure, probably with an occasional α-1,6 branch. The principal hydrolysis product of both is *maltose*, the α-linked analog of cellobiose. Amylose is crystallizable and may be separated from amylopectin by recrystallization from *n*-butanol. Degrees of polymerization are much higher for amylopectin, as a result of the generation of branches during its formation. Values as high as 800 are found for corn amylose and 3800 for potato amylose.

Starches do not form fibers and x-ray information must be obtained from powders and oriented films. For further detail the reader is referred to the treatise of Walton and Blackwell (1973, pp. 499–505).

Glycogen is very similar in structure to amylopectin but occurs in animals, where it serves as a readily mobilized storage form of glucose. The principal sites of its occurrence are the liver and in skeletal muscle. It is visible under the electron microscope in the cytoplasm of liver cells as small granules that also contain some of the enzymes responsible for its synthesis and degradation.

8.5.3. CHITIN

Chitin is very similar in structure and morphology to cellulose and might be described as 2-acetylamino cellulose:

It forms the exoskeletons of insects and many lower invertebrate animals, including crustacea, playing a role analogous to that of collagen in higher animals and cellulose in plants. It frequently occurs in combination with a large proportion of calcium carbonate and protein. There are three recognized polymorphs as observed by x-ray fiber diagrams. The most abundant of these, α-chitin, is shown in Fig. 8.36 (Carlström, 1957). The neighboring chains are antiparallel with conformations similar to that of cellulose, and are linked by hydrogen bonds between the amide groups. The other polymorphic forms have sheets of similar design, but differing in the relative directions of neighboring chains. Association with protein may take the form of a composite design consisting of chitin fibers or rods of approximately 50-nm diameter embedded in a protein matix. In other organisms, a layered structure is found.

8.5.4. MISCELLANEOUS POLYSACCHARIDES

Dextrans are a class of polysaccharides that have a predominantly 1,6 linkage between the glucose units:

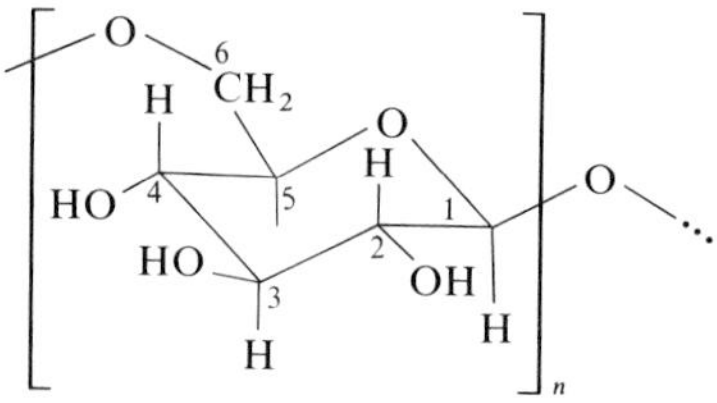

The polymer is synthesized from sucrose by the bacterium *Leuconostoc mesenteroides* and other related species under the influence of an enzyme called *dextransucrase*. This enzyme does not act by the direct condensation of glucose molecules (which is very unfavorable energetically), but by transfer of glucose units from sucrose with expulsion of fructose:

$$n\,\text{sucrose} \rightleftharpoons (\text{glucose})_n + n\,\text{fructose} \qquad (8.19)$$

This equilibrium runs very strongly to the right (Hehre, 1941). The evidence seems to indicate that the formation of each single chain is carried out rapidly by an enzyme molecule—in a manner somewhat analogous to vinyl polymerization—as the molecular weights, which may be very high, are essentially independent of conversion.

In addition to the normal links, the dextran chains contain short branches consisting mostly of a single glucosyl unit attached by a 1,3 linkage every

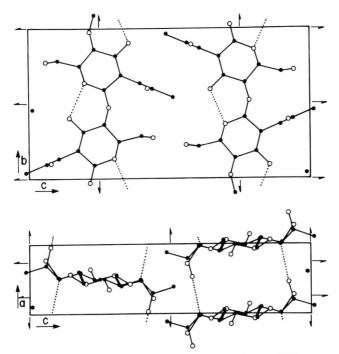

FIG. 8.36.   The structure of chitin (from Carlström, 1957).

20–30 main chain residues. There is also a possibility of the formation of long branches through the action of a rearrangement enzyme that breaks off sections of the chain and adds them back on along the chain. However, it is at least as likely that the very high molecular weights observed (of the order of 10^8) may be the result of physical aggregation.

There are a number of examples in nature of alternating AB-type polysaccharide copolymers. The mucopolysaccharide of bacterial cell walls, composed of alternating N-acetylglucosamine and N-acetylmuramic acid units joined by β-1,4 linkages, has already been discussed in Section 8.3.2.3 as the substrate of lysozyme. Additional mucopolysaccharides are given as the first six entries in Table 8.6. The *carrageenans* occur in seaweeds. *Agarose* and *porphyran* are well-characterized members of the agar family, and likewise occur in seaweeds. Both classes of polysaccharides find extensive practical use. The agar group are used in bacterial culture plates and gel electrophoresis, having the capacity to form stiff gels at low concentrations in aqueous solution. The carrageenans are used in a variety of food products to provide a smooth and palatable texture.

TABLE 8.6

Chain Structures of Some A-B Polysaccharide Copolymers[a]

Polysaccharide	Residue A	Residue B
Chondroitin	β-D-Glucopyranuronic acid	2-Acetamido-2-deoxy-β-D-galactopyranose
6-Sulfate (chondroitin sulfate C)	β-D-Glucopyranuronic acid	2-Acetamido-2-deoxy-β-D-galactopyranose 6-sulfate
4-Sulfate (chondroitin sulfate A)	β-D-Glucopyranuronic acid	2-Acetamido-2-deoxy-β-D-galactopyranose 4-sulfate
Hyaluronic acid	β-D-Glucopyranuronic acid	2-Acetamido-2-deoxy-β-D-glugopyranose
Dermatan sulfate (chondroitin sulfate B)	α-L-Idopyranuronic acid and β-D-glucopyranuronic acid	2-Acetamido-2-deoxy-β-D-galactopyranose 4-sulfate and, sometimes, 6-sulfate
Keratan sulfate	2-Acetamido-2-deoxy-β-D-glucopyranose and its 6-sulfate	β-D-Galactopyranose and its 6-sulfate
Agarose	3,6-Anhydro-α-L-galactopyranose	β-D-Galactopyranose
Porphyran	3,6-Anhydro-α-D-galactopyranose and L-galactopyranose 6-sulfate	β-D-Galactopyranose and its 6-methyl ether
κ-Carrageenan	3,6-Anhydro-α-D-galactopyranose, its 2-sulfate, and α-D-galactopyranose 6-sulfate	β-D-Galactopyranose 4-sulfate
ι-Carrageenan	3,6-Anhydro-α-D-galactopyranose 2-sulfate and α-D-galacto-pyranose 2,6-disulfate	β-D-Galactopyranose 4-sulfate
λ-Carrageenan	α-D-Galactopyranose 2,6-disulfate	β-D-Galactopyranose and its 2-sulfate
μ-Carrageenan	3,6-Anhydro-α-D-galactopyranose and α-D-galactopyranose 6-sulfate	β-D-Galactopyranose 4-sulfate
κ-Furcellaran	3,6-anhydro-α-D-galactopyranose and α-D-galactopyranose 6-sulfate	β-D-Galactopyranose, some of it possibly representing branch points, and its 4-sulfate

[a] Basic structures of some polysaccharides from animal and seaweed tissues: general structure $[A—(1 \to 3)—B—(1 \to 4)—]_n$. From Walton and Blackwell (1973, p. 506).

REFERENCES

Allen, F. S., Gray, D. M., Roberts, G. P., and Tinoco, I. Jr. (1972). *Biopolymers* **11**, 853.
Anfinsen, C. B. (1973). *Science* **181**, 223.
Astbury, W. T. (1938). *Trans. Faraday Soc.* **34**, 378.

Bamford, C. H., Elliott, A., and Hanby, W. E. (1956). "Synthetic Polypeptides." Academic Press, New York.

Beavan, G. H., Holiday, E. R., and Jope, E. M. (1950). *Disc. Faraday Soc.* **9**, 406.

Beavan, G. H., Holiday, E. R. and Jope, E. M., (1952). *Adv. Protein Chem.* **7**, 319.

Bloomfield, V. A., Crothers, D. M., and Tinoco, I. (1974). "Physical Chemistry of Nucleic Acids." Harper, New York.

Bovey, F. A. (1969). "Polymer Conformation and Configuration." Academic Press, New York. p. 110–111.

Bovey, F. A. (1972). "High Resolution NMR of Macromolecules", Academic Press, New York.

Bovey, F. A. (1975). *J. Polym. Sci. Macromol. Rev.* **9**, 1.

Bovey, F. A., and Hood, E. P. (1967). *Biopolymers* **5**, 325.

Brant, D. A. (1968). *Macromolecules* **1**, 291.

Cantor, C. R., Warshaw, M. M., and Shapiro, H. (1970). *Biopolymers* **9**, 1059–1079.

Carlström, D. (1957). *J. Biophys. Biochem. Cytol.* **3**, 669.

Carver, J. P., and Blout, E. R. (1967). "Treatise on Collagen" (G. N. Ramachandran, ed.), Vol. I. Academic Press, New York.

Carver, J. P., Schechter, E., and Blout, E. R. (1966). *J. Am. Chem. Soc.* **88**, 2550.

Cowan, P. M., and McGavin, S. (1955). *Nature (London)* **198**, 1165.

Deber, C. M., Madison, V., and Blout, E. R. (1976). *Accounts Chem. Res.* **9**, 106.

Deutsche, C. (1970). *J. Chem. Phys.* **52**, 3703.

Dickerson, R. E. (1964). *In* "The Proteins" (H. Neurath, ed.), Vol. 2, p. 634. Academic Press, New York.

Dickerson, R.E., and Geis, I. (1969). "The Structure and Action of Proteins." Harper, New York.

Dwek, R. A. (1973). "Nuclear Magnetic Resonance in Biochemistry: Applications to Enzyme Systems." Oxford Univ. Press (Clarendon), London and New York.

Förster, T. (1960). "Comparative Effects of Radiation" (M. Burton, J. S. Kirby-Smith, and J. L. Magee, eds.). Wiley, New York.

Frey-Wyssling, A., and Mühlethaler, K. (1963). *Makromol. Chem.* **62**, 25.

Gabor, G. (1968). *Biopolymers* **6**, 809.

Goodman, M., and Fried, M. (1967). *J. Am. Chem. Soc.* **89**, 1264.

Goodman, M., Davis, G. W., and Benedetti, E. (1968). *Accounts Chem. Res.* **1**, 275.

Haggis, G. H., Michie, D., Muir, A. R., Roberts, K. B. and Walker, P. M. B. (1964). "Introduction to Molecular Biology." Wiley, New York.

Harrington, W. F., and Von Hippel, P. (1961). *Adv. Protein Chem.* **16**, 1.

Hehre, E. J. (1941). *Science* **93**, 237.

Heyn, A. J. (1966). *J. Cell. Biol.* **29**, 181.

Holley, R. W. (1968). "The Molecular Basis of Life" (R. H. Haynes and P. C. Hanawalt, eds.). Freeman, San Francisco, California.

Holzwarth, G., and Doty, P. (1965). *J. Am. Chem. Soc.* **87**, 218.

Holzwarth, G., Hsu, E. C., Mosher, H. S., Faulkner, T. R., and Moscowitz, A. (1974). *J. Am. Chem. Soc.* **96**, 251.

James, T. L. (1975). "Nuclear Magnetic Resonance in Biochemistry," Chapter 7. Academic Press, New York.

Karlson, R. H., Norland, K. S., Fasman, G. D., and Blout, E. R. (1960). *J. Am. Chem. Soc.* **82**, 2268.

Kartha, G., Bello, J., and Harker, D. (1967). *Nature (London)* **213**, 862.

Katchalski, E., Sela, M., Silman, H. I., and Berger, A. (1964). "The Proteins" (H. Neurath, ed.), Vol. II, pp. 414–423.

Kendrew, J. C. (1963). *Science* **139** (Nobel Prize Address).

Kendrew, J. C. *et. al.* (1960). *Nature (London)* **185**, 422.

Kendrew, J. C., *et al.* (1961). *Nature (London)* **190**, 665.

Kendrew, J. C. *et. al.* (1970a). *Biochemistry* **9**, 3471.

Kendrew, J. C. *et. al.* (1970b). *J. Biol. Chem.* **245**, 289.

Kendrew, J. C. *et. al.* (1970c). *J. Mol. Biol.* **52**, 1.

Koenig, J. L. (1972). *Macromol. Rev.* **6**, 59–133.

Kolata, G. B. (1976). *Science* **192**, 645.

Komoroski, R. A., Peat, I. R., and Levy, G. C. (1976). "Topics in Carbon-13 NMR Spectroscopy" (G. C. Levy, ed.), Vol. 2. Wiley (Interscience), New York.

Kurtz, J., Berger, A., and Katchalski, E. (1956). *Nature (London)* **178**, 1066.

Leuchs, H. (1906). *Chem. Ber.* **39**, 857.

Liang, C. Y., and Marchessault, R. H. (1959). *J. Polym. Sci.* **37**, 385.

Lifson, S., and Roig, A. (1961). *J. Chem. Phys.* **34**, 1964.

Longworth, J. W. (1970). "Excited States of Proteins and Nucleic Acids" (F. R. Steiner and I. Weinryb, eds.), Chapter 6. Plenum, New York.

Marmur, J., and Doty, P. (1962). *J. Mol. Biol.* **5**, 109.

McDonald, C. C., and Phillips, W. D. (1969). *J. Am. Chem. Soc.* **91**, 1513.

McDonald, C. C., Phillips, W. D., and Lazar, J. J. (1967). *J. Am. Chem. Soc.* **89**, 4166.

Merrifield, R. B. (1969). *Adv. Enzymol.* **32**, 221.

Miyazawa, T. (1960). *J. Chem. Phys.* **32**, 1647.

Miyazawa, T., and Blout, E. (1961). *J. Am. Chem. Soc.* **83**, 712.

Mizushima, S. (1954). "Structure of Molecules and Internal Rotation." Academic Press, New York.

Moffitt, W. J. (1956a). *J. Chem. Phys.* **25**, 467.

Moffitt, W. J. (1956b). *Proc. Nat. Acad. Sci. U.S.* **42**, 736.

Morawetz, H. (1975). "Macromolecules in Solution," p. 255. Wiley, New York.

Moscowitz, A. (1960). "Optical Rotatory Dispersion" (C. Djerassi, ed.). McGraw-Hill, New York.

Nagayama, K., and Wada, A. (1973). *Biopolymers* **12**, 2443.

Nirenberg, M. (1973). "The Genetic Code," in Nobel Lectures: Physiology or Medicine (1963–1970), pp. 341–395. American Elsevier, New York.

Patel, D. J. (1976). Private communication.

Patel, D. J., and Canuel, L. L. (1977). *Biopolymers* **16**, 857.

Patel, D. J., Canuel, L. L., and Bovey, F. A. (1975). *Biopolymers* **14**, 987.

Pauling, L., and Corey, R. B. (1951). *Proc. Nat. Acad. Sci. U.S.* **37**, 729.

Pauling, L., Corey, R. B., and Branson, H. R. (1951). *Proc. Nat. Acad. Sci. U.S.* **37**, 205.

Perutz, M. F. (1963). *Science* **140**, 863 (Nobel Prize Address).

Pysh, E. S. (1967). *J. Mol. Biol.* **23**, 587.

Ramachandran, G. N., Ramakrishnan, C., and Sasisekharan, V. (1963). *J. Mol. Biol.* **7**, 95.

Rich, A. (1976). *Science* November 16.

Rippon, W. B., Koenig, J. L., and Walton, A. G. (1970). *J. Am. Chem. Soc.* **92**, 7455.

Roberts, G. C. K., and Jardetzky, O. (1970). *Adv. Protein Chem.* **24**, 448.

Ryle, A. P., Sanger, F., Smith, L. F., and Kitai, R. (1955). *Biochem. J.* **60**, 541.

Samejima, I., Fujii, I., and Miura, K. A. (1968). *J. Mol. Biol.* **34**, 39.

Sanger, F. (1975). *Proc. R. Soc. London Ser. B* **191**, 317.

Sasisekharan, V. (1959). *Acta. Crystallogr.* **12**, 897.

Schellman, J. A., and Oriel, P. (1962). *J. Chem. Phys.* **37**, 2114.

Schellman, J. A., and Schellman, C. (1964). *In* "The Proteins" (H. Neurath, ed.), Vol. II, Chapter 7. Academic Press, New York.

Spiro, T. G. (1974). *Accounts Chem. Res.* **7**, 339.

Stryer, L. (1975). "Biochemistry." Freeman, San Francisco, California.

Stryer, L., and Haugland, R. P., (1967). *Proc. Nat. Acad. Sci. U.S.* **64**, **719**.

Szwarc, M. (1965). *Adv. High Polym. Sci.* **4**, 1.

Takano, T., Kallai, O. B., Swanson, R., and Dickerson, R. E. (1973). *J. Biol. Chem.* **248**, 5244.

Tanzer, M. L. (1973). *Science* **180**, 562.

Teale, F. W. J. (1960). *J. Biol. Chem.* **76**, 381.

Tinoco, I. Jr., Halpern, A., and Simpson, W. I. (1962). "Polyamino Acids, Polypeptides, and Proteins" (M. A. Stahmann, ed.), Univ. of Wisconsin Press, Madison, Wisconsin.

Tinoco, I., Jr., Woody, R. W., and Bradley, D. F. (1963). *J. Chem. Phys.* **38**, 1317.

Traub, W., and Shmueli, V. (1963). *Nature (London)* **198**, 1165.

Tsuboi, M. (1962). *J. Polym. Sci.* **50**, 139.

Ullman, R. (1970). *Biopolymers* **9**, 471.

Urnes, P. J., and Doty, P. (1961). *Adv. Protein Chem.* **16**, 401.

Voet, D., Gratzer, W. B., Cox, R. A., and Doty, P. (1963). *Biopolymers* **1**, 193.

Walton, A. G., and Blackwell, J. (1973). "Biopolymers." Academic Press, New York.

Watson, J. D., and Crick, F. H. (1953). *Nature (London)* **171**, 964.

Weber, H., and Khorana, H. G. J. (1972). *J. Mol. Biol.* **72**, 219.

Wüthrich, K., (1976). "NMR in Biological Research: Peptides and Proteins." American Elsevier, New York.

Wyckoff, H. W., Hardman, K. D., Allewell, N. M., Inagami, T., Johnson, L. N., and Richards, F. M. (1967). *J. Biol. Chem.* **212**, 3984.

Yanari, S., and Bovey, F. A. (1960). *J. Biol. Chem.* **235**, 2818.

Yang, J. T. (1964). "Newer Methods of Polymer Characterization" (B. Ke, ed.), pp. 103–153. Wiley, (Interscience), New York.

Zimm, B. H., and Bragg, J. K. (1959). *J. Chem. Phys.* **31**, 526.

INDEX

A

G

H